ISBN 10: 81-7019-661-5
ISBN 13: 9788170196617

Published by
Today and Tomorrow's Printers and Publishers
4436/7, Ansari Road, Daryaganj, New Delhi - 110 002 (India)
Ph : 23242621; 23241021; Fax 23242621;

CURRENT STATUS OF RESEARCHES IN BIOSCIENCES

Editors

Prof. Prakash Chandra Joshi

Prof. (Mrs.) Namita Joshi

Dr. Reshman Yasmin

Dr. Dalip Kumar Mansotra

Today & Tomorrow's Printers and Publishers

4436/7, Ansari Road, Daryaganj, New Delhi - 110 002

Ph : 23242621; 23241021; Fax 23242621;

E-mail: ttppindia@gmail.com; info@ttppbooks.com

Web : ttpp.in

Preface

With the rapid advances in science and technology (S&T), the societal expectation are increasing, which includes, providing the basic infrastructure of assured food, water and electricity supply, sanitation, efficient transportation and communication, IT connectivity and e-governance, safety and security of all, as well as finding solutions for alleviating malnutrition and poverty offering adequate healthcare and so on.

To cope up with the increasing population in India, the only option left with us is to tap every channel of nutritional food for the security of life and sustainable development. Fortunately, we have plenty of natural resources like aquatic and open land, coastal belt and oceanic resources with rich faunal diversity.

Livestock/animal resources are an important contribution to total food production. The role of animals in food and agriculture development programmes is underrated almost everywhere throughout the increasing demand, especially in developing countries.

The focal theme of our Animal, Vetirinary and Fisheries Sciences have been well-tuned with the main focal theme of ISCA "Science and Technology: Rural India". Rural development as assumed greater importance in India today than in the earlier period in the process of development of the country and our section is playing an important role in it.

The papers and articles received in our section have been compiled in the form of books to make it more meaningful. We have published the entire proceedings in 4 Volumes. This volume contains 50 articles. We are thankful to all the authors and contributors for giving their time and writeup, who are responsible for the work presented in the paper and if any plagiarism question arises, they are answerable.

We are thankful to ISCA, Kolkata and office bears of ISC of University of Agriculture, Bangaluru (from 3-7 January 2020) for encouraging us to hold symposium of our section in the section of Animal, Veterinary and Fisheries Sciences.

The editorial boards is also thankful to Ms. Shivalika Sharma, Mr. Gaurav Pant, Mr. Hemant Singh and Ms. Meera Goswami, of Department of Zoology and Environmental Sciences, Gurukula Kangri Vishwavidyalaya,

Haridwar for their day and night work to make it possible in compiling all these research papers in the form of book.

This book will be very useful for researchers, scientists, students and all entrepreneurs who are working in the field of livestock.

Contents

Current Status of Researches in Biosciences (2020) : 1-21
Editor : Prof. P.C. Joshi, N. Joshi, Reshman Yasmin, and D. K. Mansotra
Today & Tomorrow's Printers and Publishers, New Delhi - 110 002

DESIGNER FOODS, DIETARY SUPPLEMENTS, FOR BETTER CARDIOVASCULAR HEALTH

Gundu H. R. Rao

Laboratory Medicine and Pathology, Director, Thrombosis Research, Lillehei Heart Institute, University of Minnesota

E mail:gundurao9@gmail.com

ABSTRACT

Metabolic diseases, such as hypertension, excess weight, obesity, type-2 diabetes, and vascular diseases, have increased in the incidence and prevalence, to epidemic proportions worldwide, in the last three decades. During this period, obesity has increased by two-fold, and diabetes by four-fold. Cardiovascular diseases are the number one killers and have retained this position for over 100 years. No country has reduced, reversed, or prevented the increase in the incidence of these diseases. Modern medicine has failed, to prevent the increase of these chronic diseases and has instead, opted for drug discovery and management of observed risks for these diseases. These diseases are basically life style diseases. In view of this fact, we would like to encourage holistic lifestyle, emphasize the need for robust physical activity, wellness, and heart healthy diet. In an earlier article, we discussed changing concepts of health care, which includes, heavy emphasis on physical fitness and wellness. In this article, we would discuss heart healthy diet, designer foods, dietary supplements, and complementary therapies for reduction, reversal, and prevention of metabolic risks, such as oxidative stress, chronic inflammation, altered blood flow, endothelial dysfunction, and subclinical atherosclerosis. When it comes to designer foods, examples include, designer eggs, with enhanced omega-3 fatty acids, improved variety of staple grains, such as white rice, wheat, and corn, improved animal diet, for the development of high-quality tissues. We will discuss the importance of genomic studies of animals and plants, for selective breeding and genetic engineering and use of stem cells for synthetic meat. Since we are interested in the early interventions of metabolic risks, as a major strategy for prevention of CVDs, we will discuss development of dietary supplements, and complementary therapies, using indigenous phytochemicals. Since basic cause for developing metabolic risks is altered lifestyle, we emphasize the importance of healthy life style including physical fitness, heart healthy diet and wellness.

KEYWORDS: Designer foods, Cardiovascular health, CVDs.

INTRODUCTION

Chronic metabolic diseases such as hypertension, excess weight, obesity, type-2 diabetes, and vascular diseases, have increased to epidemic proportions, worldwide in the last three decades (1-19). The high incidence of hypertension is a major global public health challenge, as it increases morbidity and mortality from cardiovascular and kidney diseases. Recent studies indicate, that close to 32% of the population (1.4 billion) had hypertension in 2010 (20). Currently, 2.1 billion people, -nearly one third of the world's population, are either obese or overweight, according to a new, first of a kind analysis of trend data, from 188 countries (21). An analysis of 45 countries, representing 90% of the population, estimates that in 2018, there was more than 500 million type-2 diabetes worldwide (22). NCD Risk Factor Collaboration group used data from 751 studies, from 146 of the 200 countries for their estimate. According to them, if post -2000 trends in the increased incidence of diabetes continue, the probability of meeting the global target of halting the rise in the prevalence of diabetes by 2025, to the level of 2010 level worldwide is lower than 1% (12). During the same period (1980-2010), in China, 7- national diabetes mellitus surveys were conducted. They found that the prevalence of diabetes had increased by 17-fold (23). Potential risk factors which contributed include, social and economic development, urbanization, dietary pattern, and Westernized lifestyle. Cardiovascular disease (CVD), remains a major cause of health loss for all regions of the world (24).

Framingham Heart Study (FHS) was initiated in 1948, by the National Heart Institute (NIH) in the town of Framingham, Massachusetts (25). Investigators sampled 2 of every three adults aged 30-59 years of age. This study is now considered one of the longest, most important epidemiological studies in the medical history. The study demonstrated a link between the modifiable risk factors such as cigarette smoking, hypertension, low-density lipoproteins, and individual's risk for developing heart disease. During the same period, Professor Ancel Keys at the University of Minnesota initiated Seven Countries Study (SCS), the first major study to investigate diet and lifestyle, along with other risk factors for cardiovascular disease. The main hypothesis of this study was, that the rate of coronary artery disease in populations and individuals, vary in relation to their physical characteristics and lifestyle, particularly in the fat composition of the diet and serum cholesterol. Exploratory studies began in Italy, Spain, South Africa and Japan. During the first phase of the study, standardized lifestyle and risk factor surveys were conducted at 5 and 10 years of follow-up. Early on in the study, Professor Keys and colleagues noticed, that across cultures in the Seven Countries, cultures differed in their diets and corresponding differences were seen in saturated fat, serum cholesterol, and coronary artery disease incidence, after 5 to 10 years follow-up (26-30).

As a continuation of the population-based CVD epidemiology and prevention project, the North Karelia project was launched in 1972, to carry out a comprehensive community-based CVD prevention program (31). By 1972, North Karelian men had earned the dubious distinction of having the highest rate of heart disease in the world. By the early 2000s, the number of deaths of Finnish men ages 35-64 from coronary artery disease had plunged down to 75%. Much of this reduction came from dramatic reduction in risk factors, like high cholesterol and smoking. Cholesterol lowering was the strongest contributor. Professor Pekka Puska the principal investigator of this probably unrepeatable successful project, summarizes his effort following way:

"The experiences and results of the North Karelia Project in Finland supports the idea, that a well-planned and determined community-based Programme, can have major impact on lifestyles and risk factors, and such a development really leads quite rapidly, to reduced cardiovascular rates in the community. Furthermore, they demonstrate, the strength of community-based approach in changing the people's risk factors as well as give practical experience in organizing such activities (32). Since then, several clinical studies have demonstrated the beneficial effects of robust management of modifiable risk factors, in the reduction of premature mortality (33-35).

Altered metabolism, leads to the development of metabolic risks such as oxidative stress, chronic inflammation, excess weight, changes in the flow dynamics of circulating blood, endothelial dysfunction, hardening of the arteries, and subclinical atherosclerosis. Metabolic diseases such as hypertension, excess weight, obesity and diabetes are basically lifestyle diseases. The progression of these diseases leads to various clinical complications including the precipitation of acute vascular events, such as heart attacks and stroke. No country has reduced, reversed, or prevented these chronic metabolic diseases. Modern medicine has failed, to prevent the increase in the incidence and prevalence of these diseases. Instead, has opted on drug discovery and management of observed risk factors. Just management of risk factors will not reduce, the serious health care burden imposed by these diseases. We and others have articulated this dilemma, and called for immediate action, by all the stakeholders. In an earlier article, we have articulated about a novel way of using the cohorts from Fitness and Wellness participants, who are motivated about their health, to develop an early diagnosis and prevention program (36). In this overview, we will describe, some novel approaches to develop healthy diets, designer foods, and phytochemical-based dietary supplements, for use as complimentary therapies.

DISCUSSION

Some of the dietary components, that contribute to the development of metabolic risks, include, excess salt, excess sugar, excess carbohydrates, and type of fats. Following large-scale studies showing, that a high salt intake leads to high blood pressure (BP), the Dietary Guidelines for Americans set the recommended sodium intake at 2300 milligrams per day. Historically, salt consumption is in use for several thousands of years. At about 1000 years ago, the salt consumption already had risen to 5gms per day. International Study on Macro/Micronutrients and Blood Pressure (INTERMAP), in 17 populations in China, Japan, the UK and the USA concluded, that "The adversary association of dietary sodium with BP, is minimally attenuated by other dietary constituents; these findings underscore the importance of reducing salt intake for the prevention and control of prehypertension and hypertension (37). The DASH-(Dietary Approaches to Stop Hypertension)-like diet, with increase in nutrient-dense intakes of vegetables, fruits, whole grains, fat-free and low-fat dairy products, legumes, unsalted nuts and seeds, seafoods and unsaturated vegetable oils- is effective in lowering BP (38-40). When it comes to dietary salt, the first thing to know is that the salt you use at the dinner table is not the biggest sodium contributor in the American diet, in fact, the majority of the sodium Americans consume, -more than 70% -is found in processed foods and restaurant meals.

Sugar consumption doesn't cause diabetes. A diet high in calories from any source (including sugar) contributes, to weight gain and weight gain increases your risk for type-2 diabetes. In people with diabetes, cells are resistant to insulin, as a result of this insulin resistance, sugar accumulates in the blood. Excess sugar and fat contribute to weight gain, thereby reducing insulin sensitivity in the body. Highly processed foods, seem to be the major source of excess sodium, added sugars and unhealthy fats. Highly processed foods contribute almost 60% of calories and 90% of the added sugars in the American diet, according to the American Heart Association. Globally, policies are shifting towards mandatory reformulation of processed foods in terms of added salt, sugar, and types of fats. According to a recent article in the British Medical Journal, 'Consensus on Salt and Health (CASH)', - only one out of the 28 food categories surveyed, are on track to meet Public Health England's (PHE) 2017 salt reduction targets (41). The food industry will also fail, to hit a PHE target to achieve 20% reduction in added sugar. Of the three nutrients, proteins, carbs and fats, - carbohydrates (carbs) have the greatest impact on blood sugar control as the body breaks them down to glucose. The rise of ultra-processed foods has coincided with growing rates of obesity and diabetes (42).

Dariush Mozaffarian, dean of Tufts University's School of Nutrition, agrees with the findings of a land mark study by the NIH, which claims that it is not salt, sugar, or fat; but ultra-processed foods that drive weight gain. One of the critiques says, that we should try to eat as much of real food as we can. He also suggests, designing processed foods that are healthy. The rise in obesity and type-2 diabetes prevalence, occurred in parallel with an increasingly industrialized food system, characterized by large scale production of high-yield, inexpensive, agricultural products (corn, soy, wheat and rice), which were processed to generate an abundance of 'added value' foods. NCD Risk Factor Collaborator Group found, that from 1975 to 2016, the number of girls with obesity increased from 5 million to 50 million, and boys from 6 million to 74 million. In other word, childhood obesity increased 10-fold worldwide since 1975 (43). Reducing excess weight, will require modifying both energy intake and energy expenditure and not simply focusing on either alone (44). Americans Consume on an average 250-300 grams of carbohydrate a day, accounting for 55% of their daily intake. Studies done in several laboratories including the Seven Countries study, have demonstrated the beneficial effects of a Mediterranean diet and low carb diets, compared to the low-fat diets. Dr. Walter Willet, Chairman of the Department of Nutrition, Harvard School of Public Health says, "Fat is not the problem." "If the Americans could eliminate sugary beverages, potatoes, white bread, pasta, white rice and sugary snacks, we would wipe out almost all the problems we have with excess weight, diabetes, and other metabolic diseases."

What really is the healthy diet? Harvard Heart Healthy plate consists of 50% fruits and vegetables, 25% of whole grains, 25% of protein with vegetable oils or monosaturated fats. The United States Department of

Agriculture (USDA), recommends a similar combination of nutrients and includes dairy products (Fig 1).

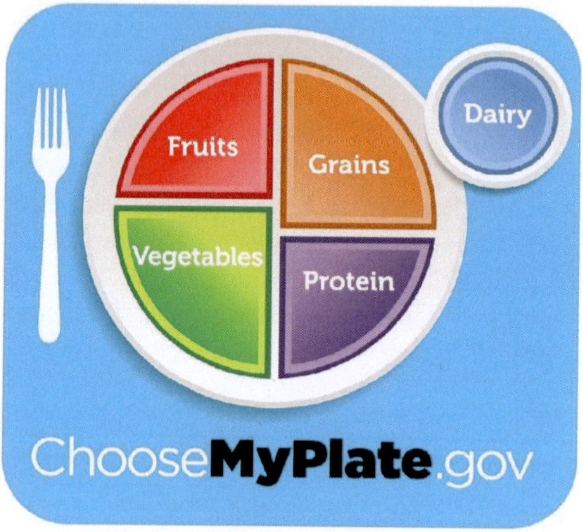

Source: U. S. Department of Agriculture

Fig 1. Healthy Diet Plate

In majority of countries, people cannot afford this type of heart healthy food plate. Major source of daily diet consists of staple grains such as, corn, wheat and rice, supplemented with a minimum of vegetables, meat, fish, poultry and dairy products, depending upon the availability and affordability. According to the Food Agriculture Organization (FAO), maize, rice and wheat, provide more than half of caloric intake. Worldwide, rice wheat, maize and to a lesser extent, sorghum and millets, are important staple foods, critical to the survival of billions of people. In 2018, corn production was 1100 metric tons (MT), wheat was 735 MT, and Rice 500 MT. Rice is the staple food of the half of the world's population. Most staple foods are inexpensive, plant-based. In recent years, there is a growing trend to develop organically grown products. Similarly, we assume that there will be a growing interest in developing designer, 'value added' healthy foods. Examples include, designer eggs, designer milk, designer grain, probiotics, designer foods enriched with micro and macro nutrients, designer fats, proteins and amino acids (45). In China, more than 3,000 varieties of health foods are available for use, in their traditional medicine practice. In this overview, we will briefly discuss some aspects of this novel approach with special reference to our area of interest; cardiometabolic diseases.

If you do an internet search on Value-added staple crops, mostly you will see genetically engineered crops, to fight some pest or the other and not staple foods that are developed for better nutritional qualities or for better health. Excess weight, obesity and type-2 diabetes are major health problems worldwide. The Danish Diet study demonstrated, that higher whole-grain intake is associated with lower risk of type-2 diabetes, among middle-aged Men and Women (46). In this study, researchers used wheat, rye, and oats as the source of whole grains. They found consistent association between high-whole grain intake and lower risk of type-2 diabetes, irrespective of the type of cereal consumed. In this study, with participants between ages 50-65, (n=55,645) the researchers found, that the highest whole grain intake among men was associated with a 34% decrease in risk of diabetes. Whole grains are rich in fiber, harder to break down, and help with insulin sensitivity. Researchers at Kawasaki University, Japan, have demonstrated that eating glutinous brown rice twice daily for 8 weeks, improves glycemic control in Japanese patients with diabetes (47). Can brown rice slow the spread of type-2 diabetes? To tackle this issue the researchers of Harvard School of Public Health (HSPH), launched a Global Nutritional and Epidemiological Transition (GNET) Working Group, a collaborative initiative, between researchers from the Departments of Epidemiology at HSPH and their colleagues in Asia, Latin America and Africa. The aim of this group is, to prevent the global diabetes epidemic by improving the carbohydrate quality of staple foods in the diets of people from around the world.

About half a century ago, researchers at the Central Food Technological Research Institute, Mysore, India, developed deodorized fish flour as well as yogurt from peanut milk as a good source of protein. Although, these products were welcome initially, did not get acceptance and as such failed to get into the main stream market. The reason I am mentioning this is, though brown rice revolution may pay a big dividend in the areas, where people are used to consume brown rice, will fail in the countries, where white rice is preferred. White rice, a common Indian staple grain, has a high glycemic index and low fiber content. High-Fiber White Rice (HFWR) was developed in India, using biochemical screening approaches and classical breeding techniques. Mohan and associates from Madras Diabetes Research Foundation (MDRF), conducted a randomized control trial. The dietary fiber content of HFWR, was five-fold higher and glycemic index of HFWR was 23% lower than white rice. They concluded, that the new HFWR variety can be considered as a potentially healthier alternative, to commercial white rice, in rice-eating populations, on account of its lower glycemic index and high fiber content (48). Similar breeding and selection processes are in progress, for other staple grains in different countries, to improve the nutritional quality of these staple foods.

Wahab and associates from Pakistan report a study, in which they selected eight different varieties of wheat for nutritional qualities. They concluded, that it is possible to select and cultivate new varieties of wheat with better nutrition qualities, enriched with preferred micro nutrients (49). Out of the ten genotypes validated, RSP-561 possessed maximum levels of starch protein, iron, zinc, soluble protein, gluten, calcium, carotenoids, and antioxidant compositions (50). Several studies worldwide have demonstrated, that improved crops can be developed by classical breeding, mutagenesis, and transgenesis (51-53). Similar studies have been done in corn breeding, to improve the nutritional quality and suitability for health benefits. Among various corn varieties, high amylose corn (HACS) has attracted the most research interest (54). In this study, particular emphasis was placed on upon the impacts of high-amylose corn flour/starch (HACF/HACS), on postprandial glycemic/insulinemic responses, insulin sensitivity, satiety/ appetite, lipid metabolism/obesity, colon/gut health and mineral absorption. In view of the fact, that these have components with slow digestive rate, HACS/HACF have been incorporated into various types of food products for feeding studies, and their effect on the postprandial glucose/insulinemic responses validated.

The principal phytochemicals present in corn seed and corn silk include polyphenols, phenolic acids, flavonoids, anthocyanins, glycosides, carotenoids and polysaccharides of biological importance (55). The presence of these chemicals makes corn a medicinal plant, which shows various biological activities particularly the antioxidant, antimicrobial, antidiabetic, anti-obesity, antiproliferative, hepatoprotective, cardioprotective, and renal protective activities. In Vietnam, 32 percent of corn grown, are genetically modified, to prevent diabetes and obesity. Global Nutrition Epidemiologic Transition Initiative has been working with 11 countries to reduce diabetes by improving the quality of staple foods, through culturally-appropriate interventions (56-61). Great progress has been made over the past decade, with respect to application of biotechnology to generate nutritionally improved crops. Biofortified staple crops such as rice, maize, and wheat harboring essential micro-nutrients to benefit the malnourished individuals as well as new varieties of crops, which have the ability to combat chronic metabolic diseases have been developed (62). Recently, a new line of biotechnology, based on gene editing has come to the forefront. Genome editing focuses on nuclease- based forms of engineering, such as the

TALENS (transcription activator like effector nucleases) or the CRISPR associated systems. These and other similar technologies, will revolutionize the way we think about enhancing food crops, to improve global nutrition and health.

Now that we have briefly discussed the role of dietary salt, sugar, carbohydrates, and some recent studies related to the improvement of dietary components, let us turn our attention to what is being done to modulate the dietary fats. Framingham heart studies, as well as many other studies, have reported the ill effects of unhealthy fat for developing excess weight, obesity, diabetes and vascular diseases, as well as their role in the precipitation of acute vascular events such as heart attacks and stroke. Since the time Bang and Dyerberg, studied the Greenland Eskimo diet and published their observations, about the rarity of ischemic heart disease in Greenland Eskimos, there is a great interest in the modifications of certain diets by enriching with Omega-3 fatty acids (63). These investigators also analyzed the plasma lipids of Eskimos and found that they had low lipid concentrations, in spite of high fat diet (Omega 3-fatty acids: Eicosapentaenoic acid (EPA) and Docosahexaenoic acid (DHA). Based on their observations they concluded, that coronary atherosclerosis, diabetes mellitus, is very rare among Eskimos (64). This revolutionary discovery at that time, inspire several leading researchers worldwide, to concentrate their work on demonstrating the health benefits of dietary Omega-3 fatty acids. In blood platelets, upon activation, arachidonic acid is released from membrane lipids and converted to a proactive Prostaglandins G_2 and H_2, which are further converted to thromboxane A_2 (Fig).

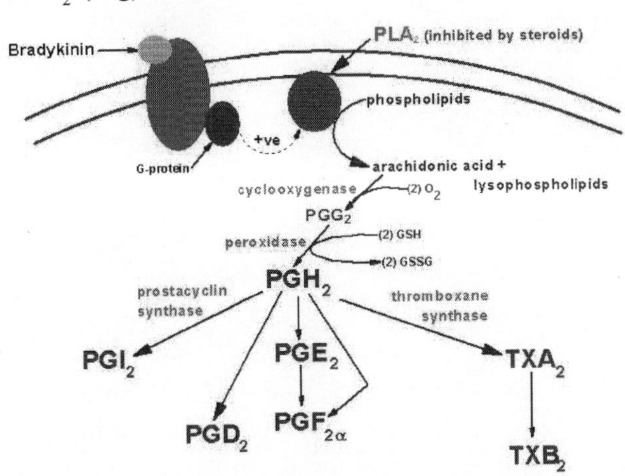

Fig 2. Arachidonic acid Metabolism via Cyclooxygenase (COX) and Prostacyclin Synthetase.

On the other hand, in the vessel wall, anti-aggregatory, vasoactive, prostacyclin (PGI_2) is produced. Following the work of Bang and Dyerberg, there was a concerted effort, to show that Omega three fatty acids are converted to their respective prostaglandins and Thromboxanes A_3 and A_4, seemed to be less aggregatory than the Thromboxane A_2.

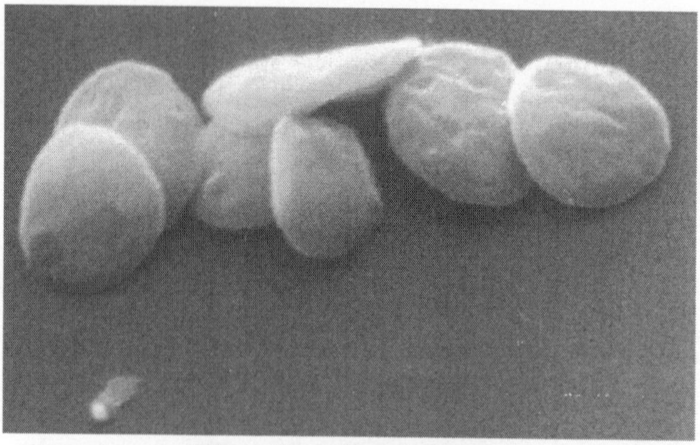

Fig 3. Unactivated Platelets.

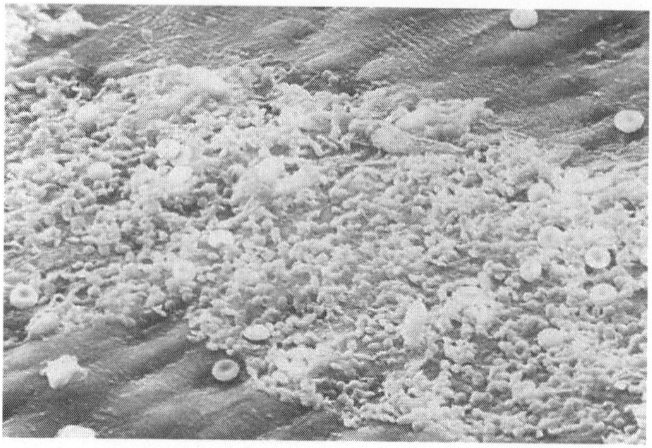

Fig 4. Activated Platelets on Vessel Wall.

Courtesy: (Late) Professor, James G White

During the same period, researchers demonstrated that the antiplatelet drug, aspirin (acetyl salicylic acid) inhibited irreversibly the key enzyme of (COX) PG synthesis. In view of this observation, various professional societies, public health policies promoted these ideas, and as such, there was a shift in the way the dietary fats were modified to meet these health guidelines. Omega three fatty acids have received a lot of attention in recent years, to the growing body of evidence that indicates, that these healthy fats help prevent range of medical conditions, including cardiovascular disease and thrombotic episodes. We have articulated in our earlier articles, the mechanisms involved in platelet activation, as well as about the meta-analysis results of multiple clinical trials on Fish oil (65-67). Naturally omega 3-fatty acid enriched foods include, mackerel, salmon, cod liver oil, herring, oysters, sardines, anchovies, and caviar.

Many Governments consider WHO dietary guidelines, to the state-of-the-art scientific evidence, translating them to regional and national dietary guidelines (68). The 2018 WHO draft guidelines (which relies on 84 randomized control trials), on dietary fatty acids and trans fatty acids, recommend reducing total intake of saturated fat, and replacing it with polyunsaturated and monounsaturated fatty acids. According to a new report, the recommendation fails to take into account, considerable evidence that the health effects of saturated fat varieties depends on the specific fatty acid and on the specific food source (68). A food-based transition of the recommendation for saturated fat intake, would avoid unnecessary reduction or exclusion of foods that are key source of important nutrients. In view of such observations and recommendations, we feel that it would be a better approach, to enrich the food with appropriate healthy fats. Having said that, it is not easy to incorporate Omega-3 three fatty acids into processed foods. Highly unsaturated fats tend to oxidize fast and are not good for incorporation into the processed foods. The expert committee strongly recommends, a more food-based translation of how to achieve a healthy diet, than consideration of the draft guidelines on the reduction in the total intake of saturated fatty acids.

By changing the diet of hens, egg producers have succeeded in improving upon the already nutritious egg. Most 'designer eggs' are enriched with omega-3 fatty acids, a heart-healthy nutrient that is otherwise found in seafood (salmon), flaxseed and walnuts (69). Russian researchers have demonstrated, that selenium-enriched chicken, pork, and meat can also be produced by using organic selenium in the diet of these animals (70). French researchers have developed, a new natural multi-enriched egg (Benefic eggs) with omega-3 fatty acids, vitamins D, & E, carotenoids, iodine, and selenium (71). The same research group demonstrated, that by feeding the animals with fish extracts or algae (oils), the level of DHA could be increased by 20-fold in fish, 7-fold in chicken, and less than 2- fold in beef (72). Some studies have shown, that it is not necessary to feed animals with linseed-supplemented diets for a long period, to have the highest increase in poly unsaturated fatty acids (PUFA) content of the products (73). Researchers from Czech Republic showed, that feeding rapeseed oil for just 4 weeks before slaughter was sufficient for improving fatty acid profile, without negative effect on the backfat (74). Danish studies with dietary manipulation showed, that it is possible to lower cholesterol in humans by providing meat from specially grown designer pigs. By increasing the content of rapeseed oil in pig feed, the researchers were able to change the fatty acid composition of the pig fat, in a way that it now had, potential to reduce blood cholesterol in human subject (75).

Just like the plant scientists, animal scientists, also have been taking advantage of the advances made in the area of biotechnology, genetic engineering, and genomics. Major source of food worldwide, is of animal origin as only 25% of the human population seems to be vegetarian. Australian researchers have developed, a high-resolution 'atlas of gene expression' in the domestic sheep (76, 77). This gene expression atlas for sheep, seems to be the largest transcriptomic dataset. Freeman and associates have described, the first genome-wide association analysis of the transcriptional landscape of 62 pig tissue/cell types (78). In India, national dairy development board has developed, a genome assembly named "NDDB-ABRO Murrah" of riverine buffaloes (May 2019). They claim, that this is the first complete parent-wise genome assembly of buffalo. Similar to the green revolution, national dairy board of India launched "Operation Flood" called white revolution, with the leadership of Dr Veghese Kurien. This development of the genome assembly, comes on the back of another successful launch of INDUSCHIP, a customized genotyping for indigenous cattle and their crosses.

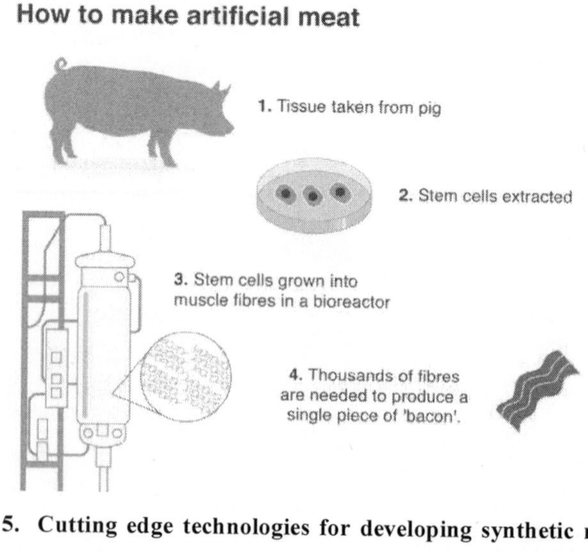

Fig 5. Cutting edge technologies for developing synthetic meat

The new start-ups are racing to develop lab-grown meat, using cutting-edge-technologies. Clean meat is the term, used for genuine animal meat grown in the laboratory using cell culture and appropriate nutrients. In August of 2013, a Dutch scientist Mark Post, presented a first lab-grown burger ever created. It costed 250,000 Euros. Mosa Meat for instance, harvests muscle tissue from a living cow with a biopsy, muscle cells are then cultured and grown to produce mass of customized designer tissue. Scientists at Israel are working in this cutting-edge technology for developing lab-grown meat.

Now that we have briefly discussed the advance made in developing designer plant and animal foods, let us touch base on another very important part of health foods; dietary supplements. Scientists from Tehran University, Iran, claim that higher intake of phytochemical-rich foods is inversely related to prediabetes (79). In this study, they demonstrated that higher dietary intake of phytochemicals, correlated with lower prediabetes- odds ratio. In countries with large diabetic population, like China and India, there is a great opportunity to develop indigenous dietary supplements, enriched with phytochemicals, known to have anti-diabetic potential (79-83). Most promising phytochemical groups are the alkaloids, glycosides, polysaccharides, and phenolics as well as flavonoids, terpenoids and steroids. If we extend this logic further, we can develop dietary supplements for intervention of most of the early metabolic risks, such as oxidative stress, chronic inflammation, excess weight, endothelial dysfunction, and subclinical atherosclerosis. For instance, Kasliwal and associates at the Medanta Hospital, New Delhi, India, have shown that just eating pistachio nuts for a few weeks, reduces or normalizes endothelial dysfunction, which is one of the earliest signs of vascular diseases (84). Dr. Jay N. Cohn, a pioneering Professor of Cardiology, Director of Rasmussen Center for Prevention of Cardiovascular Disease, University of Minnesota says, "No cardiovascular diseases without endothelial dysfunction." (85).

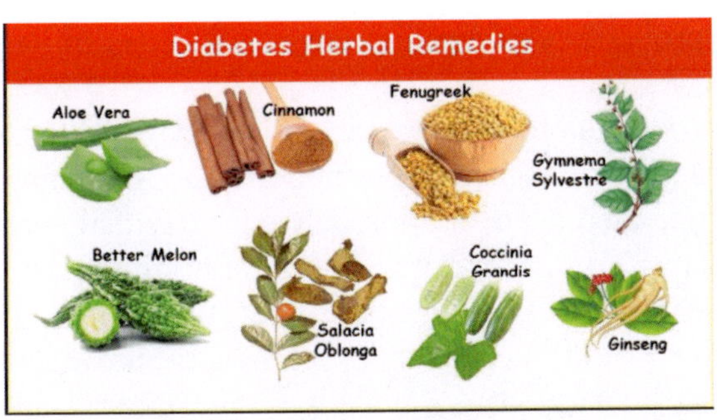

Fig 6. Common Tropical Diabetes Herbal Plants

Oxidative stress is another early metabolic risk, and at present clinicians is not treating diabetics with any drugs for this clinical condition. The most healthful diets like Mediterranean and Okinawa diets, are rich in Nrf2 raising nutrients. The transcription factor Nrf2 (nuclear factor, erythroid -2 related factor) is a master regulator of detoxification, anti-oxidant, anti-inflammatory and also possess other cytoprotective functions. Recent studies have demonstrated, that Protandim (a mixture of five phytochemicals; Ashwagandha, Indian Bacopa, Indian Green Tea, China Milk Thistle, and China Turmeric) is associated with reduction in oxidative stress and fibrosis, preservation of the right ventricular (RV) microcirculation and RV function (86-88). Since early metabolic risks such as oxidative stress, inflammation, free radical mediate pathology and ageing are associated with multiple mechanisms modulated by scores of genes and gene expressed proteins, we feel that a combination therapy using potent phytochemical extracts may be useful. Having said that, we strongly emphasize the need for a well thought-out randomized clinical trials, for validating the effectiveness of these phytochemical-based complementary therapies.

CONCLUSIONS

Metabolic diseases such as hypertension, excess weight, obesity, type-2 diabetes, and vascular disease have rapidly increased in the last three decades, to epidemic proportions worldwide. Seems like the increased incidence and prevalence of these diseases, parallels the introduction of the modern processed foods and sugary beverages. A collaborative study between the Imperial College of London and the World Health Organization demonstrated, that childhood and adolescent obesity has increased by ten-fold in the last four decades. An analysis of 45 countries representing 90% of the population, estimates that in 2018, there were more than 500 million type-2 diabetes worldwide. NCD Risk Factor Collaboration group, used data from 751 studies from 146 of the 200 countries for their estimate. According to them, if post -2000 trends in the increased incidence of diabetes continue, the probability of meeting the global target of halting the rise in prevalence of diabetes by 2025 to the level of 2010 level worldwide is lower than 1%. No country has reduced, reversed, or prevented the increase in the incidence or prevalence of chronic metabolic diseases. In an earlier article, we articulated a novel approach for early diagnosis of metabolic risks and developing appropriate intervention strategies. These chronic metabolic diseases are more or less life style diseases. In view of this observation, we would like to encourage holistic lifestyle, emphasize the need for robust physical activity, wellness and heart health diet. In this overview we have briefly discussed the concept of designer food for better cardiovascular health, with some selected examples from plant foods (cereals) and animal foods (eggs, and animal products) enriched with value added micronutrients and fats.

Food therapy of traditional Chinese medicine aims, to maintain balanced nutrition through diet. Medical diet therapy on the other hand, tries to achieve the balance of 'Yin and Yang' through the combination of nutrition and medicine. What we have discussed in this overview, is more or less a working model for how staple foods, whether from plant source (cereals, pulses and millets) or from animal sources, could be enriched with essential nutrients, balanced in terms of quality of carbohydrates, protein and fat content. Furthermore, we also have briefly discussed, the need to develop dietary supplements for complementary therapy, so that we need not depend on the mercy of pharma industries for all our medical needs. According to WHO, over 80% of the World population rely on such traditional plant based-system of medicine. Over 50,000 plant species are in use in herbal sector. The market for herbal products runs into trillions of dollars. We would like to see that the hidden potential of this untapped market is explored. There are of course, great opportunities and challenges in developing evidence-based support for these herbal products.

Currently, there are more prediabetics, than diabetics in the world. If we add the number of prediabetics and diabetics, it will be almost close to the number of hypertensives, and obese individuals (over a billion) worldwide. These three metabolic diseases, contribute significantly to the excess morbidity and mortality, associated with cardiovascular disease. Early intervention of these "at risk" population, will go a long way, in reducing the healthcare burden of these diseases. These at-risk individuals, are not covered by any prevention programs and providing affordable health care for such a large population (several billion), is almost unaffordable in many countries. We and others working in this area feel, that the major contributing cause for these metabolic risks is unhealthy life style. In view of this fact, we feel that development of a robust education program for creating awareness, and encouragement of healthy living will go a long way, to reduce the burden of these diseases. Such a possibility is documented by a Harvard study, which concluded, "Even among participants with high genetic risk, a favorable lifestyle, was associated with a nearly 50% lower relative risk of coronary artery disease, than was an unfavorable life style." Considering the rapidity with which the metabolic diseases are increasing, the time for action is now. Public health policy makers should focus on prevention, and encourage healthy diet, physical activity, and a healthy lifestyle. As a part of our prevention efforts, we have articulated the benefits of physical fitness, wellness and a healthy life style.

ACKNOWLEDGEMENTS

This article was prepared as a part of my presentation on 'this topic', for invited lectures at the Wellness Conference-2019 at Kanpur University, Kanpur, India, and at the Indian Science Congress-2020, Bengaluru, India. Figures 1, 2, 5, and 6 are from public (INTERNET) source. Author thanks the individuals, who posted these figures in the public domain.

REFERENCES

1. Afshin, A., Sur, P. J., Fay, K. A., Cornaby, L., Ferrara, G., Salama, J. S., & Afarideh, M. (2019). Health effects of dietary risks in 195 countries, 1990–2017: a systematic analysis for the Global Burden of Disease Study 2017. *The Lancet*, 393(10184), 1958-1972.

2. Saklayen, M. G. (2018). The global epidemic of the metabolic syndrome. *Current hypertension reports*, 20(2), 12.

3. Forouzanfar, M. H., Liu, P., Roth, G. A., Ng, M., Biryukov, S., Marczak, L., & Ali, R. (2017). Global burden of hypertension and systolic blood pressure of at least 110 to 115 mm Hg, 1990-2015. *Jama*, 317(2), 165-182.

4. Bloch, M. J. (2016). Worldwide prevalence of hypertension exceeds 1.3 billion. *Journal of the American Society of Hypertension: JASH*, 10(10), 753.

5. Dorans, K. S., Mills, K. T., Liu, Y., & He, J. (2018). Trends in prevalence and control of hypertension according to the 2017 American College of Cardiology/ American Heart Association (ACC/AHA) guideline. *Journal of the American Heart Association*, 7(11), e008888.

6. Bowet P, Chiolero A(2018). Prevalence and control of hypertension. *Lancet.* 392 (10155) P1305-1306.

7. Reilly, J. J., El-Hamdouchi, A., Diouf, A., Monyeki, A., & Somda, S. A. (2018). Determining the worldwide prevalence of obesity. *The Lancet,* 391(10132), 1773-1774.

8. Editorial(2018). Tackling obesity seriously: the time has come. *Lancet.* 3 (4); PE153.

9. NCD Risk Factor Collaboration (NCD-RisC)(2016). Trends in adult body-mass index in 200 countries from 1975 to 2014: a pooled analysis of 1968 population-based measurement studies with 19.2 million participants. *Lancet.* 387 (10026): P1377-1396.

10. World Health Organization: Ten-fold increase in childhood and adolescent obesity in four decades: new study by Imperial College, London and WHO. October 2017, News Release, https://www.who.int/news-room/detail/11-10-2017-tenfold-increase-in-childhood-and-adolescent-obesity-in-four-decades-new-study-by-imperial-college-london-and-who

11. Krug, E. G. (2016). Trends in diabetes: sounding the alarm. The Lancet, 387(10027), 1485-1486.

12. NCD Risk Factor Collaboration (NCD-RisC) (2016). Worldwide trends in diabetes since 1980: a pooled analysis of 751 population-based studies with 4.4 million participants. *Lancet.* http://dx.doi.org/10.1016/S0140-6736(16)00618-8

13. Collaborators, G. B. D., Forouzanfar, M. H., Alexander, L., Bachman, V. F., Biryukov, S., Brauer, M., ... & Frostad, J. J. (2015). Global, regional, and national comparative risk assessment of 79 behavioural, environmental and occupational, and metabolic risks or clusters of risks in 188 countries, 1990-2013: a systematic analysis for the Global Burden of Disease Study 2013. The Lancet, 386(10010), 2287-2323.

14. World Health Organization (2016.). Global report on diabetes. WHO, Geneva.

15. Knowler, W. C., Barrett-Connor, E., Fowler, S. E., Hamman, R. F., Lachin, J. M., Walker, E. A., & Nathan, D. M. (2002). Reduction in the incidence of type 2 diabetes with lifestyle intervention or metformin. The New England journal of medicine, 346(6), 393-403.

16. Gregg, E. W. (2017). The changing tides of the type 2 diabetes epidemic—smooth sailing or troubled waters ahead? Kelly West Award Lecture 2016. *Diabetes care,* 40(10), 1289-1297.

17. Anand, S. S., & Yusuf, S. (2011). Stemming the global tsunami of cardiovascular disease. *The Lancet*, 377(9765), 529-532.

18. Maranhão, M., Tse, T. F., Poole-Wilson, P., & de Luna, A. B. (2001). Global cardiovascular disease prevention. The Lancet, 358(9293), 1642.

19. Beaglehole R. (2001). Global cardiovascular disease prevention: time to get serious. *Lancet.* 358:661-663.

20. Mills KT, Bundy JD, Kelly TN *et al.*, (2016).Global disparities of hypertension prevalence and control: A systematic analysis of population-based studies from 90 countries. *Circ.* 134 (6): 441-450.

21. Ng M, Fleming T, Robinson M *et al.*, (2014). Global, regional, and national prevalence of overweight and obesity in children and adults during 1980-2013: a systematicanalysis for the Global Buren of Disease Study. *Lancet.* 384 (9945): P766-781.

22. Kaiser AB, Zhang N, Vander-Plujim W. (2018). Global prevalence of type2 diabetes over the next ten years (2018-2028). *Diabetes.* https://doi.org/10.2337/db18-202-LB.

23. Shen X, Vaidya A, Wu S *et al* (2016.). The diabetes epidemic in China: An integrated review of national surveys. *Endocr. Pract.* 22 (9): 1110-1129.

24. Roth, GA, Johnson C, Abajobir A *et al.*, (2017). Global, regional, and national burden of cardiovascular diseases for 10 causes,1990-2015. *J. Am. Coll. Cardiol.* 70 (1): 1-25.

25. Tsao CW, Vasan R. (2015). The Framingham Heart Study: past, present and future. *Int. J. Epidemiol.* 44(6): 1763-1766,

26. Keys A (1970). Coronary Heart Disease in seven countries. *Circ.* 41(S1): 118-139.

27. Keys A, Menotti A, Aravanis C *et al.*, (1984). The Seven Countries Study: 2,289

deaths in 15 years. *Prev. Med.* 13 (2):141-154.

28. Pett KD, Willett WC Vartiainen E *et al.*, (2017). The Seven Countries Study. *Eur. Heart J.* 38 (42): 3119-312.

29. Blackburn H, Kromhout D: The Seven Countries Study: A scientific adventure in cardiovascular disease epidemiology. http://www.epi.umn.edu/cvdepi/pdfs/SCSpapers.pdf

30. Kahn P, Willett WC, Katz DL: Ancel Keys and the Seven Countries Study: An evidence-based response to revisionist histories. A White Paper. http://www.truehealthinitiative.org/

31. Vartiainen E: The North Karelia Project: Cardiovascular disease prevention in Finland. Glob. Cardiol. Sci. Pract. 2018 (2): 13 doi: 10.21542/gcsp.2018.13

32. Pekka Puska (2002). Successful prevention of non-communicable diseases: 25-year experiences with North Karelia Project in Finland. *Publ. Health. Med.* 4 (1): 5-7.

33. Yusuf S, Hawken S, Ounpuu S *et al* (2004). Effect of potentially modifiable risk factors associated with myocardial infarction in 52 countries (the INTERHEART study): case-control study. *Lancet.* 364: 937-952.

34. Di Cesare M, Bennett JE, Best N *et al.*, (2013). The contribution of risk factor trends to cardiometabolic mortality decline in 26 industrial countries. *Int. J. Epidemiol.* 42 (3): 838-848.

35. Khera AV, Emdin CA, Drake I *et al* (2016.). Genetic risk, adherence to a healthy lifestyle, and coronary disease. *N. Engl. J. Med.* 375:2349-2358.

36. Malmefeldt E, Rao GHR (2019) Changing concepts of healthcare: Physical fitness, wellness. *EC Endocrinol. Metab.* Res. (In Press).

37. Ha, S. K. (2014). Dietary salt intake and hypertension. *Electrolytes & Blood Pressure*, 12(1), 7-18.

38. Stamler, J., Chan, Q., Daviglus, M. L., Dyer, A. R., Van Horn, L., Garside, D. B., ... & Elliott, P. (2018). Relation of dietary sodium (salt) to blood pressure and its possible modulation by other dietary factors: the INTERMAP study. *Hypertension*, 71(4), 631-637.

39. Appel, L. J., Moore, T. J., Obarzanek, E., Vollmer, W. M., Svetkey, L. P., Sacks, F. M., ... & Lin, P. H. (1997). A clinical trial of the effects of dietary patterns on blood pressure. New *England journal of medicine*, 336(16), 1117-1124.

40. Mozaffarian, D., Fahimi, S., Singh, G. M., Micha, R., Khatibzadeh, S., Engell, R. E., ... & Powles, J. (2014). Global sodium consumption and death from cardiovascular causes. New England Journal of Medicine, 371(7), 624-634.

41. Caraher, M., & Perry, I. (2017). Sugar, salt, and the limits of self regulation in the food industry.

42. Hall, K. D., Ayuketah, A., Brychta, R., Cai, H., Cassimatis, T., Chen, K. Y., ... & Fletcher, L. A. (2019). Ultra-processed diets cause excess calorie intake and weight

gain: an inpatient randomized controlled trial of ad libitum food intake. *Cell metabolism,* 30(1), 67-77.

43. NCD Risk Factor Collaboration (NCD-RisC): Worldwide trends in body mass19 index, underweight, overweight, and obesity from 1975-2016: a pooled analysis of population-based measurement studies in 128.9 million children, adolescents, and adults. Lancet 390: P2627-2642, 2017.

44. Hill, J. O., Wyatt, H. R., & Peters, J. C. (2012). Energy balance and obesity. *Circulation*, 126(1), 126-132.

45. Rajasekaran A, Kalaivani M: Designer foods and their benefits: A review. J. Food Rajasekaran, A., & Kalaivani, M. (2013). Designer foods and their benefits: A review. *Journal of food science and technology,* 50(1), 1-16.

46. Kyrø, C., Tjønneland, A., Overvad, K., Olsen, A., & Landberg, R. (2018). Higher whole-grain intake is associated with lower risk of type 2 diabetes among middle-aged men and women: The Danish Diet, Cancer, and Health Cohort. *The Journal of Nutrition,* 148(9), 1434-1444.

47. Nakayama, T., Nagai, Y., Uehara, Y., Nakamura, Y., Ishii, S., Kato, H., & Tanaka, Y. (2017). Eating glutinous brown rice twice a day for 8 weeks improves glycemic control in Japanese patients with diabetes mellitus. *Nutrition & diabetes*, 7(5), e273-e273.

48. Mohan, V., Anjana, R. M., Gayathri, R., Ramya Bai, M., Lakshmipriya, N., Ruchi, V., ... & Krishnaswamy, K. (2016). Glycemic index of a novel high-fiber white rice variety developed in India—A randomized control trial study. *Diabetes technology & therapeutics,* 18(3), 164-170.

49. Wahab S, Khan S, Khattak MK *et al* (2007). Nutritional qualities of different wheat varieties in north west frontier of Pakistan. *Sarhad J. Agri.* 23 (4): 1131-1136.

50. Mallick, S. A., Azaz, K., Gupta, M., Sharma, V., & Sinha, B. K. (2013). Characterization of grain nutritional quality in wheat. *Indian journal of plant physiology,* 18(2), 183-186.

51. Aggarwal PK, Gupta HS (2016). Enhancement of nutritional quality of cereals using biotechnological options. In: P.S Kendurkar, GP, Srivastava M, Mohan & Vajpeyi (Eds) *Proc. of ICPHT* (pp48-58).

52. Shankar KS, Kumar EN, Pusphanjali *et al* (2016). Reshaping Agriculture and Nutrition Linkages: Food and Nutrition Security. ICAR-Central Research Institute, Hyderabad, India. ISBN: 978-93-80883-42-7.

53. Lafiandra, D., Riccardi, G., & Shewry, P. R. (2014). Improving cereal grain carbohydrates for diet and health. *Journal of cereal science*, 59(3), 312-326.

54. Ai Y, Jane J. (2016) Macronutrients in corn and human nutrition. Comp Rev. *Food Sci and Food Safety.* 15 (3). https://doi.org/10.1016/j.jcs.2014.01.001

55. Nawaz H, Muzaffar S, Aslam M *et al* (2018) Photochemical composition: Antioxidant potential and biological activities of corn. DOI:10.5772/ interchopen.79648.

56. Mattei, J., Malik, V., Wedick, N. M., Hu, F. B., Spiegelman, D., Willett, W. C., ... & Global Nutrition Epidemiologic Transition Initiative. (2015). Reducing the global burden of type 2 diabetes by improving the quality of staple foods: The Global Nutrition and Epidemiologic Transition Initiative. *Globalization and health*, 11(1), 23.

57. Ley, S. H., Hamdy, O., Mohan, V., & Hu, F. B. (2014). Prevention and management of type 2 diabetes: dietary components and nutritional strategies. *The Lancet,* 383(9933), 1999-2007.

58. Swinburn, B. A., Sacks, G., Hall, K. D., McPherson, K., Finegood, D. T., Moodie, M. L., & Gortmaker, S. L. (2011). The global obesity pandemic: shaped by global drivers and local environments. *The Lancet*, 378(9793), 804-814.

59. Willett, W. C., & Stampfer, M. J. (2013). Current evidence on healthy eating. *Annual review of public health*, 34, 77-95.

60. Micha, R., Shulkin, M. L., Penalvo, J. L., Khatibzadeh, S., Singh, G. M., Rao, M., ... & Mozaffarian, D. (2017). Etiologic effects and optimal intakes of foods and nutrients for risk of cardiovascular diseases and diabetes: systematic reviews and meta-analyses from the Nutrition and Chronic Diseases Expert Group (NutriCoDE). PloS one, 12(4).

61. Singh, G. M., Danaei, G., Farzadfar, F., Stevens, G. A., Woodward, M., Wormser, D., ... & Di Angelantonio, E. (2013). The age-specific quantitative effects of metabolic risk factors on cardiovascular diseases and diabetes: a pooled analysis. *PloS one*, 8(7), e65174.

62. Hefferon, K. L. (2015). Nutritionally enhanced food crops; progress and perspectives. *International journal of molecular sciences*, 16(2), 3895-3914.

63. Bang, H. O., Dyerberg, J., & Sinclair, H. M. (1980). The composition of the Eskimo food in north western Greenland. The *American journal of clinical nutrition,* 33(12), 2657-2661.

64. Bang, H. O., Dyerberg, J., & Hjørne, N. (1976). The composition of food consumed by Greenland Eskimos. *Acta Medica Scandinavica*, 200(1 6), 69-73.

65. Rao GHR (2018). Omega 3-fatty acids: Cardiovascular Disease. Cardiology and Cardiovascular Therapy. J. Cardiovasc. Ther. 10 (5): JOCCT.MS.ID.555797.

66. Rao, G. H. (2018). Arachidonic Acid Metabolism, Thrombosis, and Stroke. *J of CardiolCardiovascTher*, 11(1).

67. Rao GHR, Kishore NP, Peller JD *et al* (1987) Influence of polyenoic acids on arachidonic acid metabolism and platelet function. In: Cardiovascular Disease (Gallo l, Ed), N.Y. Plenum Press, pp495-505.

68. Astrup A, Bertram HCS, Bonjour J *et al* (2019). WHO draft guidelines on dietary saturated and trans fatty acids: time for a new approach? BMJ 366:14137. Doi:10.1136/bmj.14137.

69. Burns-Whitmore B, Haddad E, Sabate J *et al* (2014). Effects of supplementing n-3 fatty acid enriched eggs and walnuts on cardiovascular disease risk markers in healthy free-living lacto-ovo-vegetarians: a randomized, crossover, free-living

intervention study. *Nutr. J.* 13:29, Doi: 10.1186/1475-2891-13-29.

70. Fisinin VI, Papazyan TT, Surai PF (2006). Producing selenium enriched eggs and meat to improve the selenium status of the general population. *J. Nutr. Health* 10 (5): 371-376.

71. Bourre JM, Galea F (2006). An important source of omega-3 fatty acids, vitamin D & E, carotenoids, iodine, and selenium: a new natural multi-enriched egg. *J. Nutr. Health* Aging 10 (5): 371-376.

72. Bourre JM (2005). Effect of increasing the omega-3 fatty acid in the diets of animals on the animal products consumed by human. *Med. Sci.* 21 (8-9): 773-779.

73. Kouba M, Mourot J (2011). A review of nutritional effects on fat composition of animal products with special emphasis on n-3 polyunsaturated fatty acids. *Biochimie* 93 (1): 13-17.

74. Okrouhia M, Stupka R, Citek J: *et al*: (2018). Effect of duration of dietary rapeseed and soybean oil feeding on physical characteristics, fatty acids profile, and oxidative stability of pig fat. *Animals.* 8 (11); pil: E193, dloi:10.3390/ani8110193.

75. Sandstrom B, Bugel S, Lauridsen C *et al* (2000). Cholesterol-lowering potential in human subjects of fat from pigs fed rapeseed oil. *Br. J. Nutr.* 84 (2): 143-50.

76. Clark EL, Bush SJ, McCullough MEB *et al* (2017). A high resolution atlas of gene expression in the domestic sheep (*Ovies aries*). *PLoS Genet* 13 (9): e10006997. Doi: 10.1371/journal.pgen.1006997

77. Jiang Y, Xie M, Chen W *et al* (2014). The sheep genome illuminates biology of the rumen and lipid metabolism, *Science* 344 (6188): 1168-73.

78. Freeman TC, Ivens A, Baillie JK *et al* (2012). A gene expression atlas of domestic pig. *BMC Biol.* 10:90. Doi:10.1186/1741-7007-10-90.

79. Abshirini M, Mahaki B, Bagheri F *et al* (2018). Higher intake of Phytochemical-rich foods in inversely related to prediabetes: A case-control study. *Int. J. Prev. Med.* 9:64. PMID: 30147853.

80. Parveen R, Singh N (2016). A review on antidiabetic Angiosperm plants form the regions of Uttarakhand, India. *IDOR J. of Pharmacy* 6 (10):14-61. ISSN: 2250-3013, (p)-ISSN: 2319-4219.

81. Muhammad HI, Aswami MZ, Khan NAK (2016). A review on promising phytochemical, nutritional and glycemic control studies on Moringa oleifera Lam. In tropical and sub-tropical regions. *Asian Pac J. Trop. BioMed.* 6 (10): 896-902.

82. Prabhakar, P. K., & Doble, M. (2008). A target based therapeutic approach towards diabetes mellitus using medicinal plants. *Current Diabetes Reviews*, 4(4), 291-308.

83. Grover, J. K., Yadav, S., & Vats, V. (2002). Medicinal plants of India with anti-diabetic potential. *Journal of ethnopharmacology*, 81(1), 81-100.

84. Gaikwad SB, Mohan GK, Rani MS (2014). Phytochemicals for diabetes management. *Pharmaceutical Crops* 5(suppl 1:M2): 11-28.

85. Kasliwal, R. R., Bansal, M., Mehrotra, R., Yeptho, K. P., & Trehan, N. (2015).

Effect of pistachio nut consumption on endothelial function and arterial stiffness. *Nutrition*, 31(5), 678-685.

86. Cohn, J. N., Finkelstein, S., McVeigh, G., Morgan, D., LeMay, L., Robinson, J., & Mock, J. (1995). Noninvasive pulse wave analysis for the early detection of vascular disease. *Hypertension*, 26(3), 503-508.

87. Boggard HJ, Natarajan R, Henderson R *et al* (2009). Chronic pulmonary artery pressure elevation is insufficient to explain right heart failure. *Circ*. 1201951-60.

88. Hybertson, B. M., Gao, B., Bose, S. K., & McCord, J. M. (2011). Oxidative stress in health and disease: the therapeutic potential of Nrf2 activation. *Molecular aspects of medicine*, 32(4-6), 234-246.

89. Joe M McCord, Fridovich I (1978). The biology and pathology of oxygen radicals. *Ann. Int. Med*. 89 (1): 122-27.

Current Status of Researches in Biosciences (2020) : 23-31
Editor : Prof. P.C. Joshi, N. Joshi, Reshman Yasmin, and D. K. Mansotra
Today & Tomorrow's Printers and Publishers, New Delhi - 110 002

BIOCHEMICAL CHANGES IN THE HAEMOLYMPH, SILK GLAND AND FAT BODY OF THE TIGER BAND INFECTED LARVAE OF OAK TASAR SILKWORM, *ANTHERAEA PROYLEI* JOLLY

Dalip K. Mansotra and P. C. Joshi

Department of Zoology and Environmental Sciences, Gurukula Kangri Vishwavidyalaya, Haridwar, Uttarakhand, India

Email: dalipmansotra@gmail.com

ABSTRACT

The present study was carried out to analyze the variations in the biochemical constituents like protein, carbohydrates and amino acids in different body tissues viz., haemolymph, silk gland and fat body in the healthy and Tiger band infect fifth instar larvae of the oak tasar silkworm *A. proylei* Jolly. The larvae were reared during the late spring season (2015-17) at Munsyari, Uttarakhand under poly houses and were fed with *Q. semecarpifolia*. A significant decline in the concentration of the protein, carbohydrates and amino acids was recorded in different body tissues mainly due to the incidence of a deadly viral disease commonly called as Tiger band disease. Maximum decline was recorded in the silk gland protein. As a result of this disease, the larvae becomes lethargic, stops feeding and finally leads to the death. The variations in temperature and humidity were main contributing factors responsible for the incidence of the disease in the larvae.

Keywords: Oak tasar silkworm, *Antheraea proylei* Jolly, Tiger band Disease, Biochemical constituents

INTRODUCTION

Indian sericulture industry has made significant progress during the past decades, occupying the second position after China in the global silk production. As per the data available with the International Sericulture Commission, there has been a continuous increase in the silk production in India i.e. 28,523 MT, 30,348 MT, and 31,906 MT during 2015, 2016 and 2017, respectively. However, the silk production is still low as compared with the silk production of the Chinese silk industry i.e. 1, 70, 000 MT, 1, 58, 400 MT and 1, 42, 000 MT, during 2015, 2016 and 2017, respectively ("Statistics | INTERNATIONAL SERICULTURAL COMMISSION,"). The members of the genus *Antheraea* inhabit different ecological niches, widely ranging from the tropical to temperate with transitional zones. Among all the species of oak tasar silkworms, *Antheraea proylei* Jolly is considered as the most prominent, viable and commercially exploited species in the temperate regions of the Indian sub-continent. Oak tasar silkworm being semi-domestic in nature shows weak bivoltinisim, thus depending upon the geographical and agro-climatic conditions one or more crop can be obtained per year. The biology and life table parameters are greatly affected by the topology and climatic conditions of the area, thus it shows great variations both in biology and economic yield in both the tasar rearing regions i.e. north-eastern region and north-western region. In the recent past, trends in the decline of the oak tasar productivity have been observed in different regions of the state of Uttarakhand. In some regions, complete failure of the crop has been also well recorded. Several reasons may be attributed to the downfall of the oak tasar industry, among which the incidence of disease in the silkworms are on the top. As a result of the occurrence of disease in the adult and larval stages, there has been a sharp decline in the cocoons per DFL, which is mainly due to the high rate of mortality. The insects are prone to viral infections, thus studies on their pathology have been extensively carried out, especially considering two insects as animal model viz. *Bombyx mori* and *Apis mellifera.* Hughes, (1957) reported that nearly 200 species of the Lepidoptera are affected by viral diseases.

One of the most deadly viral disease affecting the oak tasar silkworm is "Tiger Band disease", with remarkable symptoms of black spots on the infected larvae. This disease results in a high rate of mortality and causes significant loss to the rearers, thus having economic importance. Initially, Jameson, (1922) observed such kind of disease in muga silkworm. In this case, the incidence of the disease was followed by the occurrence of black spots on the larval body and finally death of the larvae. The black spots were named as "Pebrine spots" by him and also reported that the disease was non-infections in nature. But the study Shantibala *et al.*, (2018) made it clear that the disease was caused by the Group I Alphabaculovirus i.e. *Antheraea proylei* nucleopolyhedrovirus (AnprNPV). They further observed that the diseases were highly infectious in nature and thus results in the multiplication of the disease, with an increased rate of mortality.

The occurrence of the disease is mainly attributed due to the unfavorable climatic conditions. Any rise in the temperature with reduced humidity or fall in the temperature with high humidity leads to the occurrence of the disease. The manifestation symptoms comprise of the dull behavior by the larvae, stops feeding, restlessness, loss of appetite, the occurrence of black spots similar to the strips of the tiger across the body length of the silkworm. Temperature exhibits a great deal of impact on the pathogenicity and occurrence of the different disease in the silkworm races and thus affecting the sericulture industry (Rahmathulla, *et al.*, 2012; Ram, *et al.*, 2016). The viral infection affects the insect metabolism and in case of serious viral infection in the silkworm, there has been a continuous decline in the concentration of the important macronutrients which are essential for growth and development. The poor understanding of the pathology and occurrence of the disease, particularly the morphology, physiology and biochemical composition of the silkworms results in the increased rate of disease. The earlier studies have been confined to a particular race of the silkworm mostly region specific. No sequential study has been carried out in the western Himalayan region of the Uttarakhand, comparing the variations in the biochemical constituents during the progression of the tiger band disease. Thus this study was designed to study the changes in the concentration of biochemical constituents in different body tissues of the fifth instar larvae under the influence of the Tiger band disease.

MATERIAL AND METHODS

The present study on the effect of incidence of tiger band disease on the biochemical constituents of different body tissues like haemolymph, silk gland and fat body of fully grown fifth instar larvae of oak tasar silkworm, *A. proylei* Jolly was carried out at Munsyari, Uttarakhand (30°04’96"N 80°28’95"E) during the spring seasons by selecting *Q. semecarpifolia* as host plant. For the estimation of biochemical changes both healthy and diseased fifth instar larvae were collected on the eight day after 4th moulting. The quantitative estimation of protein in haemolymph, silk gland and fat body was carried out as per the standard methods described by Lowry *et al.* (1951) using crystalline bovine serum albumin (BSA) as standard. The values of the protein content have been expressed as mg/ml in haemolymph, mg/g in silk gland and fat body. The quantitative estimation of total sugar present in the haemolymph, silk gland and fat body of silkworm (healthy and diseased) were estimated by Anthrone method (Dubois *et al.*, 1956) using glucose as standard, while the total free amino acids content in the haemolymph, silk gland and fat body of healthy and diseased Vth instar larvae of oak tasar silk worm was estimated as per the standard procedure of Spies *et al.*, (1957) and modified by Singh and Agarwal (1989). The values of the carbohydrates and free amino acids content have been expressed as mg/ml in haemolymph, mg/g in silk gland and fat body.

RESULTS AND DISCUSSION

The results of the present study revealed a significant decline in the concentration of protein, carbohydrates and amino acids in the haemolymph, silk gland and fat body of the tiger band infected larvae. The concentration of haemolymph protein, carbohydrates and amino acids in the healthy fifth instar larvae was found to be ranging from 23.47-25.17 mg/ml, 10.81-11.43 mg/ml and 4.44-5.11 mg/ml, while there was a significant decline in their concentration in case of the tiger band infected larvae i.e. 13.51-14.23 mg/ml, 4.77-5.30 mg/ml and 2.82-3.96 mg/ml, respectively. This decline in the essential macromolecules results in retarded growth and development of the larvae and increased mortality. The results obtained for the haemolymph showed approximately 50 percent decline in the concentration of the protein, carbohydrates and amino acids when infected with Tiger band disease. The results of the present study are in conformation with the studies of several earlier researchers. Reeta *et al.*, (2016) observed that the tiger band disease causes a major effect on the growth and development of the oak tasar silkworm even during the favorable climatic conditions. They further observed that the incidence of disease caused major damage to productivity and lead to 70-80 % of larval and pupal mortality in North-eastern India. Since then, tiger band disease has become a great threat to the oak tasar industry both in the North-eastern and North-western regions of India. Borphukan and Bhola (2013) observed changes in the haemolymph protein and carbohydrate during the larval development in *Samia ricini* infected with *Nosema* and found that the content of protein and carbohydrate continuously decline with the progression of the infection. The content of the haemolymph protein and carbohydrate in the infected larvae was found to be 23.33 ± 1.63 mg/ml and 23.66 ± 3.67 mg/ml as compared with the healthy larvae i.e. 60.00 ± 3.68 mg/ml and 30.54 ± 1.62 mg/ml, respectively. The results obtained during the study for haemolymph protein, carbohydrates and amino acids are given in the Table 1 and Figure 1.

Table 1: Biochemical changes in the haemolymph of Healthy and diseased Fifth Instar larvae of Oak tasar silkworm *Antheraea proylei* J. fed with *Q. semecarpifolia* during spring season of 2015-16, 2016-17 and 2017-18

Year	Condition	Protein (mg/ml)	Carbohydrate (mg/ml)	Amino acids (mg/ml)
2015-16	Healthy Larvae	24.38±0.59	11.16±0.62	5.11±0.34
	Diseased larvae	13.51±0.55	5.12±0.33	3.60±0.59
2016-17	Healthy Larvae	25.17±0.86	10.81±0.63	4.44±0.71
	Diseased larvae	14.23±0.53	4.77±0.72	3.96±0.87
2017-18	Healthy Larvae	23.47±0.57	11.43±0.97	4.78±0.74
	Diseased larvae	13.90±0.60	5.30±0.62	2.82±0.60

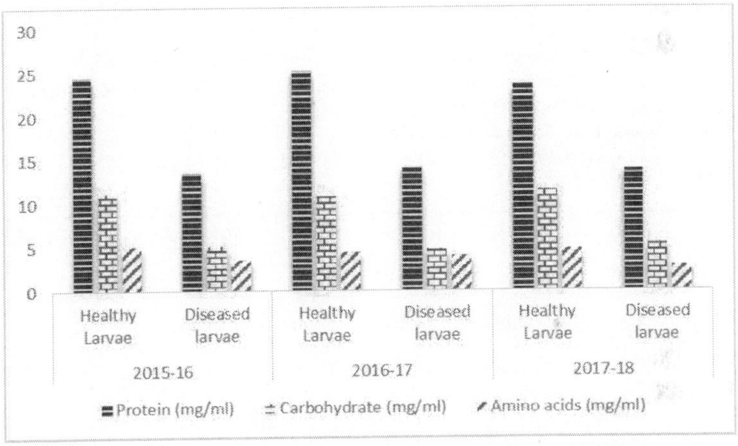

Fig. 1: Biochemical changes in the haemolymph of Healthy and diseased Fifth Instar larvae of Oak tasar silkworm *Antheraea proylei* J. fed with *Q. semecarpifolia* during spring season of 2015-16, 2016-17 and 2017-18

Similarly, the concentration of protein, carbohydrates and amino acids in the silk gland of the healthy and tiger band infected larvae showed great variations. The concentration of silk gland protein, carbohydrates and amino acids in the healthy fifth instar larvae was found to be ranging from 79.27-94.02 mg/g, 17.13-18.34 mg/g and 21.18-26.38 mg/g, while there was a significant decline in their concentration in case of the tiger band infected larvae i.e. 19.82-23.51 mg/g, 11.36-12.18 mg/g and 10.40-12.63 mg/g, respectively. Among these maximum decline was observed in the silk gland protein, followed by amino acids and carbohydrates. Rajakumari, (2005) examined the histopathological changes in the different regions of the silk gland viz. anterior, middle and posterior, in the disease infected larvae and found various cytomorphological changes such as hypertrophy, necrosis, and vacuolation. The diseased larvae became lethargic and stopped feeding, leading to starvation, which ultimately resulted in reduced secretions of enzymes in the anterior silk gland and finally in non-release of silk thread during the spinning stage. Among all the three regions of the silk gland, the middle region of the silk gland is the major site for the production and storage of sericin and fibroin. Due to the disease, the middle silk gland got greatly damaged and resulted in reduced sericin and fibroin content. The progression of the disease greatly affects the physiology and biochemistry of the silk gland. The results of the present study are well supported by the studies of earlier researchers (Gani *et al.*, 2017; Mohanta *et al.*, 2015 and Singh, *et al.*, 2011). The results obtained during the study for haemolymph protein, carbohydrates and amino acids are given in the Table 2 and Figure 2.

Table 2: Biochemical changes in the Silk gland of Healthy and diseased Fifth Instar larvae of Oak tasar silkworm *Antheraea proylei* J. fed with *Q. semecarpifolia* during spring season of 2015-16, 2016-17 and 2017-18

Year	Condition	Protein (mg/g)	Carbohydrate (mg/g)	Amino acids (mg/g)
2015-16	Healthy Larvae	79.27±0.81	18.34±0.81	21.18±0.57
	Diseased larvae	23.51±0.76	11.36±0.60	10.40±0.66
2016-17	Healthy Larvae	86.27±0.43	17.54±0.76	24.40±0.82
	Diseased larvae	19.82±0.67	11.57±0.76	12.22±0.88
2017-18	Healthy Larvae	94.02±0.85	17.13±0.50	26.38±0.61
	Diseased larvae	21.57±0.84	12.18±0.83	12.63±0.69

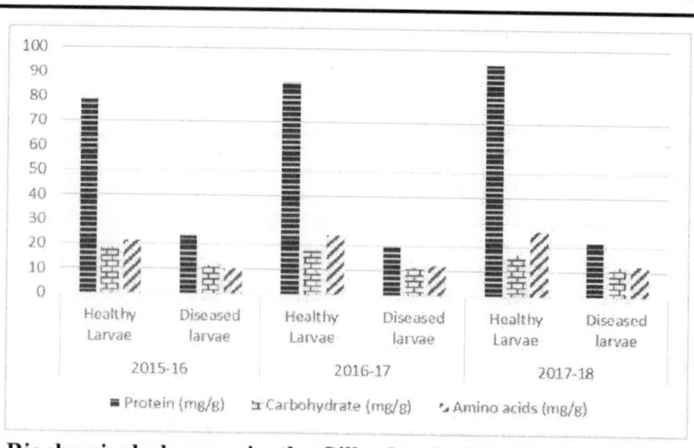

Fig. 2: Biochemical changes in the Silk gland of Healthy and diseased Fifth Instar larvae of Oak tasar silkworm *Antheraea proylei* J. fed with *Q. semecarpifolia* during spring season of 2015-16, 2016-17 and 2017-18

In case of the fat body, there was maximum decline in the concentration of protein, followed by carbohydrates and amino acids in case of the Tiger band infected larvae. The concentration of fat body protein, carbohydrates and amino acids in the healthy fifth instar larvae was found to be ranging from 61.66-66.26 mg/g, 14.61-17.18 mg/g and 17.72-18.31 mg/g, while there was a significant decline in their concentration in case of the tiger band infected larvae i.e. 15.41-16.57 mg/g, 8.92-9.70 mg/g and 11.55-13.20 mg/g, respectively. Gupta, *et al.*, (2018) studied the influence of temperature on the free amino acids contents in the fat body of the multivoltine mulberry silkworm *Bombyx mori* Linn. and observed that the free amino acids contents in the fat body are significantly influenced by the varying temperature. A maximum of 28.44 µg/mg of free amino acids was recorded from the fat body of the healthy pupae reared at a temperature of 26°C, while minimum content i.e. 10.12 µg/mg was observed in the fat body of the adult reared at a temperature of 14°C. Though the temperature was found to be more suitable during the autumn season for the better growth and development of the larvae, sudden rise in the relative humidity and rainfall increase the incidence of disease and thus results in increased mortality rate. The results of the present study are in corroboration with the findings of the earlier researchers (Chen *et al.*, 2018; Etebari, *et al.*, 2007 and Gupta *et al.*, 2018). The results obtained during the study for haemolymph protein, carbohydrates and amino acids are given in the Table 3 and Figure 3.

Table 3: Biochemical changes in the Fat body of Healthy and diseased Fifth Instar larvae of Oak tasar silkworm *Antheraea proylei* J. fed with *Q. semecarpifolia* during spring season of 2015-16, 2016-17 and 2017-18

Year	Condition	Protein (mg/g)	Carbohydrate (mg/g)	Amino acids (mg/g)
2015-16	Healthy Larvae	63.11±0.75	15.83±0.56	18.18±0.58
	Diseased larvae	15.78±0.65	9.39±0.33	13.20±0.84
2016-17	Healthy Larvae	66.26±0.85	17.18±0.67	17.72±0.42
	Diseased larvae	15.41±0.51	9.70±0.61	11.55±0.78
2017-18	Healthy Larvae	61.66±0.62	14.61±1.06	18.31±0.47
	Diseased larvae	16.57±0.56	8.92±0.56	12.94±0.30

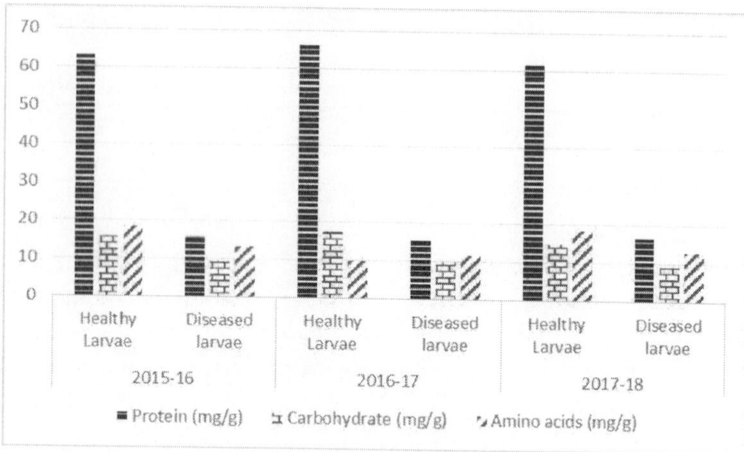

Fig. 3: Biochemical changes in the Fat body of Healthy and diseased Fifth Instar larvae of Oak tasar silkworm *Antheraea proylei* J. fed with *Q. semecarpifolia* during spring season of 2015-16, 2016-17 and 2017-18

Thus, the results of the present study showed that the incidence of the Tiger band disease in the fully grown fifth instar larvae results in decline in the concentration of the essential macronutrients i.e. proteins, carbohydrates and amino acids in different body tissues. This decline in turn results in retarded growth and development of the larvae, leading to increased mortality rate and ultimately crop loss.

REFERENCES

1. Borphukan, A., & Bhola, R. K. (2013). Hemolymph Protein and Carbohydrate Alteration during Larval Development in *Samia ricini* Infected with *Nosema*. *International Journal of Science and Research* 4(5), 4.

2. Chen, X.-D., Wang, Y.-F., Wang, Y.-L., Li, Q.-Y., Ma, H.-Y., Wang, L., Xu, S. Q. (2018). Induced Hyperproteinemia and Its Effects on the Remodeling of Fat Bodies in Silkworm, *Bombyx mori*. *Frontiers in Physiology*, 9.

3. Dubois, M., Gilles, K. A., Hamilton, J. K., Rebers, P. T., & Smith, F. (1956). Colorimetric method for determination of sugars and related substances. *Analytical chemistry*, 28(3), 350-356.

4. Etebari, K., Ebadi, R., & Matindoost, L. (2007). Physiological Changes of Silkworm (*Bombyx mori* L.) Larvae Fed on Mulberry Leaves Supplemented with Nitrogenous Compounds. *Journal of the Entomological Research Society*, 9.

5. Gani, M., Chouhan, S., Babulal, R. K., Gupta, G. K., Kumar, N. B., Saini, P., & Ghosh, M. K. (2017). *Bombyx mori* nucleopolyhedrovirus (BmBPV): its impact on silkworm rearing and management strategies. *Journal of Biological Control*, 31(4), 189–193.

6. Gupta, S. K., Gaur, K. P., & Upadhyay, V. B. (2018). Influence of temperature on

the free amino acids. *International Journal of Scientific Research.* Vol.7 (2): 35-36

7. Hughes, K. (1957). An annotated list and bibliography of insects reported to have virus diseases. *Hilgardia,* 26(14), 597–629.

8. Jameson, A. P. (1922). Report on the Diseases of Silkworms in India. Report on the Diseases of Silkworms in India.

9. Lowry, O. H., Rosebrough, N. J., Farr, A. L., & Randall, R. J. (1951). Protein measurement with the Folin phenol reagent. *Journal of biological chemistry,* 193, 265-275.

10. Luikham, R., Shantibala, T., James, T. K., and Bidyapati Devi, L. B. (2016): Characterization and evaluation of wild sericigenous insects of Manipur: exploration for conservation. *Sericologia*, 56 (1):51-66.

11. Mohanta, M. K., Saha, A. K., Saleh, D. K. M. A., Islam, M. S., Mannan, K. S. B., & Fakruddin, M. (2015). Characterization of *Klebsiella granulomatis* pathogenic to silkworm, *Bombyx mori* L. 3 Biotech, 5(4), 577–583. https://doi.org/10.1007/s13205-014-0255-4

12. Rahmathulla, V. K., Kumar, C. K., Angadi, B. S., & Divaprasad, V. (2012). Influence of weather factors on incidence and intensity of Microsporidiosis in silkworm (*Bombyx mori* L.). *J. Entomol,* 9(5), 266–273.

13. Ram, R. L., Maji, C., & Bindroo, B. B. (2016). Impact of climate change on sustainable sericultural development in India. *International Journal of Agriculture Innovations and Research*, 4(6), 2319–1473.

14. Reeta L., Shantibala T., James T. K. and Devi L. B. (2016) Characterization and evaluation of wild sericigenous insects of Manipur: Explore for conservation. *Sericologia*, 56 (1): 51-66.

15. Sheeba Rajakumari, D. V. (2005). Bacterial diseases in the silkworm *Bombyx mori* L in Tirunelveli district and management techniques to improve economic traits. Ph. D thesis submitted to M S University.

16. Singh, D. K., & Agarwal, R. A. (1989). Toxicity of Piperonyl Butoxide—Carbaryl Synergism on the Snail *Lymnaea acuminata. Internationale Revue der gesamten Hydrobiologie und Hydrographie,* 74(6), 689-699.

17. Singh, G. P., Sinha, A. K., Kumar, P. K., & Prasad, B. C. (2011). Characterization and Identification of Bacteria Infecting Indian Tropical Tasar Silkworm, *Antheraea mylitta* D. *Research journal of microbiology*. Vol. 6(12):891-897

18. Spies, J.R. 1957. Colometric procedures for amino acids, p. 467–477. *In: S.P. Clowick and N.O. Kaplan (eds.). Methods in enzymology. vol. 3. Academic, New York*

19. Statistics | International Sericultural Commission. (2019). Retrieved May 31, 2019, from https://inserco.org/en/statistics

Current Status of Researches in Biosciences (2020) : 33-39
Editor : Prof. P.C. Joshi, N. Joshi, Reshman Yasmin, and D. K. Mansotra
Today & Tomorrow's Printers and Publishers, New Delhi - 110 002

APPLICATIONS OF NANOTECHNOLOGY IN HUMAN AND VETERINARY MEDICINE

Soumendra Darbar

Bioequivalence Study Centre, Department of Pharmaceutical Technology, Jadavpur University, Raja S C Mallick Road, Kolkata-700032, W.B.

E-mail: darbarsoumendra@gmail.com

ABSTRACT

Nanomedicine is a branch of medicine that applies the knowledge and tools of nanotechnology to the prevention and treatment of life threatening diseases through inhibition of ROS. Therapeutic potential and antioxidant activity of Silver Nanoparticle coupled with *Andrographis paniculata* (AP-Ag NP) was assessed against CCl_4 induced oxidative stress at tissue level. The present study aimed to evaluate the comparative efficacy of AP-Ag NP against carbon tetrachloride (CCl_4) induced oxidative stress. Male Swiss albino mice were exposed to carbon tetrachloride (CCl_4) for 28 days and were co-administered orally with AP-Ag NP (50 mg/kg each) to evaluate effects on heme synthesis pathway, hepatic oxidative injury and antioxidant potential. The results suggest a significantly elevated ROS, decreased blood and hepatic GSH levels, superoxide dismutase and catalase activities after carbon tetrachloride (CCl_4) exposure. Co-administration of nano-antioxidants (AP-Ag NP) increased SOD, CAT and GSH levels and was also beneficial in the recovery of oxidative injury and restoring inhibited aminolevulinate dehydratase (ALAD) activity. Our results suggest that Silver Nanoparticle coupled with *Andrographis paniculata* (AP-Ag NP) have the potential antioxidant effect in mice.

KEYWORDS: AgNP; *Andrographis paniculata*; Protection; Free radicals; Lipid peroxidation

INTRODUCTION

Nanomedicine is the medical application of nanotechnology

(Wagner *et al.,* 2006). Nanomaterials synthesized through green route can be useful for both in vivo and in vitro biomedical research and applications. Chronic diseases such as diabetes, cardiovascular disease and liver fibrosis increase oxidants and decrease antioxidants in patients(Patra and Baek, 2014; Csányi and Miller, 2014). Clinical and experimental studies have shown that disturbing the balance of the oxidant–antioxidant system can contribute to the pathogenesis of liver and kidney fibrosis (Ahmadvand *et al.,* 2012; Ahmadvand *et al.,* 2013 and Salata, 2004).

Green synthesis is defined as the use of environmentally compatible materials such as bacteria, fungi and plants in the synthesis of nanoparticles (Patra and Baek, 2014). In this regard, leaf extract of *Andrographis paniculata* (commonly known as Kalmegh) a species of family Acanthaceae was used for bioconversion of silver ions to nanoparticles. The effect of concentration of metal ions and concentration of leaf extract quantity were also evaluated to optimize route to synthesise silver nanoparticle. The method applied here is simple, cost effective, easy to perform and sustainable (Gannimani *et al.,* 2014).

Radical scavengers (RS) are highly reactive species that can potentially abstract hydrogen atoms from activated bonds of biological materials under physiological environment. RS plays crucial role in neutralizing the direct attack of reactive oxygen species (ROS) that prevents a number of acute and chronic cellular implications by inhibiting the oxidative damage(Ríos-Silva *et al.,* 2014 and Csányi and Miller, 2014). So, there is a worldwide need to find out a new drug that can fight against these diseases and ameliorate the redox healing. The present study was aimed at investigating the efficacy of Silver Nanoparticle Coupled with *Andrographis paniculata* (AP-Ag NP) against CCl_4 induced oxidative injury. Liver & Kidney was selected as the major organ for investigation as it is the major target for free radical attack leading to lipid peroxidation.

MATERIAL AND METHODS

In this study we synthesized of silver nanoparticles (NPs) using *Andrographis paniculata.* Initially, 50 mM of silver nitrate ($AgNO_3$) solution was prepared. Then plant extract was added drop wise to it under stirring. The reaction mixture became yellowish to brownish and precipitate of silver was formed. The reaction mixture was left for 30 min for complete reduction to silver nitrate. Then the precipitate was collected by centrifugation at 15000 rpm for 10 min at 4 °C. Characterization of synthesized AP-Ag NP was carried out by UV-VIS spectra, XRD patterns and FTIR. In the in vivo experiment we developed liver and kidney fibrosis in animal model using CCl_4.

EXPERIMENTAL DESIGN

Groups	Treatment
Control	Normal Saline for 28 days
CCl_4 treated	Received 1:1 (v/v) CCl_4 in olive oil for 28 days
CCl_4 + AP-AgNP	Received 1:1 (v/v) CCl_4 in olive oil along with 0.50 g/kg /day for 28 days
AP-AgNP	0.50 g/kg /day for 28 days

After the experimental period liver and kidney function enzymes like AST, ALT, ALP, GGT BUN, creatinine and uric acid were measured using biochemical kits. Liver MDA content and different antioxidant enzymes like SOD, CAT, GPx and GSH were measured. Histology of the liver was done as per standard method. H&E Masson's trichrome (MT) and Sirius red (SR) staining were performed to quantify the extent of fibrotic damage in liver. Histopathological changes were examined under the microscope (Olympus BX51).

RESULTS AND DISCUSSION

In this study, AgNPs were synthesized successfully using the leaf extract of *Andrographis paniculata* as a novel bio-reductant. Plant extract being very eco-friendly and cost effective can be used for the large scale synthesis of AgNPs in nanotechnology processing industries. The present study synthesized AgNPs from *Andrographis paniculata* leaf extract seem to be promising antioxidant activity which can be useful for the development of a new drug for biomedical applications.

It was observed that AP-AgNP significantly attenuate CCl_4 induced alteration in the activities of liver function enzymes more effectively. A significant recovery of liver function enzymes was observed after co-administration of AP-Ag NP. Administration of CCl_4 significantly decreases the essential antioxidant effect in the liver. A significant recovery in GSH, CAT and SOD was observed after co-administration of AP-Ag NP. Histology strongly proved that AP-Ag NP attenuate CCl_4 induced redox healing in the *in-vivo* model.

Cellular oxidative stress was evident by elevated ROS level, reduced glutathione level, increased lipid peroxidation and impaired antioxidant defense status. Our study reported elevated ROS levels in CCl_4-treated group, suggesting free radical generation leading to oxidative stress conditions. Increased lipid peroxidation like elevated TBARS and reduced GSH, further signifies oxidative stress condition. Concomitant administration of Ag nanoparticle coupled with *Andrographis paniculata* led to pronounced recovery, suggesting it become a more effective scavenger of free radicals. ROS generation leads to impaired cellular antioxidant defense system. GSH levels decreased after CCl_4 treatment, possibly to its increased utilization in neutralizing free radicals. Glutathione is the major form of cellular glutathione and earlier reports support our results for the protective efficacy of developed silver nanoparticle. Elevated ROS levels are implicated in the damage of biological molecules such as lipids, which are altered by peroxidation. Elevation in TBARS is an indicator of lipid peroxidation under oxidative stress condition. We observed a significant elevation in TBARS level following CCl_4 exposure and back to normal level after NP administration. Interestingly the level of plasma AST, ALT, ALP and GGT were restored to normal in the animals co-exposed to AP-Ag NP.

We also determined the antioxidant profile. SOD prevents the harmful effects of superoxide ion by converting them into less toxic hydrogen peroxides which subsequently splits into nontoxic water and oxygen molecule by catalase activity. Catalase is another major antioxidant enzyme whose activity decreases during oxidative stress, leading to H_2O_2 accumulation and finally peroxidation of lipids. We observed decreased hepatic SOD activity in our study. SOD is one of the components of intrinsic antioxidant defence system, and is responsible for dissemination of Superoxide radicals. During oxidative stress the body uses its defence mechanism to minimize the process of lipid peroxidation by using the antioxidant enzymes such as SOD, thus, the activity of this enzyme become higher in early stages of insult, but if the insult continue, the enzyme become depleted which means that in advance stages of per-oxidation the activity of SOD declined. We observed a significant increase in hepatic SOD and Catalase activity on NP exposure which responded favorably to the co-administration of antioxidants.

Table 1: Effect of AP-AgNP on body weight changes on CCl_4 Exposure in mice

Groups	Initial BW (g)	Final BW (g)	BW gain or loss (g)
Control	25.14±0.13	29.37±0.11	4.23±0.042
CCl_4	24.98±0.14	26.01±0.14	1.03±0.051
CCl_4+AP-AgNP	25.22±0.11	28.99±0.12	3.77±0.039
AP-AgNP	25.31±0.12	29.48±0.11	4.17±0.044

Values are means ± SEM (n=6)

Table 3: Effect of AP-AgNP on blood biochemical and antioxidant variables on CCl_4 Exposure in mice

Blood	Control	CCl_4	CCl_4 + AP-AgNP	AP-AgNP
ROS (FIU)	412.59±14.3	736.13±17.8#	492.81±11.24	462.14±12.19
SOD (U/mg)	106.14±8.1	72.05±6.9#	98.34±7.1**	100.02±6.2**
CAT (U/mg)	232.62±21.3	161.1±19.4#	212.4±16.2**	224.9±13.1**
GSH (mg/g)	38.39±0.49	21.92±0.51#	33.12±0.81**	33.66±0.41**
MDA (nmol/g)	34.16±2.14	76.39±3.02#	40.12±3.44**	36.13±2.91**

Values are means ± SEM (n=6), a $P<0.05$, significant change with respect to control group, changes between # $P<0.001$ compared with respect to control group

** $P<0.001$ compared with respect to CCl_4 group.

Table 4: Effect of AP-AgNP on renal function test variables on CCl_4 Exposure in mice

Renal Function Test	Control	CCl_4	CCl_4 + AP-AgNP	AP-AgNP
Urea (mg/dl)	**32.52±2.5**	**64.16±3.1#**	**38.06±2.9**	**39.16±3.4**
Uric acid (mg/dl)	**1.410±.24**	**2.920±0.22#**	**1.850±0.14****	**1.510±0.31****
Creatinine (mg/dl)	**0.510±0.05**	**0.84±0.04#**	**0.56±0.04****	**0.59±0.06****
BUN (mg/dl)	**15.2±1.8**	**35.14±2.3#**	**17.03±2.1****	**16.22±1.8****

Values are means ± SEM (n=6), # $P<0.05$, significant change with respect to control group, ** $P<0.001$ significant change with respect to CCl_4 group

Characterization of Silver nanoparticles (AgNPs)

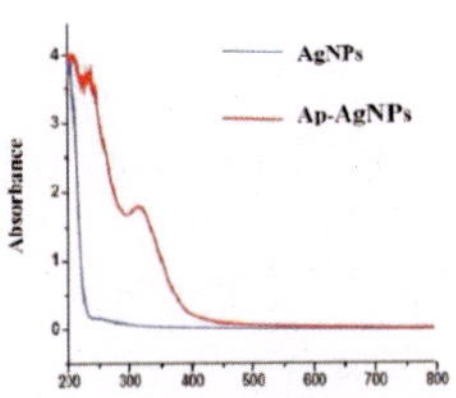

(A). UV-Vis Spectra of silver nanoparticles of *A. paniculata*

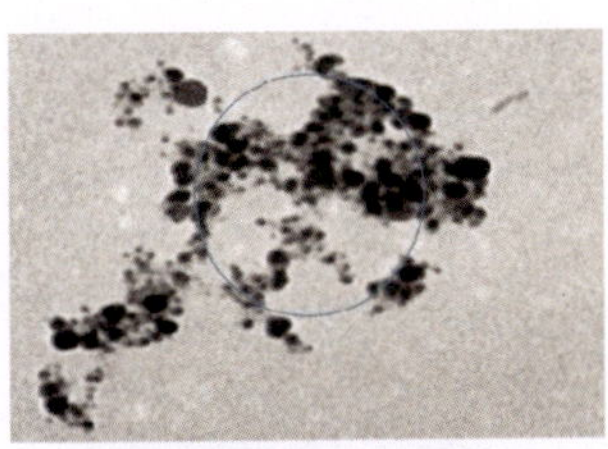

(B). Image of TEM of silver nanoparticles of *A. paniculata*

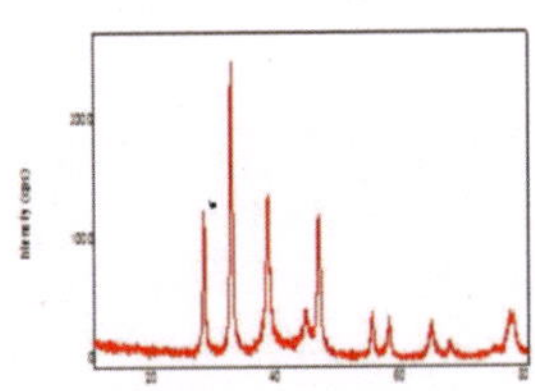

(C). 5 XRD of silver nanoparticles of *A. paniculata*

Fig. 1: Characteristics Ag nanoparticle

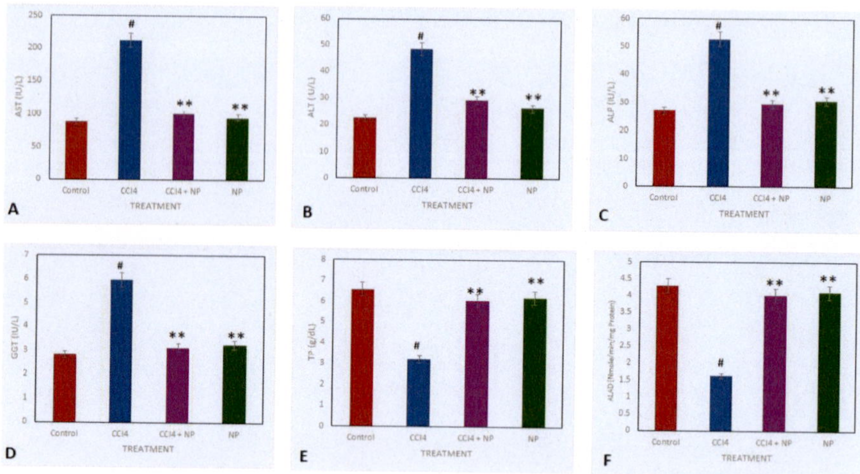

Fig. 2: Effect of AP-AgNP on Liver biochemistry on CCl_4 exposed mice

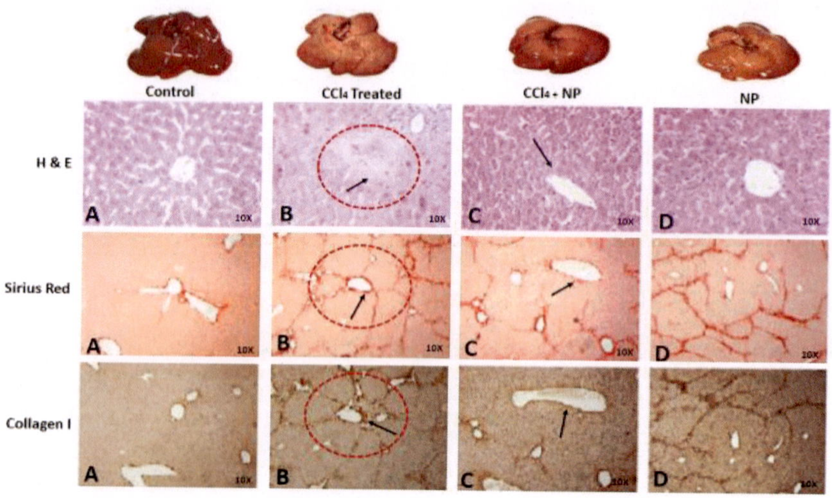

Fig. 3: Effect of AP-AgNP Histopathology of Liver on CCl_4 exposed mice

CONCLUSION

In this study, AgNPs were synthesized successfully using the leaf extract of *Andrographis paniculata* as a novel bio-reductant. Plant extract being very eco-friendly and cost effective can be used for the large scale synthesis of AgNPs in nanotechnology processing industries. The present study synthesized AgNPs from *Andrographis paniculata* leaf extract seem to be promising antioxidant activity which can be useful for the development of a new drug for biomedical applications. This biological chemistry approach towards the synthesized AgNPs through green route is highly essential effort being addressed in nanomedicine because of its varied advantages.

ACKNOWLEDGMENTS

The authors are thankful to University Grant Commission, New Delhi, India for providing financial grants.

REFERENCES

1 Ahmadvand H, Mabuchi H, Nohara A, Kobayahi J, Kawashiri MA. (2013) Effects of coenzyme Q (10) on LDL oxidation in vitro. *Acta Med Iran.*, 51(1):12-18.

2 Ahmadvand H, Tavafi M, Khosrowbeygi A. (2012) Amelioration of altered antioxidant enzymes activity and glomerulosclerosis by coenzyme Q10 in alloxan induced diabetic rats. *J Diabetes Complications,* 26 (6): 476-482.

3 Csányi G, Jr Miller FJ. (2014) Oxidative stress in cardiovascular disease. *Int J Mol Sci.*, 15(4):6002-6008.

4 Gannimani, R., Perumal, A., Krishna, S., Sershen, Mishra, A., Govender, P. (2014) Synthesis and antimicrobial activity of silver and gold nanoparticles produced using aqueous seed extract of *Protorhus longifolia* as reducing agent, *Dig. J. Nanomater. Biostructures*, 9, 1669-1679.

5 Patra, J.K., Baek, K.H. (2014) Green Nanobiotechnology: Factors Affecting Synthesis and Characterization Techniques, *J. Nanomater.* 219-230.

6 Ríos-Silva M, Trujillo X, Trujillo-Hernández B, Sánchez-Pastor E, Urzúa Z, Mancilla E, Huerta M. (2014) Effect of chronic administration of forskolin on glycemia and oxidative stress in rats with and without experimental diabetes. *Int J Med Sci.*, 11(5):448-452.

7 Salata, O. (2004) Application of nanoparticles in biology and medicine, *J Nanobiotechnol*, 2, 1-6 (2004).

8 Wagner, V., Dullaart, A., Bock, AK., Zweck, A. (2006) "The emerging nanomedicine landscape", *Nature Biotechnology*, 24 (10), 1211–1217.

Current Status of Researches in Biosciences (2020) : 41-71
Editor : Prof. P.C. Joshi, N. Joshi, Reshman Yasmin, and D. K. Mansotra
Today & Tomorrow's Printers and Publishers, New Delhi - 110 002

LIVESTOCK FOR NUTRITIONAL SECURITY, LIVELIHOOD AND PROSPERITY OF FARMERS

M. P. Yadav

IVRI Deemed University, Izatnagar, SVP University of Agriculture & Technology, Meerut

Email: yadav_mp@hotmail.com

ABSTRACT

Livestock (LS) sector, inclusive of Dairy, Poultry & Fishery, contribute significantly towards Agricultural GDP, especially in developing countries in Asia, Africa and Latin America. Besides being an integral component of national GDP and economy, food, nutritional and livelihood security, employment generation, gender equity, social security, wellbeing and cultural heritage, it serves as source of income on regular basis, and a moving bank on hooves with easy liquidity. As compared to crop agriculture, livestock is more resilient to climatic shocks. It also facilitates organic farming and natural resource management. Livestock sector contributes about 40% of the agricultural GDP globally, which may be as high as 50% or more in mountainous, semi-arid and drought prone regions. The LS promotes sustainable agriculture, rural development with climate justice and inclusive growth. It serves as cushion to sustain the agricultural GDP/GVP (gross value added). The growth of Livestock Sector was 1.6 times more than the Crop Sector and 1.3 times more than Agricultural Sector as a whole in 1990s and 2000s. During the 11th Five Year Plan, average growth of Livestock Sector was 4.15% compared to 3.6% for agriculture as a whole. Animal husbandry, has been growing at sustainable rate and is ahead among all sub-sectors of agriculture. The GVA from livestock sector contributed 4% to the overall national GVA and 26.7% to the agricultural GVA in 2014- 15. According to estimates of the Central Statistics Office (CSO), the value of output from livestock sector was about Rs. 9,17,910 crore at current prices during 2016-17 which is about 31.25% of the value of output from agricultural and allied sector. During the decade of 2004-05 to 2014-15, crop, livestock and fisheries sub-sectors had growth of 2.93%, 6.11% and 5.13% per annum, respectively. It is worth mentioning that the LS sector never experienced a negative growth during the last three and a half decades and thus serve as engine of growth for the agriculture sector as a whole, and can be relied upon for risk mitigation and minimizing the losses to the farmers even if the outcomes from crop and others sub-sectors happens to be worst as a result of drought, floods, pests or disease etc. Livestock Sector serves as a source of employment and Livelihood to 70% of the rural population in India.

Milk is the largest agricultural commodity in India. In value term, the value of milk is more than the combined value of Wheat and Rice - the two important staple crops. Comparison of value of output from Livestock and Crop Sectors from 1980-81 to 2009-10 showed that Livestock production was more stable. For sustainable agriculture, livestock is the best supplementary/complementary enterprise as the waste of one sector is used as input by the other Sector. Low input-low output-based Animal Husbandry has sustained since centuries in India by practicing Vedic Agriculture based on organic farming. In Maharashtra, it was observed that there was no farm stressed related suicides among the farmers keeping Livestock. In the present-day scenario, Livestock, Poultry, Fishery and Horticulture are best options for higher and sustainable growth in agriculture in India. For sustainable as well as more profitable farming, diversification in Agriculture in Farming System Mode preferably with Livestock component (s) is more profitable. Livestock is more gender sensitive as about 70% of total workers engaged in livestock are women.

Livestock is best suited for Inclusive growth and to reduce rural – urban disparity in incomes. Livestock distribution is more egalitarian than land distribution. Livestock growth is more Pro-Poor and reduction in rural poverty is more responsive to growth in LS Sector than in Crop Sector. Livestock contributes nearly half of the income of small holder and land less farmers, accounting for 80% of the farming families having 75% of livestock resources. States having higher growth in livestock sector in India have less rural poverty. In the last five decades, India has witnessed several agricultural revolutions, namely the Green Revolution (wheat and rice), White Revolution (milk), Blue Revolution (fish) and Pink Revolution (meat). To accomplish these, technology application, Innovation, Policy support, Public and Private Sector participation and political will played important role. The lessons learnt subsequent to green revolution need to be addressed for achieving evergreen revolution.

Various aspects pertaining to enhancing farmers' income through animal husbandry and dairying with increased and sustainable production of milk, meat, egg, fish and other animal products will be discussed with emphasis on newer technologies, innovation, best practices, strategies and policy support along with way forward. Value addition and value chains in livestock origin food, HACCP, promotion of Farmers' Company (FPCs and FPOs) and SHGs for production, value addition and marketing of livestock products, production of mozzarella cheese from buffalo milk, promotion of livestock products' exports by following suitable EXIM policy, conservation and development of indigenous milch breeds of cattle as envisaged under Rashtriya Gokul Mission project of the government of India, developing modern livestock and poultry markets throughout the country including rural areas, performance recording and identification of the animals, reduction in the unproductive animal population by using sexed semen and high yielding elite germ plasm, reorienting breeding policy and developing breeding Acts by the States and the Central government, registration of livestock breeds and protection of livestock keepers rights, addressing shortage of feed and fodder, increasing AI coverage in cattle and buffalo, production of certified hygienic semen to meet the requirement of the country, capacity building and strengthening of veterinary and extension services at the farmers door, upgrading of the slaughter houses, farmer friendly livestock insurance; stray animal problem, pro-poor livestock policies, human resource development, reaching to the farmers; policy support for credit, subsidy, price regulation; application and preparedness for future science for sustainable development of livestock and poultry sectors will be deliberated.

KEYWORDS: Livestock, Nutritional Security, GDP.

LIVESTOCK AND LIVELIHOOD

India has a large livestock population reared under varied production systems and agroclimatic conditions. As per the 19th livestock Census (2012) there are about 300 million bovines, 65.07 million sheep, 135.2 million goats and about 10.3 million pigs in the country. The country has about 58%, 15%, 18%,.7% and 5% of world's buffalo, cattle, goat, sheep and chicken population, respectively. LS sector plays a multi-faceted role in providing livelihood to more than 60% of the rural population and employs about 8% of the country's labour force (11th Five Year Plan: 2007-12 report). Besides their economic benefits and providing a steady stream of food and revenue for households, it provides employment to the family, acts as insurance during crop failures, contributes to gender equality by generating opportunities for women, generates *in situ* fertilizers for enhancing soil fertility, contributes to day-to-day expenses of the farm family, recycles waste products and residues from cropping and agro-industries and supplies energy source for cooking and thus provides a sustainable environment.

Role of Small ruminants: Small ruminants, namely goat and sheep play significant role in food and nutritional security of the rural poor and landless farmers, especially in the rainfed regions where crop production is uncertain, and rearing of large ruminants is restricted due to acute scarcity of feed and fodder. The goat is truly known as 'Poor man's cow'. Being hardy animals, they can be reared in regions that are unsuitable for other livestock production systems. As per NSSO data there are about 33 million goat farmers in the country. Goat rearing has distinct advantages over other livestock species as it requires less initial investment, low input requirement, besides having higher reproductive rate, early sexual maturity and ease in marketing. Goats can eat and thrive on all sorts of vegetation including tree leaves. In spite of being a strong contributor to the income and livelihoods of the poorest segments of the society, goat has remained neglected at various levels. India stands first in goat milk production (4.59 million tons) and contribute about 29% share globally. In view of the social acceptance of goat meat and health promoting nutraceutical value of its milk, goat farming would be ideal option for poor rural households, if common community land is developed for grazing of small ruminants. India is no. 2 in the world in goat meat production of 0.595 million tons (mt) having 12% of world share. Goat meat production has doubled (9.3% to 18.3%) in India during the last decade. The meat production from sheep and goat in India is estimated as 485.53 and 942.91 million kg, respectively, constituting 7% and 13% to the 7000 million kg of total meat produced in country (BAH&FS, 2016).

Poultry a shining sector: With about 10% growth rate is one of the fastest growing sectors of Indian agriculture. Other countries after India are Brazil, China and USA having growth rate of 7%, 2.1% and 2%, respectively. Presently, India is world's third largest egg producer (7% of world production) and 5th largest in broiler production (5% of the world). Transformation of poultry industry from a backyard activity to a major commercial activity in just about four decades is unparallel anywhere in the world. This transformation has been possible with sizeable investments in breeding, hatching, rearing and processing. Poultry industry has grown to its present status as a result of the initiative and support of private sector, diverse indigenous poultry genetic resources and needed support for veterinary health services; quality poultry feed, equipment, housing, processing infrastructure, policy support and market linkages. Egg production showed an impressive growth during the past few decades; from10 billion eggs during 1980-81 to 39.1 billion during 2000-01 and 88.1 billion in 2016-17. Andhra Pradesh is the highest egg producing state (17.78 billion) followed by Tamil Nadu (17.42 billion), Telangana (12.67 billion), West Bengal (7.64 billion), Maharashtra (5.7 billion), Haryana (5.59 billion) and Karnataka (5.56 billion).

LIVESTOCK COMMODITIES

1. Milk: India stands 1st in milk production globally since 1998. Presently, it accounts for about 20% of total world milk production. There has been a steady and remarkable rate of growth in milk production reaching 176.35 million tons (MT) in 2017-18 (DAHDF, 2017-18) as compared to 17.40 MT in 1950-51 and about 50 MT during 2000-01. Milk production increased more than 10 times between 1950-51 and 2017-18. Growth rate of milk during 2017-18 was 6.62% as compared to the preceding year. The per capita milk availability at national level during 2017-18 was 375 g/day which is well above the minimum requirement prescribed by the ICMR. However, due to wide variations in milk production in various parts of the country, per capita availability of milk varies from as low as 71 g/day in Assam to 1120 g/day in Punjab. Despite being the highest milk producer in the world, average milk production/cow/year is 50% of the global average at about 1200 kg as compared to Israel (10214 kg), US (9332 kg), UK (7101 kg), Australia (5600 kg) and New Zealand (3343 kg). Uttar Pradesh is the highest milk producing state (29.05 MT) followed by Rajasthan (22.4 MT), Madhya Pradesh (14.71 MT), Andhra Pradesh (13.7 MT), Gujarat (13.57 MT), Punjab (11.85 MT), Maharashtra (11.1 MT) and Haryana (9.81 MT). The projected demand of milk in 2030 in India is estimated to be about 185 MT. Country is expected to meet this if Dairy Sector continues to grow at the present rate of about 6 to 6.5% and the rise in number of non-productive animals is controlled. However, per animal productivity is highest in Punjab and Haryana in all three types of milch animals i.e. buffalo, crossbred and indigenous cattle breeds. This difference in productivity is mainly due to availability of feed and fodder, good animal germ plasm and better facilities of AI and health care of milch animals. As per CSO, the value of output from livestock sector was INR 9, 17, 910 crore at current prices during 2016-17 which is about 31.25% of the value of output from agriculture and allied sector. The growth rate of milk during 2017-18 was 6.62% as compared to the preceding year 2016-17.

2. Meat: India is the largest exporter of buffalo meat and third largest exporter of meat after Brazil and Australia. India started exporting meat, both fresh and frozen, since 1969. The major importing countries are Vietnam, Malaysia, Thailand, Australia, UAE, Saudi Arabia and Egypt. Among Indian States, Uttar Pradesh has emerged as the major exporter of buffalo meat, followed by Punjab and Maharashtra. Meat production showed a good growth rate during the last decade. According to the Department of Animal Husbandry, Dairying and Fisheries, the total meat production increased from 1.9 MT during 2001-02, to 7.7 MT during 2016-17 (http:/dare.nic.in and www. Dadf.gov.in). Uttar Pradesh contributes highest share (15%) followed by Maharashtra (12.1%), West Bengal (10.1%), Andhra Pradesh (9.3%) and Telangana (8.4%). In the year 2015-16, poultry contributed 3.26 MT chicken meat followed by buffalo meat (1.94 MT), mutton (0.49 MT) and Pork (0.39 MT).

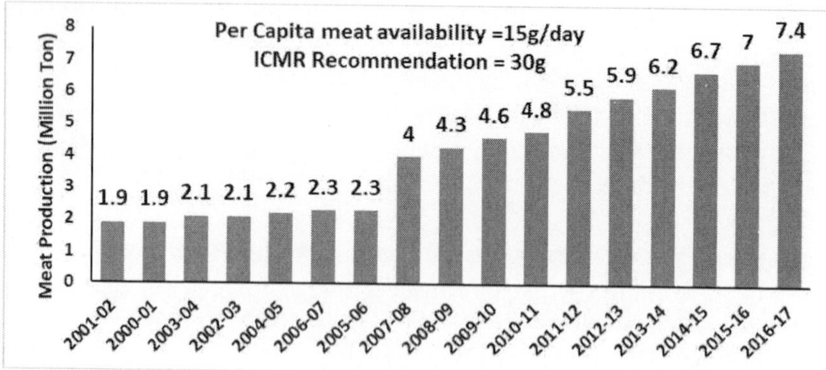

Fig. 1. Meat production in India during 2000-01 to 2016-17

Presently there exits a huge gap between the demand availability of meat. Moreover, by 2050, it is expected that the population in India would increase by 34% and to fulfil the dietary recommended levels of the livestock products as per ICMR for a population of 1.7 billion people, we would be required to produce 186.2 million tons of milk, 18.7 million tons of meat and 306 billion eggs per annum against the present availability of 176.35 MT milk, 7.7 MT meat and 95.2 billion eggs, respectively during 2017-18 (DADF, 2018-19). Control and eradication of Rinderpest (Cattle plague) in India in 2006 as a result of the mass vaccination program launched from 1954 onward till 2000 has a direct bearing on enhancing the production of both milk and meat in India and export of buffalo meat and dairy products as well as producing additional food for human consumption (Uppal, 2011).

India is second in world in goat meat (5.96 lakh ton) production sharing 12% of world share though per animal productivity of sheep and goat meat is much below the world average. During the last decade, the goat meat production has doubled (9.3% to 18.3%). The meat production from sheep and goat in India as estimated (BAH&FS 2016) is 485.53 and 942.91 million kg, respectively, that constitutes 7% and 13% contribution to 7000 million kg of total meat produced in country. Thus, the availability of meat in India is only about 15g/person/day against the ICMR recommendation of 30g/person/day. It is apparent that there exists a huge gap between demand and availability of meat. Moreover, by 2050, it is expected that the population in India would increase by 34% and to fulfil the dietary recommended levels of the livestock products as per ICMR for a population of 1.7 billion people, the livestock sector should produce 186.2 million tons of milk, 18.7 million tons of meat and 306 billion eggs per annum.

3. Eggs: Egg production showed an impressive growth during the past few decades; from 10 billion eggs during 1980-81 to 39.1 billion during 2000-01 and 88.13 billion in 2016-17 (Fig. 2).

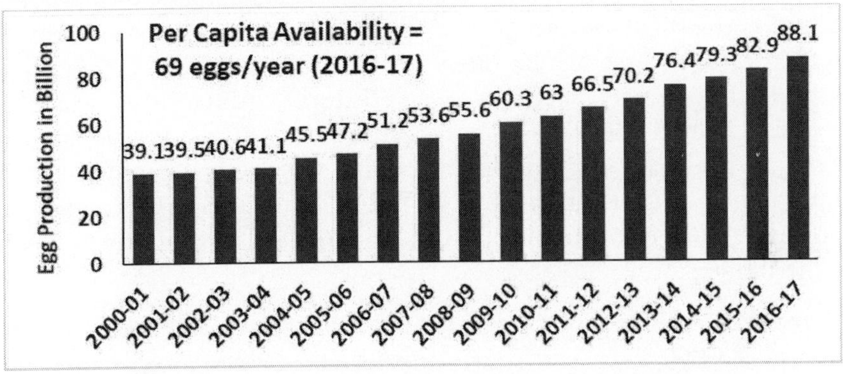

Fig. 2. Egg production in India during 2000-01 to 2015-16

At present the annual production of eggs amounts to about 24 crore eggs every day of the year. If 60% of India's population were to eat an egg every day, the production will need to be increased three-fold. Nationally, around 19% of the egg production is from backyard poultry enterprises, in which 64% are produced by indigenous (desi) fowl. Ducks contribute about 7% of the eggs. The organized and commercial sector contributes about 81% of the eggs produced. Andhra Pradesh is highest egg producing state (17.78 billion) followed by Tamil Nadu (17.42 billion), Telangana (12.67 billion), West Bengal (7.64 billion), Maharashtra (5.7 billion), Haryana (5.59 billion) and Karnataka (5.56 billion). This difference is mainly due to large number of organized poultry farms and good market chain in high egg producing states. The per capita availability (2017-18) is around 74 eggs per person per annum.

Value Addition in Livestock Food: Animals provide nutritious food (milk, meat, eggs) which are rich in high quality proteins, minerals, vitamins and micronutrients (such as zinc, iron, calcium) and thus provide nutritional security to the human population. Animal proteins are more digestible and metabolized more efficiently than plant proteins. The dietary animal proteins contain essential amino acids which are deficient in cereals. Eating even a small amount of animal product corrects amino acid deficiencies in cereal-based human diets. The newer technological innovations such as probiotic fermented dairy products, fortified milk and meat products and composite dairy food, developed by NDRI, CIRG and other institutes need to be scaled out for enhancing the farmers' income. Addition of value to dairy products ranging from dahi to highly industrialized yogurts, cheeses, powders, dairy probiotics and nutraceuticals has become imperative for the dairy industry due to increased demand. It is now a well-established fact that value addition helps to increase the profitability of the dairy enterprise and there is a huge scope for growth of the organized dairy industry due to increased demand. Buffalo milk cheeses are becoming increasingly popular throughout the world as they display typical body and textural characteristics which are unique in nature. Mozzarella cheese made from buffalo milk is the most popular cheese for pizza toppings. Thus, there is great potential and opportunity for value addition in buffalo milk to earn handsome profits. Moreover, branding of buffalo milk can help to obtain better economic returns to the farmers.

Farmers' Companies for Production of Livestock Products: For the success of FPOs in case of livestock products, the need is to create some model FPOs in the country, may be on the lines of the famous 'milk cooperative'. Livestock FPOs need an assisting hand from the government to stand on their feet. Today, many successful milk cooperatives in the country are surviving on their own while competing in the open market, but they have institutional support in the form of the National Dairy Development Board. Similarly, the government needs to create some other specific institutions for the benefit of FPOs created for production of livestock products other than milk. Formation of such SHGs and FPOs are more important in small ruminants farming, as there is complete absence of organizations of small ruminant keepers.

For supporting small scale dairy farming there is a need for the formation of farmers producer's organization, farmers, producer's companies, co-operative societies or self-help group as demonstrated by NDRI in Dairy sector in the villages. This will enhance value addition and farmers income. The producers should be linked to market. There is a need to prioritize clean milk production, traceability, adulterants detection kits and value addition of milk and milk products.

Exports of Livestock and Livestock Products: India exported Livestock and their products valued at Rs. 45776.81 crores during 2017-18. The major export was from meat and edible meat offal (58.8%). There is tremendous scope of export of meat specially buffalo meat from India, as India is having more than 50 % of World's buffalo population. There is a need to develop a breed of buffalo for Meat purpose and the slaughter system, packaging and transport of these products at par International standards. In the present situation the registered slaughter houses needs to be increased and improved to bring at par with international standards. There is plenty of scope to export Dairy and Poultry products with due value addition and at par with international norms. Presently, the export of dairy and poultry products and honey is in the tune of Rs.2364 crores. Besides the above, significant exports are made from hides, skins leather and leather products, and fish and fish products.

Value of Export and import of Livestock and Livestock Products in 2017-18 (Rs. In Crore)

Groups	Import	Export
Livestock	56.55	411.02
Meat and Edible Meat offal	29.99	26921.75
Dairy and Poultry products & Honey	302.88	2363.60
Animal fodder and Feed	3558.07	9384.01
Raw hide and skin & Leather	3967.71	5636.67
Raw wool and animal hair	2223.96	1059.76
Total	**10139.16**	**45776.81**
India's Total	***3001015.73***	***1955541.12***

Livestock Farms for Producing Elite Germplasm: Presently the number of livestock farms under the control of animal husbandry departments for Cattle, Buffalo, Sheep, Goat, Rabbit, Pig, Camel, Horses, Poultry and Duck are 132, 23, 69, 161, 34, 366, 2, 14, 244 and 32, respectively. Besides these, livestock farms under control of other organizations for cattle, buffalo, sheep, poultry and duck are 53, 12, 19, 18, 55 and 6, respectively. In spite of a large network of livestock farms, the herd/flock size in most of the farms is inadequate to take up genetic improvement programmes. For cattle, a total of 6495 gaushalas are also registered in the country. Therefore, there is a need to strengthen / establish sufficient number of farms with adequate strength of livestock (at least 1000 heads of each breed) with sufficient budget and man power to fulfil the mandate for i) Supply of germplasm of high genetic merit for genetic improvement of farmers' herds either in form of breeding males or semen/embryos, ii) To act as a model demonstration unit of animal husbandry activities through training to the farmers, iii) To act as conservation unit of a particular breed.

Livestock and Poultry Markets: There are about 2,000 markets for live animals throughout India, falling under the jurisdiction of state governments. These are managed by local bodies such as Municipal Corporations and Gram Panchayats. Most of these markets are short of basic infrastructure and marketing facilities. In handling live animals, the role of available infrastructure is important to comply with animal welfare and to maintain the health of the animals. Further, lack of basic facilities in markets, for farmers as well as for animals also discourages producers to bring their animals to the markets. A considerable proportion of live animals, mainly small ruminants (sheep & goat) are exchanged amongst livestock farmers themselves and between farmers and intermediary traders. Intermediaries assemble animals from farmers for further sale to larger traders as well as to slaughter houses and butchers, exhibiting varied market chains. Sometimes, small producers assemble their produce to sell collectively to large buyers. Butchers-cum-retailers in small towns too procure live animals directly from producers. Unlike in case of fruits and vegetable farmers, direct sales to end consumers are rare.

Ease of livestock movement, including the documentation to ensure safe transit, is also an aspect that livestock markets are expected to facilitate. Prices of live animals, especially ruminants, are negotiated by buyers and sellers, taking into consideration the animal characteristics, however, no harmonized standard of the identified parameters is uniformly applied.

The marketing of milk and milk products is comparatively organized. It is estimated that 48% of milk consumed at production points and out of the remaining 52%, dairy cooperatives handle 16%, private organized sectors 14% and unorganized sector 70%. As per NCDFI, there are at present 27 State Milk Federation, 218 district milk union, 1.77 lakhs village milk societies and 163 lakh dairy farmers members.

No doubt there is a well-established marketing and supply chain network for export of meat, however, the domestic marketing system does not demonstrate equal quality and standardization in these products. About half of the total meat production comes from un-registered, make-shift slaughter houses. Marketing and transaction costs of livestock products are high up to 15-20% of the sale price. Bovine meat output in India, the second largest bovine meat producer in Asia, increased by 1.2 percent in 2017 (2.55 million tonnes), suggesting that the ban on selling cattle for slaughter and the crackdown on illegal meat establishments have not interrupted the existing supply channels (FAO, 2018).

Marketing of poultry live birds is more organized where the producers have integrated their output into the supply chain of the poultry processing industry. Bulk of trade in live broilers and eggs takes place between producers and traders, directly or indirectly through commission agents, in designated markets or at the farm gates. In some states, poultry co-operatives also facilitate marketing, but on a limited scale.

Contract farming in poultry has emerged in big way for taking the poultry farming down to the path of industrial farming. Producers associations like Broiler Producers Marketing Association (BROMARK) set the daily harvest prices depending on the market. The majority of domestic consumers prefer poultry meat from small butchers, and reports indicate that more than 90 per cent of chicken is sold freshly butchered at retail level. This allows smallholding farmers to continue trade in the live animals with small town butchers and wet market retailers.

Animal identification and performance recording: Presently we do not have an effective mechanism to identify the animals and their performance recording is lacking except small scale level progeny testing programs at some places. The elite germ plasm of all dairy breeds of cattle and buffaloes needs to be identified, subjected to performance recording and putting systematic breeding program in place so that more numbers of elite germ plasm are obtained and male offspring from these elite females can be used for future artificial breeding purpose. The large number of non-descript and non-characterized livestock population need to be worked out and subjected to performance recording.

Biodiversity in Indian Livestock: India has a large livestock population and mega biodiversity. Presently there are184 registered indigenous breeds comprising of cattle (43), buffalo (16), goat (34), sheep (43), camel (9), donkey (2), horse & pony (7), poultry (21), pig (8) and Yak (1). Out of total cattle population, 69.7% are non-descript and only 11.6% are recognized indigenous breeds. Among the buffalo population, 56.1% are non-descript, while only 30.2% confirm to registered indigenous breeds. Despite these large number of breeds in livestock and poultry, considerable livestock and poultry population exist in the country deserving a breed status yet to be recognized.

Conservation and improvement of indigenous breeds: The exotic male germ plasm of cattle was imported and cross-bred with indigenous cows which led to increased crossbred population and milk production. Population of crossbred cows increased at the annual growth rate of about 7.5% during 1982–2003 and about 7% during 2007-2012 compared to about

0.1% and 0.17% for indigenous cows, respectively. This threatens the loss of our valuable biodiversity in indigenous dairy animal breeds. Indigenous animals have the advantage of sustaining productivity under low inputs and are also known for their drought tolerance and disease resistance. Some indigenous cattle breeds, namely Gir, Sahiwal, Rathi, Red Sindhi, Thaprparkar, are having good potential of milk production with lactation milk yield varying between 1200-3000 liters or higher. Such breeds need to be promoted and may also be used for grading up of non-descript cattle in different parts of the country for improving their milk productivity. Besides the regular programmes for conservation and improvement of indigenous cattle, the strength of having several Gaushalas can be converted into opportunity by involving them as focal point for development and conservation of indigenous breeds in their respective region. Similar policy can be put in vogue for upgrading the non-descript buffaloes by crossing with elite germplasm of dairy buffalo breeds.

Genomic based selection needs to be initiated for higher coverage and better accuracy. In the meantime, for sheep and goat husbandry, community - based breeding program should be implemented on priority. *In-situ* and *ex-situ* conservation programs of important genetic merit elite germplasm of livestock and poultry should be undertaken in mission mode. Special focus is needed for conservation of migratory breeds of sheep for future. Further, scrub bull castration is urgent need of the day. Breeder Associations should be formed and National Gene Bank of livestock should be strengthened.

Breeding Policy and Breeding Act: Livestock is a State subject and all the states are developing their breeding policies for livestock. The breeding policies are developed as per broader guidelines of Government of India and as per the need/capacity of the state. These are periodically reviewed and improved. All the states are required to formulate their Livestock Breeding policy and review the same after every 5/10 years. Most of the states have developed Breeding Policy for bovines (cattle and buffalo) only. It is therefore suggested that the breeding policy should be developed for all the species and under different production systems and agro-climatic zones.

Standards of semen/embryo production in terms of veterinary norms and genetic merit of breeding males should be checked by the government through legislative procedure. Further, the technical ability of AI workers is to be certified by government so as to avoid any type of long-term problem to animals in terms of disease and poor genetic improvement. Although

Bovine Breeding Policies are in place, but Bovine Breeding Acts need to be formulated in each state. Punjab has established its Bovine Breeding Act and other states may also develop similar mechanism.

Registration of Livestock Breeds and protection of Livestock Keepers' Rights: Presently different Livestock species are having a big proportion of non-descript population. As per the Breed Survey-2013 of Government of India, the proportion of non-descript population is 59.3%, 43.4%, 61.3%, 38.7%, 73.1%, 28.2% an 80.4% in cattle, buffalo, goat, sheep, pig, camel and Horses/ponies, respectively. Therefore, there is a need to complete the inventories of livestock by characterization and registration as breeds. ICAR has initiated this process in 2008 but the registered breeds are not published under any gazette notification of the government. Therefore, there should be a legitimate authority to register and conserve all kinds of domestic animal diversity as is being done by PPVF&R for plant varieties. Such Act may also protect the Livestock Keepers' Rights in the country.

CONSTRAINTS AND CHALLENGES FOR LIVESTOCK DEVELOPMENT

1. Shortage of breeding bulls and semen: Major constraints in livestock and animal husbandry development include shortage of breeding bulls and quality semen, shortage of feed and fodder, suboptimum outreach of veterinary health services, lack of quality slaughter houses; inadequate incentives for credit, market linkage, insurance etc. As per 2012 livestock census, among the total breedable bovine milch animals' population, a large proportion comprising of 1.6 million crossbred cattle, 6.03 million indigenous/ non-descript cattle and 4.6 million buffaloes have not calved even once. This population accounts to nearly 10% of the total breedable milch animals' population. Conception rate through artificial insemination in buffaloes is very less when compared with cattle. Delayed age at sexual maturity (31-33 months in Murrah buffaloes compared to 18-19 months in Mediterranean buffaloes), and poor expression of heat symptoms still remains as major issues in achieving high reproductive efficiency in buffaloes. Summer infertility characterized by high incidence of silent heat (even up to 70%) is serious problem in buffaloes. On a conservative estimate, the country is losing 20-30 million tons of milk annually on account of anestrus and repeat breeding in cattle and buffaloes which translates to a loss of nearly INR 40,000-50,000 crores annually.

Artificial insemination for breeding purpose was started in 1950s but its coverage and conception rate are still unsatisfactory. India has one of the largest breeding infrastructures in the world (51 frozen semen stations, 3321 bulls and 94,688 Artificial Insemination Centres) with total production of about 90 million frozen semen straws per year. However, this could cover hardly 25-30% of the breedable population. To achieve the national target of 50% AI coverage by 2021-22, the quality semen production must reach to 140 million doses. The major limiting factor in achieving the required numbers of frozen semen straws production is the availability of adequate number of quality bulls. In sheep and goat husbandry also, the availability of quality breeding males is very limited and the situation is further amplified by the poor quality semen produced by the breeding male. Goat keepers are unwilling to keep bucks due to high rearing cost. Further, there is no perfection in AI with frozen semen use in goat industry.

For enhancing the reproductive performance, the proven assisted reproductive technologies need to be employed under field conditions. In developed countries, the third-generation reproduction technologies, such as *in-vitro* embryo production, are being applied at farmer level. Whereas, in India, still we are mainly depending on the first-generation reproduction technologies, namely semen cryopreservation and artificial insemination. Nevertheless, we need to take full advantage of the potential of these technologies to multiply superior germplasm. Promising reproductive bio techniques like multiple ovulation and embryo transfer (MOET), ovum pick-up and *in vitro* fertilization need to be utilized to maximum extent at least for breeding bull production. MOET, sexed semen and sexed embryos should be used for faster multiplication of superior germplasm of elite dairy female animals. The technology protocols of AI in goat, sheep, poultry and pigs need to be upgraded, strengthened and out scaled.

To bridge the gap between demand and availability of male germplasm, a project/scheme to rear "the male calf to bull" should be initiated by State Govt./ Central Govt. Similar schemes are also needed for goat and sheep husbandry for the improved breeds in their breeding track.

2. Feed and fodder Shortage: Shortage of feed and fodder is one of the major constraints in livestock farming. The inadequacy of feed, both in quantity and quality, remains biggest handicap, particularly during the dry season. The area under fodder crops in India has stagnated at about 8.5-9.0 million hectares during the past decade which is about 4.6% of the total cultivated area. The projected demand for dry fodder, green fodder and concentrate for 2020 is 468, 213 and 81 MT on dry matter basis whereas the availability is estimated to stand at 417, 138 and 44 MT leaving a short fall of 11%, 35% and 45%, respectively. Although significant quantities of crop residues are produced, they don't meet the nutritional quality required for dairy cows. Scarcity of feed-fodder for sheep and goats and raw material, particularly maize at reasonable cost for poultry is also a big issue. In view of the wide spread deficiencies of trace elements in the soil, feed and fodder, supplementation of mineral mixture is inevitable to achieve optimum health, reproduction and production of livestock and poultry.

At least 10% of irrigated land should be used for fodder production. There should be GIS based record for area under fodder cultivation and the estimates should be made on regular basis by the NSSO or any other government organization. There is urgent need to calculate the total requirement of feed and fodder in all the States of the country. Quality forage seed production chain must be given high priority.

In order to meet the nutritional requirements of animals fed on crop residues and roughages, there is a need to increase the bioavailability of nutrients in the feeds and fodders using chemical, biological and biotechnological approaches. We also need to have enriched feed/fodder banks like food/seed banks as technologies for compaction, enrichment and fortification are now available.

Keeping in view the shortage of feed and fodder, a very well-planned programme to upscale and out scale the available technologies of "Precision Feeding" (bypass protein and fat), alternate feed and fodder resources, area specific mineral mixtures need to be formulated and implemented. Availability of Maize for poultry feed is to be assured at affordable cost. The poultry producers may be encouraged to procure maize directly by using contract farming.

3. Suboptimum outreach of Veterinary health services: Livestock sector is facing challenges of increased incidence of emerging and re-emerging animal diseases and vulnerability to exotic diseases. Further, it is also presumed that many diseases may surge upon during changed climatic scenario. The outreach of veterinary health care services to the livestock farmers is also low. Lack of awareness and timely non-availability of inputs for preventive measures lead to high incidence of diseases and epidemics. In India, it has been estimated that brucellosis causes a loss of at least Rs. 35.0 crore every year, udder infections (mastitis) about Rs. 6,053.21 crore, and for FMD about Rs. 23,000 crore loss every year. Small, marginal and unorganized landless livestock farmers are most sufferers due to infectious diseases. Infectious disease status also affects the export potential of the livestock industry. Milk and milk products, meat, skin and hides, eggs and fish are not accepted by the disease-free importing countries. If FMD is controlled in India, the milk production can be increased by at least 5% annually and the export of meat could be enhanced by 3-5 times of the present level. In sheep and goat husbandry, the very high mortality can be reduced with prophylactic care, vaccination and improved method of goat rearing. PPR or goat plague is the most important disease of sheep and goats causing an economic loss to the tune of Rs. 1800 million/annum. The mass scale use of PPR vaccine has resulted in reduction of >75% disease incidence.

Although vaccination is the most promising way to control several diseases, it is not 100-percent effective in preventing some of the diseases. Thus, vaccination coupled with good husbandry practices is essentially required for cost effective control of infectious diseases. A comprehensive animal health cover system is possible only if there are adequate facilities for prompt diagnosis of livestock diseases. Facilities for specific and general disease diagnosis need to be strengthened by introducing good laboratory practices. There is a need to put an effective system in place for ensuring quality control of vaccines and diagnostics produced/imported. There is also a need to improve existing vaccines, diagnostics, therapeutics in terms of efficacy, thermostability, long lasting immunity, and easy delivery using advanced knowledge for use in National control programmes. Vaccines and diagnostics need to be produced in required quantities and made available at affordable cost. Quarantine facilities need to be further strengthened and zoo-sanitary and quarantine procedures need to be strictly followed to prevent ingression of exotic and transboundary diseases. Mechanism of Emergency Preparedness against emerging and exotic diseases also needs to be put in place. A well-planned and operational livestock disease control programme involving PPP mode could be an option for capacity building in the domain of vaccines and diagnostics to ward off huge economic losses due to changing climatic conditions and newer emerging, re-emerging and Transboundary diseases. Emphasis should also be given to the prevention and control of zoonotic diseases.

Launching of systematic disease control / eradication programmes for OIE listed diseases along with effective disease surveillance on the lines of Rinderpest eradication programme, and strict enforcement of Sanitary and Phytosanitary (SPS) conditions in processing the livestock products are critical in promoting the international acceptance and export.

A comprehensive informative package about disease awareness, epidemiology, surveillance, management and control measures need to be developed for education at farmers level for effective and timely control of the common diseases of cattle, buffalo, sheep, goats, pigs, horses and poultry. Required quantity of vaccines against common diseases of cattle, buffalo, sheep, goats, pigs, horses and poultry need to be arranged either by import or manufacturing indigenously.

4. Upgradation of Slaughter Houses: India has more than 1176 slaughter houses besides 75 modern abattoirs. As majority of slaughter houses are operated and managed by municipalities in cities and panchayats in rural areas, standard practices are often lacking and investment opportunities for improvement of infrastructure are difficult. As a result, there is low recovery rate coupled with wastage of by-products like blood, skins, tallow etc. The meat slaughter houses are basically linked to hide production and the leather industry. Schemes to improve slaughter houses framed by the Central and State governments in the past have been effective in refining the basic infrastructure like building, lighting, water supply, drainage arrangement etc. but slaughter houses which required repositioning under different schemes have not been effective primarily as the traders' associations resisted the centralized process of slaughter house. In the current system, small and medium traders bring their own animals for slaughter and they take back the meat and all by-products with them. To make this sector dynamic, merely targeting large slaughter houses will not serve. Addition of value at the local level will be more important, as it will limit and reduce the transport of animals and environmental pollution, and help animal owners to earn more income by providing them uninterrupted admittance to local market. Since at the industrial scale, slaughter houses have been mainly successful in the private sector, it is a necessity to improve and enhance the private participation,

To prevent long distance transportation of live animals and birds and associated losses due to shrinkage, stress, sickness and death, it is suggested to construct one hygienic slaughterhouse at district level or in each cluster of 2-4 districts with all the facilities for efficient utilization of edible and in-edible by-products and effluent treatment plant. Instead of one big slaughterhouse in Metropolitan cities, it would be ideal to establish few modern slaughterhouses surrounding the city in peri-urban areas so that the small meat traders (with 3-5 animals) need not to transport the animals for long distance.

5. Inadequate Policy Support (Credit, incentives, price regulation, rebate, insurance and governance): The institutional mechanisms to protect animals against risk are not strong enough. Livestock extension has remained grossly neglected in the past. Only about 5% of the farm households in India access information on livestock technology. These facts indicate an apathetic outreach of the financial and information delivery systems (Planning Commission, 2012). Livestock sector did not receive the policy and financial attention it deserved in terms of policy support with regard to credit, insurance, rebate, incentives and price regulation. The sector received only about 12% of the total public expenditure on agriculture and allied sectors, which is disproportionately lesser than its contribution to agricultural GDP. The budget allocation to livestock sector must be more than 30% of agriculture, matching to its contribution in agricultural GDP. The sector too has been neglected by the financial institutions also in granting credit. The share of livestock in the total agricultural credit has hardly ever exceeded 4% in the total (short-term, medium-term and long-term).

Currently, only 6% of the animal heads (excluding poultry) are provided insurance cover. As per 19th Census (2012), there were 122.9 million cows and 92.5 million female buffaloes in the country. In 2014-15, 14.80 lakh animals were insured. However, the number came down to 7.65 lakh in 2015-16 and 7.44 lakh in 2016-17 (Lok Sabha Question). Thus, it is clear that the number of insured animals is less than a miniscule. The animal husbandry sector should be treated at par with agriculture sector, in all terms including credit, investments, incentives for electricity and water and other charges.

The benefit of food processing sector should also be extended to dairy processing and meat processing sectors. The dairy co-operative societies should be exempted from the purview of income tax and CESS. In the list of food commodities covered under FTA with Australia and New Zealand, milk and milk products should not be included.

Govt. Policies and Programmes: Under the XII Five Year Plan, Department of AHD, Ministry of agriculture, Govt. of India introduced several important policy interventions for livestock sector in Mission mode. Amongst these, two important programs include i). "Rashtriya Gokul Mission" for Bovines and, ii). National livestock Mission for Non-bovine livestock species. National Livestock Mission and Rashtriya Gokul Mission (RGM) should address the problem of

- Availability of elite germplasm of high genetic merit for livestock and poultry breeds
- Problem of Feed and Fodder Resources for genetically improved livestock and poultry including grazing land
- Health management including Vaccination, De-worming, disease

diagnosis and treatment

- Credit, Insurance and marketing facilities to the livestock farmers

1. Rashtriya Gokul Mission: It is the part of mega programme "National Programme for Bovine Breeding and Dairy Development (NPBBDD)". Major components of NPBBDD include

(i) National programme for Bovine breeding;

(ii) National Programme of Dairy Development and

(iii) Rashtriya Gokul Mission.

"Rashtriya Gokul Mission" has been started with the objective of development and conservation of indigenous breeds. The objective of the scheme is to provide breeding services at the farmers' doorstep for genetic improvement of cattle and buffaloes, establishment of multi-purpose AI technicians for rural India (MAITRIs) and streamlining of liquid nitrogen storage and distribution system are inbuilt in RGM. Two new "National Kamdhenu Breeding Center" (one in Madhya Pradesh and one in Andhra Pradesh) have been established for which an amount of Rs. 50 Crore has been released.

2. National Mission on Bovine Productivity: Government of India has initiated National Mission on Bovine Productivity scheme in 2016-17 for a period of 3 years with a financial allocation of Rs. 825 crores. The scheme has been currently subsumed under the umbrella scheme "Rashtriya Gokul Mission". The scheme aims towards faster genetic improvement of bovines for milk production in the country by use of advanced breeding techniques.

3. Dairy Processing and Infrastructure Development Fund: Under this scheme, Dairy Processing and Infrastructure Development Fund (DIDF) of Rs.10,881crores has been set up. Loan assistance is provided through NABARD at the subsidized interest rate of 6.5% per annum to the milk co-operative institutions. It is envisaged to provide the benefit of the scheme for 3 years (2017-18 to 2019-20). Under this scheme, 95 lakh dairy farmers will be benefited from additional milk procurement. Additional milk chilling capacity of 140 lakh liters per day and additional milk processing capacity of 126 lakh liters per day will be set up. A target of establishing 28,000 Bulk Milk Cooler (BMC) in 50,000 villages has been envisaged.

4. National Livestock Mission (NLM): The NLM has been formulated by subsuming and modifying 7 Centrally Sponsored and 7 Central Sector Schemes of non-bovine livestock species of Government of India. It commenced in 2014-15 with major objectives to bridge the gap in demand and availability of feed and fodder, conservation and improvement of indigenous breeds, higher productivity and production in a sustainable and environment friendly manner, better availability of quality animal products to consumers; and overall socio-economic upliftment of livestock farmers. Under NLM, there is need to augment the efforts of the states to allow small and marginal farmers to gain better price realization, access to markets, technical support and improved technologies for value addition. The scope and coverage of livestock insurance has been increased from 300 districts to all the 719 districts of the country. Coverage of livestock insurance has been extended from 2 to 5 milch animals/ other animals or 50 small animals per farm house hold. It is suggested to include marketing of livestock and livestock produce in NLM or in RGM.

5. Animal Husbandry Infrastructure Development Fund: In the Union Budget 2018, a provision for around Rs. 2450 crore has been made for establishment of Animal Husbandry Infrastructure Development Fund (AHIDF). This will empower small and marginal farmers and entrepreneurs, especially women, self Help groups and weaker sections to avail latest infrastructure facilities and to get better remuneration for their produce.

Human Resource Development for service to Farmers:

(i) For effective transfer of technologies and implementation at the level of end users, adequate number of trained manpower and training facilities need to be ensured. Networking of research institutions with KVKs and other local bodies may be an option in this regard.

(ii) In addition to graduates in veterinary, animal husbandry and dairy sciences, there is urgent need of regulation and policy of training for para-veterinarians, inseminators, vaccinators, technology agents and imparting diplomas in dairy processing units and other relevant areas.

(iii) Animal waste (dung, urine and poultry excreta) management and manure production require utmost attention. In addition to cow dung and urine, the excreta of other animal species should be utilized for farm yard manure or vrmiculture production. The manure produced from goat dung is very rich source of NPK, and has long lasting effect on soil fertility and minimizing soil erosion.

(iv) Establishing Special Animal Product Economic Zones (SAPZ) on the lines of "Special Economic Zones" and value addition in order to promote the export of livestock products is the need of the hour. The SAPZs may be linked with export-based processing plants, compound feed industry, regulated animal market, veterinary polyclinics, semen bank etc.

(v) Continuous Veterinary Education (CVE) program, on the pattern of Continuous Medical Education (CME) should be implemented in all the States of the country.

Future Science for Sustainability of Livestock Sector

1. Development of "electronic nose" for oestrus detection in cattle and buffaloes.
2. Cow side tests for diagnosis of sub-clinical mastitis, early pregnancy and metabolic disorders need to be developed.
3. The puberty age of cattle and buffalo is quite long. Hence, need based R & D should be taken up on priority for reducing age at puberty in both male and female bovines.
4. Biotechnological tools for prediction of fertility of bull during calf-hood should be worked out. It will encourage farmers to rear male calves.
5. To reduce the number of unproductive and low producing dairy animals, the technologies for sexed semen and sexed embryos should be developed for Indian breeds of cattle and buffalo.
6. To reduce the gap between demand and availability of male germ plasm rearing of male calves to become bulls, with good management practices should be encouraged.
7. In future, we need to develop suitable tools for selection of animals not only on the basis of production and productivity but also based on "input efficiency" and health parameters.
8. Development of fool proof workable technologies for environment friendly animal waste disposal is needed most.
9. Modern science should be applied for increasing the shelf life of milk and milk products, meat and meat products, egg and egg products for better profitability of the farmers.
10. To Explore the health attributes of milk, especially of non-bovine origin, particularly camel milk, goat milk, sheep milk, mare and donkey milk) on priority.

Challenges faced and Way Forward: Agriculture is the back bone of developing countries in Asia, Africa and Middle East countries for their GDP/economy, livelihood support, employment and cultural heritage. Agriculture and livestock have direct bearing for some of the sustainable developmental goals (SDGs).

The Livestock sector faces several challenges as described elsewhere in this document. These need to be addressed by keeping the livestock farmers, environment, human and animal health and welfare in the centre so as to increase the production, productivity, income and overall profitability through application of best practices, innovations, technology and market linkages, and pro-poor policies. The future path needs to be planned for finding implementable solutions to achieve sustainable growth in agriculture sector with climate justice. Livestock and animal farming (inclusive of poultry and fishery) are equally important for crop agriculture, food and nutritional security, inclusive growth.

Contribution of Livestock sector to National GDP: Globally livestock sector contributes around 40% of the agricultural GDP. In India it varies between 26% to 31% based on the fluctuations in crop agriculture on account of monsoon aberrations. In mountainous and semi- arid/arid regions or areas lacking in irrigation facilities and having scant rains, contribution of livestock to agricultural GDP may be as high as 50% or more. In India, the share of livestock in gross value of output and growth of agricultural sector between TE 1992-93 and TE 2008-09 was 23.3% and 26.8%, respectively, with annual growth rates of 3.6% and 2.7% in 2000s for livestock sector and agricultural sectors, respectively. Share of livestock in agricultural growth was 31.1% and 36.3% in 1990s and 2000s, respectively (Birthal and Negi, 2012). The rate of growth in livestock sector was 1.6 times the growth in the crop sector and 1.3 times than the overall growth in agricultural sector. During the 11th Plan, average growth of livestock sector was 4.15% as compared to 3.6% for agriculture as a whole. Though the contribution of agriculture to national GDP declined from 36.4% in 1982-83 to 14.1% in 2012-13, the contributions of livestock sector to agriculture GDP increased more than 28% and is likely to further increase in future (BAHS, 2014). The growth value of output from livestock sector was about 4.8% during XI Plan. Thus, over the years, livestock sector has been acting as a buffer to sustain the agricultural GDP. Milk accounting for more than two-thirds of the value of livestock output, is now the largest agricultural commodity in India. The value of milk has been larger than the combined value of wheat and rice, the main staple serials in the country. Livestock being more resilient to climatic shocks, acts as a buffer to the losses in crop production. Thus, livestock sector provides a cushion to agricultural growth and food security.

Livestock provide Sustainable Growth: A comparison of the value of output from livestock and crop sectors from 1980-81 to 2009-10 indicated that the livestock production was relatively more stable compared to crop production, and hence serves as a sustainable source of agricultural growth (Birthal and Negi, 2012).

For sustainable crop agriculture, livestock/animal farming serves as the best supplementary and complementary enterprise. Wastes from livestock, particularly dung and urine, are helpful in promoting organic agriculture, restoring soil organic carbon, soil biota and fertility, besides retaining more water in the soil. Livestock production and agriculture are intimately linked with each other as both utilize the waste products of the other for the survival, growth and production of plants and human being. Low input and low output-based animal husbandry, generally practised in the developing countries, provide livelihood and employment to sizeable rural population.

Livestock sector as Job Provider: A large section of small, marginal, medium and landless farmers does not have sufficient work all the year around. Rearing of livestock species for food (milk, eggs, meat), leather, draught power for agricultural operations, transport etc. provide part time work and indirect income from the animal produce to men and women alike, thus strengthening nutrition, employment and social securities. In India, 16.44 million workers as per usual status were engaged in the activities related to the farming of animals, mixed farming, aquaculture and fishing (NSS 68th Round: July 2011-June 2012).

Technology, Innovation and Best Practices for Livestock Development: Improved technologies, innovation, best practices and policy support and planning play crucial role in enhancing productivity and profitability of livestock. Adoption of artificial insemination (AI) using superior germplasm, sexed semen/embryos, balanced nutrition, both quantitatively and qualitatively, embryo transfer, MOET, area specific mineral mixture (ASMM) feeding; cross breeding, balanced total ration, green fodder and hydroponic fodder/feed; timely vaccination/medication against killer/economic infectious and non-infectious diseases are some of the technologies, adoption of which has the potential to enhance the production of milk, eggs, meat, fish, wool and skin and hides. Simple technologies and best practices in animal feeding and reproduction, such as chaffing of green and dry fodder, soaking of straws in water overnight or treating with 4% urea or ammonia; grinding, soaking of grains in warm water, not only make the feed more palatable and digestible but also save energy used by the animal in chewing the hard feed and fodder. Innovative feeding practices using secondary plant compounds (condensed tannins, saponins), feeding of essential oils and select spices reduce methane production by the ruminants, and act as anti-parasitic to reduce parasite load. The feed energy which was wasted in the form of methane is thus saved and used by the animal for higher production of milk, meat, wool etc.

Experience with AI in India has shown that the success rates are much below the expected ones. One of the reasons for this is the deficiencies of micro nutrients in feed and fodder due to excessive mining of nutrients from soil for intensive crop agriculture practised using high yielding varieties of wheat, rice and other crops since 1970s onward. Feeding of ASMM has found to be a cheap and affordable solution not only to cure repeat breeding and infertility in bovines, it also ensures better growth and early maturity of calves. Silent heat in high producing crossbred cows and buffaloes create problems in judging the right time for insemination. An innovative technology, named cystoscope, developed at Indian Veterinary Research Institute (IVRI) to know the ideal time for insemination by examining the fern pattern of cervical mucus, has been found to significantly improve the conception rates. This can be tested in rural background without the need of laboratory or electricity with the help of battery-operated equipment costing around INR 1000.

Feeding of concentrate: Farmers practice innovative practices for feeding concentrates by mixing these with chaffed green/dry fodder rather than feeding of concentrates and fodder separately as practiced in some dairy farms. High milk yielding cows are known to have higher incidence of clinical as well as sub clinical mastitis leading to significant loss in milk production. One of the reasons for this is attributed to the fact that the high yielding cows tend to sit on floor immediately after milking due the drain of calcium, phosphorus and other nutrients in the milk. Therefore, concentrate is offered to the animals after milking so that they remain standing for about half an hour to avoid bacterial contamination of the udder from the floor through the open teat canal. Feeding of concentrates in the form of pellets is a good practice to ensure balanced ration free from moulds and mycotoxins.

Livestock keeping reduces Farmers Stress by Ensuring Inclusive Growth: About 80% of livestock farmers in India are small, marginal, medium or landless. Livestock driven agricultural growth is considered as pro-poor. Christiaensen *et al.,* (2006) provided a global perspective on the role of agriculture on poverty reduction and observed that agricultural growth has a larger impact on poverty reduction than non-agricultural growth. It is believed that 1% growth in agriculture is equal to 2% industrial growth to achieve inclusive growth. Birthal and Taneja (2006) reported that reduction in rural poverty was more responsive to growth in livestock sector than growth in the crop sector. It is also reported that farmers' suicides were more in the states where monocropping was practised and the farmers were not keeping livestock. In the present scenario, livestock, poultry, fishery and horticulture are best options for higher and sustainable inclusive growth in agriculture. Post green revolution in India, intensive crop production using high yielding varieties of wheat and rice has resulted in degraded lands on account of over mining of micronutrients as a result of unbalanced use of fertilizers. Monocropping with high yielding varieties has also reduced the biodiversity, thus losing the genetic variability, including resistance to abiotic and biotic stresses. Diversification of agriculture is the key to overcome these problems. Diversification of agriculture in farming system mode, preferably with livestock component (piggery, goatery, duckery, poultry, fish etc.) is an effective way of accelerating agricultural growth and reducing rural poverty. In India, lately livestock accounts for a larger share of the value of agricultural output than food grains. In TE 2009-10, the output from livestock was valued at Rs. 2,130 billion (at 2004-05 prices) which is about 18% more than the value of food grains (Birthal and Negi, 2012). In 1980s, the value of livestock output was 28% less than the value of food grains.

Increased demand for Animal Protein and Animal origin Food: The demand of food products of animal origin (milk, egg, meat, fish) is increasing as a result of sustained and economic growth and expanding urban population, particularly in developing countries. High rise in the prices of food commodities of animal origin, namely milk, egg, meat, fish) in recent years are mainly responsible for inflation, indirectly affecting the prices of other commodities. Though India stands 1st in milk production, 2nd in aquaculture and fish production, 3rd in egg production and 5th in meat production globally, we are able to meet only about half of the requirement of protein of animal origin for our population. For an adult person weighing about 60 kg, out of the total 60 g protein requirement, 25 g should be of animal origin (milk, egg, meat, fish) having higher biological value.

Livestock Boosts Agricultural Growth and Promote Gender & Social Equity: Contribution of livestock to agricultural GDP has been increasing consistently from 1970s onward. In the decadal estimates made between 1970-80, 1980-90, 1990-2000, 2007-10 the growth in the livestock sector in

India has been higher than the growth in the crop sector. The 'State of Agriculture' document (NAAS, 2009) states that "Rapid growth in livestock sector has led not only to boost agricultural growth but also to reduce rural poverty and promote rural equity. The distribution of livestock is more egalitarian than that of land. The smallholders and landless rural population together control 75 percent of the livestock resources. Livestock is thus an important source of livelihood for smallholder and the landless, and the sector's rapid growth benefits the poorest households the most. Evidence shows that livestock contributes nearly half of the total income of the smallholders." It further documents, "Livestock sector also seems to promote gender and social equity. About 70 percent of the total workers engaged in the livestock sector are women. The participation of women in other activities including agriculture is low compared to that of animal husbandry. Further, a majority of workers engaged in livestock sector belong to socially and economically backward communities. Scheduled tribes, scheduled castes and other backward castes together constitute about 70 percent of the persons employed in livestock sector." There is also evidence to suggest that the States (Punjab, Andhra Pradesh, Haryana, Rajasthan, Kerala, Gujarat and Karnataka) having higher growth in livestock sector have less rural poverty as compared to the states (Orissa, Bihar, Uttar Pradesh, Maharashtra and West Bengal) having low growth in livestock sector (NAAS, 2016).

Livestock sector has the Potential to make India a developed Country: Economists have suggested that if India is to become a developed country, it should have two-digit (10% or more) GDP growth rate for more than a decade and to achieve this national GDP, the agricultural GDP should grow minimum 4%. Since the beginning of the Ninth Five-Year Plan in 1997, India's policymakers have been targeting 4% growth for the agricultural sector but so far it has been elusive. Considering the higher growth potential in livestock and fisheries, an overall growth of 6 to 7 percent per annum was envisaged during the 11th Five-Year Plan for the livestock sector as a whole, with milk group achieving a growth of 5% per annum and meat and poultry achieving a growth of 10% per annum. Against these targets, the growth value of output from the livestock sector during 11th Plan was about 4.8% per annum and from fisheries it was about 3.6% per annum. It is thus apparent that for agriculture to achieve a growth rate of 4% per annum; livestock, poultry and fishery should grow between 6 to 8% per annum. This can be achieved provided more funds for R & D and policy support is provided to these enterprises.

RECOMMENDATIONS AND ACTION PLAN

General: The livestock sector needs to equip for providing technological backstopping along with efficient input delivery system, facilitating favourable market and marketing network. Availability of superior germ plasm in required numbers, quality inputs, such as, frozen semen, feed and fodder, vaccines, diagnostics and other health measures; application of reproductive biotechnological tools, automation for large scale livestock farms, value addition and ensuring the quality of the LS produce are some of the areas needing concerted efforts in terms of research, technology, and development are required. Salient recommendations / action plans are given below:

(i) Reliable and proven reproductive bio techniques, such as multiple ovulation and embryo transfer (MOET), ovum pick-up and *in vitro* fertilization need to be utilized to maximum extent at least for production of breeding bull. Besides, MOET and sexed semen and sexed embryos should be used for faster multiplication of superior germplasm of elite female animals.

(ii) The protocols of AI technology in goat, sheep, pig and poultry need to be upgraded, strengthened and out scaled.

(iii) To overcome the gap between demand and availability of male germplasm, a well-structured project/scheme to rear the male calf to bull should be initiated by State Govt./ Central Govt. Similar schemes are also required for goat and sheep husbandry to improve the existing breeds in their breeding track. Such schemes already exit in some states,

(iv) Genomic selection in breeding programs need to be initiated for higher coverage and better accuracy. For sheep and goat husbandry, a community-based breeding programs should be implemented on priority. Castration of scrub bulls is urgent need of the day. In-situ and ex-situ conservation programs of important genetic merit livestock and poultry should be undertaken in mission mode. Special focus is required for conservation of migratory breeds of sheep for future.

(v) Veterinarians may be slowly dissociated from routine AI activities, which can very well be taken up by trained para-veterinary staff and inseminators. The services of veterinarian should be utilized for their expertise in specialized needs like precision livestock production, treatment and health cover management.

(vi) A well-planned and operational livestock disease control program in PPP mode could be an option to ward off huge economic losses due to changing climatic conditions and the challenge due to newer emerging and re-emerging diseases. Emphasis should also be given for prevention and control of zoonotic diseases. A comprehensive package about disease awareness with respect to epidemiology, surveillance, management and control measures need to be developed for education at farmers' level to limit the disease incidences. Veterinary health service being very poor for small ruminants, need special attention for upgradation.

(vii) Required quantity of vaccines against common diseases of cattle, buffalo, sheep, goats, pigs, horses and poultry need to be arranged (either by import or by manufacture in India).

(viii) Considering our large livestock population, minimum of 10% of irrigated land should be used for fodder production. There should be GIS based record for area under fodder cultivation. There is urgent need to establish a mechanism to calculate the total requirement of Feed and Fodder through periodic surveys. Quality forage seed production chain need to be maintained.

(ix) Keeping in view the shortage of feed and fodder, a very well planned "Program" to upscale and out scale the available technologies of "Precision Feeding" (bypass nutrients), alternate feed and fodder resources, area specific mineral mixtures are to be formulated and implemented keeping in view the regional variations. Availability of Maize for poultry feed, should be assured at affordable cost. The poultry producers may be encouraged to procure maize directly through contract farming.

(x) The technologies for fortification of crop residues for making nutritive fodder for animals should be out scaled to farmers' door.

(xi) The program of "Balancing of Ration" should be expanded in all the districts of the country.

(xii) To ensure the quality of feed, the standard and specificity of their ingredients should be set by BIS / FSSAI. Separate specifications are needed for Fodder Blocks making units and

Feed manufacturing units to have separate identity for providing subsidy as there is a need for greater subsidy for fodder block units due to relatively higher cost involved for processing, transport and establishment.

(xiii) Licensing of fodder block units should not be coupled with feed manufacturing units to minimize misuse of the subsidy.

(xiv) A greater subsidy is justified for fodder block units as relatively higher cost is involved for processing, storage, transport and establishment.

(xv) Roughage should be an essential component (about 30%) in the feed block units for granting subsidy.

(xvi) Promoting semi intensive goat management system, to enhance the profitability of goat farmers is need of the day. In this regard, rejuvenation of grazing land/common property resources may be taken on priority.

(xvii) The biggest handicap of non-availability of quality fodder seeds should be addressed on priority with the involvement of private sector.

(xviii) All the benefits and concessions of agriculture sector must be extended to livestock and fishery sector. Further, the benefit of food processing sector should also be extended to dairy processing and meat processing sectors. The dairy co-operative societies should be exempted from the purview of income tax and CESS. In the list of food commodities covered under FTA with Australia and New Zealand, milk and milk products should not be included.

(xix) To prevent long distance transportation of live animals and birds and associated losses due to shrinkage, stress, sickness and death, it is suggested to construct one hygienic slaughterhouse at district level or in each cluster of 2 or more districts with all the facilities for efficient utilization of edible and in-edible by-products and effluent treatment plant. Instead of one big slaughterhouse in Metropolitan cities, it is suggested to establish few modern slaughterhouses surrounding the city in peri-urban areas so that the small meat traders (with 3-5 animals) need not to transport the animals for long distance.

(xx) In the currently approved food parks/mega food parks there are hardly any meat/poultry processors due to restrictions from other food commodity entrepreneurs about inclusion of meat and poultry. To address this issue, Exclusive ANIMAL PROTEIN food parks/food hubs for meat/fish/poultry/egg need to be considered.

(xxi) Establishing the Special animal product economic zones and value addition in order to promote the export of livestock products is the need of the hour. The "Special Animal Product Zones (SAPZ)" on the lines of "Special Economic Zones" are to be set up and may be linked with export based processing plants, compound feed industry, regulated animal market, veterinary polyclinics, semen bank etc.

(xxii) Formation of farmers producer organization (PFO), co-operative societies or self-help groups in the villages will enhance farmers income and value addition. The producers should be linked to market. Further, clean milk production, traceability, adulterants detection kits and value addition of milk and milk products, meat and meat products, egg and egg products should be prioritized.

(xxiii) Continuing Veterinary Education (CVE) program, on the pattern of Continuing Medical Education (CME) should be implemented in all the states on utmost priority so as to acquaint field veterinarians with latest knowledge.

REFERENCES

1. BAH&FS (2016). Basic Animal Husbandry and Fishery Statistics. Published by the Department of Animal Husbandry, Dairying and Fisheries, Ministry of Agriculture & Farmers Welfare, GOI, New Delhi.

2. Birthal, P. S. and Taneja, V.K. (2006). Livestock Sector in India: Opportunities and Challenges for Smallholders. In: P. S. Birthal, V. K. Taneja and W Thrope (eds.), Smallholder Livestock Production in India: Opportunities and Challenges, National Centre for Agricultural Economics and Policy Research (NCAP), New Delhi, and International Livestock Research Institute (ILRI), Nairobi.

3. Birthal, P. and Negi, D.S. (2012). Livestock for higher, sustainable and inclusive agricultural growth. *Economics & Political Weekly Supplement*. June 30, 2012. Vol XLVII Nos. 26 & 27: 89-99.

4. Christiaensen, L., Demery, L. and Kuhl, J. (2006). The Role of Agriculture in Poverty Reduction: An Empirical Perspective, World Bank Policy Research Working Paper 4013, World Bank, Washington DC.

5. DAHDF (2017-18). Annual Report. Department of Animal Husbandry, Dairying and Fisheries, Ministry of Agriculture & Farmers Welfare, GOI, New Delhi.

6. DAHDF (2018-19). Annual Report. Department of Animal Husbandry, Dairying and Fisheries, Ministry of Agriculture & Farmers Welfare, GOI, New Delhi, pp 1-192.

7. NAAS (2009). Agriculture Sector: Status and Performance. In: State of Agriculture, Chapter 1 (ed.) M. Rai *et al.* Published by National Academy of Agricultural Sciences (NAAS), New Delhi, pp 1-34.

8. NAAS (2016). Breeding Policy for Cattle and Buffalo in India, Policy Paper No. 82, National Academy of Agricultural Sciences, New Delhi: 36 p.

9. Uppal P.K. (2011). National testimonies under the Global rinderpest Eradication Programme. Project report (Project no. GCP/GLO/302/EC) submitted to FAO, June 2011.

Current Status of Researches in Biosciences (2020) : 73-83
Editor : Prof. P.C. Joshi, N. Joshi, Reshman Yasmin, and D. K. Mansotra
Today & Tomorrow's Printers and Publishers, New Delhi - 110 002

ASSESSMENT OF ENDPOINT BIOMARKERS, THE MICRONUCLEI, FOR DNA DAMAGE

Sunil P. Trivedi

Environmental Toxicology & Bioremediation Laboratory, Department of Zoology, University of Lucknow, Lucknow-226007, India

Email: trivedi_sp@lkouniv.ac.in

ABSTRACT

The micronuclei induction happens in cells of organisms exposed to genotoxicants. The micronucleus test (MNT) reveals genotoxic damage in cells at interphase stage. This is an alternative and viable technique to chromosomal aberration test (CAT), because it examines interphase cells while CAT reflects metaphase stages of the cells. Further, MNT allows faster evaluation of genotoxicity. Furthermore, the damage analysis is less subjective and more susceptible to automation. Chromosomal DNA damage tests are crucial part of genetic toxicology because chromosomal mutations are highly significant events in carcinogenesis. The Micronucleus test is one of the preferred method for assessing chromosomal damage as it allows both chromosome loss and chromosome breakage to be identified reliably and quickly. This method is now being widely applied to many types of cells to monitor genetic damage among populations, investigate chemicals for genotoxic potential and for particular purposes such as predicting the sensitivity of tumor radiation and carcinogenesis. In its current form, MN analysis shows the measurement of genotoxicity and cytotoxicity in the form of chromosome destruction, chromosome loss, chromosome rearrangement, inhibition of cell division, and apoptosis. Therefore, MNT provides multiple and complementary measurements of genotoxicity and cytotoxicity which can be easily achieved in a single system. Among aquatic organisms, fish have the ability to metabolize, concentrate and accumulate waterborne pollutants quickly. They have a broad range of sensitivity against a variety of aquatic pollutants and widely preferred as an appropriate model for genotoxicity assessment. Thus, they can be effectively used as a subject to determine the distribution, toxic effects and the molecular mechanism of action of toxicants in aquatic environment.

KEYWORDS: Micronucleus; Clastogens; Aneugens; Genotoxicity; Carcinogenesis; Apoptosis.

INTRODUCTION

Micronuclei (MN) are extra nuclear bodies consisting of impaired chromosome pieces or entire chromosomes that were not fused into the nucleus after cell division. MN can form by deformities in the cell repair system, accretion of DNA damages and chromosomal aberrations. Many genotoxic pollutants may induce MN formation prompting to cell death, genomic instability and cancer progression (Luzhna *et al*., 2013). The routine methods employed for assessment of genotoxic impact of toxicants in the life form utilize a variety of genetic end points. DNA damage testing at the chromosome level is a significant piece of genetic toxicology on the grounds that chromosomal transformations are significant events in carcinogenesis. MN can be build up during anaphase from the rest of the chromosome or chromatid pieces brought about by damage to DNA. Micronuclei measurement is one of the significant assessment strategies for deciding chromosome damage (Fenech, 2008) and in the investigation of genotoxicity incited by different toxicants (Ratn *et al*., 2017). On the other hand, Cytokinesis block micronucleus assay (CBMN) is a comprehensive variant of MN test. One of the advantages of this test is the analysis of toxicant responses at metaphase stage. However, scoring of micronuclei in the interphase is in fact a lot simpler and quicker than the scoring of chromosomal variations during metaphase. The MN test in aquatic organisms likewise has the capability for recognizing genotoxic substances present in aquatic media. Aside from the biochemical parameters, micronuclei acceptance is additionally a delicate pointer of DNA damage and, in this way; it may be adequately utilized to have looks at genotoxicity. MN examination is routinely utilized for the evaluation of genotoxicity incited by toxins in different organisms, carcinogenic substances and development of cancer and apoptotic factors responsible for cell death.

Water pollution is a serious and growing problem. Although there are legal provisions that limit emissions/discharge of toxic chemicals, aquatic pollution still exists (Fleeger *et al*., 2003). Apart from fresh water habitats, marine ecosystem also serves as global sink for many pollutants originating from natural and anthropogenic sources. In fact, marine pollution is becoming increasingly attractive to the scientific community and society. Chemical pollutants with genotoxic and carcinogenic potential in the water are of great concern, because they threaten life in water and also on Earth. Fishes are perfect models for the study of genotoxic potential of contaminants present in water samples (Yadav and Trivedi, 2009). This consideration has sparked interest in the development of techniques and bio-indicators for monitoring genomic damage due to dangerous aquatic pollutants.

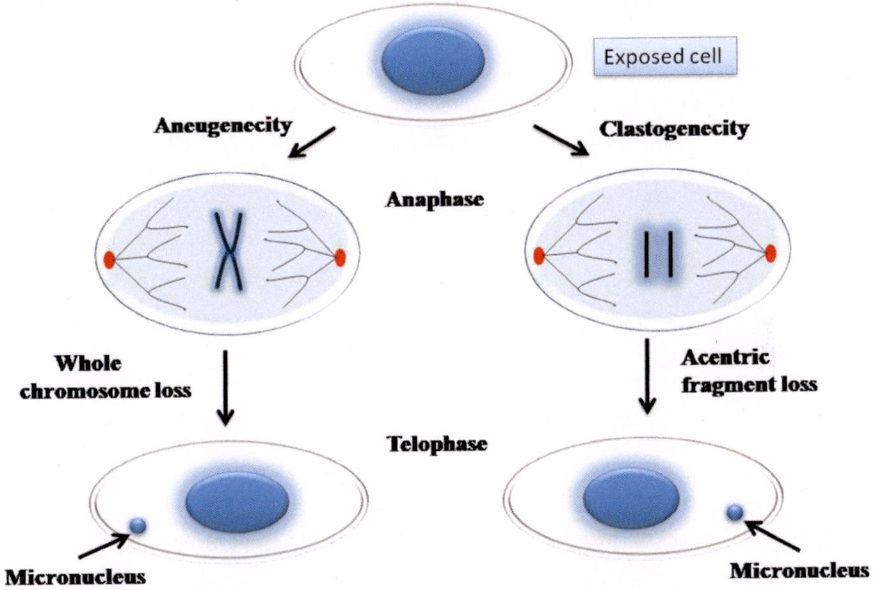

Fig 1. Diagrammatic representation of micronuclei formation

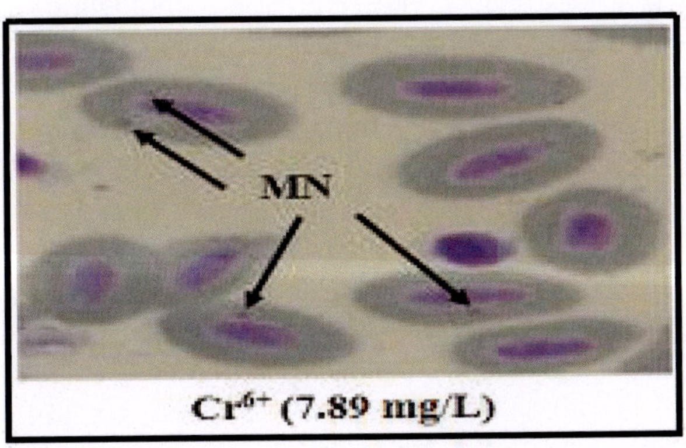

Fig 2. Represents MN induction in erythrocyte of fish, *Channa punctatus* exposed to sub-lethal concentration of Cr (VI) (Awasthi *et al.*, 2019).

ORIGIN OF MICRONUCLEUS

Micronuclei Induction due to Oxidative Stress: Oxidative stress is a condition associated with an imbalance between reactive oxygen species (ROS) and antioxidants. Increased level of ROS causes lipid, protein and DNA damage. Oxidative stress is associated with various diseases. Moreover, ROS molecules also have effects on immune defense as well as on signaling molecules and on maintaining physiological processes. Findings of Awasthi *et al.*, (2018) signifies the role of oxidative stress in induction of micronuclei on exposure to Cr^{6+} in *Channa punctatus.* Additionally, the prolonged exposure of arsenic induces the high frequency of MN in the fish due to oxidative stress (Dwivedi *et al.*, 2017). Moreover, a positive correlation between MN % and ROS level was observed in the blood of fish indicating the direct link of DNA damage with oxidative stress (Ratn *et al.*, 2018). Oxidative stress mediated induction of micro nucleated erythrocytes following the exposure to dimethoate has also been documented in fish by Ali *et al.*, (2014). Ratn *et al.*, 2017 have elaborated that the DNA damage in erythrocytes of fish can be attributed through indirect ROS binding with the DNA .

Micronuclei induction due to defects in the Dna Repair Machinery: A little extent of MN may just emerge from unrepaired double strand DNA breaks, but this only happens when DNA damage load surpasses the repair capacity of the cell within a specified time interval. Topoisomerase II is widely known enzyme liable for the DSBs resulting induction of MN due to failure of DNA repair (Awasthi *et al.*, 2019). The penchant for misrepair of DNA breaks is increased if the error-free homologous recombinational DNA repair pathway is nonfunctional due to fault in genes such as BRCA1 and BRCA2. Furthermore, DNA breaks, which lead to MN formation, may be left unrepaired if repair enzymes in the nonhomologous end joining pathway are defective (Hartlerode and Scully, 2009; O'Donovan and Livingston, 2010). Other processes that could lead to MN formation from mismatch DNA repair involve simultaneous excision repair of damaged DNA (e.g. 8-oxo-deoxyguanosine) or integration of inappropriate bases in DNA (e.g. uracil) that are closely located on opposite complementary DNA strands. Such simultaneous excision repair events, particularly if the gap-filling step is not completed, lead to DNA double-strand breaks and MN formation (Bull and Fenech, 2008; Dianov *et al.*, 1991; Fenech and Crott, 2002). In fact, this mechanism can be exploited to greatly enhance the lymphocyte MN assay responses to genotoxic agents that mainly induce DNA adducts. This increased sensitivity is accomplished by changing excision-repairable DNA abrasion into DNA strand breaks and therefore MN induction by treatment with cytosine arabinoside (during G1 phase of the cell cycle), which constrain the gap-filling step of excision repair (Fenech, 1993). Recently, it has been demonstrated that MN can arise from damaged chromosome material when nucleoplasmic bridge (NPB) are framed, extended and broken during telophase (Hoffelder *et al.*, 2004). Furthermore, the study of Awasthi *et al.*, (2019) illustrated a significant induction in MN frequency which occurs due to the DSBs in the chromium exposed fish, *Channa punctatus.*

There are many molecular methods that could assess chromosomal malsegregation at anaphase bringing about MN induction. One of the components that may prompt MN formation is hypomethylation of cytosine in centromeric and pericentromeric repeat sequences. For example, classical satellite repeats at pericentromeric areas and higher order repeats of satellite DNA in centromeric DNA (Pironon *et al.*, 2010).Various molecular events are there to induce MN formation and a key feature of MNT is that it permits detection of different molecular mechanisms that lead to chromosome damage and instability.

Micronuclei induction due to chromosomal aberrations: In recent years micronucleus test has emerged as a significant biomarker of genotoxicity with in various organisms (Çakal Arslan *et al.*, 2010). It makes possible to determine the remaining chromosomes and broken chromosomes which are failed to incorporate in nucleus. MNT has several advantages *viz.,* easy to manage, doesn't need to count the chromosomes to observe the chromatids and chromosomal damage difficult to detect and observe in the metaphase stage, presenting more objective results than other tests in identifying chromosomal damage, possibility to count thousands of cells, and fast preparation of smear (OECD, 2014). The MN test initially started to use as a test to identify chemical carcinogens in human cells in the 1970s and since then, it has been employed in a vast variety of organisms in order to determine cytogenetic damage (Schmid, 1975; Venier *et al.*, 1997). Further, for detection of aneugenic and clastogenic effects of harmful chemicals, micronuclei assay has been routinely used because they are formed at the cell division in anaphase stage when whole chromosome or fragment of chromosome breaks due to the lack of centromere (Prashad *et al.*, 2017). The chromosomal loss is due to eugenic event related to the spindle apparatus, while the chromosome breakage occurs due to clastogenic events. These inclusions can be visualized in any type of cell, both somatic and germinal. Therefore, the MNT can be performed in any active tissue. The rise in MN count is symptomatic of the numerical and structural chromosomal damage produced by various agents in cells. Aneuploidy stimulating agents encourages centromeric cleavage errors and malfunctioning of spindle strands, while clastogenes play a role in MN formation by forming chromosomal breaks (Bolognesi and Fenech, 2012).

OUTCOME OF MICRONUCLEI INDUCTION

Genomic instability: Micronuclei induction in peripheral blood lymphocytes is enormously used as a biomarker of genome instability in organisms. Ovarian cancer cells can form micronuclei that undergo vigorous DNA metabolism and generates single strand DNA breaks, leading to destabilizing genomic structures and activating genetic variations (Tang *et al.*, 2018). Increased level of micronuclei causes genetic instability which is a predictive biomarker of response to A siDNA in breast cancer cell lines (Jdey *et al.*, 2017). Furthermore, evaluation and examination of bladder cancer patients shows efficacy of micronucleus assay as a biomarker of genomic instability. MN assay, particularly in urothelial exfoliated cells, may be used to identify individuals who are at high risk of developing cancer due to extreme level of genomic instability (Podrimaj-Bytyqi *et al.*, 2018).

Cancer development: MN induction is considered as the initial stage in the development of genomic instability and tumorigenesis. Individuals inclined to cancer tend to form MN more widely than persons without hereditary history. Therefore, MN screening may serve as a valuable procedure in identifying many diseases, including cancer. Cancer treatment therapies also rely on the ability of a substance to cause profound damage and induce death in malignant cells (Luzhna *et al.*, 2013). Aberrant rearrangements can take place when DNA in micronuclei is damaged and gets reincorporated into the primary nucleus, a phenomenon referred to as chromothripsis where one or multiple chromosomes achieve dozens to hundreds of clustered rearrangements in a single event (Stephens *et al.*, 2011). Chromothripsis is frequent in cancer patients and is usually combined with poor prognosis (Rode *et al.*, 2016; Stephens *et al.*, 2011). One of the recent studies shows that the chromothripsis involves DNA damage in micronuclei followed by reincorporation into a daughter nucleus, where random re-location can takes place (Ly *et al.*, 2017). Relationship between MN induction and cancer development is supported by a number of studies *viz.,* the high frequency of this biomarker in untreated cancer patients and in subjects affected by cancer-prone congenital diseases (Fenech, 2002). The increased MN frequencies in oral mucosa, served as a surrogate biomarker of cancer in clinical chemoprevention trials (Schooten *et al.*, 2002). There exists a correlation between genotoxic MN-inducing pathogens and carcinogenicity, e.g. ionizing and ultraviolet radiation (Bettega *et al.*, 2003).

Apoptosis and cell death: Inappropriate separation of many chromosomes causes rapid cell death. The mitotic catastrophe involves the formation of giant cells with many micronuclei or two nuclei. Micronuclear cells are non-viable and results from the formation of nuclear envelopes around chromosome clusters or chromosome fragments during catastrophic mitosis. Damage to mitotic DNA can also cause cytokinesis failure which causes binucleation.

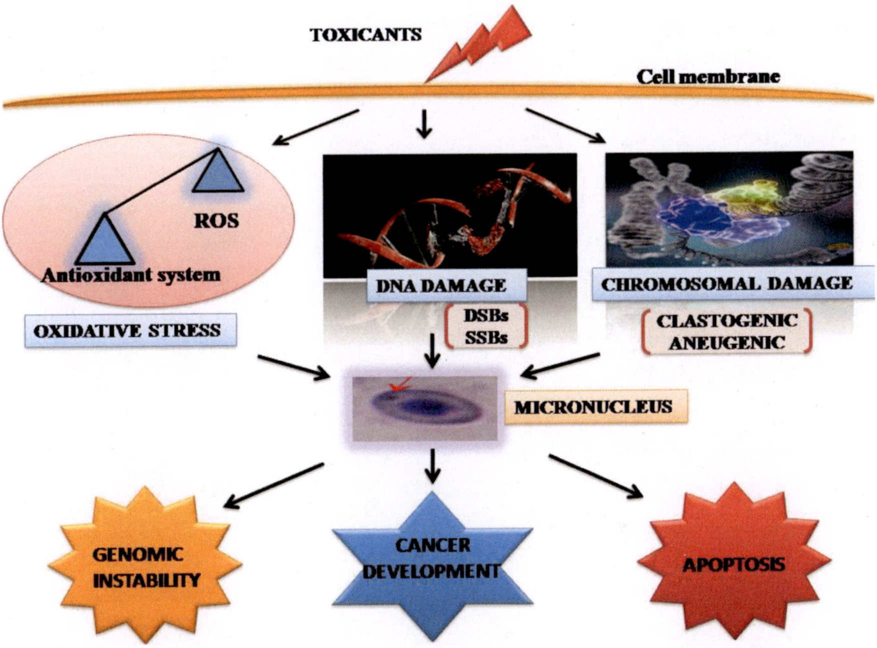

Fig 3. Diagrammatic representation of induction of micronuclei and its possible outcome.

Cells that contain two nuclei are restricted in G1 as they are dependent on p53 (Ricci *et al.*, 2006). The transcriptional alteration in p53 gene that forms a bridge between oxidative stress and DNA damage initiates the apoptosis (Awasthi *et al.*, 2018). Utani *et al.*, (2010) found that the presence of micronuclei associated with apoptosis and low hydroxylurea concentrations effectively induced multipolar mitosis, which produced lagging chromosomes or chromatin bridges in addition to poly-nuclear cells that were closely related to apoptosis. Ratn *et al.*, (2017) studied MN induction and subsequent differential expression of tumor suppressor gene p53, apoptosis protease activating factor-1 (*apaf-1*) and catalase (*cat*) caused by DNA injury in *Channa punctatus*. Furthermore, *Nox-1* gene plays an important role in cell signaling which combines the three key processes *viz.,* oxidative stress, DNA damage and apoptosis, responsible for tissue injury (Awasthi *et al.*, 2018). They concluded that the oxidative stress plays a pivotal role in instigating DNA strand breaks resuling in induction of MN. In assessing genotoxicity, it is important to consider not only DNA damage, but also apoptotic and cytostatic effects, all of which may be interrelated. H_2O_2 levels induce MN frequency, suggesting that apoptosis plays an important role in removing cells with DNA damage (Fenech *et al.*, 1999). Micronucleus analysis can be used for measurement of damaged chromosomes, chromosome loss, and integrated apoptotic effects. In addition, this system provides a better way to assess the relative importance of aforementioned events when cells are stressed by different drugs or xenobiotics.

CONCLUSION

Development of the micronucleus test as a stable technique is helpful for identification of once dividing cells by their binuclear emergence. The development of MN analysis has evolved into a complete cytomic technique, which, in addition to MN, also allows for quantification of stable nuclear fission rates as well as two types of cell death-necrosis and apoptosis. In addition, the relevance and validation of the MN test was significantly enhanced by prospective epidemiological studies showing that MN biomarkers in binucleotide lymphocytes quickly predict cancer risks. This attainment is expanded by the widespread use of MN assays for both *in vitro* and *in vivo* genotoxicity studies. This allows a nearly harmonious approach to assess genotoxic damage in humans, because the same system can be used not only for *in vitro* testing but also as a measure of *in vivo* induced DNA damage. Under the exposure of different water borne toxicants, the formation of free radicals (ROS) in fish is obvious. This generates consequent oxidative stress resulting in detrimental effects in the oxidation of cell constituents, including DNA in fish. Damaged DNA reflects in grater MN induction. Thus, assessment of this endpoint biomarker can be effectively utilized in routine bio-monitoring of aquatic regimes contaminated with potential genotoxicants.

REFERENCES

1. Ali, D., Kumar, G., Kumar, S., Ahmed, M., Gend Kumar, P., (2014). Chemical Speciation & Bioavailability Evaluation of genotoxic and oxidative stress response to dimethoate in freshwater fish *Channa punctatus* (Bloch) Evaluation of genotoxic and oxidative stress response to dimethoate in freshwater fish *Channa punctatus* (Bloch). *Chem. Speciat. Bioavailab.* 26. https://doi.org/10.1080/09542299.2014.11073965

2. Awasthi, Y., Ratn, A., Prasad, R., Kumar, M., Trivedi, A., Shukla, J.P., Trivedi, S.P., (2019). A protective study of curcumin associated with Cr6+ induced oxidative stress, genetic damage, transcription of genes related to apoptosis and histopathology of fish, *Channa punctatus* (Bloch, 1793). *Environ. Toxicol. Pharmacol.* 71, 1–10. https://doi.org/10.1016/j.etap.2019.103209

3. Awasthi, Y., Ratn, A., Prasad, R., Kumar, M., Trivedi, S.P., (2018). An in vivo analysis of Cr 6+ induced biochemical, genotoxicological and transcriptional profiling of genes related to oxidative stress, DNA damage and apoptosis in liver of fish, *Channa punctatus* (Bloch, 1793). *Aquat. Toxicol.* 200, 158–167. https://doi.org/10.1016/j.aquatox.2018.05.001

4. Bettega, D., Calzolari, P., Doneda, L., Belloni, F., Tallone, L., Redpath, J.L., (2003). Differential effectiveness of solar UVB subcomponents in causing cell death, oncogenic transformation and micronucleus induction in human hybrid cells. *Int. J. Radiat. Biol.* 79, 211–216. https://doi.org/10.1080/0955300031000075345

5. Bolognesi, C., Fenech, M., 2012. Mussel micronucleus cytome assay. *Nat. Protoc.* 7, 1125–1137. https://doi.org/10.1038/nprot.2012.043

6. Bull, C., Fenech, M., (2008). Genome-health nutrigenomics and nutrigenetics: nutritional requirements or 'nutriomes' for chromosomal stability and telomere maintenance at the individual level. *Proc. Nutr. Soc.* 67, 146–156. https://doi.org/10.1017/S0029665108006988

7. Çakal Arslan, Ö., Parlak, H., Katalay, S., Boyacioglu, M., Karaaslan, M.A., Guner, H., (2010). Detecting micronuclei frequency in some aquatic organisms for monitoring pollution of Izmir Bay (Western Turkey). *Environ. Monit. Assess.* 165, 55–66. https://doi.org/10.1007/s10661-009-0926-5

8. Dianov, G., Timehenko, T., Sinitsina, O., Kuzminov, A., Medvedev, O., Salganik, R., (1991). Repair of uracil residues closely spaced on the opposite strands of plasmid DNA results in double-strand break and deletion formation. *MGG Mol. Gen. Genet.* 225. https://doi.org/10.1007/BF00261686

9. Dwivedi, S., Kumar, M., Trivedi, S.P., (2017). Mitigating potential of Melissa officinale against As 3+-induced cytotoxicity and transcriptional alterations of Hsp70 and Hsp27 in fish, Channa punctatus (Bloch). *Env. Monit Assess* 189, 1–15. https://doi.org/10.1007/s10661-017-6002-7

10. Fenech, M., (2008). The Micronucleus Assay Determination of Chromosomal Level DNA Damage. pp. 185–216. https://doi.org/10.1007/978-1-59745-548-0_12

11. Fenech, M., (2002). Chromosomal biomarkers of genomic instability relevant to cancer. *Drug Discov. Today* 7, 1128–1137. https://doi.org/10.1016/S1359-6446(02)02502-3

12. Fenech, M., (1993). The Cytokinesis-Block Micronucleus Technique and Its Application to Genotoxicity Studies in Human Populations, Environmental Health Perspectives Supplements.

13. Fenech, M., Crott, J.W., (2002). Micronuclei, nucleoplasmic bridges and nuclear buds induced in folic acid deficient human lymphocytes—evidence for breakage–fusion-bridge cycles in the cytokinesis-block micronucleus assay. *Mutat. Res. Mol.*

Mech. Mutagen. 504, 131–136. https://doi.org/10.1016/S0027-5107(02)00086-6

14. Hartlerode, A.J., Scully, R., (2009). Mechanisms of double-strand break repair in somatic mammalian cells. *Biochem. J.* 423, 157–168. https://doi.org/10.1042/BJ20090942

15. Jdey, W., Thierry, S., Popova, T., Stern, M.-H., Dutreix, M., (2017). Molecular and Cellular Pathobiology Micronuclei Frequency in Tumors Is a Predictive Biomarker for Genetic Instability and Sensitivity to the DNA Repair Inhibitor AsiDNA. https://doi.org/10.1158/0008-5472.CAN-16-2693

16. Luzhna, L., Kathiria, P., Kovalchuk, O., (2013). Micronuclei in genotoxicity assessment: from genetics to epigenetics and beyond. *Front. Genet.* 4, 1–17. https://doi.org/10.3389/fgene.2013.00131

17. Ly, P., Teitz, L.S., Kim, D.H., Shoshani, O., Skaletsky, H., Fachinetti, D., Page, D.C., Cleveland, D.W., (2017). Selective Y centromere inactivation triggers chromosome shattering in micronuclei and repair by non-homologous end joining. *Nat. Cell Biol.* 19, 68–75. https://doi.org/10.1038/ncb3450

18. O'Donovan, P.J., Livingston, D.M., (2010). BRCA1 and BRCA2: breast/ovarian cancer susceptibility gene products and participants in DNA double-strand break repair. *Carcinogenesis* 31, 961–967. https://doi.org/10.1093/carcin/bgq069

19. Oecd, (2014). OECD Test Guideline 487.

20. Pironon, N., Puechberty, J., Roizès, G., (2010). Molecular and evolutionary characteristics of the fraction of human alpha satellite DNA associated with CENP-A at the centromeres of chromosomes 1, 5, 19, and 21.

21. Podrimaj-Bytyqi, A., Borovečki, A., Selimi, Q., Manxhuka-Kerliu, S., Gashi, G., Elezaj, I.R., (2018). The frequencies of micronuclei, nucleoplasmic bridges and nuclear buds as biomarkers of genomic instability in patients with urothelial cell carcinoma. *Sci. Rep.* 8, 1–9. https://doi.org/10.1038/s41598-018-35903-5

22. Prashad, R., Kumar, M., Trivedi, S.P., (2017). Antigenotoxic effect of turmeric powder extract curcumin against chromium trioxide induced genotoxicity in fish *Channa punctatus*. *J. Entomol. Zool. Stud.* 5, 89–94.

23. Ratn, A., Awasthi, Y., Kumar, M., Singh, S.K., Tripathi, R., Trivedi, S.P., (2017). Phorate induced oxidative stress, DNA damage and differential expression of p53, apaf-1 and cat genes in fish, *Channa punctatus* (Bloch, 1793). *Chemosphere*. https://doi.org/10.1016/j.chemosphere.2017.05.008

24. Ratn, A., Prasad, R., Awasthi, Y., Kumar, M., Misra, A., Trivedi, S.P., (2018). Zn 2+ induced molecular responses associated with oxidative stress, DNA damage and histopathological lesions in liver and kidney of the fish, *Channa punctatus* (Bloch, 1793). *Ecotoxicol. Environ.* Saf. 151, 10–20. https://doi.org/10.1016/j.ecoenv.2017.12.058

25. Rode, A., Maass, K.K., Willmund, K.V., Lichter, P., Ernst, A., (2016). Chromothripsis in cancer cells: An update. *Int. J. Cancer* 138, 2322–2333. https://doi.org/10.1002/ijc.29888

26. Schmid, W., (1975). The micronucleus test. *Mutat. Res. Mutagen. Relat. Subj.* 31,

9–15. https://doi.org/10.1016/0165-1161(75)90058-8

27. Schooten, J.V.F., Nia, B.A., Flora, D.S., Agostini, D.F., Izzotti, A., Camoirano, A., Balm, J.. A., Dallinga, W.J., Bast, A., Haeman, M.M.R.G., Veer, V.L., Baas, P., Sakai, H., Zandwijk., (2002). Effects of oral administration of N-Acetyl-L-cysteine: A multi-biomarker study in smokers. *Cancer Epidemiol. Biomarkers Prev.* 11, 167–175.

28. Stephens, P.J., Greenman, C.D., Fu, B., Yang, F., Bignell, G.R., Mudie, L.J., Pleasance, E.D., Lau, K.W., Beare, D., Stebbings, L.A., McLaren, S., Lin, M.-L., McBride, D.J., Varela, I., Nik-Zainal, S., Leroy, C., Jia, M., Menzies, A., Butler, A.P., Teague, J.W., Quail, M.A., Burton, J., Swerdlow, H., Carter, N.P., Morsberger, L.A., Iacobuzio-Donahue, C., Follows, G.A., Green, A.R., Flanagan, A.M., Stratton, M.R., Futreal, P.A., Campbell, P.J., (2011). Massive Genomic Rearrangement Acquired in a Single Catastrophic Event during Cancer Development. *Cell* 144, 27–40. https://doi.org/10.1016/j.cell.2010.11.055

29. Tang, Z., Yang, J., Wang, X., Zeng, M., Wang, J., Wang, A., Zhao, M., Guo, L., Liu, C., Li, D., Chen, J., (2018). Active DNA end processing in micronuclei of ovarian cancer cells. *BMC Cancer* 18, 1–10. https://doi.org/10.1186/s12885-018-4347-0

30. Venier, P., Maron, S., Canova, S., (1997). Detection of micronuclei in gill cells and haemocytes of mussels exposed to benzo[a]pyrene. *Mutat. Res. Toxicol. Environ. Mutagen.* 390, 33–44. https://doi.org/10.1016/S0165-1218(96)00162-0.

31. Yadav, K.K., Trivedi, S.P., (2009). Sublethal exposure of heavy metals induces micronuclei in fish, *Channa punctata. Chemosphere* 77, 1495–1500. https://doi.org/10.1016/j.chemosphere.2009.10.022.

Current Status of Researches in Biosciences (2020) : 85-95
Editor : Prof. P.C. Joshi, N. Joshi, Reshman Yasmin, and D. K. Mansotra
Today & Tomorrow's Printers and Publishers, New Delhi - 110 002

PREVALENCE OF GASTROINTESTINAL HELMINTH PARASITES OF PIGS SLAUGHTERED AT LUCKNOW, UTTAR PRADESH (INDIA)

Kamal Jaiswal*, Suman Mishra, Amit Kumar Singh, Divyam Srivastava and Anjum Bee

Department of Zoology, Babasaheb Bhimrao Ambedkar University, Lucknow- 226025

Email: drkamaljaiswal@gmail.com

ABSTRACT

Helminthiasis, caused by helminth parasites, is one of the most common infections affecting health of pigs and indirectly small livestock holders are in great economic loss. According to month, age and gender wise prevalence of gastrointestinal helminth parasites of *Sus scrofa* was studied from June, 2018 to November, 2018 in the Parasitology lab of Zoology Department, Babasaheb Bhimrao Ambedkar University, Lucknow, Uttar Pradesh. In this study, 147 gastrointestinal tract of carcass pig were examined, out of which 80 gastrointestinal tracts of pigs were found to be infected with helminth parasites. The overall prevalence of parasites was 54.42%. During the study period, *Ascaris suum* (32.50%) was the most prevalent parasite than the *Strongyloides* spp (20%), *Fasciolopsis* spp (15.00%), *Taenia solium* (6.25%). Month wise prevalence of gastrointestinal parasite of pigs was higher in the month of August (73.91%) and low in November (43.47%), age-wise prevalence was higher in adult group (3.0-4.0 year) (70.00%), than the piglets (<1.0 year) (37.93%). Females (56.25%) were more prone to infection than the males (52.23%) during the study period. So it is suggested that deworming is necessary to avoid the risk of pig endoparasites and the types of anthelmintic drugs should be changed annually.

KEYWORDS: Helminthiasis, *Sus scrofa*, parasites, anthelmintic drugs, national economy

INTRODUCTION

Pigs belong to genus *Sus* under family Suidae also named as swine/ hogs, which have been reared for food, leather and related products. Commonly large proportion of pigs are rearing in free ranging system in developing countries like in India, in this system freely they can feed faecal matter, raw garbage and kitchen waste therefore are more prone to parasitic infections (Weka & Ikeh, 2009 and Tiwari *et al.*, 2009). Generally the common weight of pig carcass in China is 76kg, 84kg in USA and 90kgs in Germany, but in India it is up to 35kgs as a whole (ICAR Handbook of Animal Husbandry, 2012). Since ancient time different parasites are causing to pigs which can harbour to human also. A lot of literature exists regarding porcine taeniasis and cysticercosis, but little literature available on the gastrointestinal parasites of pigs in the geographical conditions of India. Way of rearing is an important factor related to the risk of gastro-intestinal parasitism (GIP) (Nansen & Roepstorff, 1999). A parasite is an organism that has sustained contact with another organism to the detriment of the host organism (Weng, 2005). A parasite can also be defined as any living organism which can gain the food from the host and can live on the surface or inside the host animal (Henderson, 1990). Gastrointestinal parasites can be categorized as nematodes (round worms), Trematodes (flukes), cestodes (Tapworms), and protozoans (Brander, 1982). Malnutrition and un-thriftiness are responsible for the high parasitic load leading to poor weight gain in India (Yadav & Tandon, 1989, Ebibeni, 2013). In India, about 90% proportion of pig are reared by poor families having reduced hygiene level (Laha, 2014, NABARD). Parasites of pig cause major economic loss globally as a consequence of reduced feed conversion and weight gains and the condemnation of affected organs found after slaughtered (Boes, 2000; Joaching, 2001). Pigs are infected with gastrointestinal parasite through oral ingestion of helminthes eggs or cysts from pig's environment. Once ingested, the parasite competes with infected pig for nutrients, and pigs that are heavily infected grow slower, and become less resistant to disease and stress (Weng, 2005). In some cases, internal organs are damaged due to internal parasite which may result in condemnation of such organs as meat. Although many schemes are being run by the government to promote pig rearing but piggery sector suffers from setbacks due to execution and development of policies. This often led to great nutritional and economic loss (Harper, 2004). Keeping in mind the above facts, the present study was carried out to generate data on prevalence of gastrointestinal helminth parasites in pigs of Lucknow region.

MATERIALS AND METHODS

Study area: The study was conducted in and around Lucknow which stands at an elevation of approximately 123 meters (404 ft) above sea level. Lucknow district covers an area of 2,528 square kilometres (976 sq mi) (Lucknow District Population Census, 2011) (Lucknow (District, Uttar Pradesh, India) 2014). Bounded on the east by Barabanki, on the west by Unnao, on the south by Raebareli and in the north by Sitapur, Lucknow sits on the northwestern shore of the Gomti River. This city has a humid subtropical climate with cool, dry winters from mid-November to February and dry, hot summers with thunderstorms from late March to June. The rainy season is from July to September when the city gets an average rainfall of 896.2 millimetres (35.28 in) from the south-west monsoon winds, and occasionally frontal rainfall will occur in January. In winter, the maximum temperature is around 25 °C (77 °F) and the minimum is in the 3 °C (37 °F) to 7 °C (45 °F) range (Lucknow District, Uttar Pradesh, India 2012).

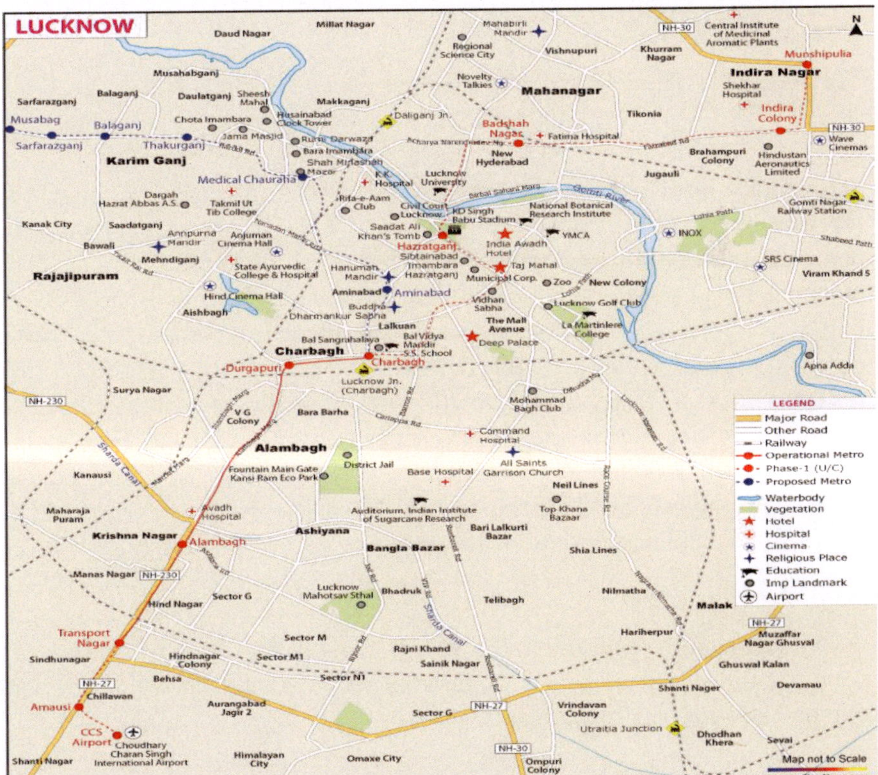

(Source- https://www.google.com/search?q=lucknow+city+map&tbm)

Study period- Study was conducted during June 2018 to November 2018 in Lucknow Uttar Pradesh, to determine the prevalence of gastrointestinal helminth parasites of *Sus scrofa* (Pig).

Parasitological examination- During the present study faecal samples was collected with respect to age, month and gender from the pigs brought for slaughter. All the samples were brought to the laboratory of Department of Zoology, BBAU, Lucknow and examined thoroughly. Adult worms like nematode, cestode and trematodes were also collected from the same host and preserved in Lectophenol and cornoys fixative respectively, and eggs of helminth detected by a direct smear method (Soulsby, 1986 and Cheesbrough, 2000). Slightly placed the sample on the clean slide with the help of glass rod and made a smooth smear. After that carefully mixed the 1% normal saline and placed a glass cover slip on it and observed under the light microscope for the presence of gastrointestinal helminths. The adult worms and eggs of helminths were identified according to the keys described by Soulsby (1982), Chandler *et al.,* (1961), Cheesbrugh (2000),

Definitions: The prevalence of parasites was calculated using the formula-

$$\text{Prevalence} = \frac{\text{Total number of hosts infected}}{\text{Total number of hosts examined}} \times 100$$

Data analysis: The most common measurements of parasite population levels in hosts are prevalence (Bush *et al.,* 1997). Prevalence refers to the percentages of organisms infected by a particular species of parasite. The nomenclature used to define ecological parameters is in consistency with that of Margolis, (1982). The information obtained from laboratory test and observation was entered on the IBM SPSS version 20. Chi-square ($\chi 2$) test was used to analyze the sample data. Chi-square test was used to assess whether there exist statistically significant difference in gastrointestinal nematode infection between months, sex and age. A statistically significant association between variables was considered to exist if the calculated p-value is less than 0.05 with 95% confidence level.

RESULTS

During the present study, 147 gastrointestinal tract of carcass pig were examined, out of which 80 gastrointestinal tracts were infected with one or more species of parasites.

Table1. Overall prevalence of gastro intestinal helminths in *Sus scrofa*

Total no of hosts examined	No of infected hosts	Prevalence of infection
147	80	54.42%

The overall prevalence was 54.42% (Table 1), in terms of species wise prevalence the highest prevalence was of *Ascaris suum* (32.50%) as compared to *Strongyle* spp (20.00%), *Taenia solium* (6.25%), *Fasciolopsis* spp (15.00%), and Mixed infection (26.25%) (Table 2).

Table 2: Species-wise prevalence of gastro intestinal helminths in *Sus scrofa*

S.N	Helminths Species	Total no of infected intestines	% Prevalence
1.	*Ascaris sum*	26	32.50%
2.	*Strongloids* spp	16	20.00%
3.	*Taenia solium*	5	6.25%
4.	*Fasciolopsis* spp	12	15.00%
5.	Mixed infection	21	26.25%
Total	**147**	**80**	**54.42%**

During the study period, month-wise prevalence was higher in the month of August (73.91%) and lowest in November (43.47%) (Table 3).

Table 3: Month wise prevalence of gastro intestinal helminths in *Sus scrofa*

Months	Temperature		Humidity		Hosts Examined		
	Max. (°C)	Min. (°C)	Max. (%)	Min. (%)	Total no. of hosts examined examined	Total no. of infected hosts	% of infected hosts
June	43	24	100	25	27	13	48.14%
July	41	25	100	35	25	13	52.00%
August	38	24	100	56	24	17	70.83%
September	37	22	100	42	23	16	69.56%
October	37	17	100	24	24	11	45.83%
November	34	12	100	28	24	10	41.66%
							Ç² = 7.974 **P= >0.05**

Table 4: Age-wise prevalence of gastro intestinal helminths in *Sus scrofa*

S.No.	Age-Group (years)	No. of hosts examined	Total no. of Infected hosts examined	% Prevalence
1.	<1.0	29	11	37.93%
2.	1.0-2.0	29	13	44.82%
3.	2.0-3.0	30	17	56.66%
4.	3.0-4.0	30	21	70.00%
5.	>4.1	29	18	62.06%
				$\chi^2 = 5.019$
				P= >0.05

Present study showed that all age groups were infected with gastrointestinal parasites where adult group (3.0-4.0 year) (70.00%) were found to have higher rate of infection, whereas the infection was found to be low in piglets (<1.0 year) (37.93%) as shown in Table no. 4.

Table 5: Gender-wise prevalence of gastro intestinal helminths in *Sus scrofa*

Gender	No. of examined	No. of infected host	% Prevalence
Male	67	35	52.23%
Female	80	45	56.25%
			$\chi^2 = 0.262$ P= >0.05

Out of 147 specimens of *Sus scrofa* examined during the present study, 67 were males and 80 were females. Gender wise prevalence was higher in females (56.25%) as compared to males (52.23%) (Table 5).

DISCUSSION

Pig production can be a profitable enterprise but gastrointestinal parasites are a limiting factor for piggery production (Nsoso *et al.*, 2000). Several parasites of pig are zoonotic in nature that can cause to human beings (FAO, 2008; Spencer, 2010). During the study, we studied on the prevalence of gastrointestinal helminth parasites in pigs kept by the farmers in Lucknow India. The overall prevalence of gastrointestinal helminth parasites was 54.42% during study period. The overall prevalence of intestinal helminth parasites in pigs was more or less similar to studies conducted in Mumbai (51.11%) (Sachin Dadas *et al.,* 2016), Zimbabwe

(58.7%) (Marufu *et al.*, 2008). In contrary, higher prevalence of intestinal parasites in pigs had been recorded in Bangladesh (96.4%) (Dey *et al.*, 2014), Southeastern part of Uganda (94%) (Waiswa *et al.*, 2007), Burkina (92.7%) (Tamboura *et al.*, 2006) and Kabale Uganda (91%) (Nissen *et al.*, 2011), Kenya (83%) (Obonyo *et al.*, 2012, Nigeria (80.4%), (Sowemimo *et al.*, 2012), Korea (73.4%) (Ismail *et al.*, 2010). These recorded results showed variation due to geographical and climatic conditions, way of farm running, status of health and nutrition of the pigs, way of sample collection and investigation, and differences in the number of samples analysed.

The overall prevalence (32.50%) of *A. suum* during the study was more or less similar (32.59%) to a study conducted by Sachin Dadas *et al.*, 2016 in Mumbai, 40% by Tamboura *et al.*, (2006) in Burkina Faso, 50.9% by Dey *et al.*, (2014) in Bangladesh and 54.6% by Nsoso *et al.*, 2000 in Botswana. Though, deviations in the season and geography can cause a variation in the results of *A. suum,* which help in the development and proliferation of helminth parasites. Kagira (2010) and Obonyo *et al.*, (2012) reported that some factors like humid-temperature, long-lasting wet farm conditions, and unhygienic environments are favourable for the development of parasites. The eggs of parasites can survive in unfavourable climatic conditions and they can be viable and infective for a long time (Roepstorff & Nansen, 1998). According to Leman *et al.* (1986), milk spot of liver in the growing pigs are due to transmission of *A. suum* larva. It is reported that 40% reduction in weight and 25% reduction in feed conversion efficiency is due to the infection by *A. suum* (Polley and Mostert,1980).

Higher prevalence was recorded for 3-4 years (70.00%) and 62.06% in >4 years old pigs while piglets recorded the lowest (<1 year) (37.93%). Similarly the highest prevalence of intestinal helminth parasites in growers and/or adult pigs was also reported in the studies of Dey *et al.*, (2014); Dutta *et al.*, (2005); Nsoso *et al.*, (2000); Roepstorff & Nansen, (1998); Jufare *et al.*, (2015) in Nordic countries, Botswana, India, Bangladesh and Ethiopia respectively. Generally heavy infestations in growing and adult pigs kept in outdoors responsible for bloody diarrhoea (Roepstorff & Nansen, 1998).

Contrary, Obonyo (Obonyo *et al.*, 2012) reported the lowest prevalence of helminth infections in adult pigs in the District of Homabay, Kenya. Bugg *et al.*, (1999) reported the highest prevalence of infection in the piglets of Western Australia and some other researcher also reported the same prevalence of infection in the Grenada region of West Indies, Chongqing region of China, and Ibadan region of Nigeria (Tiwari *et al.*,

2009, Lai *et al.,* 2011, Sowemimo *et al.,* 2012). During the present study prevalence rate of infection was higher in females (56.25%) than the males (52.23%). Similar prevalence of infection was found in the female pigs of Bangladesh and Ethiopia reported by Dey *et al.,* (2014) and Jufare *et al.*, (2015) respectively. Additionally, various reports from Nigeria and Burkina Faso concluded that females are more prone to the helminth parasites than males (Obonyo *et al.*, 2012, Tamboura *et al.*, 2006), but Kagira (2010) and Sowemimo *et al.*, (2012) reported that male pigs are more infected in the Kenya and Nigeria, respectively. In females, various factors like hormonal imbalance, gravidity, parturition, lactation and stress leads to poor immunity and inclination of pathogens responsible for the higher prevalence of infection (Lloyd, 1983; Swai *et al.*, 2010).

CONCLUSION

The present study exposed that pigs were highly susceptible to parasitic infection and affected by different helminth parasites in different stages of their life and different climatic conditions existing in the study area. Helminthes cause a higher risk to flourishing pig farming industry as well as meat consumers by transmitting the infection. Month-wise, age-wise and gender-wise prevalence of helminth parasite was studied in Lucknow region and it can be concluded that helminthes are a serious threat to pig farming. According to present study it can be concluded that various species of gastrointestinal pig parasites are occur in the study area. For in-depth studies of parasitic infection in this area, survey should be done during all four seasons and sample size should be large to determine the parasitic load. This will help to execute suitable strategies against unprofitability of piggery enterprises. Every effort should be beneficial for smallholder pig farmers to get higher profit for their products and pork should be clean and safe for the consumers. It is very needed that the formulation of appropriate and cost-effective strategies and drugs to control the gastrointestinal helminth parasite in pigs to get the benefit of pig rearing.

ACKNOWLEDGMENTS

Thanks our Hon'ble Vice-Chancellor Prof. Sanjay Singh and other authorities of the University for Necessary Infrastructural Facilities to carry out the research work.

REFERENCES

1. Boes, J., A. L. Willingham, P. Nansen (2000). Prevalence and distribution of pig helminthes in Dongtin region China. *Journal of Helminthology*., 74: 45-52.
2. Dadas S., S. Mishra, V. Jawalagatti, S. Gupta, T.S. Vinay, J Gudewar (2016).

Prevalence of gastro-intestinal parasites in pigs (*sus scrofa*) of mumbai region. International Journal of Science, *Environment and Technology.*, Vol. 5, No 2, 2016, 822 – 826

3. Brander, G. C. (1982). Veterinary applied pharmacology and the erapentics. Fourth Edition Pp. 470 – 471

4. Bugg, R.J., I. D. Robertson, A.D. Elliot, R.C. Thompson (1999). Gastro-intestinal parasites of urban pigs in Perth, *Western Australia, Veterinary Journal.*, 157, 295–301.

5. Bush AO., K.D. Lafferty, J.M. Lotz, A.W. Shostak (1997). Parasitology meets ecology on its own terms: Margolis *et al.*, revisited. *J Parasitol.*, 83: 575-583.

6. Chandler, A. C., C.P. Read (1961). Introduction to Parasitology. John Wiley and sons Inc. *London 10th edition.*, pp 140-145.

7. Cheesbrough, M. (2000). District laboratory practice in tropical countries, part 2. *Press sundicate of the University of Cambridge U.K.*, Pp 320 – 321.

8. Dey, T.R., A.R. Dey, N. Begum, S. Akther, B.C. Barmon (2014). Prevalence of endo parasites of pig at Mymensingh, Bangladesh, *Journal of Agriculture and Veterinary Science.*, 7(4), 31–38.

9. Dutta, S., J.D. Ghosh, N.K. Sasmal, G.S. Mukherjee (2005). Prevalence of gastrointestinal parasites affecting pig farms of West Bengal, India, *Journal of Veterinary Parasitology.*, 19, 23–26.

10. Ebibeni N., J.K. Chamuah, O.K. Raina, A. Sakhrie, Perumal, P.D. Borkotoky, N. Lily (2013). Prevalence of gastrointestinal parasites of pigs in Dimapur district (Nagaland). *Journal of Veterinary Parasitology.*, 27: 57-58.

11. Food and Agricultural Organization (FAO) (2008). Guidelines for humane handling, transport and slaughter of livestock, viewed 11 November 2015, from http://www.fao.org/docrep/011/i0291e/i0291e00.htm

12. Harper, A. (2004). Internal parasites control in pigs. Unpublished data. University of Missouri pp. 1-6.

13. Henderson, D. C. (1990). The Veterinary sheep famers. Farming press books Pp 477.

14. https://www.google.com/search?q=lucknow+city+map&tbm

15. ICAR, Handbook of animal husbandry (2012) ICAR publication, New Delhi.

16. Ismail, H. A., H. K. Jeon, Y. M. Yu, C. Do, Y. H. Lee (2010). Intestinal parasite infections in pigs and beef cattle in rural areas of Chungcheongnam-do, Korea, *Korean Journal of Parasitology.*, 48, 347–349.

17. Joaching, A., N. Dulmar (2001). Occurrence of helminthes in the pig fattening units with different management system in Northern Germany. *Veterinary Parasitology.*, 9: 135-146.

18. Jufare, A., N. Awol, F. Tadesse, Y. Tsegaye, B. Hadush (2015). Parasites of pigs in two farms with poor husbandry practices in Bishoftu, Ethiopia, *Onderstepoort*

Journal of Veterinary Research., 82(1), 1–5.

19. Kagira, J.M. (2010). Characteristics of production system, epidemiology and control of pig parasites in Busia District, Kenya, Ph.D Thesis, University of Nairobi, Kenya.

20. Laha, R., M. Das, A. Goswami, B. Sailo, B. K. Sharma, D. Gangmei, E. Ngullie (2014). Prevalence of gastrointestinal parasitic infections in pigs of North Eastern Region of India. *Indian Journal of Hill Farming*. 27(1): 110-117.

21. Leman, A. D., B. Straw, R. D. Glock, W. L Mengeling, R.H.C. Penny, E. Scholl (1986). Diseases of swine, 6th edn., pp. 336, Iowa State University Press, Ames, IO.

22. Lloyd, S. (1983). Effect of pregnancy and lactation upon infection, *Veterinary Immunology Immunopathology* 4, 153–176.

23. Lucknow (District, Uttar Pradesh, India) – population statistics, map and location". City population. 10 January 2014. Retrieved 4 August 2014.

24. Lucknow District Population Census 2011, Uttar Pradesh literacy sex ratio and density". Census 2011.Retrieved 4 August 2014.

25. Lucknow Minimum Temperature". The Times of India. 29 December 2012.

26. Margolis L, W. Esch J.C. Holmes, A.M. Kuris, G. A. Schad. (1982). The use of ecological terms in parasitology (Report of an Ad Hoc Committee of theAmerican Society of Parasitologists). *J Parasitol.*, 68: 131-133.

27. Marufu, M.C.M., P. Chanayiwa, M. Chimonyo, E. Bhebhe. (2008). Prevalence of gastro-intestinal nematodes in Mukota pigs in a communal area of Zimbabwe', *African Journal of Agricultural Research.*, 3, 91–95.

28. Nansen, P. A. Roepstorff (1999). Parasitic helminths of the pig: factors influencing transmission and infection levels. *International Journal for Parasitology.*, 29(6): 877-891.

29. Nissen, S., Paulsen, I., Nejsum, P., Olsen, A., Roepstorff, A., Rubaire, A. *et al.*, 2011, Prevalence of gastrointestinal nematodes in growing pigs in Kabale District, Uganda', *Tropical Animal Health and Production* 43, 567–572.

30. Nsoso, S.J., K.P. Mosala, R.T. Ndebele, S.S. Ramabu (2000). The prevalence of internal and external parasites in pigs of different ages and sexes in Southeast District, Botswana, *Onderstepoort Journal of Veterinary Research.*, 67, 217–220.

31. Obonyo, F.O., N. Maingi, S.M. Githigia C.J. Nganga (2012). Prevalence, intensity and spectrum of helminths of free range pigs in Homabay District, Kenya, Livestock Research.

32. Opara, M., N. Ibekwe, J. Azubuike, C. Okoli (2006). Occurrence of parasitic helminths among free-range pigs in five local government areas of Owerri zone, Imo State, Nigeria', *International Journal of Natural and Applied Sciences.*, 2(1), 72–77.

33. Polley, L.R., P.E Mostert (1980). Ascaris suum in Saskatchewan pigs: An abattoir

survey of prevalence and intensity of infection', *Canadian Veterinary Journal.*, 21(11), 307–309.

34. Roepstorff, A., P. Nansen (1998). Epidemiology, diagnosis and control of Helminth Parasites of Swine, FAO Animal Health Manual, 3, Rome, Italy.

35. Souslsby, E. J. L. (1982., Helminths, arthropods and protozoa of domesticated animals Bailerier Tindel and cox limited London seventh edition pp 517 –755.

36. Sowemimo, A.O., Asaolu, S.O., Adegok, F.O. & Ayanniyi, O.O., 2012, 'Epidemiological survey of gastro-intestinal parasites of pigs in Ibadan, Southwest Nigeria', *Journal of Public Health and Epidemiology.*, 4, 294–298.

37. Spencer, T. (2010). Zoonotic diseases of pigs, SAPPO: News and industry articles, viewed 09 August 2015.

38. Swai, E.S., E.J. Kaaya, D.A. Mshanga, W.E. Mbise (2010). Survey on gastro-intestinal parasites of non-descript dogs in and around Arusha Municipality, Tanzania', *International Journal of Animal and Veterinary Advances.*, 3(2), 63–67.

39. Tamboura, H.H., H. Banga-Mboko, D. Maes, L. Youssao, A. Traore, B. Bayala (2006). Prevalence of common gastro-intestinal nematode parasites in scavenging pigs of different ages and sexes in Eastern Centre Province, Burkina Faso, Onderstepoort

40. Tiwari, K.P., A. Chikweto, G. Belot, G. Vanpee, C. Deallie, G. Stratton (2009) Prevalence of intestinal parasites in pigs in Grenada, West Indies', *West Indian Veterinary Journal.*, 9(1), 22–27.

41. Waiswa, C., M. Joseph, K. James, J. Oweikanga (2007). Prevalence of endoparasitic infections in pigs kept in South Eastern Uganda, *Veterinary Medicine.*, 39, 377–380.

42. Weka, R.P., E.I. Ikeh (2009). Seroprevalence of cysticercosis and intestinal parasitism in pigs in Jos Metropolis. *Journal of Animal and Veterinary Advances.*, 8(5): 883-887.

43. Weng, Y. B., Y.J. Hu, Y. Li, B.S. Li, D.H. Lin, Xie, R.B. Gasser, X.Q. Zhu (2005). Survey of intestinal parasites of pigs from intensive farms in Guangdong province, people's Republic of China. *Journal of Veterinary Parasitology.*, 127: 333-336.

44. Yadav, A.K., V. Tandon (1989). Nematode parasite infections of domestic pigs in subtropical and high rainfall area of India. *Veterinary Parasitology.*, 31: 133-139.

Current Status of Researches in Biosciences (2020) : 97-101
Editor : Prof. P.C. Joshi, N. Joshi, Reshman Yasmin, and D. K. Mansotra
Today & Tomorrow's Printers and Publishers, New Delhi - 110 002

PROSPECTS OF IVF IN CATTLE AND BUFFALOES

Shyam Zawar

JK Trust and Chief Scientist, JK Bova Genix, Ist Floor, Sapphire Building, Pokhran Road No. 1, Jekegram, Thane (West), Maharashtra – 400606

Email: Shyam.Zawar@raymond.in

ABSTRACT

IVF is one of the Assisted Reproductive Technologies; which can be effectively used for rapid genetic improvement of Indigenous Cattle. OPU (Ovum Pick Up) has been considered as the most flexible and repeatable technique to produce embryos from any given donor. Unlike Multiple Ovulation and Embryo Transfer, OPU does not interfere with the normal reproduction and production cycles of the donor. Any female starting from 6 months of age to third month of pregnancy and 2-3 weeks after calving could be a donor. In OPU-IVF technology, the oocytes are gently aspirated from ovarian follicles by means of a fine needle, guided by an ultrasound scanner probe placed in the vagina. The recovered oocytes are then matured and fertilized in laboratory. Further maturation and culture takes place in incubator for 7 days, resulting in the production of viable IVF embryos which can be transferred into recipient cows. From a Gir Donor cow named "Radha", JK Trust produced 14 IVF calves in a year using surrogate mothers and from another Gir Cow named "Gauri", 56 IVF pregnancies were established in 6 months. Further, for the first time in the country, nine IVF pregnancies are established in buffaloes at the doorsteps of farmers. This pioneering work was undertaken in a Buffalo Farm near Pune in Maharashtra and the calves are expected to be born in August 2020. With this, we have shown the practical application of IVF in Indian field conditions, for rapid multiplication of superior genetics among Cattle and Buffaloes.

KEYWORDS: IVF, OPU, Cattle and Buffaloes.

INTRODUCTION

Dairy Sector plays an important role in India's economy, with an overall contribution of 4.11% to the total GDP during 2012-13. According to the Livestock Census 2012, there are 190.9 million cattle in India, but the alarming factor is that the livestock population has decreased by 4% during the period from 2007 to 2012, even while the demand for milk and milk products has increased at the rate of 12.6% on year on year basis. Though India is the world's top milk producer (contributing close to 20% of total global milk production), the average milk productivity of Indian cattle is around 3.5 litres per animal per day which is a matter of concern for sustaining Dairy farming as a stand alone commercial activity. The low production average is mainly contributed by the large population of Indigenous and non-descript cattle population.

Indigenous cattle breeds are known for disease resistance, heat tolerance, work-capacity, ability to withstand natural calamities, conversion of low-quality forages, utilization of bio-mass, suitability to organic and natural farming and adjustment to local eco-systems. But, low milk production of the Indigenous cattle prevents dairy farmers from taking up commercial dairy farming of indigenous cows. Therefore, the population of Indigenous cattle breeds has been reducing drastically over the last few decades due to uncontrolled crossbreeding and interbreeding with exotic cattle breed for rapid increase in milk production. Unless and until, strategies are adopted to increase the productivity of the indigenous cattle with more focused genetic selection, the scenario of Indigenous cattle will remain unchanged and in the coming days many indigenous cattle breeds will become history. There is huge scope for improving the productivity of Indigenous cattle through interventions in animal genetics and adopting better management practices.

Ovum Pick Up and In Vitro Embryo Production (OPU-IVEP), also termed as IVF technology is one of the Assisted Reproductive Technologies; which can be effectively used for genetic improvement of the Indigenous Cattle to bring about rapid genetic improvement with desired genetic traits. An added advantage of using IVF technology is that it is possible to collect oocytes from open cows, pregnant cows, virgin heifers, as well as problematic females that have had difficulty in conventional breeding attempts, but known for higher value genetic merit; thus increasing the overall base for getting oocytes from valuable and the rare animals for production of IVF embryos. With the use of OPU-IVF technology, we can rapidly breed the available cows with higher milk production and develop a large population of high milk yielding cows in a very short span of time resulting in overall improvement in milk productivity.

In India, till date, we have not focused much on propagating the genetics of superior female animals. With the breakthrough technology of IVF and ET, we can multiply genetics of the superior females, keeping the male side also very strong with usage of sexed semen of superior bulls. Ultimately, we can produce 10-25 calves from a single donor cow in a year through surrogacy which is not possible with traditional breeding methods. Thus, selective breeding program for milk productivity enhancement of the indigenous cattle can be achieved at a very faster rate.

With the use of OPU and IVF technology, it is now possible to harvest immature oocytes once in two weeks and subsequently these oocytes can be developed into Embryos. By this way, the female with superior genetics can be used to its fullest potential and a farmer can create a herd of his choice by IVF technology.

There may be some female animals with very good production characteristics, but may be having dysfunctional reproductive anatomy which makes conceiving and giving birth difficult for them. There may be some animals which can be stimulated with hormones, but have difficulty in fertilization. Similarly there may be good animals which cannot carry a baby to full term due to sickness or injury which makes full term pregnancy difficult for them. In all these cases, we can harvest the oocytes and subject it to IVF, with the aim of doing an ET in surrogate mothers. Another advantage of IVF is that oocytes can also be aspirated from pregnant animals. In short, the generation gap can be shortened and we can put into use the best out of best animals to be propagated for producing future generations. This will lead to a continuous up-gradation of animals.

In OPU-IVF technology, the oocytes are gently aspirated from the ovarian follicles by means of a fine needle, guided by an ultrasound scanner probe placed in the vagina. The recovered oocytes are then matured and fertilized in the laboratory. Further maturation and culture takes place in an incubator for approximately 7 days resulting in the production of viable IVF embryos which can be transferred into recipient cows.

OPU has been taken as a most flexible and repeatable technique to produce embryos from any given live donor. Unlike Multiple Ovulation and Embryo Transfer (MOET), OPU does not interfere with the normal reproduction and production cycles of the donor. Any female starting from 6 months of age to the third month of pregnancy and even soon after calving (2-3 weeks) could be a suitable donor. It is more cost effective to use IVF on low embryo production donors or females the owner would like to keep in the production cycle. Additionally, IVF can reduce donor boarding costs and semen costs by utilizing a single straw of semen to fertilize multiple oocytes, with the opportunity to use more than one sire on a single donor aspiration. It has been shown to be a feasible and practical alternative to the conventional multiple ovulation and embryo transfer (MOET) program and it is being more and more used for commercial applications across the globe.

We, at JK Trust have launched JK BovaGenix Initiative in 2016 with the objective of bringing about rapid genetic improvement in Indian dairy Cattle and Buffaloes by use the of Advanced Assisted Reproductive Technologies and we have successfully demonstrated that OPU-IVF technology can be effectively used in Indian condition for rapid propagation and upgradation of the High Genetic Merit Indian Cattle. We produced India's First IVF Male Calf (named as Krishna) in 2017 and India's first IVF female Calf from Frozen IVF Embryo at our Gopalnagar Facility in 2016.

From our experience in OPU-IVF Technology, we can confidently assume that around 10 Oocytes can be recovered from one OPU and may expect around 2-3 viable IVF embryos; which may result into one IVF pregnancy. Thus, there is potential to produce more than 12 calves in a year from a Donor cow, which is not possible through conventional breeding practices.

We have one Gir Donor cow namely "Radha" from whom we had performed 24 OPUs and produced 99 embryos in a span of 20 months. All those embryos had been transferred to suitable recipients and today we are proud to share that we have 14 IVF calves from "Radha".

From a Gir Donor Cow named "Gauri", 56 IVF pregnancies are established in a span of 6 months and calves are expected to be born from January to June 2020.

Further, for the first time in the country, nine IVF pregnancies are established in buffaloes at the doorsteps of farmers. This pioneering work was undertaken in a Buffalo Farm near Pune in Maharashtra and the calves are expected to be born in August 2020.

JK Trust is providing Cattle IVF services at the doorsteps of the progressive Dairy farmers of various states of the country through Mobile ET-IVF Vans specially designed for this purpose. Till date, we have successfully provided IVF and ET services to 43 farmers of 10 states namely Maharashtra, Gujarat, Chhattisgarh, Punjab, Tamil Nadu, Kerala, Andhra Pradesh, Haryana, Uttarakhand and Rajasthan. As on 31st October 2019, we have produced 3,279 IVF Embryos and established 308 IVF Pregnancies. Further, 132 IVF calves are born.

Benefits of OPU-IVF to a dairy farmer

1. A farmer can form a herd of his choice by propagating the genetics of a cow & a bull of his choice; at a faster pace.
2. The average cow will only produce only a few calves in her entire lifetime, but this technology can potentially allow a genetically superior cow to produce 10 to 25 (and sometimes even more) female calves within one year.
3. The biggest advantage of the IVF technology is that it helps us to use not only the genetics of the best bulls available, but also taps the genetic merit of the females to the fullest potential.
4. Eggs (oocytes) from sick, injured or aged female with high genetic merit can also be used in this technology.
5. Oocytes from pregnant donors can still be collected throughout the first trimester of pregnancy.
6. In addition, the use of sexed semen allows the farmer to selectively produce female calves only.

IVF is one of the best techniques to speed up genetic advancement in indigenous breeds of cattle and it will become a mainstream breeding tool over the next few years. IVF Technology will definitely help us to develop our indigenous breeds to their fullest genetic potential as this technology will help us to multiply the cows with superior genetics at a rapid pace.

CONCLUSION

In vitro production of bovine embryos is an efficient tool to produce animals of higher genetic merit. In fact its usage has been continually increasing, mainly in developing countries of South America and Asia. However it's noteworthy that several factors can influence the success of in vitro embryo production and increase or decrease the embryo yield and gestational viability. Differences between donors, sperm, and breeds should be taken into account in regard to *in vitro* embryo production. Using breeds adapted to the environment will certainly overcome the effects of heat on oocyte quality and embryo yield.

Current Status of Researches in Biosciences (2020) : 103-126
Editor : Prof. P.C. Joshi, N. Joshi, Reshman Yasmin, and D. K. Mansotra
Today & Tomorrow's Printers and Publishers, New Delhi - 110 002

EVALUATION AND MANAGEMENT OF NON-COMMUNICABLE DISEASE IN RELATION WITH RURAL DEVELOPMENT: AN EXPERIMENTAL STUDY ON DIABETES WITH *ACALYPHA INDICA*

Kesireddy Sathyavelu Reddy*, Singamala Hari Prasad, Ganjikunta Venkata Subbaiah, Sahukari Ravi, Borreddy Venu Gopal Reddy, Bhasha Shanmugam and Phatan Usnan Taj

Division of Molecular Biology and Ethno pharmacology, Department of Zoology, Sri Venkateswara University, Tirupati – 517 502, India.

Email: sathyakreddy@hotmail.com*

ABSTRACT

70% of Indian population resides in rural, they experience higher rate of type 2 diabetes than urban residents resulting in not only healthcare burden but also socio-economic burden of the country. The present study which has an impact on rural health made an attempt to investigate the anti-diabetic activity of polyphenol rich fraction of *Acalypha indica* in diabetic liver of rats. In this study, blood glucose, Insulin, liver biomarkers, lipid metabolic profile and histopathological changes were measured. The obtained results were shown abnormal changes of the induced diabetes was reversed by treatment with polyphenol rich fraction of *Acalypha indica*.

KEYWORDS: Rural population, *Acalypha indica*, Polyphenolic rich fraction, Diabetes, Liver markers, Lipid profile.

INTRODUCTION

The achievement of the Millennium Development Goals is at the centre of sustainable development. Sustainable rural development is vital to the economic, social and environmental viability of nations. A healthy and dynamic agricultural sector is an important foundation of rural development, generating strong linkages to other economic sectors. The traditional approach has been to regard health activities as a small component of rural development programmes. Under the Primary Health Care approach health is seen as the lever for rural development. An India today has over twice the odds of dying of non-communicable diseases (NCDs), giving rise to a high mortality rate in the country [1-3]. As 70% of the Indian population resides in rural and rural healthcare as one of the biggest challenges faced by the Health Ministry of India thus puts the rural population at a higher risk [4]. NCDs share common behavioural risk factors such as unhealthy diet, lack of physical activity, use of tobacco and alcohol. In India, Twelfth Five Year Plan (2012-2017) is going on. Department of AYUSH was created under the ministry of health and family welfare where traditional system of medicine like Ayurveda, Unani, Sidha and Homeopathy was clubbed together to give acronym AYUSH.Though considerable progress has been made in documenting identity and quality standards of herbal medicines, scientific validation of AYUSH principles, remedies and therapies has not progressed. Similarly, the National Medical Plans Board has supported many projects for conservations, cultivation and storage of medicinal plants. Diabetes mellitus (DM) is a metabolic disorder which was associated with hyperglycemia results in lack of insulin or insulin resistance.[5,6] DM is most prevalent disease affecting 25% of the population in Worldwide and it is predicted to be rise to 552 million by 2030.[7] Sustained hyperglycemic condition could ensure the diabetic complications beside the dysfunction or damage of insulin depended organs due to the generation of high amount of free radicals.[8,9] Etiology of liver diseases is an important in the incidence of DM. As a consequence, liver is the large and complex insulin depended organ which play a major role in glucose homeostasis and involved in carbohydrates and lipid metabolism. During loss of insulin, liver could alleviate to ensure the glycogenolysis which often hypertriglyceridemia and hypercholesterolemia [10, 11]. Chronic mild elevations of transaminases are frequently found in diabetes due to oxidative injury of hepatocytes by free radicals [12]. Consequently, this could lead to develop the micro and macro vascular diseases which are the major causes of morbidity and death.[13,14] Therefore, it is immediate search for more effective and safer anti-diabetic drugs. Currently, for the diabetes management, number of anti-diabetic drugs including biguanide, thiazolidinedione, and sulfonylurea were used. However, the usage of these agents were restricted due to several considerable side effects.[15] Many herbal medicinal plants and their bioactive constituents have been recommended to treat diabetes and its complications from hundreds of years throughout the World [16] Plants mostly contain carotenoids, flavonoids, phenols, tannins, terpenoids, alkaloids, and glycosides, are often have anti-diabetic properties. [17, 18] *Acalypha indica (Ai)* is extensively grown in tropical India, Sri Lanka, South Africa and Pakistan. It has anti-venom, wound healing, antioxidant, anti-inflammatory [19] diuretic, antibacterial [20] and antidiabetic properties. *Ai* has rich in polyphenolic compounds (phenols, tannins and flavonoids) and other phytochemicals. However, to the best of our knowledge, hepatoprotective activity under diabetic condition by polyphenolic rich fraction of *Ai* leaves remains to be ascertained. In order to obtain more knowledge about *Ai*, this study was made to attempt an investigation on therapeutic potentiality of polyphenolic rich fraction of leaves against STZ induced liver damage in rats.

MATERIALS AND METHODS

Chemicals: Polyvinyl polypyrrolidine (PVPP), Streptozotocin were purchased from Sigma Aldrich, Ultra-sensitive ELISA kit for rat insulin (Linco Research, Inc., St. Charles, MO), liver biomarkers, Bilirubin test kits were purchased from Recon Diagnostics, Ltd., Vadodara, India. All other chemicals used in the study were analytical grade and obtained from standard commercial suppliers.

Plant material: Leaves of *Ai* were collected from Raghavarajapuram, Kodur, Kadapa, Andhra Pradesh, India. The identity of the plant (*Ai)* was confirmed by Taxonomist of Department of Botany, Sri Venkateswara University, Tirupati, India and deposited a voucher specimen in the same department. According Paulo method [21] Polyphenolic-rich fraction (PLF) was prepared.

Preparation of Polyphenolic-rich fraction: Polyphenolic-rich fraction (PLF) was prepared according to the method [21] of briefly, dried leaves powder macerated overnight in dichloromethane then filtered with Whatt man No.1 filter paper. The filtrate was separated to get the defatted residue. Further, this residue macerated in methanol/water (70:30 v/v) for a day and filtered. The filtrate was concentrated in rotary evaporator until complete evaporation of methanol from the mixture. The obtained water extract adjusted to pH 4 with the help of 1N HCl. to 100 ml of water extract, 5 g of PVPP (poly vinyl poly pyrrolidine) was added as adsorbent. PVPP adsorbed

Ai 'OH' group contained moieties have reside in residue of filtration. This PVPP-OH group mixture was mixed with acetone/water (70:30 v/v) and stirred for a while then again filtration was done. Finally, filtrate (acetone/water) subjected to dry where acetone evaporated and polyphenols exist in water. The complete dry form of polyphenols have stored in refrigerator for future use.

Quantitative estimation of phenols: The quantitative estimation of phenols present in the separated fraction was assessed using Folin–Ciocalteu method [22]. Briefly, the reaction mixture made with 1 ml of fraction (0.5 mg)/ Gallic acid (Different concentrations include 5, 10, 15, 20 and 25 μg/ml), 5 ml of double distilled water and 1.5 ml of the Folin-Ciocalteu reagent. Total mixture kept in room temperature for 5 min. To this mixture, 2 ml of Na_2CO_3 solution was added and final volume made up to 15 ml with distilled water. Now, incubated for another 30 min and reaction was read at 765 nm in spectrophotometer. The total phenolic content of fraction calculated from calibration curve of Gallic acid. The phenolic content of fraction expressed in Gallic acid equivalent/gram weight of fraction.

OH group phytochemicals (polyphenols) identification: The obtained polyphenols mixture from PVPP procedure has been subjected to HR-LC/Q-TOF/MS to explore individuals from concoction. We followed the same method for this analysis which is available in our previous publication.[23] All the obtained phytochemical compounds have drawn in Marvin sketch, a free tool for chemical structure calculations.

Animals: Male Wistar rats (200–220 g) were purchased from the Indian Institute of Science, Bengaluru, India. The study was carried out according to the CPCSEA principles guided for the care and use of experimental animals. The animals received a standard pellet diet, water ad libitum, and were maintained under laboratory environmental conditions (23 ± 2^{0}C, 40–60% relative humidity and 12 h of light–dark cycle).

Induction of diabetes: One week acclimatized healthy animals were fasted overnight and diabetes was induced by a single intra peritoneal injection of freshly prepared Streptozotocin (STZ) (50 mg/kg b.w) in 0.1 M cold citrate buffer (pH 4.5).The animals were considered as diabetic when their blood glucose levels reached above 250 mg/dL on the third day after STZ injection.

Grouping of animals: The rats were divided into five groups of six rats each and the oral treatment was given every day for 1 month.

Group I, Normal Control (NC): Rats received 0.9% saline and fed with normal diet.

Group II, Diabetic Control (DC): Streptozotocin (STZ 50 mg/kg b.w.) was given intra peritoneally for the induction of diabetes.

Group III, Polyphenolic fraction treatment (Pt): Rats received Polyphenolic fraction (100 mg/kg b.w.) orally for a period of month.

Group IV, Diabetics + Polyphenolic fraction treatment (Di + Pt): Diabetic rats received Polyphenolic fraction, as described in group III, for a month treatment.

Group V, Diabetics + Glibenclamide treatment (Di + Glbt): Diabetic rats received Glibenclamide (20 mg/kg b.w.) for one month treatment.

Collection of blood and analysis: After completion of treatment period, blood was collected in EDTA coated vials through retro orbital sinus puncture under mild ether anaesthesia from over-night fasted animals. Plasma was obtained by centrifugation at 3000 rpm for 10 min at 4^0C and stored at -20^0C until estimation of liver biomarkers and lipid profile.

Estimation of blood glucose levels: Blood glucose was measured by using a one touch glucometer (Accucheck). After 24 h of the last dose, blood glucose levels were measured from overnight fasted rats in each group.

Estimation of Insulin levels in plasma: Insulin levels were analyzed in all groups before and after the treatment. Insulin was measured by ELISA assay (Linco, Millipore).

Determination of liver biomarkers in plasma

Estimation of Alkaline Phosphatase (ALP): Plasma ALP activity was measured by Recon Diagnostics kit was followed by Bassey *et al* [24] method and was modified by Wright *et al* [25]. Briefly, 10 µl of plasma was mixed with 500 µl of the reagent. Initially, after 3 minutes the absorbance was read at 405 nm. The mean absorbance per every minute was considered and the final values were expressed in U/L.

Estimation of Alanine transaminase (ALT) activity in Plasma: Plasma ALT activity was measured by Recon Diagnostics kit was followed by IFCC the method [26]. Briefly, 50 µl of sample was added to1 ml of working reagent and incubated at 37^0C for 60 s. The absorbance was read at 340 nm continuously for 2 min at an interval of 30 s. The values were expressed in U/L.

Estimation of Aspartate transaminase (AST) activity in plasma: Plasma AST activity was measured by Recon Diagnostics kit was followed by IFCC the method. Briefly, 0.05 μl of sample was added to 1 ml of working reagent and incubated at 37^{0}C for 60s. The absorbance was read at 340 nm for 2 min with an interval of 30s. The values were expressed in U/L.

Estimation of γ -Glutamyltransferase (GGT) activity in plasma: Plasma GGT activity was measured by Recon Diagnostics kit was followed by Szasz method [27]. Briefly, γ-Glutamyl-p-nitroanilide in ammediol-HCl buffer at pH 8.2 used as substrate and was added to 0.1 ml of plasma. The absorbance was read at 405 nm in UV/VIS spectrophotometer.

Estimation of Bilirubin levels in plasma: Plasma bilirubin levels were measured by Bio vision assay kit was followed by Mauro method [28]. Briefly, 50 μl of sample was added to sodium nitrite & sulfanilic acid solutions. Pink colored azocompund was read at 546 nm in UV/VIS spectrophotometer.

Determination of lipid profile in plasma: Triglycerides (TGs), total cholesterol (TC) levels were measured using commercially available kits (Merck India Ltd), where as high density lipoproteins (HDL) levels were estimated enzymatically by colorimetric kits (Merck India Ltd) and low density lipoproteins (LDL) were calculated by Friedewald method.[29]

Histopathology: The liver tissues were isolated from sacrificed animal groups followed by immediate fixation in 10% formalin for 24 hours. Then the tissues were further processed with washing, dehydrated with the help of increased concentration of alcohols. The dehydrated tissues were embedded in paraffin wax then cut into sections about 5 μ thick using Microtome. All the sections deparafinized in xylene and hydrated with reverse process of alcohols treatment on glass slides. Initial staining with Hematoxylin and counter stain with eosin were done for tissues fixed on slides.

RESULTS

Quantity of polyphenols and Chemical composition of PLF of *Ai* leaves: Quantitative estimation of phenols in the fraction revealed that it has 368 mg/gram of Gallic acid equivalent. The phytochemical profile analysis of PRF of *Ai* using HR-LC-MS revealed 59 phenolic compounds as represented in Table-1 and Figure 1 including their name, retention time (RT), molecular formula, molecular weight and % peak area. These structures strongly supported the quantity of phenols present in fraction measured from Folin-Ciocalteu method.

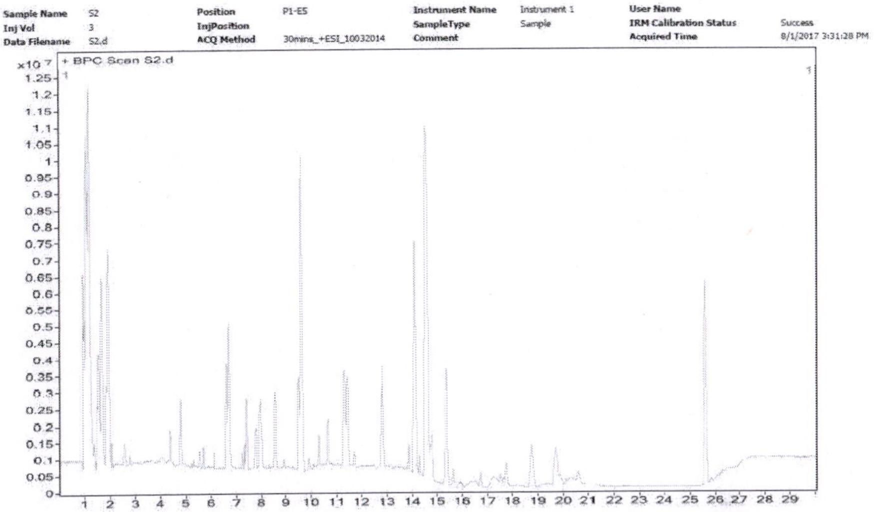

Figure-1: Methanolic extract of PLF of Ai leaves GC-MS chromatograph

Effect of PLF of *Ai* on blood insulin and glucose levels: As shown in Figure-2 the plasma insulin levels in diabetic rats were significantly decreased compared with normal rats. After one month treatment with PLF of *Ai* (40 mg/kg), plasma insulin levels of diabetic rats were significantly increased compared with diabetic control rats. In normal and PLF of *Ai* treated rats there was no effect on insulin levels. The hypoglycemic effect of PLF of *Ai* was examined by measuring the blood glucose levels from 0 to 4 weeks in treatment period. As shown in Figure-2 the blood glucose levels of diabetic control group was significantly increased ($p<0.01$) compared with normal control rats. Administration of PLF of *Ai* significantly decreased the plasma glucose levels in diabetic rats compared with diabetic control rats, while in normal and PLF of *Ai* treated rats there was no effect on the plasma glucose levels.

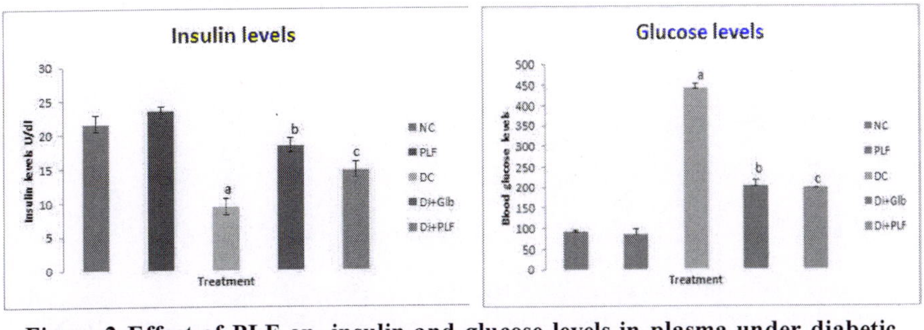

Figure-2 Effect of PLF on, insulin and glucose levels in plasma under diabetic condition. The groups which not have same alphabet letter are significant (P<0.01) to the control. Values are means ± S.D for six rats per group.

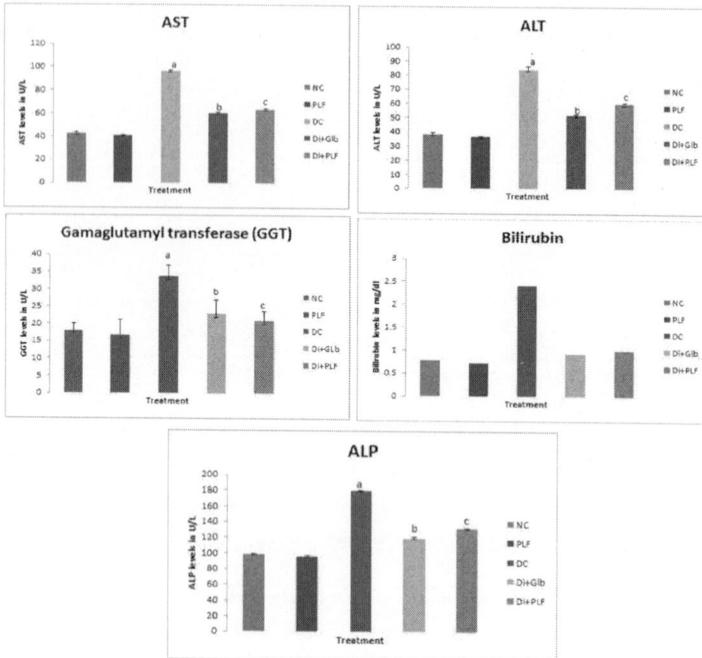

Figure-3 Effect of PLF on AST, ALT, GGT, Bilirubin and ALP levels in plasma under diabetic condition. The groups which not have same alphabet letter are significant (P<0.01) to the control. Values are means ± S.D for six rats per group.

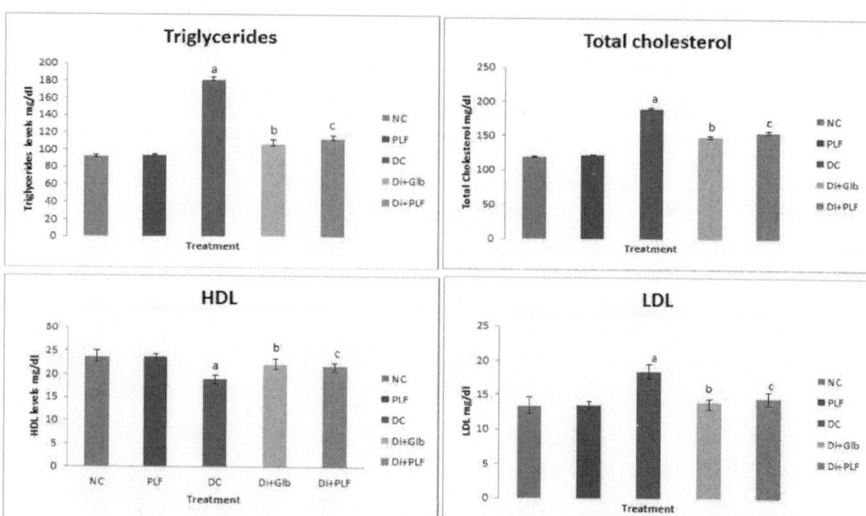

Figure-4 Effect of PLF on LDL, HDL, Total cholesterol and Triglycerides levels in plasma under diabetic condition. The groups which not have same alphabet letter are significant (P<0.01) to the control. Values are means ± S.D for six rats per group.

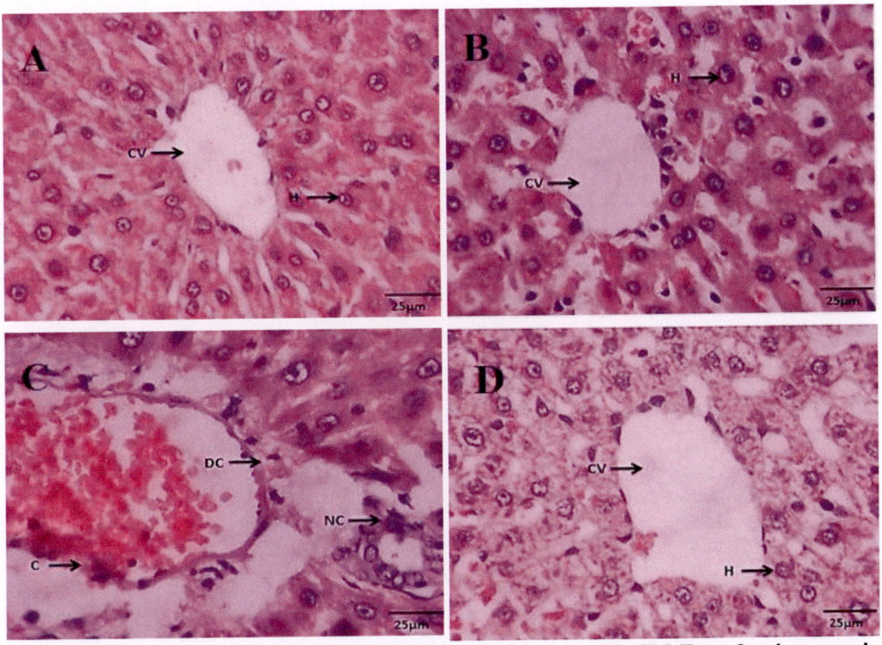

Figure-5 Photomicrograph sections of livers stained with H&E and microscopic magnification (40X). A: Cyto-architecture of liver tissue with normal hepatocytes (H) and Central vein (CV).B: Normal cyto-architecture maintained by polyphenols treatment as like normal control, C: Degenerative changes in central vein having congestion (C), vacuolization and necrotic changes (NC), D: Regenerative changes in liver tissue due to Glibenclamide treatment having normal hepatocytes (H) and central vein, E: Regenerative changes in liver tissue due to polyphenolic fraction treatment having normal hepatocytes (H) and central vein like normal control.

Table-1 Poly phenols of methanolic extract of *Acalypha indica* leaves

S.No	Compound Name	RT	Formula	Mass	% area
1	2-Furoic acid	0.946	C5 H4 O3	112.013	1..07
2	2-Butanone, 4-[6-(sulfooxy)-2-naphthalenyl]	0.951	C14 H14 O5 S	294.0549	1.43
3	Lyxosylamine	0.957	C5 H11 N O4	149.0658	1.97
4	Canavanine	0.96	C5 H12 N4 O3	176.0891	1.21
5	2-Naphthaleneacetic acid, 6-hydroxy	0.961	C12 H10 O3	202.0657	1.55
6	Taurine	0.963	C2 H7 N O3 S	125.0137	2.24
7	Choline	1.088	C5 H14 N O	104.1093	0.89
8	ArgSer	1.093	C9 H19 N5 O4	261.1412	0.89
9	3,5-Pyridinedicarboxylic acid,2-(hydroxymethyl)-6-	1.129	C17 H16 N2 O8	376.0952	3.66

	methyl-4-(3-nitrophenyl)-, 5-(2-hydroxyethyl) est				
10	1,4 Dideoxy 1,4 Imino D Arabintol	1.139	C5 H11 N O3	133.074	3.66
11	Bis (2-hydroxypropyl) amine	1.196	C6 H15 N O2	133.1102	2.15
12	Ala His Ala	1.267	C12 H19 N5 O4	297.1385	1.61
13	Carnitine	1.384	C7 H16 N O3	162.1107	1.25
14	1-Aminocyclohexane-carboxylic	1.45	C7 H13 N O2	143.0946	1.07
15	D tiapride	1.501	C15 H24 N2 O4S	328.1464	0.44
16	O-Acetylserine	1.509	C5 H9 N O4	147.0511	0.38
17	Indospicine	1.516	C7 H15 N3 O2	173.1136	0.47
18	Asn Met Asp	1.572	C13 H22 N4 O7S	378.123	0.62
19	Arg Pro	1.582	C11 H21 N5 O3	271.1616	0.71
20	Nalorphine	1.587	C19 H21 N O3	11.1539	0.80
21	N-Acetyl-b-neuraminic acid	1.623	C11 H19 N O9	309.1086	0.80
22	Epiafzelechin trimethylether	1.679	C18 H20 O5	316.1349	1.23
23	3-Hydroxynorvaline	1.703	C5 H11 N O3	133.0737	0.67
24	Probenecid	1.887	C13 H19 N O4S	285.1036	0.53
25	4-Hydroxy-L-threonine	1.888	C4 H9 N O4	135.0524	1.95
26	CysGlnGlu	1.941	C13 H22 N4 O7S	378.1228	0.67
27	L-2-Aminoadipic acid	2.074	C6 H11 N O4	161.0665	2.18
28	Tromethamine	2.643	C4 H11 N O3	121.0743	0.58
29	Tetranor-PGEM	4.406	C16 H22 O7	326.1435	0.58
30	3-Methylamino-1, 2-propandiol	4.824	C4 H11 N O2	105.0794	0.83
31	Glu Pro Pro	4.976	C15 H23 N3 O6	341.1583	0.31
32	Pro Glu Pro	5.566	C15 H23 N3 O6	341.1585	0.31
33	Beta-nonylenic acid	5.822	C9 H16 O2	156.1129	0.29
34	Homocysteinepenicillamine disulfide	6.141	C9 H18 N2 O4 S2	282.0727	0.38
35	Glutathione, oxidized	6.6	C20 H32 N6 O12S2	612.1625	1.16
36	D-Pantetheine 4'-phosphate	6.617	C11 H23 N2 O7 P S	358.0974	1.34
37	Dihydrorobinetin	6.701	C15 H12 O7	304.0549	1.52

38	1-Methylinosine	6.779	C11 H14 N4 O5	282.0974	1.52
39	5-(4-Hydroxy-3-methoxyphenyl)-5-phenylhydantoin	7.252	C16 H14 N2 O4	7.252	0.40
40	8R-hydroxy-nonanoic acid	7.421	C9 H18 O3	174.1259	1.05
41	Val Pro Lys	7.773	C16 H30 N4 O4	342.227	0.58
42	Pro Glu Ser	8.069	C13 H21 N3 O7	331.1384	0.80
43	Thr Phe Asn	8.315	C17 H24 N4 O6	380.1642	0.80
44	Avocadene Acetate	8.922	C19 H36 O4	328.2688	0.89
45	SNAP (s-nitroso-nacetylpenicillamine	9.52	C7 H12 N2 O4 S	220.0533	1.03
46	Dihydroergocornine	9.596	C31 H41 N5 O5	563.3118	1.03
47	Idebenone Metabolite (QS-4)	9.6	C13 H16 O6	268.094	0.67
48	4-Methyl-decanoic acid	11.3	C11 H22 O2	186.1615	1.01
49	4,7-Dioxo-octanoic acid	11.412	C8 H12 O4	172.0734	1.03
50	(E)-2-Methylglutaconic acid	11.412	C6 H8 O4	144.0425	1.07
51	Ephedrine	12.763	C10 H15 N O	165.1153	1.12
52	Pro Ala Gly	12.804	C10 H17 N3 O4	243.1251	1.12
53	3-Methyl-tetradecanedioic Acid	14.279	C15 H28 O4	272.1981	2.20
54	Lactone of PGF-MUM	14.76	C16 H24 O5	296.1593	3.17
55	Ambroxol	15.376	C13 H18 Br2 N2 O	375.9793	1.10
56	4-Chloro-N1-methyl-N1-(4-carboxy-2-hydroxy-2-methylbutyl)-mbenzenedisulfonamide	17.743	C13 H19 Cl N2 O7 S2	414.0314	0.17
57	Trandolapril glucuronide	19.723	C30 H42 N2 O11	606.2782	0.89
58	Ecgonine-methylester	25.608	C10 H17 N O3	199.1206	1.91
59	Capryloylglycine	25.634	C10 H19 N O3	201.1361	1.91

Effect of PLF of *Ai* on liver biomarkers in plasma (AST, ALT, ALP, GGT and Bilirubin): Figure-3 showed that the effect of PLF of *Ai* on plasma AST, ALT and ALP levels in normal and diabetic rats. Plasma AST, ALT and ALP levels were significantly increased after injection of STZ to the rats when compared with the normal rats while treated the diabetic rats with PLF of *Ai* significantly decreased the plasma AST, ALT and ALP levels. However, in normal and PLF of *Ai* treated rats there was no effect on the plasma AST, ALT and ALP levels. The plasma GGT and Bilirubin levels in diabetic group were significantly increased compared with normal control group shown in Figure 3. A significant decrease in GGT and Bilirubin levels were observed in diabetic group when treated with PLF of *Ai* while PLF of *Ai* treatment to the normal rats were not shown any significant changes in GGT and Bilirubin levels compared with the normal control rats.

Effect of PLF of *Ai* on lipid profile in plasma (Total Cholesterol, Triglycerides, HDL and LDL): As shown in (Figure-4), Plasma Total Cholesterol and Triglycerides levels were significantly increased after the administration of STZ compared with normal rats. Administration of PLF of *Ai* significantly decreased the plasma Total Cholesterol and Triglycerides levels in diabetic rats compared with diabetic control rats, while in normal and PLF of *Ai* treated rats there was no effect on the plasma Total Cholesterol and Triglycerides levels. Plasma LDL levels were increased whereas HDL levels decreased after administration of STZ to the normal rats when compared with normal control group rats. However, LDL levels decreased whereas HDL levels increased in diabetic rats after treatment with PLF of *Ai* compared with diabetic control rats.

Histopathology: Examined liver in normal control rats revealed that sections having normal cyto-architecture without any pathological changes in central vein and hepatocytes. The group of rats which received STZ induces the complication include degenerative changes (severe damage) in liver tissue with an evidence of congestion. The polyphenolic rich fraction treated rats have not shown any pathological alterations in hepatocytes and central vein of liver. The treatment of polyphenolic rich fraction of *Ai* to diabetic rats confined that it has potential hepatoprotective property with an evidence of regeneration of liver tissue to normal.

Table 2: Biological activities of phenolic compounds of *Acalypha indica*

S.No	Name of the compound	Compound structure	Biological activity
1	1,4 Dideoxy 1,4 Imino D Arabintol		1.17-beta-hydroxysteroid dehydrogenase-Inhibitor 2.Aryl-Hydrocarbon-Hydroxylase-Inhibitor 3.Testosterone-Hydroxylase-Inducer 4.Increase Aromati Amino Acid Decarboxylase Activitity 5.Arachidonic acid-Inhibitor Inhibit Production of Uric Acid 6. Urinary-Acidulant

2	2-Hydroxy-3-(4-methoxyethyl-phenoxy)-propanoic acid	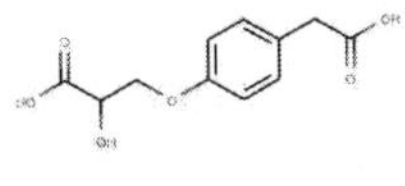	1.Arachidonic acid-Inhibitor 2.17-beta-hydroxysteroid dehydrogenase-Inhibitor 3.Inhibit Production of Uric Acid
3	2-methylglutaconic acid	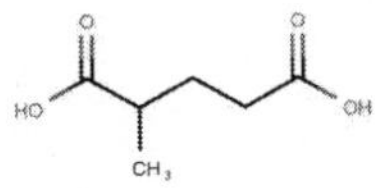	1. Arachidonic acid-Inhibitor 2. Inhibit Production of Uric Acid 3. Anticancer (Esophagus) 4.Cytochrome-P450-2E1-Inhibitor 5.Stimulate Epinephrine Production
4	2-Naphthaleneacetic acid, 6-hydroxy	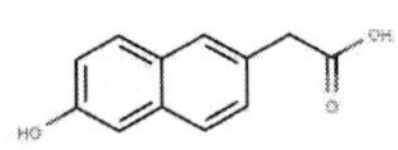	1.17-beta-hydroxysteroid dehydrogenase-Inhibitor 2.Aryl-Hydrocarbon-Hydroxylase-Inhibitor 3.Testosterone-Hydroxylase-Inducer 4.Increase Aromatic Amino Acid Decarboxylase Activity 5.Arachidonic acid-Inhibitor 6.Inhibit Production of Uric Acid 7.Urinary-Acidulant
5	3,5-Pyridinedicarboxylic acid,2-(hydroxymethyl)-6-methyl-4-(3-nitrophenyl)-, 5-(2-hydroxyethyl) est	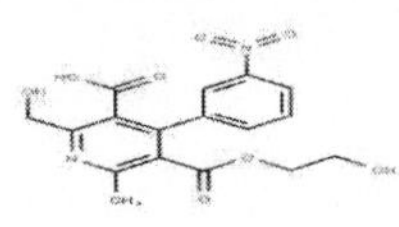	1.Estrogen-Agonist 2.Estrogenic 3. Increase Aromatic Amino Acid Decarboxylase Activity 4.Arachidonic acid-Inhibitor 5.Inhibit Production of Uric Acid 6.Urinary-Acidulant
6	3-Methyl-tetradecanedioic	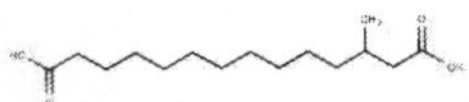	Catechol-O-Methyl-Transferase-Inhibitor
7	4,7-Dioxo-octanoic acid	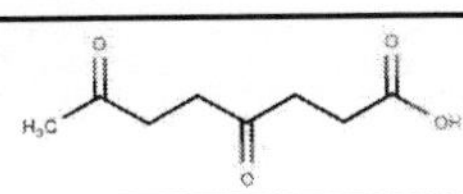	1.Arachidonic acid-Inhibitor 2.Inhibit Production of Uric Acid
8	4-Chloro-N1-methyl-N1-(4-carboxy-2hydroxy 2methylbutyl) m-benzenedisulfonamide	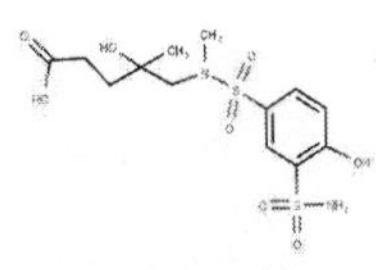	1.Antitumor (Nasopharynx) 2.Increase Natural Killer (NK) Cell Activity 3. Inhibit Production of Tumor Necrosis Factor 4.NADH-Oxidase-Inhibitor 5.Nephroprotective 6.Neuromuscular-Blocker 7.Neuroprotective 8.NF-kB-Inhibitor 9.NO-Scavenger

			10.Stimulate Norepinephrine Production
9	4-Hydroxy-L-threonine	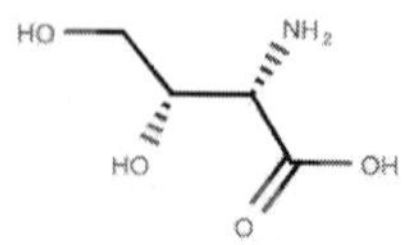	1.Anti-LDL 2.Anticancer (Liver) 3.Antioxidant (LDL) 4.Decrease Endothilial Leukocyte Adhesion 5.Inhibit Lipoxygenase 6.Lipase-Inducer 7.Lipolytic 8.Lyase-Inhibitor
10	4-Methyl-decanoic acid	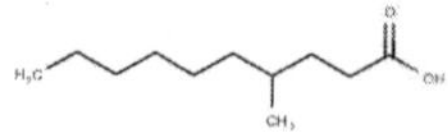	1.Arachidonic acid-Inhibitor 2.Inhibit Production of Uric Acid 3.Catechol-O-Methyl-Transferase-Inhibitor
11	5-(4-Hydroxy-3-methoxyphenyl)-5-phenylhydantoin	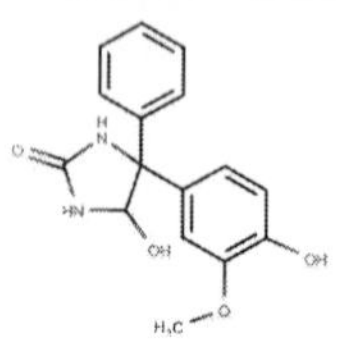	1.17-beta-hydroxysteroid dehydrogenase-Inhibitor 2.Aryl-Hydrocarbon-Hydroxylase-Inhibitor 3.Testosterone-Hydroxylase-Inducer
12	8R-hydroxy-nonanoic acid	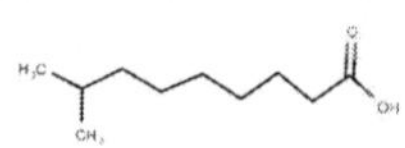	1.Testosterone-Hydroxylase-Inducer 2.5-Alpha-Reductase-Inhibitor 3.17-beta-hydroxysteroid dehydrogenase-Inhibitor 4.Free-Radical Scavenging 5.Beta-2-Receptor-Agonist 6.HIV-RT-Inhibitor 7.RNA-Inhibitor 8.Suppress HMG-CoA Reductase Activity 9.T-suppressor cell ratio
13	Ala His Ala	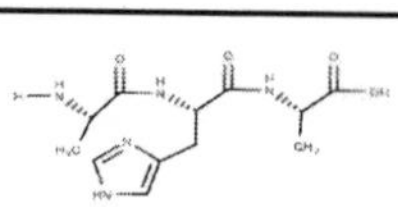	1.Histamine-Inhibitor 2.Histidine-Decarboxylase-Inhibitor 3.Histidine-Kinase-Inhibitor
14	Arg Pro	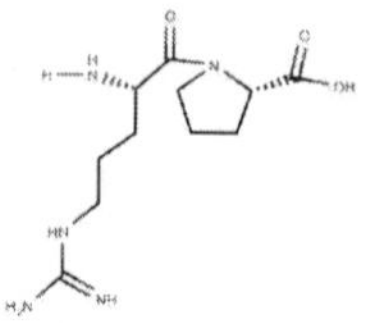	1.Anticancer (Prostate) 2.Antitumor (Breast) (Lung) (Prostate) 3.Calcium Provider 4.DNA-Protective 5.Endothelium-Derived Relaxing 6.Factor Promoter

			7.Increase Prolactin Production 8.Inhibit Production of Uric Acid 9.Proinflammatory 10.Prooxidant 11.Prostaglandin Synthase-Inhibitor 12.Protein-Kinase-C-Inhibitor 13.Prothrombogenic 14.Provide Vit D 15.Provide Vit K 16.Provide Zinc 17.Ulcer Proctective
15	Asn Met Asp	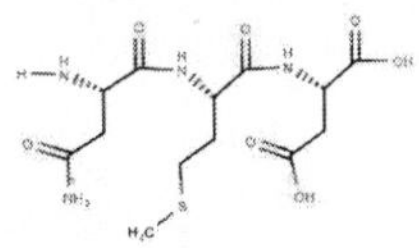	1.Catechol-O-Methyl- 2.Transferase-Inhibitor 3.Metal-Chelator 4.Regulate Calcium Metabolism 5.Methyl-Guanidine-Inhibitor
16	Beta-nonylenic acid	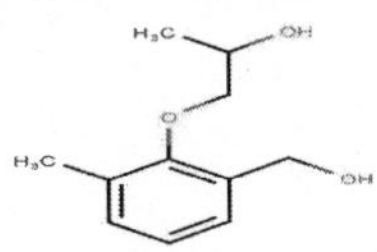	1.Arachidonic acid-Inhibitor 2.Inhibit Production of Uric Acid 3.Beta-2-Receptor-Agonist 4.Beta-Galactosidase-Inhibitor 5.AntiTGF-beta
17	Choline	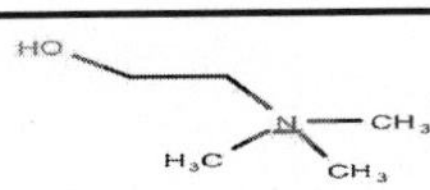	1.Acetyl-choline-antagonist 2.Cholinergic 3.Choline-Sparing
18	D-Pantetheine 4'-phosphate	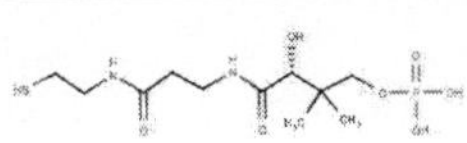	1.17-beta-hydroxysteroid dehydrogenase-Inhibitor 2.Alcohol-Dehydrogenase-Inhibitor 3.Anticancer (Duodenum) 4.Circulatory-Depressant 5.Coronary-Dilator 6.Decarboxylase-Inhibitor 7.Decrease Endothilial Platelet Adhesion 8.Decrease Endothilial Leukocyte Adhesion 9.Decrease Epinephrine Production 10.Decrease Glutamate 11.Oxaloacetate Transaminase 12.Decrease Lactate/Pyruvate Ratio 13.Decrease Norepinephrine Production 14. Decrease Oxalate Excretion Detoxicant (Alcohol) 15.DNA-Protective 16.Dopamine-Adenylate-Cyclase-Inhibitor

			17.Dopamine-Blocker 18.Increase Superoxide Dismutase Activity 19.Increase Vit D Bioavailability 20.Increase Aromatic Amino Acid 21.Decarboxylase Activity 22.Succinic-Dehydrogenase-Inhibitor
19	L-2-Aminoadipic acid	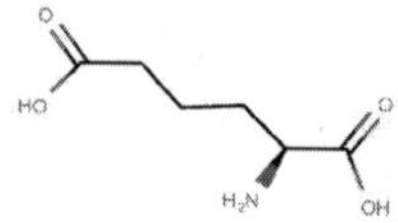	1.Anti-LDL 2.Anticancer (Liver) 3.Antioxidant (LDL) 4.Decrease Endothilial Leukocyte Adhesion 5.Inhibit Lipoxygenase 6.Lipase-Inducer 7.Lipolytic 8.Lyase-Inhibitor
20	Lactone of PGF-MUM	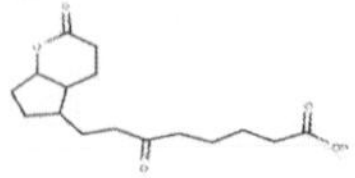	1.Inhibit Production of Tumor 2.Necrosis Factor 3.Inhibit Production of Uric Acid
21	N-(2- hydroxyethyl) stearamide		1.Antitumor (Nasopharynx) 2.Increase Natural Killer (NK) 3.Cell Activity 4.Inhibit Production of Tumor 5.Necrosis Factor 6.NADH-Oxidase-Inhibitor 7.Nephroprotective 8.Neuromuscular-Blocker 9.Neuroprotective 10.NF-kB-Inhibitor 11.NO-Scavenger 12.Stimulate Norepinephrine Production
22	N-Acetyl-b-neuraminic acid	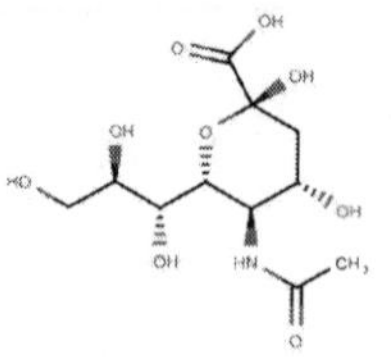	1.Arachidonic acid-Inhibitor 2.Inhibit Production of Uric Acid 3.Inhibit Acetyl Coenzyme A 4.Decrease Norepinephrine Production 5.Increase NK Cell Activity 6.Inhibit Production of Tumor- 7.Necrosis-Factor 8.NADH-Oxidase-Inhibitor 9.Neurodepressant 10.Neuroprotective 11.Neurotoxin

			12.NO-Inhibitor 13.11B-HSD-Inhibitor 14.17-beta-hydroxysteroid dehydrogenase-Inhibitor 15.Adrenergic-Blocker 16.Antitumor (Bladder) 17.Bacteriophagicide 18.Balance Prolactin Production 19.Beta-Galactosidase-Inhibitor 20.Bronchostimulant 21.Fat-Burning 22.DNA-Binder 23.Increase Boron Bioavailability 24.P450-2B1-Inhibitor
23	O-Acetylserine	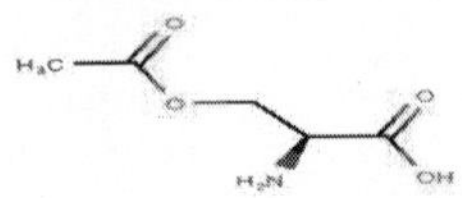	1.Aldehyde-Oxidase-Inhibitor 2.Anticancer (Oral) 3.Antitumor (Ovary) 4.Decrease Glutamate Oxaloacetate Transaminase 5.Inhibit Production of Tumor- 6.Necrosis-Factor 7.Inhibit Production of Uric Acid 8.NADH-Oxidase-Inhibitor 9.Nitric-Oxide-Synthase-Inhibitor 10.Ornithine-Carboxylase-Inhibitor 11.Osmoregulator 12.Ovulation-Stimulant 13.Xanthine-Oxidase-Inhibitor Oxidant
24	Pro lle	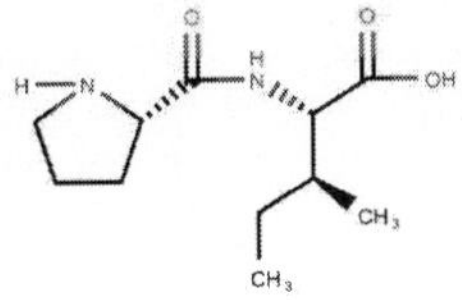	1.Anticancer (Prostate) 2.Antitumor (Breast) (Lung) (Prostate) 3.Calcium Provider 4.DNA-Protective 5.Endothelium-Derived Relaxing Factor Promoter 5.Increase Prolactin Production 6.Inhibit Production of Uric Acid 7.Proinflammatory 8.Prooxidant 9.Prostaglandin Synthase-Inhibitor 10 Protein-Kinase-C-Inhibitor 11Prothrombogenic 12.Provide Vit D 13.Provide Vit K 14.Provide Zinc 15.Ulcer Proctective

25	Tiapride	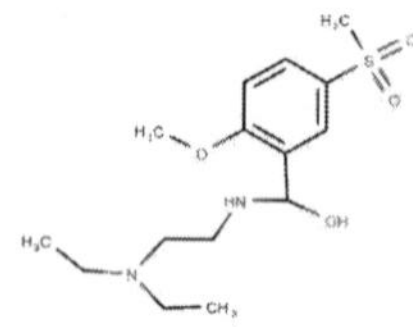	1.17-beta-hydroxysteroid dehydrogenase-Inhibitor 2.Alcohol-Dehydrogenase-Inhibitor 3.Anticancer (Duodenum) 4.Circulatory-Depressant 5.Coronary-Dilator 6.Decarboxylase-Inhibitor 7.Decrease Endothilial Platelet Adhesion 8.Decrease Endothilial Leukocyte Adhesion 10.Decrease Epinephrine Production 11.Decrease Glutamate 12.Oxaloacetate Transaminase 13.Decrease Lactate/Pyruvate Ratio 14.Decrease Norepinephrine Production 15.Decrease Oxalate Excretion Detoxicant (Alcohol) 16.DNA-Protective 17.Dopamine-Adenylate-Cyclase-Inhibitor 18.Dopamine-Blocker 19.Increase Superoxide Dismutase Activity 20.Increase Vit D Bioavailability 21.Increase Aromatic Amino Acid Decarboxylase Activity 22.Succinic-Dehydrogenase-Inhibitor
26	Trans-aconitate	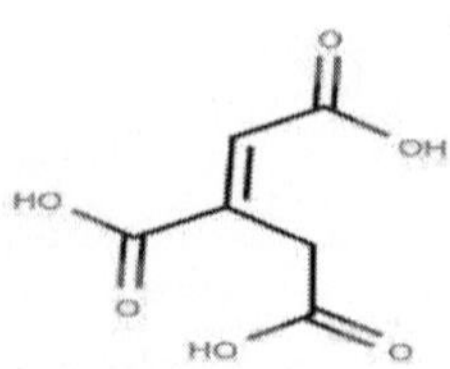	1.Catechol-O-Methyl-Transferase-Inhibitor 2.Decrease Glutamate Oxaloacetate Transaminase 3.Decrease Glutamate Pyruvate Transaminase 4.Glucosyl-Transferase-Inhibitor 5.Increase Glutathione-S-Transferase (GST) Activity 6.Increase Glyoxalate Transamination 7.Reverse-Transcriptase-Inhibitor

DISCUSSION

Polyphenols are commonly found in both edible and non-edible plants but in both cases they have many biological properties like anti-lipid peroxidation and antioxidant activities. Crude extracts of vegetables, herbs, fruits, cereals and other sources are having rich in phenolic content and also are increasing interest in pharmacy for treating various radicals associated diseases.[30] In this study, separated and quantified phenolic compounds of *Ai* revealed that it has rich in phenolic compounds (Figure 1). Further, phenolic profile analysis in HR-LC-MS explored huge number of OH group contained phytophenols. This phenolic data supports that the *Ai* might be act as a strong antioxidant and anti-lipid peroxidation agent. Polyphenols bioactive substances are the reducing agents that can able to diminish the oxidative stress by their radical-scavenging characteristics.[31] The previous experimental and epidemiological evidence suggested that dietary polyphenols play a role in ameliorating atherosclerosis and anti-inflammatory activity.[32,33] However, the lack of insulin or insulin resistance alleviates the hyperglycaemia and hyperlipidemia due to destruction of β cells in pancreas seemed under STZ induced intoxication in experimental animals, meanwhile it has also been accompanied with the liver dysfunction.[34,35,36] In order to that, controlling of glucose levels in plasma is main concern in the diabetes management to restrict the further development of diabetic complications[35,36]. Therefore, the available diabetic agents were shown hypoglycemic effects meanwhile the usages of these agents were restricted due to several considerable side effects.[37]

In the present study, *Ai* leaves of PLF shown 59 different polyphenols bioactive compounds. (Table 1) Among those 26 phenolic compounds shown different biological activities (Table 2). Our reports clearly demonstrated that, treatment with *Ai* leaves of PLF at dose 40 mg/kg shown hypoglycaemic, hypolipidemic and hepatoprotective activity under diabetic condition by decrease the glucose, lipid profiles and liver biomarkers in plasma. Persistent chronic high glucose level in the blood was a sign of diabetes, which was found in the present study in STZ induced diabetic rats when compared with normal rats.[38] In diabetes management, reduction of the blood glucose level is the primary therapeutic goal.[34] However, treatment with PLF of *Ai* considerably reduced the blood glucose levels in diabetic rats compared with diabetic control rats conceivably may due to potentiating of insulin secretion from existing β cells in pancreas or may enhance the conversion of blood glucose to glycogen by the liver.[39] Thus, the PLF of *Ai* potentially had shown hypoglycemic activity under diabetic condition. Improvement of insulin secretion and insulin sensitivity of the tissue was main strategy in diabetes treatment in consequence of insulin levels declined due to malfunction of pancreas.[40] In the present study obtained results shown , plasma insulin levels were significantly decreased in diabetic rats compared with normal rats, while plasma insulin levels of diabetic rats considerably enhanced when treated with PLF of *Ai* may be virtue of its hyper insulin activity to be lift the insulin secretion by stimulation or regeneration of â cells in pancreas.[41]The earlier reports have been demonstrated that diabetes was associated with liver damage, caused by elevation of liver biomarkers such as AST, ALT, GGT, bilirubin and ALP in plasma.[10] Hyperglycemia and over insulinization are believed to be metabolic preconditions in liver disease. However, Plasma enzyme activities can be used as useful biomarkers for monitoring the cytotoxicity of xenobiotics including STZ.[31]. In our study, STZ induces hepatic tissue damage which is one of the typical changes in diabetes as evidenced by elevation of plasma AST, ALT, GGT, bilirubin and ALP levels in diabetic rats compared with normal control rats. Conversely, treatment of the diabetic rats with PLF of *Ai* caused reduction in the levels of AST, ALT, GGT, bilirubin and ALP in plasma compared with the diabetic control rats. In accordance with this reports, other workers have been confirmed that anti-diabetic activity of medicinal plants especially polyphenols antioxidant activity that could protect the liver from oxidative injury initiated by hyperglycemic induced free radicals.[42,34] Thus, PLF of *Ai* administration to diabetic group of rats significantly declined liver biomarkers suggesting its hepatoprotective nature in diabetic condition. In diabetes, liver dysfunction and hyperlipidaemia accompanied with hyperglycemia.[43] Hyperglycemia accelerates the conversion of free fatty acids into triglycerides in the liver.[44] As a result of its increased the LDL and decreases the HDL in plasma to contributes immensely to accelerate the micro and macro vascular diabetic complications.[45]. In our study, similar consequences were observed in diabetic rats which shown increased plasma total cholesterol, triglycerides, LDL levels and decreased HDL levels. However, when treated the diabetic rats with PLF of *Ai*, plasma total cholesterol, triglycerides, and LDL levels significantly decreased and increased HDL levels were found may be due to its polyphenols compounds to control the glucose levels in blood by virtue of its hypoglycemic activity which in turn to regulate the lipid metabolism in liver.[46] Thus, our results clearly demonstrated that PLF of *Ai* could be able to retrieve the lipid profile to normal level under diabetic condition by asset of its hypoglycemic and hypolipidemic activity.

CONCLUSION

The overall outcomes of the current study indicate that PLF of *Ai* has remarkable hypoglycaemic, hypolipidemic and hepatoprotective activities in STZ induced diabetic rats. These properties seem to be due to the presence of bioactive compounds. Hence, the fraction can be used as herbal drug to treat diabetes. Further studies are essential to isolate polyphenolic compounds and evaluate in clinical level to clarify its possible mechanism of action at molecular level in the prevention of diabetes.

Conflict of interest: The authors declare no conflicts of interest.

ACKNOWLEDGEMENTS

All authors express sincere gratitude to the University Grants Commission New Delhi, India for the financial assistance. The first author for the award of UGC-BSR-Faculty Fellowship (No. F.18-1/2011 (BSR) dated 24-11-2017), the second author for the award of BSR Fellowship (No: F.25-1/2013-14 (BSR)/7-160/2007).

REFERENCES

1. Tappy, L., & Minehira, K. (2001). New data and new concepts on the role of the liver in glucose homeostasis. *Current Opinion in Clinical Nutrition & Metabolic Care*, 4(4), 273-277.

2. IFCC (International Federation of Clinical Chemistry), (1980). Physicochemical Quantities and Units in Clinical Chemistry. *J Clin Chem & Clin Biochem*, 18: 829-854.

3. Mauro P, Renze B, Wouter W. Enzymes, (2006) In: Tietz text book of clinical chemistry and molecular diagnostics. Carl AB, Edward R, David EB. 4th edition, *Elsevier,* 604-616.

4. Nicolle, E., Souard, F., Faure, P., & Boumendjel, A. (2011). Flavonoids as promising lead compounds in type 2 diabetes mellitus: molecules of interest and structure-activity relationship. *Current medicinal chemistry*, 18(17), 2661-2672.

5. Ahmadian, M., Suh, J. M., Hah, N., Liddle, C., Atkins, A. R., Downes, M., & Evans, R. M. (2013). PPAR³ signaling and metabolism: the good, the bad and the future. *Nature medicine*, 19(5), 557-566.

6. 70%Indians live in rural areas: Census. Available at:http://www.business-standard.com/article/economy-policy/70-indians-live-in-rural-areas-census-11071500171_1.html. Accessed on 23rd December 2016.

7. Bailey, C. J., & Day, C. (1989). Traditional plant medicines as treatments for diabetes. *Diabetes care,* 12(8), 553-564.

8. Bessey, O. A., Lowky, O. H., & Brock, M. J. (1946). A method for the rapid determination of alkaline phosphatase with five cubic millimeters of serum. *Journal of Biological Chemistry*, 164, 321-329.

9. De Silva, D. D., Rapior, S., Hyde, K. D., & Bahkali, A. H. (2012). Medicinal mushrooms in prevention and control of diabetes mellitus. *Fungal Diversity,* 56(1), 1-29.

10. Devaraj, S., Jialal, I., & Vega-Lopez, S. (2004). Plant sterol-fortified orange juice

effectively lowers cholesterol levels in mildly hypercholesterolemic healthy individuals. *Arteriosclerosis, Thrombosis, and Vascular Biology*, 24(3), e25-e28.

11. Friedewald, W. T. Levy. RI & Fredrickson, DS (1972) Estimation of the concentration of low density lipoprotein cholesterol in plasma, without the use of the preparative ultracentrifuge. *Clin Chem,* 18(6), 499-502.

12. El Gharras, H. (2009). Polyphenols: food sources, properties and applications–a review. *International Journal of Food Science & Technology,* 44(12), 2512-2518.

13. Gutierrez, R. M. P., & Navarro, Y. T. G. (2010). Antioxidant and hepatoprotective effects of the methanol extract of the leaves of *Satureja macrostema. Pharmacognosy Magazine,* 6(22), 125-131.

14. Hassan, S. K., El-Sammad, N. M., Mousa, A. M., Mohammed, M. H., Hashim, A. N. E., Werner, V., ... & Nawwar, M. A. E. M. (2015). Hypoglycemic and antioxidant activities of *Caesalpinia ferrea* Martius leaf extract in streptozotocin-induced diabetic rats. *Asian Pacific Journal of Tropical Biomedicine*, 5(6), 462-471.

15. India's Speedy–And Ominous–Disease Transition, June 2015. Available at: http://www.indiaspend. com/cover-story/indias-speedy-and-ominous-disease-transition-64225. Accessed on 6th January 2017.

16. Kameswara Rao, B., Rajasekhar, M. D., Sreelatha, A., & Apparao, C. (2006). Treatment of diabetes mellitus: Plant drugs vs oral hypoglycemic agents and insulin. *Recent Progress in Medicinal Plants,* 14, 279-296.

17. Kanchana, D., Jayanthi, M., Usharani, G., Saranraj, P., & Sujitha, D. (2013). Evaluation of plant growth promoting substance production by Azospirillum sp. isolated from rhizosphere of chilli (*Capsicum annuum* L.). *Int. J. Microbiol. Res,* 4(3), 300-304.

18. Mass spectrometry, (2010) . *J Chromatogr A,* 121732 58–3268

19. Murthy, N. S., Nandakumar, B. S., Pruthvish, S., George, P. S., & Mathew, A. (2010). Disability adjusted life years for cancer patients in India. *Asian Pac J Cancer Prev*, 11(3), 633-640.

20. Magalhães, P. J., Vieira, J. S., Gonçalves, L. M., Pacheco, J. G., Guido, L. F., & Barros, A. A. (2010). Isolation of phenolic compounds from hop extracts using polyvinylpolypyrrolidone: Characterization by high-performance liquid chromatography–diode array detection–electrospray tandem mass spectrometry. *Journal of Chromatography* A, 1217(19), 3258-3268.

21. Ponugoti, B., Dong, G., & Graves, D. T. (2012). Role of forkhead transcription factors in diabetes-induced oxidative stress. *Experimental diabetes research*, 2012.

22. Pooja, V. S., & Samanta, K. C. (2011). Hypoglycemic activity of methanolic extract of *Tectona grandis* linn. Root in alloxan induced diabetic rats. *Journal of Applied Pharmaceutical Science*, 1(04), 106-109.

23. Prasath, G. S., & Subramanian, S. P. (2013). Fisetin, a tetra hydroxy flavone recuperates antioxidant status and protects hepatocellular ultrastructure from hyperglycemia mediated oxidative stress in streptozotocin induced experimental diabetes in rats. *Food and Chemical Toxicology,* 59, 249-255.

24. Rains, J. L., & Jain, S. K. (2011). Oxidative stress, insulin signaling, and diabetes. *Free Radical Biology and Medicine*, 50(5), 567-575.

25 Ramasamy, R., Yan, S. F., & Schmidt, A. M. (2012). The diverse ligand repertoire of the receptor for advanced glycation endproducts and pathways to the complications of diabetes. *Vascular Pharmacology*, 57(5-6), 160-167.

26. Ravi, S., Shanmugam, B., Subbaiah, G. V., Prasad, S. H., & Reddy, K. S. (2017). Identification of food preservative, stress relief compounds by GC–MS and HR-LC/Q-TOF/MS; evaluation of antioxidant activity of *Acalypha indica* leaves methanolic extract (in vitro) and polyphenolic fraction (in vivo). *Journal of Food Science and Technology,* 54(6), 1585-1596.

27. Saravanan, G., Ponmurugan, P., Kumar, G. P. S., & Rajarajan, T. (2009). Antidiabetic properties of S-allyl cysteine, a garlic component on streptozotocin-induced diabetes in rats. *J Appl Biomed,* 7, 151-159.

28. Seebaluck, R., Gurib-Fakim, A., & Mahomoodally, F. (2015). Medicinal plants from the genus Acalypha (Euphorbiaceae)–A review of their ethnopharmacology and phytochemistry. *Journal of Ethnopharmacology,* 159, 137-157.

29. Shah, B., & Mathur, P. (2005). Multi-centric study on risk factors for Non Communicable diseases in India. Final Report. NICD-ICMR.

30. Gorinstein, S., Leontowicz, M., Leontowicz, H., Jastrzebski, Z., Drzewiecki, J., Namiesnik, J., ... & Trakhtenberg, S. (2006). Dose-dependent influence of commercial garlic (*Allium sativum*) on rats fed cholesterol-containing diet. *Journal of Agricultural and Food Chemistry*, 54(11), 4022-4027.

31. Singh, J., & Kakkar, P. (2013). Modulation of liver function, antioxidant responses, insulin resistance and glucose transport by *Oroxylum indicum* stem bark in STZ induced diabetic rats. *Food and chemical Toxicology*, 62, 722-731.

32. Szasz, G. (1969). A kinetic photometric method for serum y-glutamyl transpeptidase. *Clinical chemistry,* 15(2), 124-136.

33. TTaha, H., Arya, A., Paydar, M., Looi, C. Y., Wong, W. F., Murthy, C. V., ... & Hadi, A. H. A. (2014). Upregulation of insulin secretion and downregulation of pro-inflammatory cytokines, oxidative stress and hyperglycemia in STZ-nicotinamide-induced type 2 diabetic rats by *Pseuduvaria monticola* bark extract. *Food and Chemical Toxicology,* 66, 295-306.

34. Tahrani, A. A., Piya, M. K., Kennedy, A., & Barnett, A. H. (2010). Glycaemic control in type 2 diabetes: targets and new therapies. *Pharmacology & Therapeutics,* 125(2), 328-361.

35. Kondeti, V. K., Badri, K. R., Maddirala, D. R., Thur, S. K. M., Fatima, S. S., Kasetti, R. B., & Rao, C. A. (2010). Effect of *Pterocarpus santalinus* bark, on blood glucose, serum lipids, plasma insulin and hepatic carbohydrate metabolic enzymes in streptozotocin-induced diabetic rats. *Food and Chemical Toxicology*, 48(5), 1281-1287.

36. Vishwakarma, S. L., Rakesh D, S., Rajani, M., & Goyal, R. K. (2010). Evaluation of effect of aqueous extract of *Enicostemma littorale* Blume in streptozotocin-induced type 1 diabetic rats. *Indian J. Exp. Biol,* 48 (1) 26–30.

37. Whiting, D. R., Guariguata, L., Weil, C., & Shaw, J. (2011). IDF diabetes atlas: global estimates of the prevalence of diabetes for 2011 and 2030. *Diabetes Research and Clinical Practice,* 94(3), 311-321.

38. Wiernsperger, N. F. (2003). Oxidative stress as a therapeutic target in diabetes: revisiting the controversy. *Diabetes & Metabolism,* 29(6), 579-585.

39. Wright, P. J., Leathwood, P. D., & Plummer, D. T. (1972). Enzymes in rat urine: alkaline phosphatase. *Enzymologia*, 42(4), 317-327.

40. Xiang, F. L., Lu, X., Strutt, B., Hill, D. J., & Feng, Q. (2010). NOX2 deficiency protects against streptozotocin-induced ^{2}-cell destruction and development of diabetes in mice. *Diabetes*, 59(10), 2603-2611.

41. Zimmet, P., Alberti, K. G. M. M., & Shaw, J. (2001). Global and societal implications of the diabetes epidemic. *Nature*, 414(6865), 782-787.

42. Zimmet, P., Alberti, K. G. M. M., & Shaw, J. (2001). Global and societal implications of the diabetes epidemic. *Nature*, 414(6865), 782-787.

43. Girija, K., Lakshman, K., Udaya, C., Sachi, G. S., & Divya, T. (2011). Anti–diabetic and anti–cholesterolemic activity of methanol extracts of three species of Amaranthus. *Asian Pacific Journal of Tropical Biomedicine,* 1(2), 133-138.

44. Adiels, M., Westerbacka, J., Soro-Paavonen, A., Häkkinen, A. M., Vehkavaara, S., Caslake, M. J., ... & Borén, J. (2007). Acute suppression of VLDL 1 secretion rate by insulin is associated with hepatic fat content and insulin resistance. *Diabetologia*, 50(11), 2356-2365.

45. Mooradian, A. D. (2009). Dyslipidemia in type 2 diabetes mellitus. *Nature Reviews Endocrinology*, 5(3), 150-159.

46. Abdultawab, H. S., & Ayuob, N. N. (2013). Can garlic oil ameliorate diabetes-induced oxidative stress in a rat liver model? A correlated histological and biochemical study. *Food and Chemical Toxicology,* 59, 650-656.

Current Status of Researches in Biosciences (2020) : 127-131
Editor : Prof. P.C. Joshi, N. Joshi, Reshman Yasmin, and D. K. Mansotra
Today & Tomorrow's Printers and Publishers, New Delhi - 110 002

ANDROID BASED MOBILE APP. IN SHEEP AND GOAT FARMING: AN INNOVATIVE TECHNOLOGY

S. Senthilkumar

Department of Veterinary and Animal Husbandry Extension Education, Veterinary College and Research Institute, Ramayanpatti, Tirunelveli, Tamil Nadu

Email: usveteng@gmail.com

ABSTRACT

Based on the information needs of sheep and goat farmers, mobile based application software was developed in android platform. App. contents were organised into breeding, feeding, management, disease control, marketing and others sections. It has separate icons for special options like Push notification, WhatsApp, FAQ, MAP and Search. We can send important notification like forecasting, deworming and vaccination schedule to end-users in a single stroke by push notification option. Share many innovative ideas and experiences in sheep and goat farming via WhatsApp option. FAQ options are helpful to entrepreneurs. We can locate the veterinary hospital and dispensary via MAP option.

KEYWORDS: ICT, Mobile App., Sheep and Goat farming

INTRODUCTION

Among various Information and Communication Technologies (ICTs), mobile telephony has emerged as the smart technology of choice of the majority of the urban and even the rural masses (Ansari and Pandey, 2013). As such, mobile phones have been regarded as the widely accessed tool among the farmers for communication and also accessing agriculture-related information. India is the second largest smart phone user country in the world having 220 million smart phone users base with 80 million users in rural India. The cost of the smart phone device is becoming affordable and their feature allows creation of a variety of practical applications. Henceforth it has a wide scope and potential of faster information dissemination, thus making the farmer empowered with information. Based on these facts, a research project on "Development of Mobile Based Technology Transfer Application System to Empower the Small Ruminant Farmers in Tirunelveli District" sponsored by the NABARD, Chennai was implemented at Department of Veterinary and Animal Husbandry Extension Education, Veterinary College and Research Institute, Tirunelveli during 2017 – 2019.

METHODOLOGY

An exploratory research design was used in this study. Tirunelveli district is selected as study area since, NABARD Project on development of mobile app. in small ruminant was operated in Tirunelveli District which has sheep and goat population of about 3,03,105 and 3,30,230 respectively (19th Livestock Census, 2012). Identification of needs of small ruminant farmers means that the level of information needed by the farmers as perceived by the researcher, extension personnel and small ruminant farmers on various small ruminant farming activities. The primary data were collected from the selected respondents by using well-structured pre-tested interview schedule. A schedule of information needs contained 48 items on various aspects of small ruminant farming like breed & breeding, feeding, management, disease control and marketing. The different need items were identified in consultation with the subject matter specialists and secondary data on small ruminant rearing. The responses for each of the 48 items were measured on a five point continuum i.e., most needed, more needed, needed, less needed and least needed. The mean was calculated and the item above the mean was considered as the identified needs for the development of mobile application. Suitable contents with relevant photographs and videos on breeds and breeding, feeding, management, disease control and marketing were developed in consultation with subject matter specialists of Veterinary College and Research Institute, Tirunelveli. The contents pertinent to the information needs were incorporated as an Android based Mobile Application Systems in English and Tamil languages with suitable text, video and audio etc.

RESULTS AND DISCUSSION

The sheep and goat rearing activity was classified under five major heads i.e breed & breeding, feeding, management, disease control and marketing. Senthilkumar *et al.*, (2019) reported that a total of 24 items were considered as the identified needs that include topics on breed &

breeding (three), feeding (five), management (five), disease control (six) and marketing (five). Based on the needs identified and prioritized, the content were organised, script prepared both in Tamil and English languages, video clipping, photos collected and mobile based application software was developed in android platform. Technological booming in this Information Era, native and hybrid based mobile applications software is commonly used in the development of mobile app. Among these, Android based mobile application software was selected since android is one of the smart phone operating system whose most applications are available freely on an android market. An android user can easily and freely download android applications. Android is an operating system for mobile device and also a platform to develop key application for the smart phone. Java programming language is used to develop android application by using Android SDK tools and API (Patel *et al.,* 2014).

Structure of the Mobile App: Based on the information needs of sheep and goat farmers, mobile based application software was developed in android platform. App. contents were organised into main and sub-menu. Main menu has breeding, feeding, management, disease control, marketing and others sections. Under Breed Sub-menu, content were organised into breeds and breeding management. Breeds section has Exotic, Indian and Tamil Nadu types of breeds. Breeding management section has details on traits of breeding and types of breeding. Feeding sub-menu has green fodder, dry fodder, concentrate feed, feeding management and grazing management sections. Under concentrate feeding, preparations of concentrate feed, quantities of concentrate feed, were included. Under management sub-menu, floor space requirement, dentition, method of rearing, culling of animal, transport of goat / sheep, farm appliances were provided. Disease control sub-menu has bacterial diseases, viral diseases, parasitic diseases, deworming and vaccination schedule, ethno-veterinary medicine and first aid-medicine, etc. Marketing sub-menu comprises of marketing channel, slaughtering goat, value added meat products, bankable project, livestock shandy etc. Other sub-menu has contact details of TANUVAS units and Animal Husbandry Department.

Special Features: This app. has separate icons for special options like Push notification, WhatsApp, FAQ, MAP and Search. We can send important notification like forecasting, deworming and vaccination schedule to all end-users in a single stroke by using push notification option. By using WhatsApp option, we will share many innovative ideas and experiences in sheep and goat farming. FAQ options are obviously helpful to new entrepreneurs. We can locate the veterinary hospital, dispensary, college etc. by using MAP option. By using search option, we can find exact content what we need which saves time too.

Web Link: The developed mobile app. uploaded and is available in Google Play Store which could be downloaded and installed to retrieve the information both in online and offline mode. Following link can be used to retrieve:

Product utility

- Provides remote access to extension services: Meet the immediate needs of farmers as they change their production and livelihood systems.
- Can be used by end users anytime and from anywhere which will ultimately save money, time and effort in reaching a larger population.
- This software will help to impart the scientific knowledge of diseases that affect small ruminants to farmers in local languages and at the farmer's level of understanding to reduce economic losses due to diseases.
- This application is built to break the literacy barrier and deliver the information.

ACKNOWLEDGEMENTS

The author is thankful to National Bank for Agriculture and Rural Development (NABARD), Chennai for financial assistance for the study. The facilities provided by the Vice-Chancellor, the Registrar, the Director of Research, TANUVAS, Chennai and the Dean, VCRI, Tirunelveli are sincerely acknowledged.

REFERENCES

1. Ansari, M.A and Neha Pandey. (2013). Assessing the potential and use of mobile phones in Agriculture. *Karnataka Journal of Agricultural Science,* 26(3): 388-392.

2. 19th Livestock Census. (2012). Department of Agriculture, Animal Husbandry and Fisheries, Government of India.

3. Patel B Vimal, Rahul G. Thakkar and Hardikkumar V. Desai. (2014). An Android application for farmers for Kharif and Rabi crop diseases information. *International Journal of advanced research in computer science and software engineering*, 4 (10) : 788 - 791

4. S. Senthilkumar, C. Manivannan and S. Sureshkannan. (2019). Mobile App. in Sheep and Goat Farming. Compendium - International Seminar on "Innovative

Extension Management for Uplifting Livelihood of Farmers – Status, Initiatives and Way Forward" June 27 & 28, 2019, Dept. of VAE, MVC, Chennai / 56-3 / 418 – 242 (ISBN No. : 978-93-85418).

Current Status of Researches in Biosciences (2020) : 133-140
Editor : Prof. P.C. Joshi, N. Joshi, Reshman Yasmin, and D. K. Mansotra
Today & Tomorrow's Printers and Publishers, New Delhi - 110 002

PREVALENCE OF GASTRO-INTESTINAL NEMATODES IN *CAPARA HIRCUS* (GOAT) IN LUCKNOW REGION, UTTAR PRADESH, INDIA

Kamal Jaiswal, Suman Mishra and Jyoti Pandey

Department of Zoology, Babasaheb Bhimrao Ambedkar University, Lucknow-226025,

Email Id: drkamaljaiswal@gmail.com, jyotiacu@gmail.com

ABSTRACT

Nematodes are a serious menace for small ruminants, prevalent all over the world. They cause nematodiasis which is an important pathogenic disease in livestock animals causing severe economic loss. In this study, 355 Gastro-Intestinal (GI) tracts of freshly slaughtered goats were investigated from November 2014 to October 2016 for the prevalence of GI nematodes. Out of these 231 (65.1 %), gastrointestinal tracts were found to be positive. Four nematode species were identified namely *Haemonchus* spp. (62.3 %), *Oesophagostomum* spp. (36.1 %), *Bunostomum* sp. (27.9 %), *Trichuris* sp. (56.3 %). Month-wise prevalence was found significantly (Pd"0.05) highest in September; *Haemonchus* spp. (88.5 %), *Oesophagostomum* spp. (65.4 %), *Bunostomum* sp. (61.5 %), *Trichuris* sp.(80.8 %) whereas lowest in December; *Haemonchus* spp. (11.3 %), *Oesophagostomum* spp. (11.3 %), *Bunostomum* sp. (7.4 %), *Trichuris* sp.(33.3 %). Seasonal incidence was found significantly (Pd"0.05) highest for *Haemonchus* spp. (Monsoon = 83.1%; Summer = 61.0 %; Winter= 42.9 %) through all over the season followed by *Trichuris* sp.(Monsoon = 69.5 %; Summer= 55.9 %; Winter= 43.7 %), *Oesophagostomum* spp. (Monsoon = 51.7 %; Summer= 33.0 %; Winter= 23.5 %) and *Bunostomum* sp. (Monsoon = 42.4%; Summer= 27.1 %; Winter= 14.3 %).

KEYWORDS: Nematode, small ruminants, Gastro-Intestinal tracts, Prevalence

INTRODUCTION

In the Indian agriculture system, goat livestock holds an important niche for sustainable growth and development of the remote rural population of India which progressively used to enhance the socio-economic condition as well as help in the household livelihood strategies. Goat rearing has a significant advantage over other livestock due to their less effort requirement, minimum initial investment, ease marketing, and early sexual maturity (Pandey, 2019).

Besides, their valuable importance, goat livestock had been suffering from gastrointestinal Nematodiasis which is responsible for reduced animal performance, low productivity, involuntary culling, coarse hair coat, loss of appetite, anaemia, weight loss, rapid breathing, submandibular oedema (bottle jaw), decreased milk production (Lebbie *et al.,* 1994; Badran *et al.,* 2012; Risso *et al.,* 2015), and mortality and morbidity in heavily infected animals (Negasi *et al.,* 2012) resulting in severe economic losses.

However, the data relating to epidemiology of gastrointestinal nematodes in goats is meagre from Lucknow region of Uttar Pradesh (UP) and therefore it becomes necessary to develop data on epidemiology of gastrointestinal nematodes in goats which will be very helpful in formulating an appropriate strategic parasite control program used further for the execution of future control measures. Thus, the present study is performed to investigate the incidence of GI nematodes in Lucknow region.

MATERIALS AND METHODS

The prevalence study was conducted by collecting 355 gastrointestinal tracts of host animal in every month from the different slaughtered houses of Lucknow region through all the three seasons of (November 2014 to October 2016) the year. The gastrointestinal tracts were immediately transported to the parasitology laboratory of Dept. of Zoology (Formerly dept. of Applied Animal Sciences), Babasaheb Bhimrao Ambedkar University, Lucknow, Uttar Pradesh. Each part of the gastrointestinal tract was examined carefully for the presence of nematode parasites by following the method described by Cable, (1958); Mayer and Olsen, (1975). Collected parasites were counted and transferred to fresh normal saline (0.9%). The incidence of different gastrointestinal nematodes with relation to month and season were analysed by employing the Chi-square test.

RESULTS AND DISCUSSION

A total of 355 gastrointestinal tracts were collected and examined during the study period (November 2014 to October 2016) for the infestation and noted that 231 gastrointestinal tracts were found to be positive with different nematode species showing 65.1 % prevalence (Table 1).

Table 1: Overall prevalence of nematode parasites in *Capra hircus*

Total Number of intestines examined	Number of the infected intestines	% Prevalence
355	231	65.1

Four species of nematode parasites were identified namely *Haemonchus* spp. 62.3 %, *Trichuris* sp. 56.3 %, *Oesophagostomum* spp. 36.1 %, *Bunostomum* sp. 27.9 %. (Table2). All the parasites differed significantly (Pd"0.05) with their prevalence.

Table 2: Prevalence of different nematode parasites in *Capra hircus*

Nematode species	Total Number of intestines examined	Number of the infected intestines	% Prevalence
***Haemonchus* spp.**	355	221	62.3
***Oesophagostomum* spp.**	355	128	36.1
***Bunostomum* sp.**	355	99	27.9
***Trichuris* sp.**	355	200	56.3
X^2		114.110 (Pd≤0.05)	

In the month–wise prevalence study of different nematode species, data revealed that the nematode parasite namely *Haemonchus* spp., *Trichuris* sp., and *Oesophagostomum* spp. showed a significant (Pd"0.05) increase in the prevalence rate from July (84.4 %, 68.8 % and 43.8 % respectively) to September (88.5 %, 80.8 % and 65.4 %) except *Bunostomum* sp. which showed significantly (Pd"0.05) higher prevalence rate in the month of September (61.5 %) followed by July (43.4 %) and August (42.4 %) whereas all these parasite had shown a significant (Pd"0.05) decrease in the prevalence rate in the month of December (Table 3).

Table 3: Month-wise prevalence of nematode parasites in *Capra hircus*

Month	Total number of intestine examined	% Prevalence of nematode parasites (number of infected intestines)			
		Haemonchus spp.	*Oesophago-stomum* spp.	*Bunostomum* sp.	*Trichuris* sp.
Nov	30	50 (15)	20 (6)	10 (3)	43.3 (13)
Dec	27	11.1 (3)	11.1 (3)	7.4 (2)	33.3 (9)
Jan	31	45.2 (14)	22.6 (7)	9.7 (3)	48.4 (15)
Feb	31	61.3 (19)	38.7 (12)	29.0 (9)	48.4 (15)
March	27	66.6 (18)	40.7 (11)	33.3 (9)	40.7 (11)
April	30	60.0 (18)	33.3 (10)	33.3 (10)	56.7 (17)
May	32	59.4 (19)	34.4 (11)	25.0 (8)	68.8 (22)
June	29	58.6 (17)	24.1 (7)	17.2 (5)	55.2 (16)
July	32	84.4 (27)	43.8 (14)	43.4 (11)	68.8 (22)
August	33	87.9 (29)	51.5 (17)	42.4 (14)	69.7 (23)
Sep	26	88.5 (23)	65.4 (17)	61.5 (16)	80.8 (21)
Oct	27	70.4 (19)	48.1 (13)	33.3 (9)	59.3 (16)
X^2		60.643 (Pd≤0.05)	31.012 (Pd≤0.05)	37.320 (Pd≤0.05)	24.958 (Pd≤0.05)

In season-wise prevalence study, it was recorded that all nematode parasites showed a significant (Pd"0.05) disparity throughout all the three seasons and in this respect the highest incidence rate was noted down in the monsoon season (*Haemonchus* spp. 83.1 %, *Trichuris* sp. 69.5 %, *Oesophagostomum* spp. 51.7 %, *Bunostomum* sp. 42.4 %,) followed by summer (*Haemonchus* spp. 61.0 %, *Trichuris* sp. 55.9 %, *Oesophagostomum* spp. 33.0 %, *Bunostomum* sp. 27.1 %,) and winter (*Haemonchus* spp. 42.9 %, *Trichuris* sp. 43.7 %, *Oesophagostomum* spp. 23.5 %, *Bunostomum* sp. 14.3 %, (Table 4).

Table 4: Season-wise prevalence of different nematode parasites in *Capra hircus*

Nematode species	% Prevalence (Number of infected intestine)			X^2(P value)
	Winter N=119	Summer N=118	Monsoon N=118	
***Haemonchus* spp.**	42.9 (51)	61.0 (72)	83.1 (98)	40.849 (Pd≤0.05)

***Oesophagostomum* spp.**	23.5 (28)	33.0 (39)	51.7 (61)	21.079 (Pd≤0.05)
***Bunostomum* sp.**	14.3 (17)	27.1 (32)	42.4 (50)	23.294 (Pd≤0.05)
***Trichuris* sp.**	43.7 (52)	55.9 (66)	69.5 (82)	16.037 (Pd≤0.05)

Study of relevant factors impelling the prevalence rate of nematode infection i.e., the geography of the grazing land, vegetation, agro-climatic factor viz., temperature, soil's nature, relative humidity, availability of water sources, development of resistance due to recurrent use of anthelmintics, unhygienic field management and the rainfall play an important role to determine type and levels of nematode parasite infestation (Pandey, 2019).

The results of the present study revealed 65.1 % of the overall prevalence of gastrointestinal nematode and the findings are at par with the results of Tripathi and Subedi, 2015, where they reported a 65.3% overall prevalence rate of nematodes. These findings were also in agreement with the results of Javaid *et al.* (2018) who divulged the high infection rate of nematode i.e., 42.67 %. The higher prevalence of nematodes could be associated with the fact that these parasites have a relatively short life cycle. Moreover, nematode parasites are capable of developing resistance against modern synthetic drugs faster than any other helminth species (Radostits *et al.,* 1994; Katoch *et al.,* 1999).

In the present study a total of four nematode species namely *Haemonchus* spp., *Oesophagostomum* spp., *Bunostomum* sp., and *Trichuris* sp. were identified. These nematode parasites were also reported earlier in goat by many researchers (Raza *et al.,* 2012, Raza *et al.*, 2014, Barua *et al.*, 2015, Amran *et al.,* 2018). The prevalence rate of this study noted highest for *Haemonchus* spp. 62.3 % followed by *Trichuris* sp. 56.3 % and the results are supported by Lone *et al.,* 2012 wherein the highest prevalence rate was recorded for *Haemonchus* sp. (82 %) followed by *Trichuris* sp. (74 %) in Kashmir. Somehow similar results were also reported by (Raza *et al.,* 2014) at Pakistan, and reported highest prevalence (12.8 %) for *Haemonchus* sp., followed by *Trichuris* sp. (7.2 %). The present study also revealed the most dominating and repeatedly witnessed species were *Haemonchus* spp. and the same finding was also reported by other researchers (Ahmed and Ansari, 1987; Gupta *et al.,* 1987; Tsotetsi and Mbati, 2003; Bakunzi *et al.,* 2013). The highest incidence rate of *Haemonchus* spp. might be due to the capability of adult female species to produce thousands of eggs per day in their short duration of the lifecycle, which can initiate abrupt larval pasture contamination as well as the ability to take advantage of favourable climatic conditions (Grant, 1981; Roeber *et al.,* 2013).

In the present research, monthly fluctuations in the prevalence of gastrointestinal nematodes were recorded and higher incidence rate of was observed in the month of July (84.2 %), August (87.9 %) and September (88.5 %) and lowest in the month of December (37 %) and January (48.4 %). These results are almost in the agreement with the findings of Moiloa, (2017), who observed the monthly prevalence of gastrointestinal nematodes in Maseru district and Quthing district of Lesotho, and reported the lowest prevalence rate in the month of December (54.9 %) and highest in the month of July (98.0 %) in Maseru district whereas found lowest in the month of October (64.20 %) and highest in the month of September (79.60 %) in Quthing district. These results are also supported by Tripathi (1970) and Maske *et al.* (1990) in goats from Uttar Pradesh and Maharashtra, respectively.

The seasonal prevalence of all nematode species were observed higher in the monsoon season (*Haemonchus* spp., 98 %; *Oesophagostomum* spp., 51.7 %; *Bunostomum* sp., 42.4 % and *Trichuris* sp., 69.5 %) followed by summer (*Haemonchus* spp., 61.0 %; *Oesophagostomum* spp., 33.0 %; *Bunostomum* sp., 27.1 % and *Trichuris* sp., 55.9 %) and winter (*Haemonchus* spp., 42.9 %; *Oesophagostomum* spp., 23.5 %; *Bunostomum* sp., 14.3 % and *Trichuris* sp., 43.7 %). These results were supported by the findings Pathak and Pal (2008) and reported high prevalence of nematode parasites during monsoon season (*Haemonchus* spp., 29.72 %; *Oesophagostomum* spp., 32.43 %; *Bunostomum* sp., 5.4 % and *Trichuris* sp., 27.07 %) followed by summer (*Haemonchus* spp., 25.0 %; *Oesophagostomum* spp., 34.37 %; *Bunostomum* sp., 5.40 % and *Trichuris* sp., 27.02 %) and winter (*Haemonchus* spp., 21.05 %; *Oesophagostomum* spp., 21.05 %; *Bunostomum* sp., 5.26 % and *Trichuris* sp., 21.05 %). The high incidence of parasites in the Monsoon season (July, August, and September) might be due to the favourable condition of these months i.e. peak temperature, relative humidity and rainfall for survival, development, and propagation of nematode larvae in the pasture, which leads to higher infection on the pasture animals.

CONCLUSION

Keeping in view the present results, it can be concluded that the high incidence rate of gastrointestinal Nematodiasis needs to be examined intermittently in the goat population. Further, effective control measures should be taken under consideration like rotational grazing arrangement, pasture management, applying planned deworming, and conducting extension programs to educate the farmers regarding the appropriate use of anthelmintics to check the incidence of the parasite.

REFERENCES

1. Ahmed, M. and Ansari, J.A. (1987). Prevalence of gastrointestinal nematodes of sheep and goats in Aligarh (India). *Indian Veterinary Medical Journal,* 11: 165–170.

2. Al-Amran, M., Yadav, S.K., Akter, F., Sarkar, S., Hossain, M. A., Joy, S.M. and Samrat, Al-A.K. (2018). Prevalence of gastrointestinal parasitic infections in different existing goat breeds in different districts of Bangladesh. *The Journal of Advances in Parasitology*, 5(1): 11-21.

3. Badran, I., Abuamsha, R., Aref, R., Alqisi, W. and Alumor, J. (2012). Prevalence and diversity of gastrointestinal parasites in small ruminants under two different rearing systems in Jenin district of Palestine. *An-Najah University Journal of Research*, 26: 1-18.

4. Bakunzi, F.R., Nkomo, L.K.., Motsei, L.E., Ndou, R.V. and Nyirenda, M. (2013). A survey on anthelmintic resistance in nematode parasites of communally grazed sheep and goats in a rural area of North West province, Republic of South Africa. *Life Sciences Journal*, 10 (2): 391-393.

5. Barua, C.C., Hazorika, M., Saleque, A., Bora, P. and Das, P. (2015). Prevalence of gastrointestinal helminth infections in goats at goat research station, Byrnihat. *International Journal of Advanced Research in Biological Sciences,* 2: 297–305.

6. Cable, R.M. (1958). An Illustrated Laboratory Manual of Parasitology, 4[th]ed., Burges Publishing Co., Minneapolis 15, Minnesota, USA, pp:156.

7. Grant, J.L. (1981). The epizootiology of nematode parasites of sheep in a high-rainfall area of Zimbabwe. *Journal of South African Veterinary Association,* 52: 33–37.

8. Javaid, M., Rashid, M., Nabi, B., Mudasir, M., Bhat, M., Wani, N. and Bhat, M. (2018). Prevalence of Gastrointestinal Nematodes in Sheep and Goats Reared by Nomads of Jammu Region of Jammu and Kashmir. *International Journal of Livestock Research*, 8(8): 219-225.

9. Katoch, R., Mandial, R.K. and Nagal, K.B. (1999). Outbreak of *Haemonchus contortus* infection in sheep of Himachal Pradesh. *Indian Veterinary Journal,* 76: 932–933.

10. Lebbie, S.H.B., Rey B. and Irungu, E.K. (1994). Small ruminant research and development in Africa. Proceedings of the 2 Biennial Conference of the African Small Ruminant Research Net Work. ILCA, pp: 1-5.

11. Maske, D.K., Bhilegaonkar, N.G. and Sardey, M.R. (1990). Prevalence of parasitic infection in domestic animals at Nagpur. *Journal Veterinary Parasitology,* 4(2): 23–

25

12. Mayer, M.C. and Olsen, O.W. (1975). Essentials of Parasitology. 2nd ed., Wm. C. Brown Co. Iowa, pp: 1-303.

13. Negasi, W., Bogale, B. and Chanie, M. (2012). Helminth parasites in small ruminants: prevalence, species composition and associated risk factors in and around Mekelle town, Northern Ethiopia. *European Journal of Biological Research,* 4(3): 91–95.

14. Pandey, J. (2019). Studies on prevalence of gastrointestinal helminths parasites of *Capra hircus* (L) and evaluation of selected ethnoveterinary plants for anthelmintic activity (Unpublished doctoral Thesis). Babasaheb Bhimrao Ambedkar University, Lucknow, Uttar Pradesh.pp:

15. Pathak, A.K., and Pal, S. (2008) Seasonal prevalence of gastrointestinal parasites in goats from Durg district of Chhattisgarh. *Veterinary World,* 1(5): 136–137

16. Radostits, O.M., Blood, D.C. and Gay, C.C. (1994). Diseases caused by helminth parasites: A text book of diseases of cattle, sheep, pigs, goats and horses. 8th ed., London, Balliere Tindall. *Veterinary medicine*, pp: 1223–1230.

17. Raza, M.A., Younas, M. and Schlecht, E. (2014). Prevalence of gastrointestinal helminths in pastoral sheep and goat flocks in the Cholistan desert of Pakistan.*The Journal of Animal and Plant Sciences,* 24(1): 127-134.

18. Raza, M.A., Arshad, H.M., Ayaz, M.M., Murtaza, S., Bachaya, H.A. and Basit, A. (2012). Gastrointestinal helminthiasis in goats at the localities of Jatoi, Punjab, Pakistan. *Science International (Lahore)*, 24:171-175

19. Risso, A., Kessler, J.D., Soriano, V.S., Nunes, M.L.A., Machado, G., Langaro, A., Rossetto, R., Zuffo, T., Dallago, M. and Castro, P. (2015). Influence of pathological conditions caused by gastrointestinal parasites infection on pregnant ewe's behavior. *Acta Scientiae Veterinariae,* 43, pp: 1283.

20. Roeber, F., Jex, A.R. and Gasser, R.B. (2013). Impact of gastrointestinal parasitic nematodes of sheep, and the role of advanced molecular tools for exploring epidemiology and drug resistance-an Australian perspective. *Parasites Vectors,* 6(1): 1–13.

21. Tripathi, J.C. (1970). Seasonal variation in egg output of gastrointestinal nematodes of goats. *Indian Journal of Veterinary Sciences,* 40(1): 46–59.

22. Tripathi, R.P. and Subedi, J.R. (2015). Seasonal prevalence of gastro-intestinal helminth parasites of goats (*Capra* sp.) of Shivraj municipality-13 Kapilvastu, Nepal. *Nepalese Journal of Zoology,* 3: 63-67.

23. Tsotetsi, A. and Mbati, P. (2003). Parasitic helminths of veterinary importance in cattle, sheep and goats on communal farms in the Northeastern Free State, South Africa. *Journal of the South African Veterinary Association,* 74(2): 45–48.

Current Status of Researches in Biosciences (2020) : 141-156
Editor : Prof. P.C. Joshi, N. Joshi, Reshman Yasmin, and D. K. Mansotra
Today & Tomorrow's Printers and Publishers, New Delhi - 110 002

A NEW KNOCKDOWN RESISTANCE (*KDR*) MUTATION F1534L IN *AEDES AEGYPTI* ASSOCIATED WITH INSECTICIDE RESISTANCE

Neera Kapoor[1] and Taranjeet Kaur[2]

[1]*School of Life Sciences, Indira Gandhi National Open University, Maidangarhi, Delhi- 110068.*

[2]*National Institute of Malaria Research, Sector 8, Dwarka, Delhi-110077, India*

Email: neerakapoor@ignou.ac.in

ABSTRACT

The control of *Aedes aegypti* borne-infections mainly dengue, chikungunya, yellow fever and Zika virus relies mainly on vector control measures in the absence of specific drugs or vaccines available against these infections. Emergence of insecticide resistance in *Ae. aegypti* may pose serious threat to the success of insecticide-based vector control programme. Here, we report the presence of multiple knockdown resistance (*kdr*) mutations present in an Indian *Ae. aegypti* population including a new mutation F1534L (not reported earlier in *Ae. aegypti*) which is associated with DDT and pyrethroid resistance. DNA sequencing of partial domain II, III and IV of the voltage gated sodium channel (VGSC) performed on *Ae. aegypti* collected from Bengaluru, India, revealed the presence of four *kdr* mutations, i.e., V1016G and S989P in domain II and two alternative *kdr* mutations F1534C and F1534L in domain III. Allele specific PCR assays (ASPCR) were developed for the detection of *kdr* mutations V1016G and S989P while a PCR-RFLP based strategy was adopted for the genotyping of all three known mutations in domain III (F1534L, F1534C and T1520I). Genotyping of 572 *Ae. aegypti* samples collected in 2014 and 2015 revealed a moderate frequency of V1016G/S989P (18.27%) and F1534L (17.48%), a relatively high frequency of F1534C (50.61%) and absence of T1520I in the population. Mutations V1016G and S989P were in complete linkage disequilibrium while they were having negative linkage disequilibrium with *kdr* alleles F1534C and F1534L. The new mutation F1534L showed significant protection against permethrin, deltamethrin and DDT whereas F1534C showed protection against permethrin and DDT but not against deltamethrin.

KEYWORDS: *Aedes aegypti,* Insecticide Resistance, Knockdown resistance, Pyrethroid, Voltage gated sodium channel

INTRODUCTION

Aedes aegypti is now a widely distributed mosquito species in tropical and subtropical regions and is responsible for the transmission of several arboviral infections in human mainly dengue, chikungunya, yellow fever and Zika virus. These arboviral infections are increasingly becoming a global health concern due to their rapid geographical spread and high disease burden [1]. The ancestral form of *Ae. aegypti* was found in Africa which used to feed on nonhuman primates and after its domestication, *Ae. aegypti* has now expanded from Africa and colonized most of the pan tropical world [2]. During last five decades, there have been an unprecedented emergence of epidemic arboviral diseases [3] and approximately half of the world population is under the threat of dengue [4, 5]. In India, dengue and chikungunya are main arboviral infections with recent introduction of Zika virus (ZIKV). Recently there had been an outbreak of Zika infections in Jaipur city [6], where *Ae. aegypti* has been incriminated as Zika vector [7].

In absence of specific vaccines or drugs for the treatment of *Aedes*-borne infections, vector control and personal protection remain only measures to contain the spread of these arboviral infections. Use of insecticide-based control of this vector is a common approach used worldwide. Pyrethroid group of insecticides is of special interest which is being extensively used in long lasting insecticidal nets (LLIN), space spray as well as in household repellent and personal protection. Pyrethroids are preferred group of insecticides due to their low mammalian toxicity, degradability in nature and rapid knockdown effect on insect [8]. However, extensive use of pyrethroids in public health and also in agriculture sector has led to the emergence of resistance against these insecticides in many disease vectors including *Ae. aegypti.* Several reports of pyrethroid resistance in *Ae. aegypti* have been recorded from different parts of world mainly in Southeast Asia, Latin Americas and Africa, with such reports not available from India until the year 2015 [9].

Understanding the mechanisms of insecticide resistance in vector populations is crucial for an effective insecticide resistance management. One of the known mechanisms of insecticide resistance in mosquitoes against pyrethroids and DDT is knockdown resistance (*kdr*) which is conferred by alteration in the target site of action, i.e., the voltage gated sodium channel (VGSC) resulting from non-synonymous mutations. Several *kdr* mutations have been reported in *Ae. aegypti* in different parts of world, amongst which mutations at three loci i.e., Iso1011 (I '!M/V) and Val1016 (V'!G/I) in domain II and F1534 (F'!C) in domain III are commonly associated with pyrethroid resistance [10-18]. Presence of such *kdr* mutations in the Indian subcontinent was screened in just one northern Indian population (Delhi) where we reported the presence of F1534C and a novel mutation T1520I [18]. This study reports a new mutation F1534L, co-occurring with mutations F1534C, S989P and V1016G, in a southern Indian field population of *Ae. aegypti*. The new mutation F1534L showed positive association with DDT and pyrethroid resistance.

MATERIAL AND METHODS

Mosquito collection: Immatures (eggs, larvae and pupae) of *Ae. aegypti* were collected from domestic and peri- domestic breeding sites from Basavangudi area of Bengaluru city (77° 56-572 E, 12° 92-952 N) during years 2014 and 2015. Oral informed consent was obtained from the owners of the houses for collection of immatures at residential premises. Immatures were reared in the laboratory till the emergence into adult (F_0). In addition, F_1 progenies were also obtained. To get F_1 progenies, mosquitoes were fed on blood through artificial membrane (Parafilm®) and eggs were obtained after 72 hours of blood feeding. Eggs were allowed to hatch in water and larvae were reared in enamel trays with a supplementation of fish-food till pupation. Pupae were removed from the tray and placed in a bowl containing water inside an insect cage (measuring 30 cm X30 cm X 30 cm) for their emergence into adult. Adults were fed on 10% glucose soaked on cotton pad. Insectary was maintained at a temperature of 27±°C, relative humidity (RH) 60-70% and photoperiod of 14h:10h (light:dark) ratio.

Exposure of insecticide to mosquitoes (Bio-assay): Two-to four-days old sugar fed adult *Ae. aegypti* female mosquitoes (F_0, and F_1), were exposed to 0.05% deltamethrin-, 0.75% permethrin- or 4% DDT-impregnated papers (supplied by WHO collaborative centre, Vector Control Research, Universiti Sains, Malaysia) for one-hour following WHO's standard insecticide-susceptibility test guidelines [35]. Following exposure of insecticide, they were transferred to recovery tubes and mortalities were recorded after 24 hours of recovery. Individual dead and alive mosquitoes were kept in 1.5 mL micro centrifuge tubes with a piece of silica gel for DNA isolation and at -20 °C. All bioassays were carried out in a laboratory maintained at 25°C and RH 60-70%.

DNA isolation and sequencing: DNA from individual mosquitoes was isolated following the method described by Livak *et al.,* [36] after removing 1/3rd of the posterior abdomen that carries spermatheca (to avoid contamination of sperm from mating male partner) and stored at 4°C. Some of the mosquitoes were sequenced for domain II, III and IV of the VGSC. A common PCR protocol and PCR conditions were used for amplification of partial domain II, III and IV of VGSC. The PCR reaction (25 μl) contained 1X buffer, 1.5 mM $MgCl_2$, 200 μM of each dNTP, 0.2 μM of each primer and 0.625 units of taq polymerase (AmpliTaq Gold, Invitrogen corporation, USA). The PCR conditions were: an initial denaturation at 95°C for 3 min followed by 35 cycles each of denaturation at 95°C for 15s, annealing at 55°C for 15s and extension at 72°C for 30s with a final extension at 72°C for 7 min. PCR products were purified using ExoSAP-IT PCR Product Clean up (Exonuclease I - Shrimp Alkaline Phosphatase, Thermo Fisher Scientific Inc) and were sent to Macrogen Inc, South Korea, for sequencing. The sequence chromatograms were edited using Finch TV ver 1.5.0 (http://www.geospiza.com/).

Cloning and sequencing: It was not possible to identify the correct amino acid codon through DNA sequencing in samples which were found heterozygous for two nucleotide base positions i.e., first and second codon of the F1534 residue. We therefore cloned and sequenced five such heterozygous samples to identify the correct codon. PCR product was amplified using primers AekdrF and AekdrR and high fidelity taq DNA polymerase (Phusion® High-Fidelity DNA Polymerase) purified using QIAquick PCR Purification Kit (Qiagen Inc.) and cloned in pGem-T Easy Vector Systems (Promega Corporation) as per vendor's protocol. Competent *E. coli* cells were transformed with the recombinant DNA and grown on LB-Agar plate. Positive clones were selected by blue/white screening and amplified using universal primers SP6 and T7. The PCR product of individual clone was sequenced at Macrogen, S. Korea, using primers SP6 and T7. Sequences were aligned using Mega5 [37].

Genotyping of domain II *kdr* alleles (V1016G and S989P): For genotyping of *kdr* alleles at residue S989 and V1016, separate allele-specific PCR assays were designed. The PCR conditions for both PCR were identical. The PCR was carried out using Thermo Scientific Dream Taq Hot Start DNA Polymerase ready reaction mixture. The thermal cycling conditions of PCR were: initial denaturation at 95°C for 3 min followed by 20 cycles, each of denaturation at 95°C for 15 s, annealing for 30°C at temperature starting from 65°C with an increment of -1°C each cycle and extension at 72°C for 60 s. In remaining 15 cycles, the cycling conditions were similar except annealing temperature was kept constant at 45°C. The primers used for S989-ASPCR were 1.0 μM of Aed2F, 1.5 μM of Aed2R, 0.6 μM of SSR and 0.45 μM of PPF. For V1016-ASPCR, primers used were 1.2 μM of Aed2F, 2.25 μM of Aed2R, 0.45 μM of VVF and 1.2 μM of GGR. The sizes of universal amplicon and allele- specific amplicons are as follows: universal amplicons: 620/636 bp, S989: 129 bp, 989P: 525/549 bp, V1016: 209 bp and 1016G: 446/462 bp. PCR products were electrophoresed on 2.0% agarose gel containing ethidium bromide and visualized under UV in gel documentation system.

Genotyping of domain III *kdr* alleles (T1520I, F1534C and F1534L): Earlier we reported a PCR-RFLP for the genotyping T1520I and F1534C, where a single PCR product amplified using primers AekdrF and AekdrR were subjected to RFLP using two restriction enzymes *BsaB*I and *Ssi*I respectively [9]. In order to include RFLP for identification of new mutation F1534L, we searched for F1534L-specific restriction enzyme site using an online tool available at http: //insilico.ehu.es/restriction/two_seq. We found a unique restriction site *Eco88*I in the DNA sequence with F1534L and therefore for genotyping of all *kdr* alleles present in domain III, PCR-RFLP method was modified. In modified procedure additional restriction-digestion was performed in a separate tube with *Eco88*I in addition to *BsaB*I and *Ssi*I. Additional restriction-digestion reaction mixture (20μl) contained 5 μl of PCR-amplified product, 2 units of *Eco88*I enzyme and 1X buffer (Thermo Fisher scientific). This was incubated for 4 hours or overnight at 37°C following electrophoretic run on 2.5% agarose gel. The RFLP product was visualized in gel documentation system. The criteria for scoring of F1534 alleles were modified which are presented in Table 3, while the criterion for scoring of T1520 alleles remains unchanged.

Statistical analysis: Hardy-Weinberg equilibrium (HWE) of *kdr*-alleles in a population was tested using Fisher's exact test. Phasing of haplotypes and estimation of their frequencies were performed using Arlequin 3.5 software [38]. Association of *kdr* genotypes with phenotype resistance was tested using co-dominant model and test of significance was estimated using Fisher's exact test and odds ratio (OR).

The Wright's inbreeding coefficient was calculated using formula F = (He-Ho)/He, where 'He' is expected heterozygosity and 'Ho' is observed heterozygosity.

RESULTS AND DISCUSSION

Identification of *kdr* mutations in *Ae. aegypti* population: DNA sequencing of partial domain II, domain III and domain IV of the VGSC of *Ae. aegypti* population collected from Bengaluru, India (77° 56-572 E, 12° 92-952 N), revealed the presence of four mutations i.e., S989P and V1016G in domain II and F1534C and F1534L in domain III. No mutation was found in domain IV. The mutation F1534L is being reported for the first time in Indian *Ae. aegypti*. Mutations S989P and V1016G present in domain II were due to substitution on first codon (TCC'!CCC) and second codon (GTA'!GGA), respectively. In most of the cases, the identification of S989 and V1016 mutations were based on 1X sequencing data where forward sequence was used for identification of S989 alleles and reverse sequence was used for V1016 alleles. This was due to presence of ambiguous sequence in downstream sequence resulting from multiple indels present in intron between these two *kdr* locus. In course of study a total of 294 samples were sequenced for partial domain II of which 178 were homozygous wild for both residues (SS at residue S989 and VV at residue V1016), 92 were heterozygous (SP and VG) and 24 were mutant homozygous (PP and GG). Other two alternative mutations F1534C and F1534L present in domain III were due to T>C substitution on the first position of the codon, leading to Phe (TTC)'!Leu (CTC) mutation, and T>G substitution on the second position of the codon leading to Phe (TTC)'!Cys (TGC) mutations, respectively. Out of the 27 individuals sequenced for domain III, one was homozygous wild for FF (TTC) at residue F1534, seven were homozygotes for CC (TGC), four were homozygotes for LL (CTC), four samples were heterozygotes for each of FC and FL and seven were having mixed bases at first and second position of the codon, i.e., with YKC, which could be either heterozygote for LC (CTC+TGC) or FR (TTC+CGC). The latter combination was ruled out as sequencing of 15 cloned PCR products, from five such samples (having sequence YKC), revealed the presence of two haplotypes, one with CTC (L) and another with TGC (C). We also observed that haplotype with F1534L mutation had a restriction site for *Eco88*I (5'-C"!YCGRG-3'). Therefore, all the seven heterozygote samples with the sequence YKC were subjected to PCR-RFLP with *Eco88*I and all were partially cleaved indicating the presence of LC (CTC+TGC) heterozygote. DNA sequencing of 12 samples for partial domain IV revealed absence of any non-synonymous mutation including D1794Y reported elsewhere [19]. Additionally, 25 samples were checked for presence of D1794Y using PCR-RFLP following Chang *et al.*, [19] and none were found positive.

Development of PCR-based assay for genotyping of *kdr* alleles: For genotyping of F1534-*kdr* alleles, we modified PCR-RFLP developed by Kushwah *et al.*, (2015) [18] where an additional restriction enzyme *Eco88*I was used for identification of new allele F1534L. For genotyping of S989- and V1016-*kdr* alleles, we developed allele-specific PCRs (ASPCR) for each locus. Detailed methods are described in section Material and Methods. These methods were validated through DNA sequencing. The genotyping results of PCR- RFLP based methods developed were well in agreement with DNA sequencing results of corresponding samples. The genotyping result of 27 samples that were sequenced for partial domain III of the VGSC (FF=1, CC=7, LL=4, FL=4, FC=4, LC=7) and 294 samples sequenced for domain II (SS/VV=178, PP/GG=24, SP/VG=92) matched with DNA sequencing results.

Genotyping of *kdr* alleles and their linkage association: Genotyping result of *kdr* alleles at loci F1534, S989 and V1016 carried out on 572 field collected F_0 populations are shown in **Table** 1. The overall frequencies of F1534 (F), F1534C (C), F1534L (L), S989 (S), S989P (P), V1016 (V) and V1016G (G) were 0.32, 0.51, 0.17, 0.82, 0.18, 0.82 and 0.18, respectively, in collections carried out in year 2014 and 2015. The T1520I was absent in this population. It was observed that the frequency of observed F1534- *kdr* genotypes are not in Hardy-Weinberg equilibrium (HWE) while S989 and V1016-*kdr* genotypes were in HWE in pooled samples. Analysis of inbreeding coefficient for F1534-*kdr* alleles revealed positive value indicating presence of fewer of heterozygotes than expected (F=0.09). Additionally, 242 F_1 individuals were also genotyped for *kdr* alleles. Thus, a total of 814 samples (F0 and F1 combined) were genotyped for *kdr* alleles present in domain II and III. Estimation of gametic phase based on Gibbs sampling strategy implemented in Arlequin 3.5 [36] revealed the presence of only 4 haplotypes namely, FSV (wild), CSV, LSV and FPG(underlined are mutated alleles). This indicates negative linkage disequilibrium of mutant *kdr* alleles (P and G) present in domain II with mutant *kdr* alleles in domain III (C and L). The two mutations in domain II, i.e., 989P and 1016G were in complete linkage disequilibrium.

Genetic association of *kdr* alleles with phenotype insecticide resistance: Since mutations present in domain II (S989P and V1016G) were always found in association with wild form in domain III (F1534) and have been reported to have protection with insecticides, we removed data of individual having haplotype FPG from analysis of genetic association of F1534-*kdr* mutations with phenotypic insecticide resistance. Similarly, data of individual with CSV and LSV haplotypes were removed while analysing association of S989P/V1016G *kdr* mutations with phenotypic insecticide resistance. The analysis of genetic association of F1534-*kdr* alleles with phenotype insecticide resistance is shown in Table 2. Statistical analysis revealed that allele F1534L showed strong protection against permethrin ($P < 0.0001$), moderate protection against deltamethrin ($P < 0.01$) and very low protection against DDT ($P < 0.05$). Other allele 1534C showed strong protection against permethrin ($P < 0.0001$), low protection against DDT ($P < 0.001$) and no protection against deltamethrin. Mutations 989P and 1016G together have strong protection to permethrin and we could not establish their role against deltamethrin and DDT due to low frequency of alleles at S989 and V1016 locus.

Aedes aegypti, a primary vector of dengue, yellow fever, chikungunya, Zika and other arboviral infections has attained global importance due to invasion of this species in different parts of the world from Africa. Currently, with no specific treatment or vaccine available to control transmission of these arboviral infections, vector control remains the mainstay in public health. One of the available potential methods for control of this vector is the use of pyrethroid group of insecticides for space spray and personal protection measure. Development of resistance against these insecticides is major concern for success of such application. It is therefore desirable to have knowledge on underlying mechanism(s) of insecticide resistance for an effective insecticide resistance management.

Reduced sensitivity of VGSC—the target site for DDT and pyrethroids, is one of the mechanisms of resistance in insects due to conformational changes in protein resulting from one or more mutations leading to amino acid substitution, commonly referred to as knockdown resistance (*kdr*). In *Ae. aegypti,* several such mutations are documented, of which F1534C, S989P and V1016G are widely reported *kdr* mutation and is known to confer resistance against DDT and pyrethroids [20,21,22]. This study reports the presence of four *kdr* mutations viz. S989P, V1016G, F1534C and F1534L with the latter mutation is being reported for the first time in *Ae. aegypti*. Although its presence in *Aedes albopictus* has been reported [23,24,25]. The new mutation F1534L has shown significantly higher protection against permethrin (type I pyrethroid), deltamethrin (type II pyrethroid) and DDT in *Ae. aegypti*. The discovery of new *kdr* mutations associated with insecticide resistance is of global concern as this may hamper pyrethroid-based vector control and personal protection measures.

Earlier we conducted a survey of *kdr* mutations on Delhi population in year 2014 [18], where we found F1534C mutation along with a novel mutation T1520I linked to F1534C but did not find the three mutations being reported in this study, i.e., S989P, V1016G and F1534L. Similarly, a novel mutation T1520I reported in Delhi was absent from Bengaluru. To ensure that we did not skip detection of the other three mutations (S989P,V1016G and F1534L) in Delhi population during previous study, we genotyped 184 *Ae. aegypti* samples collected from Delhi in August 2018. We did not find any of these three mutations in Delhi population. Thus, there is a contrast difference in distribution of *kdr* alleles in two different geographical locations which are approximately 1700 km apart. In another part of India (West Bengal, eastern India), there is report of presence of three mutations, F1534C, T1520I and V1016G but the presence of S989P was not investigated in this study [26].

For monitoring of *kdr* mutations in field condition, we developed a highly specific PCR- RFLP-based assay for simultaneous detection of all the three mutations (total five alleles) reported in domain III-S6 of the VGSC at locus F1534 and T1520. The PCR-RFLP for the identification of all five alleles is advantageous over other PCR-based methods being highly specific due to high sequence-specificity of restriction enzymes. Additionally, in this PCR- RFLP assay, unlike other assays, a single PCR is required for genotyping of all five alleles present at locus F1534 and T1520. For genotyping of domain II-S6 mutations, we developed two allele-specific PCR (ASPCR) assays, one each for S989P and V1016G mutations. Our ASPCRs for genotyping of S989P and V1016G alleles are advantageous over other available PCR assay for these mutations, as our PCR need single assay for each locus while method described by Stenhouse *et al.*, (2013) [27] need two PCR assays for single locus. However, ASPCR being based on single base mismatch is prone to non-specific extension and high degree of optimization is required. ASPCR is sensitive to change of type of reagents and PCR thermal conditioning. Discrepancies were noticed with such PCR based genotyping of F1534 *kdr*-alleles in an earlier study [28].

Co-occurrence of F15134C with S989 and V1016 may be of serious concern in case all three mutations are present on same haplotype. It has been shown through the site directed mutagenesis that such combination (F1534C+S989P+V1016G) may result in a seriously high degree of resistance against permethrin and deltamethrin (1100- and 90-fold, respectively) [29]. In this study, we found that 989P, 1016G and 1534C/ 1534L were found together in a single mosquito but always in heterozygous condition. Phasing out of haplotypes revealed that there are just four haplotypes FSV, FPG, LSV and CSV present in this population. Thus, 989P and 1016G are always found together but never with 1534C or 1534L. In such a population, a single recombination may lead to production of haplotype CPG or LPG which may have greater impact on insecticide resistance phenotype. Unlike our finding, a small proportion of mosquitoes in Myanmar have shown to have 1534C homozygotes with 1016G homozygotes (2.9%, double mutant) and with homozygous 989P+1016G+1534C (0.98%, triple mutant) suggesting presence of CPG haplotype [30]. Occurrence of such haplotype with three mutations may be selected in presence of insecticide pressure and may pose serious threat to insecticide-based vector control programme.

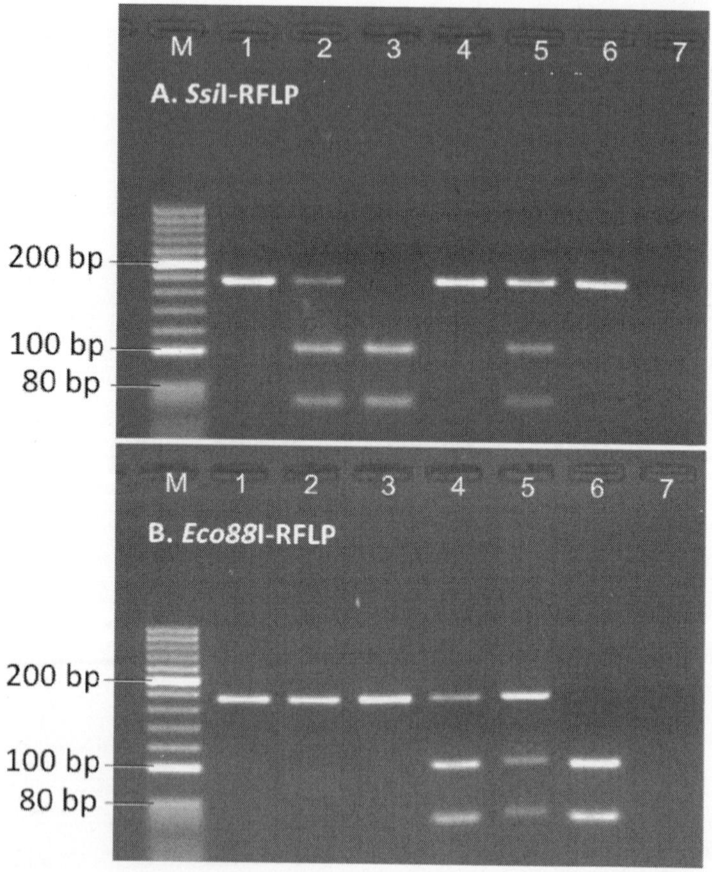

Figure 1 Gel photographs of (A) *Ssi*I- and (B) *Eco88*I-PCR-RFLP digests showing banding pattern of the different F1534-*kdr* genotypes. Lane M represents 20 bp ladder, lanes 1 to 6 represents genotypes FF, FC, CC, FL, LC and LL respectively; lane 7 represents negative control.

Table 1: Genotyping result of different *kdr* alleles

Collection date	N	*kdr* genotypes												Allele frequencies							*P*HWE		
		F1534						S989			V1016			F1534			S989		V1016		F 1534	S989	V1016
		FF	FC	CC	FL	LL	LC	SS	SP	PP	VV	VG	GG	F	C	L	S	P	V	G			
Oct-Nov 2014	148	12	26	20	20	8	62	120	20	8	120	20	8	0.24	0.43	0.33	0.88	0.12	0.88	0.12	0.00	0.00	0.00
Mar-Apr 2015	424	61	151	110	22	0	80	270	135	19	270	135	19	0.35	0.53	0.120	0.80	0.20	0.80	0.20	0.00	0.92	0.92
Total	572	73	177	130	42	8	142	390	155	27	390	155	27	0.32	0.51	0.17	0.82	0.18	0.82	0.18	0.00	0.09	0.09

P_{HWE}= test of significance for deviation from Hardy-Weinberg equilibrium

Table 2: Distribution of *kdr* alleles in dead and alive mosquitoes (F_0 and F_1) following one-hour exposure to the insecticide impregnated papers and test of genetic association of *kdr*–alleles with insecticide resistance

Insecti cide		Alleles at locus					Test of genetic association					
		F1534*			S989/V1016#		Fisher's exact test			OR (fiducial limit at 95% CI)		
		F	C	L	S/V	P/G	F vs C	F vs L	S/V vs P/G	F vs C	F vs L	S/V vs P/G
DDT 4%	Alive	44	149	23	22	16						
	Dead	12	8	0	5	2	p <0.001	p <0.05	NS	5.08 (1.95-13.21)	Infinity	1.82 (0.31-10.59)
PER 0.75%	Alive	25	236	89	5	22						
	Dead	38	74	24	16	6	p <0.0001	p <0.0001	p <0.0001	4.85 (2.75-8.56)	5.64 (2.86-11.09)	11.73 (3.04-45.27)
DEL 0.05%	Alive	77	190	75	31	10						
	Dead	46	79	17	24	2	NS	p <0.01	NS	1.44 (0.92-2.25)	2.64 (1.39-5.00)	3.87 (0.77-19.35)

*Excluding individuals having mutant alleles 989P/1016G; #excluding individuals having mutant alleles 1534C/L

OR= Odd Ratio; CI= confidence interval; DDT=dichloro diphenyl trichloroethane; PER=permethrin; DEL= deltamethrin

Table 3. Criteria for scoring of F1534 alleles in PCR-RFLP assays

F1534 genotypes	Size of PCR-RFLP bands in	
	*Eco88*I-RFLP	*Ssi*I-RFLP
FF	171	171
FC	171	171, 103 and 68
FL	171, 103 and 68	171
CC	171	103 and 68
CL	171, 103 and 68	171, 103 and 68
LL	103 and 68	171

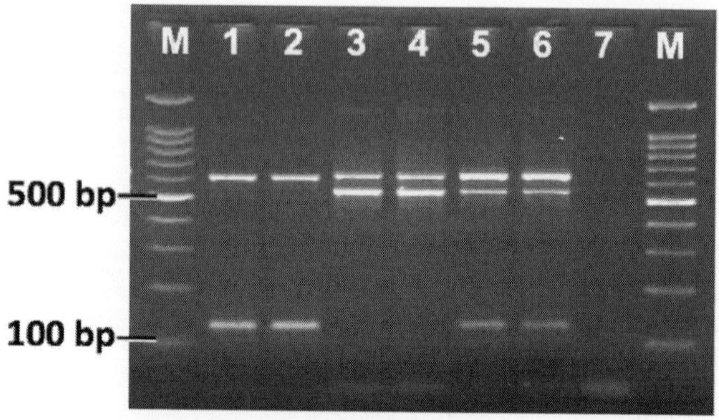

Figure 2: Gel photographs of allele-specific PCR for identification of S989-*kdr* alleles. Lane M represents 100 bp ladder, lanes 1 and 2 represents genotype SS, 3 and 4 PP, 5 and 6 SP and lane 7 represents negative control.

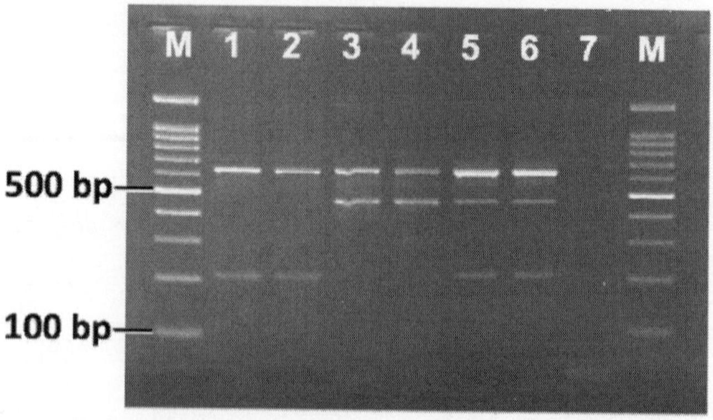

Figure 3: Gel photographs of allele-specific PCR for identification of V1016-*kdr* alleles. Lane M represents 100 bp ladder, lanes 1 and 2 represents genotype VV, 3 and 4 GG, 5 and 6 VG and lane 7 represents negative control.

In this study we found 100% linkage association of 1016G with 989P [22,30,31,32]. Though, such association has been shown in several studies, the frequency of V1016G is reported higher than S989P where V1016G mutation may be found alone but S989P was always linked with V1016G [27,29,30,32]. Similar unidirectional linkage has been shown in domain III in VGSC of *Ae. aegypti* population, where a novel mutation T1520I (reported in Delhi, India) was always found with F1534C and was suggested to be a compensatory mutation [18].

A disturbing fact we recorded in this study was the non-compliance of HWE for F1534-*kdr* mutations in Bengaluru. Similar departure from HWE for this locus was also noted in our previous study of Delhi-population [18] and in Grand Cayman Island [33]. It was interesting to note that another mutation T1520I that was linked to F1534C, was in compliance with HWE. While we don't have an explanation to this, probably remains this may be due to the presence of multiple VGSC gene or gene duplication as proposed by Martin *et al.*, [34]. This will need a further investigation.

In conclusion, this study for the first time reports the presence of F1534L mutation in an *Ae. aegypti* population which is associated with pyrethroid resistance alongside presence of three other mutations F1534C, S989P and V1016G. Molecular methods were developed for monitoring of all these *kdr* mutations. Owing to the protection conferred by F1534L mutation, it is of paramount importance to screen other populations of *Ae. aegypti* for this mutation besides *Ae. albopictus* which also plays important role in transmission of several arboviral infections.

REFERENCES

1. Al Nazawi, A. M., Aqili, J., Alzahrani, M., McCall, P. J., & Weetman, D. (2017). Combined target site (kdr) mutations play a primary role in highly pyrethroid resistant phenotypes of *Aedes aegypti* from Saudi Arabia. *Parasites & Vectors,* 10(1), 161.

2. Amelia-Yap, Z. H., Chen, C. D., Sofian-Azirun, M., & Low, V. L. (2018). Pyrethroid resistance in the dengue vector *Aedes aegypti* in Southeast Asia: present situation and prospects for management. *Parasites & Vectors,* 11(1), 1-17.

3. Beaty, B. J., Black, W. C., Eisen, L., Flores, A. E., García-Rejón, J. E., Loroño-Pino, M., & Saavedra-Rodriguez, K. (2016). The intensifying storm: domestication of *Aedes aegypti,* urbanization of arboviruses, and emerging insecticide resistance. In Global Health Im-pacts of Vector-Borne Diseases: Workshop Summary, National Academies Press.

4. Bhatt, S., Gething, P. W., Brady, O. J., Messina, J. P., Farlow, A. W., Moyes, C. L., & Myers, M. F. (2013). The global distribution and burden of dengue. *Nature,* 496(7446), 504-507.

5. Brady, O. J., Gething, P. W., Bhatt, S., Messina, J. P., Brownstein, J. S., Hoen, A. G., ... & Hay, S. I. (2012). Refining the global spatial limits of dengue virus transmission by evidence-based consensus. *PLoS Negl Trop Dis,* 6(8), e1760.

6. Brengues, C., Hawkes, N. J., Chandre, F., McCarroll, L., Duchon, S., Guillet, P., & Hemingway, J. (2003). Pyrethroid and DDT cross resistance in *Aedes aegypti* is correlated with novel mutations in the voltage gated sodium channel gene. *Medical and Veterinary Entomology,* 17(1), 87-94.

7. Chang, C., Shen, W. K., Wang, T. T., Lin, Y. H., Hsu, E. L., & Dai, S. M. (2009). A novel amino acid substitution in a voltage-gated sodium channel is associated with knockdown resistance to permethrin in *Aedes aegypti. Insect Biochemistry and Molecular Biology,* 39(4), 272-278.

8. Chang, C., Huang, X. Y., Chang, P. C., Wu, H. H., & Dai, S. M. (2012). Inheritance and stability of sodium channel mutations associated with permethrin knockdown resistance in *Aedes aegypti. Pesticide Biochemistry and Physiology,* 104(2), 136-142.

9. Dev, V., Khound, K., & Tewari, G. G. (2014). Dengue vectors in urban and suburban Assam, India: entomological observations. WHO South-East Asia *Journal of Public Health,* 3(1), 51-59.

10. Excoffier, L., & Lischer, H. E. (2010). Arlequin suite ver 3.5: a new series of programs to perform population genetics analyses under Linux and Windows. *Molecular Ecology Resources*, 10(3), 564-567.

11. Harris, A. F., Rajatileka, S. & Ranson, H. (2010). Pyrethroid resistance in *Aedes aegypti* from Grand Cayman. *Am. J. Trop. Med. Hyg.* 83, 277–284.

12. Hirata, K., Komagata, O., Itokawa, K., Yamamoto, A., Tomita, T., & Kasai, S. (2014). A single crossing-over event in voltage-sensitive Na+ channel genes may cause critical failure of dengue mosquito control by insecticides. PLoS Negl Trop Dis, 8(8), e3085.

13. Kawada, H., Oo, S. Z. M., Thaung, S., Kawashima, E., Maung, Y. N. M., Thu, H. M., ... & Minakawa, N. (2014). Co-occurrence of point mutations in the voltage-gated sodium channel of pyrethroid-resistant Aedes aegypti populations in Myanmar. PLoS Negl Trop Dis, 8(7), e3032.

14. Kushwah, R. B. S., Dykes, C. L., Kapoor, N., Adak, T., & Singh, O. P. (2015). Pyrethroid-resistance and presence of two knockdown resistance (kdr) mutations, F1534C and a novel mutation T1520I, in Indian Aedes aegypti. PLoS Negl Trop Dis, 9(1), e3332.

15. Leta, S., Beyene, T. J., De Clercq, E. M., Amenu, K., Kraemer, M. U., & Revie, C. W. (2018). Global risk mapping for major diseases transmitted by *Aedes aegypti* and *Aedes albopictus. International Journal of Infectious Diseases*, *67*, 25-35.

16. Li, Y., Xu, J., Zhong, D., Zhang, H., Yang, W., Zhou, G., ... & Yan, G. (2018). Evidence for multiple-insecticide resistance in urban *Aedes albopictus* populations in southern China. *Parasites & Vectors,* 11(1), 4.

17. Livak, K. J. (1984). Organization and mapping of a sequence on the Drosophila melanogaster X and Y chromosomes that is transcribed during permatogenesis.

Genetics, 107(4), 611-634.

18. Marcombe, S., Farajollahi, A., Healy, S. P., Clark, G. G., & Fonseca, D. M. (2014). Insecticide resistance status of United States populations of Aedes albopictus and mechanisms involved. *PloS one*, 9(7), e101992.

19. Martins, A. J., Brito, L. P., Linss, J. G. B., Rivas, G. B. D. S., Machado, R., Bruno, R. V., ... & Peixoto, A. A. (2013). Evidence for gene duplication in the voltage-gated sodium channel gene of Aedes aegypti. *Evolution, medicine, and public health*, 2013(1), 148-160.

20. Martins, A. J., Lima, J. B. P., Peixoto, A. A., & Valle, D. (2009). Frequency of Val1016Ile mutation in the voltage gated sodium channel gene of *Aedes aegypti* Brazilian populations. *Tropical Medicine & International Health*, *14*(11), 1351-1355.

21. Martins, A. J., de Andrade Lins, R. M. M., Linss, J. G. B., Peixoto, A. A., & Valle, D. (2009). Voltage-gated sodium channel polymorphism and metabolic resistance in pyrethroid-resistant *Aedes aegypti* from Brazil. *The American journal of tropical medicine and hygiene*, 81(1), 108-115.

22. Plernsub, S., Saingamsook, J., Yanola, J., Lumjuan, N., Tippawangkosol, P., Sukontason, K., & Somboon, P. (2016). Additive effect of knockdown resistance mutations, S989P, V1016G and F1534C, in a heterozygous genotype conferring pyrethroid resistance in *Aedes aegypti* in Thailand. *Parasites & Vectors*, 9(1), 417.

23. Plernsub, S., Saingamsook, J., Yanola, J., Lumjuan, N., Tippawangkosol, P., Walton, C., & Somboon, P. (2016). Temporal frequency of knockdown resistance mutations, F1534C and V1016G, in *Aedes aegypti* in Chiang Mai city, Thailand and the impact of the mutations on the efficiency of thermal fogging spray with pyrethroids. *Acta tropica,* 162, 125-132.

24. Rajatileka, S., Black IV, W. C., Saavedra-Rodriguez, K., Trongtokit, Y., Apiwathnasorn, C., McCall, P. J., & Ranson, H. (2008). Development and application of a simple colorimetric assay reveals widespread distribution of sodium channel mutations in Thai populations of *Aedes aegypti. Acta tropica*, 108(1), 54-57.

25. Saavedra Rodriguez, K., Urdaneta Marquez, L., Rajatileka, S., Moulton, M., Flores, A. E., Fernandez Salas, I., & Ranson, H. (2007). A mutation in the voltage gated sodium channel gene associated with pyrethroid resistance in Latin American *Aedes aegypti*. *Insect molecular biology*, *16*(6), 785-798.

26. Saha, P., Chatterjee, M., Ballav, S., Chowdhury, A., Basu, N., & Maji, A. K. (2019). Prevalence of kdr mutations and insecticide susceptibility among natural population of *Aedes aegypti* in West Bengal. *Plos one,* 14(4), e0215541.

27. Singh, H., Singh, O. P., Akhtar, N., Sharma, G., Sindhania, A., Gupta, N., & Valecha, N. (2019). First report on the transmission of Zika virus by Aedes (Stegomyia) aegypti (L.)(Diptera: Culicidae) during the 2018 Zika outbreak in India. Acta tropica, 199, 105114.

28. Srisawat, R., Komalamisra, N., Eshita, Y., Zheng, M., Ono, K., Itoh, T. Q., & Rongsriyam, Y. (2010). Point mutations in domain II of the voltage-gated sodium

channel gene in deltamethrin-resistant Aedes aegypti (Diptera: Culicidae). Applied Entomology and Zoology, 45(2), 275-282.

29. Stenhouse, S. A., Plernsub, S., Yanola, J., Lumjuan, N., Dantrakool, A., Choochote, W., & Somboon, P. (2013). Detection of the V1016G mutation in the voltage-gated sodium channel gene of *Aedes aegypti* (Diptera: Culicidae) by allele-specific PCR assay, and its distribution and effect on deltamethrin resistance in Thailand. *Parasites & Vectors*, 6(1), 253.

30. Tamura, K., Peterson, D., Peterson, N., Stecher, G., Nei, M., & Kumar, S. (2011). MEGA5: molecular evolutionary genetics analysis using maximum likelihood, evolutionary distance, and maximum parsimony methods. *Molecular Biology and Evolution,* 28(10), 2731-2739.

31. Wilder-Smith, A., Gubler, D. J., Weaver, S. C., Monath, T. P., Heymann, D. L., & Scott, T. W. (2017). Epidemic arboviral diseases: priorities for research and public health. *The Lancet Infectious Diseases*, *17*(3), e101-e106.

32. World Health Organization, (1998). Test procedures for insecticide resistance monitoring in malaria vectors, bio-efficacy and persistence of insecticides on treated surfaces. Report of the WHO Informal Consultation, 28 September (1998, WHO/ HQ, Geneva, World Health Organization, HO/CDS/CPC/MAL/98.12.

33. World Health Organization. (1989). *The use of impregnated bednets and other materials for vector-borne disease control: a report of the WHO* (No. WHO/VBC/ 89.981, Rev. 1. Unpublished). World Health Organization.

34. Xu, J., Bonizzoni, M., Zhong, D., Zhou, G., Cai, S., Li, Y., ... & Duan, J. (2016). Multi-country survey revealed prevalent and novel F1534S mutation in voltage-gated sodium channel (VGSC) gene in Aedes albopictus. PLoS neglected tropical diseases, 10(5), e0004696.

35. Yadav, P. D., Malhotra, B., Sapkal, G., Nyayanit, D. A., Deshpande, G., Gupta, N., ... & Mourya, D. T. (2019). Zika virus outbreak in Rajasthan, India in 2018 was caused by a virus endemic to Asia. *Infection, Genetics and Evolution,* 69, 199-202.

36. Yanola, J., Somboon, P., Walton, C., Nachaiwieng, W., Somwang, P., & Prapanthadara, L. A. (2011). High throughput assays for detection of the F1534C mutation in the voltage gated sodium channel gene in permethrin resistant *Aedes aegypti* and the distribution of this mutation throughout Thailand. *Tropical Medicine & International Health,* 16(4), 501-509.

37. Yanola, J., Somboon, P., Walton, C., Nachaiwieng, W., & Prapanthadara, L. A. (2010). A novel F1552/C1552 point mutation in the *Aedes aegypti* voltage-gated sodium channel gene associated with permethrin resistance. *Pesticide Biochemistry and Physiology*, 96(3), 127-131.

Current Status of Researches in Biosciences (2020) : 157-167
Editor : Prof. P.C. Joshi, N. Joshi, Reshman Yasmin, and D. K. Mansotra
Today & Tomorrow's Printers and Publishers, New Delhi - 110 002

EVALUATION OF KETAMINE AND KETOFOL ANESTHESIA IN DOGS UNDERGOING OVARIOHYSTERECTOMY

Kalaiselvan, E., M. Hoque*, S .K. Maiti and Rohit Kumar

Division of Surgery, ICAR-Indian Veterinary Research Institute, Izatnagar-243122 (UP)

Email: mhoq61@yahoo.com

ABSTRACT

In this study twelve female dogs for elective ovariohysterectomy were randomized for administration of ketamine (group A) and ketofol 1:1 (group B) at the dose rate of 300 µg per kg per minute during maintenance premedicated with atropine-dexmedetomidine and butorphanol. Physiological and haemodynamic observations were recorded at 0 min and 15, 20, 30, 45, 60, 75, 90, 105, 120 min. There was a significant change in physiological, haemodynamic observations. Dose of induction of group B (3.7±1.57 mg/kg) is lower than group A (5.47±1.02 mg/kg) and quality of recovery is smooth in group B than group A. It is reported that addition of ketamine in propofol as a 1:1 mixture of ketofol reduces the need of propofol for induction. Both ketamine and ketofol as a CRI-TIVA maintains stable basic vital functions during anaesthesia. Despite there is significant edge over advantage of ketofol in physiological observations as compared to ketamine in terms of maintaining vital functions merely baseline and clinically acceptable range, recovery time and recovery quality. In addition administration of oxygen indicated for both the groups due to decreased oxygen saturation of haemoglobin.Hence it is concluded that ketofol 1:1 having advantages while using CRI than ketamine alone.

KEYWORDS: Ketamine, Propofol, Ketofol, Constant rate infusion, Dogs, Anesthesis, Ovariohysterectomy.

INTRODUCTION

Inhalant anaesthetic pollution is a common greenhouse gas emission with global health problem. Despite prevention of greenhouse emission is impractical, can be minimised by maintaining anaesthesia using intravenous anaesthetic alone as a total intravenous anaesthesia (TIVA). Appropriate premedication with the use of anaesthetic agent will preserve vital functions. Propofol is a sedative hypnotic agent commonly used in both human and animals. Bradycardia, apnea, hypotension and vascular pain are its adverse effects, but it produces smooth induction (Sano *et al.*, 2003). Ketamine is an injectable anaesthetic which dissociates thalamic and limbic system. But adverse effects are catalepsy and emergence reactions during recovery period. Ketofol is a mixture of ketamine and propofol, gained popularity in human procedural sedation, emergency departments and pediatric anaesthesia (Mostafa *et al.*, 2015). Ketamine–propofol mixture has cumulative advantages than disadvantages of ketamine and propofol used alone (Yan *et al.*, 2015). For maintenance of anaesthesia using continuous constant rate of infusion (CRI) produces stable anaesthesia than intermittent bolus administration (Müller *et al.,* 2011). Most of the comparative studies pertaining to ketofol were compared with propofol (Kennedy and Smith, 2015; Saikia *et al*., 2016 and Martinez Taboada and Leece, 2014). However, studies concentrated on CRI of ketofol as TIVA comparision to ketamine are not available. Hence the objective of our study is to compare changes of vital functions and induction dose reduction by TIVA using CRI of ketamine and ketofol 1:1.

MATERIALS AND METHODS

The research work was done on twelve client owned healthy dogs irrespective of breed, age and body weight which were presented to Teaching Veterinary Clinical Complex, IVRI, Izatnagar for elective ovariohysterectomy. Study approval was obtained from competent authority at Indian Veterinary Research Institute (IVRI), Izatnagar. Written informed consent was obtained from the dog owners. Pre-operative heart rate (HR), respiratory rate (RR) and rectal temperature (RT) were recorded at 0 min (before administration) of drug atropine sulphate (Tropine; Neon Laboratories, Palghar, Thane, India) at the dose rate of 0.04 mg per kg body weight on right lumbar epaxial muscle intramuscularly. Followed by at 5 min dexmedetomidine (Dextomid; Neon Laboratories, Palghar, Thane, India) at the dose rate of 15 ug/kg body weight) and butorphanol (Butodol; Neon Laboratories, Palghar, Thane, India) at the dose rate of 0.2 mg/kg body weight on left epaxial muscle intramuscularly. At 15 min dogs were induced by ketamine (Ketmine 50; Themis Medicare Limited, Uttarakhand, India) as Group A and ketofol 1:1 (preparation procedure in appendix) single syringe mixture as Group B. Induction protocol as follows: induction drug calculated 1 mg per kg body weight. From that half of the dose infused in line over 15 seconds, followed by wait 30 seconds till abolition of pedal reflex. If pedal reflex was intact 50 % of half of dose infused till abolishment. The total dose was calculated as dose of induction. The dogs were intubated and allowed to spontaneously breathe room air immediately after induction, constant CRI of anaesthetic at the dose rate of 300 μg per kg per min till last skin sutured along with maintenance fluid normal saline by micro infusion set. Anaesthetic depth was monitored at least once every minute and if a given dog's plane of anaesthesia was deemed too light for the patient to remain intubated, then a bolus of 50% of induction dose was administered using the syringe port. Physiological observations heart rate, respiratory rate, rectal temperature, arterial blood pressures {systolic arterial pressure (SAP), diastolic arterial pressure(DAP) and mean arterial pressure(MAP)}, oxygen saturation of haemoglobin (SpO2), duration of anaesthesia (time of complete abolition of pedal reflex to reappearance of pedal reflex in minutes), recovery time (the time from the end of administration of constant rate infusion to time of appearance of pedal reflex in minutes), sternal recumbency time (the time from the end of administration of constant rate infusion to the spontaneous regaining of sternal recumbency in minutes) and standing time (the time from the end of administration of constant rate infusion to spontaneous regaining of standing position in minutes) at 0 min and 15, 20, 30, 45, 60, 75, 90, 105 and 120 min after administration of drug. Quality of recovery was evaluated by simple descriptive score scale (table 1). The data were analysed by repeated measures *t*- test to within group between time intervals and one way ANOVA to compare between groups at different time periods.

Table 1 Simple descriptive scale for quality of recovery

1	Very smooth: no excitement, vocalization, trembling or vomiting. no convulsions
2	Quite smooth: A little excitement, no paddling, vocalization, trembling or vomiting. no convulsions
3	Moderately smooth: Excitement, some paddling, vocalization, trembling or vomiting observed. Convulsions
4	Poor: not smooth and with excitement, paddling, vocalization, trembling or vomiting observed. Convulsions

RESULTS AND DISCUSSION

Animals of Group B showed 32.35 % significant ($P<0.05$) reduction in dose of induction as compared to group A. There was additional dose requirement in one of the cases within group A at 47 min and 53 min, while in one of the cases within group B at 47 min and 69 min for maintenance of anaesthesia (Table 2).

Table 2: Dose of induction mg/kg

Group	Dose of induction (milligram/kg)
A	5.47±1.02[a]
B	3.7±1.57[b]

Values with different alphabets differ significantly ($p<0.05$) at corresponding intervals.

In the current study equal concentration mixture of ketamine and propofol ratio 1:1 result 32.35% reduction in dose rate of induction than of ketamine for complete loss of pedal reflex. The dose of induction in its present study with ketofol 1:1 was lower than unpremedicated experimental study (Kennedy and Smith, 2015). The present finding of induction dose in ketofol 1:1 ratio was similar to acepromazine and methadone premedicated healthy dogs (Martinez Taboada and Leece, 2014). To cornering global green gas emissions and the impact on atmosphere by reduce usage (Feldman, 2012) and explore to better intravenous anaesthetics with low dose requirement and less impact on patients health status (Vollmer *et al.*, 2015).

There was a significant drop of heart rate during induction in group B and increased heart rates during CRI maintenance noticed in both the groups. In most of the time periods during maintenance group A had significantly higher heart rate than group B (Table- 3). This may be due to highly lipid soluble propofol reaches one arm brain circulation faster than ketamine (Dinis-Oliveira, 2018). Thus makes drop of heart rate during induction or first loading dose. But during CRI infusion ketamine predominates propofol leads to increase of heart rate. The increase heart rate in both the groups of present study is due to Ketamine which increases sympathetic nervous system activity, resulting in increased cardiovascular functions (Adams *et al.*, 1977).

The non-significant fall of respiratory rate noticed in both the groups during and post-induction and postoperative period. But during maintenance group A had significantly higher respiratory rate than group B (table 4).

Table 3: Showing heart rate(beats/min) at different intervals.

Group	T0	T15	T20	T30	T45	T60	T75	T90	T105	T120
A	98.17 ± 4.37	$109.17\pm3.47^{a**}$	$174.17\pm5.96^{a**}$	$170.67\pm4.99^{a**}$	$171.5\pm2.29^{**}$	$165\pm3.31^{a**}$	$158\pm2.63^{**}$	109.17 ± 5.42^{a}	$76\pm3.55^{a*}$	$79.33\pm2.73^{a*}$
B	98.50 ± 5.43	$77.17\pm2.6^{b*}$	$124.33\pm3.7^{b**}$	$134\pm3.28^{b**}$	$163\pm4.55^{**}$	$149.67\pm3.92^{b**}$	$147.83\pm4.13^{**}$	$148\pm3.76^{b**}$	$151.67\pm4.81^{b**}$	$149.5\pm2.29^{b**}$

Values with different alphabets differ significantly ($p<0.05$) at corresponding intervals.

*Differ significantly ($p<0.05$) from base value. ** Differ significantly ($p<0.01$) from base value.

Table 4: Showing respiratory rate (per min) at different time intervals

Group	T0	T15	T20	T30	T45	T60	T75	T90	T105	T120
A	32.67 ± 4.6	$15.00\pm1.06^{**}$	$15.17\pm1.66^{**}$	$21.50\pm1.5^{a*}$	25.17 ± 1.96^{a}	24.17 ± 0.91^{a}	22.67 ± 2.11^{a}	24.17 ± 2.26	23.17 ± 2.23	21.67 ± 1.23^{a}
B	30.83 ± 3.34	$17.50\pm1.88^{**}$	$12.33\pm1.09^{**}$	$13.00\pm1.18^{b**}$	$19.50\pm1.52^{b**}$	$14.83\pm1.17^{b**}$	$17.17\pm1.49^{b**}$	$18.83\pm2.6^{*}$	23.00 ± 1.34	33.00 ± 2.77^{b}

Values with different alphabets differ significantly ($p<0.05$) at corresponding intervals.

*Differ significantly ($p<0.05$) from base value. ** Differ significantly ($p<0.01$) from base value

In both the groups' minimal depression of respiratory drive during induction and up to 20 min were recorded. This may be due to dose dependant depression (Sano *et al.*, 2003 and Kennedy and Smith, 2015) produced by both anaesthetic agents while administering as loading dose. But during maintenance of CRI ketamine alone preserves respiratory drive with mild respiratory depression than CRI of equal ketamine and propofol. At 120 min ketofol makes respiratory rate merely equal to base line and higher than ketamine group which may be due to rapid clearance of propofol and low level of ketamine in the body. The chain of reduction in rectal temperature of both the groups as observed in this study is common (Table 5). Because in general during general anaesthesia animal is unable to maintains it body temperature (Sessler, 2008).

During induction and post induction time period, there was significant decrease in systolic blood pressure in both the groups as compared to pre values. But during maintenance and post-maintenance up to 120 min significant rise in systolic blood pressure as compared to 0 min was noticed (Table 6).

Both the groups showed increase of diastolic blood pressure as compared to pre value, however, the changes were non-significant. (Table 7).

Group A had higher and significant difference in mean arterial blood pressure during maintenance of CRI than group B. At the same time both the group showed crescendo of mean blood pressure as compared to pre and during induction values (Table 8).

The base line value of an animal's blood pressure shows higher than reference range. Since in our study many healthy individuals may be quite anxious to new environment and procedures which may be attributed to it. Arterial pressure is a product of stroke volume and systemic vascular resistance. Hypotensive effect of propofol wanes the haemodynamics into lower than baseline, especially arterial blood pressure during induction. The decrease of heart rate while induction an additional evidence of arterial baroreceptor reflex mechanism blunting by sudden loading of a propofol (Ebert *et al.*, 1992 and Sato *et al.*, 2005). Thus in CRI of ketamine alone and propofol and ketamine mixture maintains arterial baroreceptor reflex mechanism in turns increases sympathetic flow causing merely hyperhaemodynamic state. Although both the groups show increased mean arterial pressure, both were in acceptable range and ketamine has higher MAP than ketofol 1:1. Our MAP results contrast with Kennedy and Smith, (2015), this is attributed by objective and use of study was different in

Table 5: Showing rectal temperature (°F) at different time intervals

Group	T0	T15	T20	T30	T45	T60	T75	T90	T105	T120
A	101.75±0.75	101.53±1.13	100.17±0.8*	99.27±0.66**	98.93±0.76**	98.15±1.1**	97.6±1.09**	97.18±1.17**	96.97±0.92**	95.93±1.31**
B	102.17±0.6	100.88±1.12**	99.82±1.1**	99.08±1.06**	99.2±0.48**	98.02±1.53**	97.43±1.8**	96.75±2.63**	96.03±2.72**	95.75±1.89**

Values with different alphabets differ significantly ($p<0.05$) at corresponding intervals.
*Differ significantly ($p<0.05$) from base value. ** Differ significantly ($p<0.01$) from base value .

Table 6: Showing systolic blood pressure (mm of Hg) at different time intervals

Group	T0	T15	T20	T30	T45	T60	T75	T90	T105	T120
A	150.33±7.64	190.67±7.31[a]**	172.33±7.12[a]*	179.50±8.07*	190.83±3.98**	183.50±4.4*	179.00±5.84*	168.67±4.12	162.33±4.92	160.83±3.95
B	151.83±4.58	130.50±3.99[b]**	147.17±3.66[b]	169.33±4.52	177.83±4.24*	175.00±3.78*	163.50±4.65	156.67±5.15	158.83±4.69	158.67±5.32

Values with different alphabets differ significantly ($p<0.05$) at corresponding intervals.
*Differ significantly ($p<0.05$) from base value. ** Differ significantly ($p<0.01$) from base value

Table 7: Showing diastolic blood pressure (mm of Hg) at different time intervals

Group	T0	T15	T20	T30	T45	T60	T75	T90	T105	T120
A	104.67±5.91	112.17±5.86	114.83±4.82*	119.17±3.85*	124.17±4.38*	121.00±3.52	112.00±2.39	112.67±1.74	103.83±2.57[a]	111.17±5.08
B	100.17±2.63	106.67±2.96**	113±3.53*	120.33±3.34**	125.17±2.82**	124.33±2.47**	117.17±4.24*	112.33±1.43**	116.17±4.56[b]	112.5±4.42

Values with different alphabets differ significantly ($p<0.05$) at corresponding intervals.

*Differ significantly ($p<0.05$) from base value. ** Differ significantly ($p<0.01$) from base value

Table 8: Showing mean blood pressure (mm of Hg) at different time intervals

Group	T0	T15	T20	T30	T45	T60	T75	T90	T105	T120
A	128.50±6.81	125.33±4.1^{a}	137.83±3.71	156.33±5.48^{a*}	157.33±4.3^{a**}	155.83±5.36^{a**}	152.50±3.82^{a*}	150.33±4.49*	143.33±3.31	145.67±2.36^{a}
B	118.33±2.35	117.50±5.63^{b}	125.00±4.01	134.00±3.61^{b*}	142.67±2.79^{b**}	140.00±2.22^{b**}	137.00±3.84^{b*}	131.50±2.97*	133.67±5.53	129.67±3.95^{b}

Values with different alphabets differ significantly ($p<0.05$) at corresponding intervals.

*Differ significantly ($p<0.05$) from base value. ** Differ significantly ($p<0.01$) from base value

Table 9: Showing Pulse oximetry oxygen (%) saturation

Group	T0	T15	T20	T30	T45	T60	T75	T90	T105	T120
A	99±0.63	94.33±1.97**	90.33±3.72**	92.17±3.25*	92.17±4.79	90.17±4.62**	89.83±3.6**	92.67±5.01*	93±4.6*	91±11.01
B	99.33±0.52	93.83±2.32**	91.83±1.47**	90±1.9**	90.83±4.67**	93.83±3.19	91±4.01*	94.33±2.66	93.83±4.22	95.33±1.97

Values with different alphabets differ significantly ($p<0.05$) at corresponding intervals.

*Differ significantly ($p<0.05$) from base value. ** Differ significantly ($p<0.01$) from base value

terms our study done for surgical procedure ovariohysterectomy under premedicated anaesthetic trial than neither surgical procedure nor premedication.

Ketamine and ketofol 1:1 showed significant decrease of Sp02 readings, except 2 out of 9 in groups A and 3 out of 9 in group B. But there was no significance between two groups at any time period (Table 9).

Decreased pulse oximetry values in both groups were due to depressive actions of combinations balanced anaesthetic and premedications. Because ketamine has been reported to act on μ opioid receptors and its effect is similar to μ opioid drugs, which enhances respiratory depression of propofol administration (Hamza *et al.,* 1989 and Lamont and Mathews, 2007). In addition to our study we used butorphanol that can be a cumulative effect on respiratory depression. Similar finding of ketofol 1:1 on SpO2 obtained by other studies (Martinez Taboada and Leece, 2014 and Kennedy and Smith, 2015)

Animals of Group B had good recovery quality (Table 10) as well as faster recovery time and sternal recumbency time than group A. but standing time was longer in group B than group A (Table 11).

Table 10: Showing quality of recovery

Group	Recovery quality (Median range)
A	3 (2)[a]
B	1 (1)[b]

Values with different alphabets differ significantly ($p<0.05$) at corresponding intervals.

Table 11: Showing recovery time, sternal recumbency time and standing time

Group	Recovery time	Sternal time	Standing time
A	58.83±1.96[a]	68.50±1.26[a]	87.00±1.79[a]
B	25.33±1.56[b]	40.00±1.15[b]	95.17±3.4[b]

Values with different alphabets differ significantly ($p<0.05$) at corresponding intervals.

Recovery and sternal recumbency time of ketofol 1:1 was faster than ketamine alone constant continuous rate infusion. Thus similar results were obtained in comparison with bolus administration of propofol vs ketofol (Shinde *et al.*, 2018 and Thejasree *et al.*, 2018). In our study ketamine and ketofol evidenced as stated above the same but quite longer recovery time may be due to CRI infusion of ketamine with different premedication, altered state of mentation of psychoactive effects of ketamine and biologically active ketamine metabolites (Zanos *et al.*, 2018). Standing time was lesser in ketamine than ketofol 1:1. Thiss may be attributed to pshycostimulatory effect of ketamine on locomotor system during post-anaesthetic phase. Similar finding is reported in rats (Leung and Baillie, 1986).

It is concluded that addition of ketamine in propofol as a 1:1 mixture of ketofol reduces the need of propofol for induction. Both ketamine and ketofol as a CRI-TIVA maintains stable basic vital functions during anaesthesia. Despite there is significant edge over advantage of ketofol in physiological observations as compared to ketamine in terms of maintaining vital functions merely baseline and clinically acceptable range, recovery time and recovery quality. In addition administration of oxygen indicated for both the groups due to decreased oxygen saturation of haemoglobin.

REFERENCES

1. Adams, H.R., Parker, J.L. and Mathew B.P. (1977). The influence of ketamine on inotropic and chronotropic responsiveness of heart muscle. *J Pharmacol Exp Ther.* 201(1):171–83.

2. Dinis-Oliveira, R.J. (2018). Metabolic Profiles of Propofol and Fospropofol: Clinical and Forensic Interpretative Aspects. *BioMed Research International.* https://doi.org/10.1155/2018/6852857.

3. Ebert, T.J., Muzi, M., Berens, R., Goff, D. and Kampine, J.P. (1992). Sympathetic responses to induction of anesthesia in humans with propofol or etomidate. *Anesthesiology, 76*(5): 725-733.

4. Feldman, J.M. (2012). Managing fresh gas flow to reduce environmental contamination. *Anesthesia and Analgesia, 114*(5), 1093-1101.

5. Hamza, J., Ecoffey, C. and Gross, J.B. (1989). Ventilatory response to CO2 following intravenous ketamine in children. *Anesthesiology,* 70 (3): 422-425.

6. Kennedy, M.J. and Smith, L.J., (2015). A comparison of cardiopulmonary function, recovery quality, and total dosages required for induction and total intravenous anesthesia with propofol versus a propofol ketamine combination in healthy Beagle dogs. *Veterinary Anaesthesia and Analgesia, 42*(4): 350-359.

7. Lamont, L.A. and Mathews, K.A. (2007). Opioids, nonsteroidal anti-inflammatories, and analgesic adjuvants. Lumb and Jones' Veterinary Anesthesia and Analgesia. pp. 241-271.

8. Leung, L.Y. and Baillie, T.A. (1986). Comparative pharmacology in the rat of ketamine and its two principal metabolites, norketamine and (Z)-6-hydroxynorketamine. *.Journal of Medicinal Chemistry, 29*(11): 2396-2399

9. Martinez Taboada, F. and Leece, E.A. (2014). Comparison of propofol with ketofol, a propofol ketamine admixture, for induction of anaesthesia in healthy dogs. *Veterinary Anaesthesia and Analgesia, 41*(6): 575-582.

10. Mostafa Mohamad, A.M., Mostafa, M.G., Ahmed Ahmed, A.I. and Fathy Golnar, M. (2015). Ketofol for Procedural Sedation and Analgesia in Children with Acute

Lymphoblastic Leukemia. *J Anesth Clin Res*, *6*(515):.2.

11. Müller, K., Holzapfel, J. and Brunnberg, L. (2011). Total intravenous anaesthesia by boluses or by continuous rate infusion of propofol in mute swans (Cygnus olor). *Veterinary Anaesthesia and Analgesia*, *38*(4):286-291.

12. Saikia, B., Sarma, K.K., Kalita, D., Sarmah, B.K., Phukan, A. and Goswami, S. (2016). Propofol, Ketamine and ketofol as Total intravenous Anaesthesia in Dogs. *Sch. J. Agric. Vet. Sci*, *3*(2): 100-102.

13. Sano, T., Nishimura, R., Mochizuki, M., Hara, Y., Tagawa, M. And Sasaki, N., (2003). Clinical usefulness of propofol as an anesthetic induction agent in dogs and cats. *Journal of Veterinary Medical Science*, *65*(5): 641-643.

14. Sato, M., Tanaka, M., Umehara, S. and Nishikawa, T. (2005). Baroreflex control of heart rate during and after propofol infusion in humans. *British Journal of Anaesthesia*, *94*(5): 577-581.

15. Sessler, D.I. (2008). Temperature monitoring and perioperative thermoregulation. *Anesthesiology*, *109*(2): 318-338.

16. Shinde, P.R., Chepte, S.D., Thorat, M.G., Raulkar, R.V., Sajid Ali, S., Fani, F.A., Anam, A.D., Bhave, N.P and Vaidya, S.R. (2018). Clinical efficacy of ketofol and propofol in dog. *International Journal of Science, Environment and Technology,* 7(6): 1949 – 1953.

17. Thejasree, P., Veena, P., Dhanalakshmi N. and K. Veerabrahmaiah. (2018). Evaluation of Propofol and Ketofol Anaesthesia Following Atropine, Diazepam and Fentanyl Premedication in Dogs.*International Journal of Current Microbiology and Applied Sciences,* 7 (11): 3130-3137

18. Vollmer, M.K., Rhee, T.S., Rigby, M., Hofstetter, D., Hill, M., Schoenenberger, F. and Reimann, S. (2015). Modern inhalation anesthetics: Potent greenhouse gases in the global atmosphere. *Geophysical Research Letters*, *42*(5): 1606-1611.

19. Yan, J.W., McLeod, S.L. and Iansavitchene, A. (2015). Ketamine propofol versus propofol alone for procedural sedation in the emergency department: a systematic review and meta analysis. *Academic Emergency Medicine*, *22*(9): 1003-1013.

20. Zanos, P., Moaddel, R., Morris, P.J., Riggs, L.M., Highland, J.N., Georgiou, P., Pereira, E.F., Albuquerque, E.X., Thomas, C.J., Zarate, C.A. and Gould, T.D. (2018). Ketamine and ketamine metabolite pharmacology: insights into therapeutic mechanisms. *Pharmacological Reviews*, *70*(3): 621-660.

Current Status of Researches in Biosciences (2020) : 169-180
Editor : Prof. P.C. Joshi, N. Joshi, Reshman Yasmin, and D. K. Mansotra
Today & Tomorrow's Printers and Publishers, New Delhi - 110 002

MESENCHYMAL STEM CELLS USE AS A REGENERATIVE MEDICINE FOR TREATMENT OF DISEASES IN LIVESTOCK

Dhruba Malakar*, Hruda Nanda Malik, Dinesh Kumar, Sikander Saini, Vishal Sharma, Abhishek Thakur, Amit Kumar, Kamlesh Kumari Bajwa, Ramandeep Singh and Satish Kumar

Animal Biotechnology Centre, National Dairy Research Institute, Karnal-132001, India

**Corresponding author email: dhrubamalakar@gmail.com*

ABSTRACT

Mastitis, metritis and hoof wound are the most expensive dairy diseases as the animals are suffering long time and rendering them useless and unproductive. Mastitis is caused drastic milk production of cows and has heavy economical loss of the farmers. Metritis and lameness reduce the fertility performance like prolonged interval from calving to first service and lower reproductive efficiency in the dairy cow specially crossbred cattle. These diseases really have a very high incidence, even in well-managed dairy farms throughout the world. The present study was performed to treat mastitis, metritis and hoof wound with adipose tissue derived allogenic mesenchymal stem cells (MSCs). We were isolated adipose tissues aseptically from cattle and buffalo from fat pad of tail head region rump site by liposuction methods. The adipose tissues were aseptically processed as per the protocol developed in our laboratory for isolation, characterization of MSCs from goats and dogs. We were cultured the adipose tissues derived MSCs of cattle, buffaloes and mice for treatment of mastitis, metritis and hoof wound of cattle and buffalo and artificial tibial fracture of mice. MSCs were characterized with alkaline phosphatase; different molecular markers like CD29, CD44, CD166, CD99 and CD34 were expressed whereas no expression was observed for CD34, CD45 and CD71 markers in adipose tissue-derived MSCs. The MSCs were cultured in vitro condition in CO_2 incubator during 2 weeks for confluency of cells for treatment of mastitis, metritis and hoof wound cattle and buffaloes. The characterized MSCs were cryopreserved into LN2 for further revival of cells in culture condition and used these cells for treatment of diseased animals. The cultured MSCs were aseptically isolated from cultured flask and injected around 10^7cells/animal in the site of wound areas. We have already treated the mastitis, metritis and hoof wound more than 800 cattle, bulls and buffaloes with MSCs at NDRI cattle yard, Gausala Hospital, Shrikishna Gosala and Radhakrishna Gosala at Karnal Pathmera Gaosala at Rajasthan and all the patients were cured completely one an average 30 days. Tibial bone fractures of mice were treated with allogenic and xenogeneic MSCs cured in 35 days. The photographs and videos were taken of all the treated animals for the records. The videos and photographs were arranged and edited for presentation. In conclusion that the mesenchymal stem cells therapy is very much possible to cure the mastitis, metritis and hoof wound cattle and buffaloes permanently and completely within a short period of time.

KEYWORDS: Mastitis, Metritis, Wound, Mesenchymal stem cells, Cattle and Buffalo.

INTRODUCTION

Reproductive efficiency is one of the major factors that mostly affects productivity and profitability of the farmers of our country. Mastitis, metritis and wound are the most expensive dairy diseases and animals are suffering long time failure of milk production, conception rate and huge economic loss of our country. There are specific treatment which may not give permanent cure of the mastitis and metritis suffering animals and cost of treatment is also very high as per the available reports.

After FMD, mastitis may be the second most important disease in dairy cattle and buffaloes in India as huge economic losses of the farmers. In India the annual economic loss was of Rs. 7165.51 crores due to mastitis (PDADMAS, 2011) and Rs. 66180 (Singh *et al.*, 2014). Total economic loss due to subclinical mastitis has been estimated for 57.93 per cent (4151.16 crores) (PDADMAS, 2011). It has also reported annual loss per lactation due to mastitis of Rs. 1592 crores in India (Sinha and Thombhare, 2013). The overall morbidity rates of mastitis, in cattle and buffaloes were 15.5%. Total losses due to mastitis per lactation in nondescript cow, crossbred cow and buffalo were Rs. 868.34, Rs. 1,314.10 and Rs. 1, 272.36 respectively. Mastitis is dangerous disease in bovine in India as huge economic losses. It has an economic impact as it reduces fertility and milk yield and is associated with an increase in culling rates as main consequence of endometritis. It is reported to have an incidence of 10% to 15% percent in dairy herds and total cost of £160 per case (WikiVet 2019). It was estimated the cost of days open beyond 80 to 110 days post partum $2.00 to $2.25 per day (Bretzlaff, 1987) loss. Studies in the USA show that costs related to mastitis are approximately U$ 200 per cow/year. This gives an annual loss of 2 billion dollars to dairy industry (Bogni *et al.,* 2011). According to another study, the estimated loss due to mastitis in cows was almost 700 kg in first lactation and 1,200 kg in the second or higher lactation (Wilson *et al.,* 2004). Similarly the prevalence of metritis has been found sporadically and reduces conception rate of animal rendering huge economic loss of our country.

Most important causes of subfertility in dairy cows are reported to be metritis and endometritis. The bacterial infection of uterus may occur, during or immediately after parturition, coitus or while carrying out artificial insemination (Chaudhari *et al.,* 2017). Inflammation of the uterus due to bacterial infections slows down the process of involution in the uterus and delays the onset of activities of the ovaries leading to economic loss due to systemic illness, loss of milk and meat production and marked drop in fertility (Gilbert 2011, Deori and Phookan 2015). In India reported a higher prevalence rate of metritis as 28.90% in first calvers and 38.93% in all calves in Karan fries cows (Balasundaran *et al.,* 2011). Endometritis is one of the conditions of highest impact on dairy cows. It reduces fertility and increases the production costs. The metritis affects 10 to 20 per cent of dairy cows in Canada and US (Liang, 2013, Dubuc *et al.,* 2010). It has an economic impact as it reduces fertility and milk yield and is associated with an increase in culling rates as main consequence of endometritis. The cost of each case of metritis is approximately US$329–386, due to antibiotic treatment and the detrimental effects of metritis on reproductive performance, milk production, and survivability (Overton and Fetrow, 2008).

Lameness is the third most expensive dairy disease, following mastitis and reproductive failure (Juarez *et al*., 2003). The problems in the hoof lead to lameness in the animals rendering them useless. Many studies have demonstrated that lameness reduces the fertility performance like prolonged interval from calving to first service (Barkema *et al*., 1994), and lower reproductive efficiency (Somers *et al*., 2015) in the dairy cows. Lameness has a very high incidence, even in well-managed farms (Sanders *et al.,* 2009). The prevalence of lameness in the USA dairy herd in the free stall housing averages around 25% (Di Liang, 2013). The lameness costs include treatment costs, labour, discarded milk, reduced milk yield, increased culling risk, extended CI, veterinarian service fees, and extra services. The total costs were average $421.53 per case. In India at NDRI-Karnal, (Singh *et al.*, 2011) compared the production performance of 96 lame cows with 67 healthy Karan Fries crossbred cows. A significant total loss of 498.95 kg of milk yield was observed during a period of 305 days. Effect of lameness (hoof disorders) on productivity of Karan Fries crossbred cows was reported in Animal Science Journal (2011).

According to the International Society of Cellular Therapy (ISCT), Mesenchymal stem cells adhere to a plastic surface and they express their surface markers CD44, CD73, CD90, and CD105, and they are unable to express of hematopoietic markers CD14, CD34, CD45, CD11b/CD79, and CD19/HLA-DR. MSCs have the ability to differentiate into osteoblastic, adipocytic, neurocytic and chondrocytic lineages. MSCs can be isolated from bone marrow, adipose tissue, umbilical cord blood, nerve tissue, and dermis. MSCs can be administered both systemically and locally for the treatment of different wounds. It have been shown to express low levels of long-term incorporation into healing wound areas they release of trophic mediators, rather than a direct structural contribution. They generally release of vascular endothelial growth factor (VEGF), stromal cell-derived factor-1, epidermal growth factor, keratinocyte growth factor, insulin-like growth factor, and matrix metalloproteinase-9. They promote new vessel formation, recruit endogenous progenitor cells, and direct cell differentiation, proliferation, and extracellular matrix formation during wound repair (Duschera *et al.*, 2016).

MSCs have also shown key immunomodulatory properties though the secretion of interferon-λ, interleukin-1α, tumor necrosis factor-α interleukin-1β and nitric oxide synthase. MSCs secret prostaglandin E2 and regulates fibrosis as well as inflammation, promoting tissue healing. Not only that, MSCs had bactericidal properties through the secretion of antimicrobial factors and by upregulating bacterial killing and phagocytosis by immune cells. MSC is the major stem cells for cell therapy, have been used in treatment of animals. From animal models to clinical trials, MSCs have afforded promise in the treatment of many diseases, primarily tissue injury and immune disorders. MSCs for cell therapy have been shown to be safe and effective (Xin *et al.*, 2013). Cell migration and invasion are processes that offer rich targets for intervention in key physiologic and pathologic phenomena such as wound healing and cancer metastasis. Effective use of several principle formats for studying cell motility: scratch assays, transmembrane assays, microfluidic devices and cell exclusion zone assays may be done (Hulkower and Herber 2011).

Keeping in view that the higher economic loss was observed due to bovine mastitis, metritis and hoof wound and these diseases could be cured with treatment of mesenchymal stem cells. The present investigation was proposed to isolate, in vitro culture and characterization of mesenchymal stem cells from adipose tissue of bovine and mice. The in vitro cultured mesenchymal stem cells were used for treatment of mastitis, metritis and hoof wound healing of bovine and tibial bone healing of mice.

MATERIALS AND METHODS

MSCs exist in all tissues and can be easily derived from bone marrow, umbilical cord, fat tissue, liver, intestine, lung etc. (Bianco *et al.*, 2008). Among other source of MSCs, Adipose tissue derived mesenchymal stem cells is a prime source of cell therapy by virtue of its easy availability, enormous expandability, ease of isolation (Gimble *et al.*, 2007). The adipose tissues derived adult mesenchymal stem cells (MSCs) were culture for cattle, buffaloes, and mice for treatment of mastitis, metritis, and hoof wound of cattle and buffalo and tibial fracture of mice.

Adipose tissues were collected from fat pad of tail head region rump site of cattle and buffalo by liposuction methods as well as slaughterhouse of Kolkata and Delhi and transferred in transport medium containing antibiotic solution. The adipose tissues were minced into very small pieces and incubated in digestion media containing DMEM F/12, 1% type 1 collagenase enzyme for four hour under standard culture condition.

The dissociated cells were filtered in 41μm filter to remove undigested fat tissue. The cell suspension was centrifuged at 1000 rpm for 10min and pellet was seeded in 25 cm^2 culture flask in growing media under standard culture condition. Next passage, at 80% confluence, cells were trypsinized and centrifuged at 1000 rpm for 10 min and re-seeded in 25 cm^2 culture flask at density $2x10^5$ cells/cm^2.

MESENCHYMAL STEM CELL CULTURE FROM ADIPOSE TISSUE

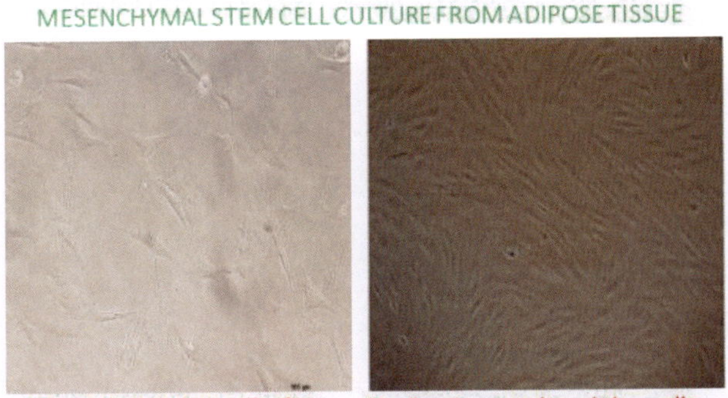

Mesenchymal stem cells after 5 days of culture (400X) of cattle

Confluent mesenchymal stem cells after 10 days of culture (100X) of cattle

Cryopreservation of mesenchymal stem cells for long time for further use: The adipose tissues derived in vitro cultured mesenchymal stem cells at 80% confluence cells were trypsinized and centrifuged at 1000 rpm for 10 min and mixed with cryopreserved medium with 10% DMSO and serum at density $1 x 10^6$ cells and put into 2 ml cryovials. The cryovials were kept at 4° C for 4 h and the -80° C overnight. Then the cryovials were shifted into liquid nitrogen tank for further use. The cryopreserved cells was thawed at 37° C water bath and removed the cryopreserved medium with centrifugation at 1000 rpm for 5 min. The pellet cells were further in vitro cultured for 3 days. We observed the cells were grown faster forming confluency more than 80% for 3 days

Cryopreservation of MSCs and further culture

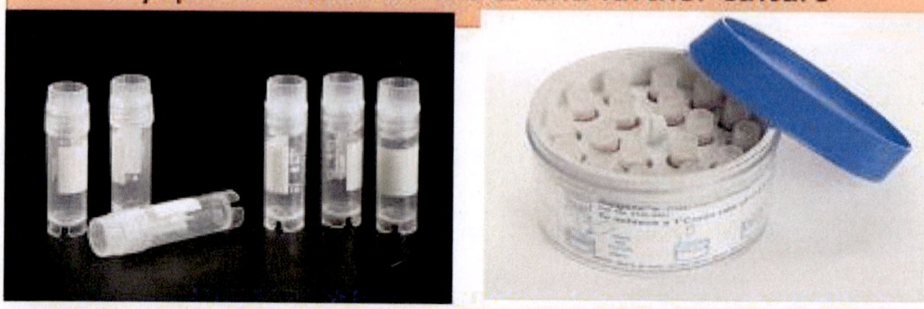

Cryopreservation of MSCs using Mr. Frosty™ Freezing Container

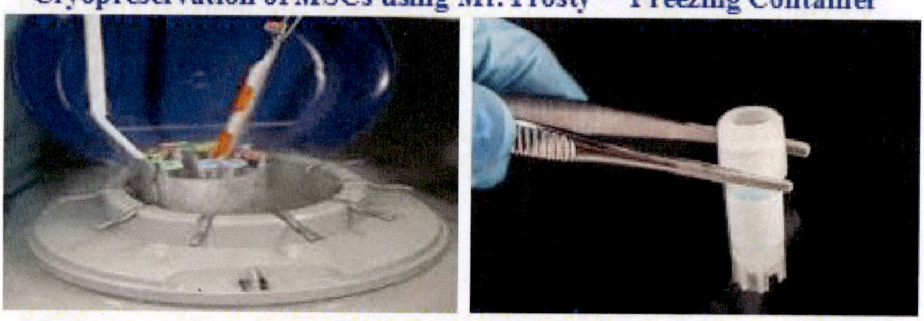

MSCs kept into LN2 tank and thawing at 37°C water for 30 second.

Revival of MSCs after freezing

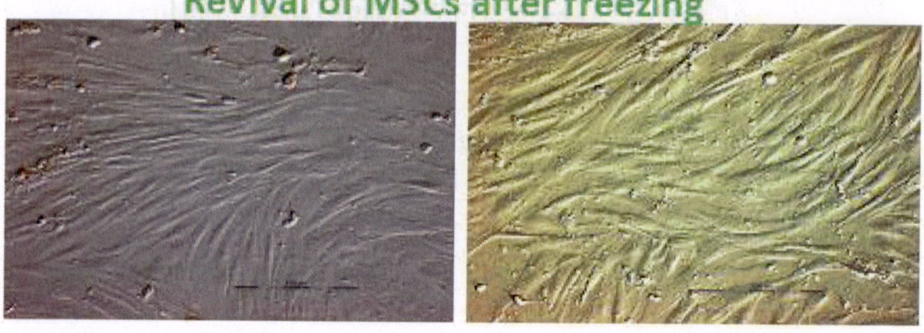

Characterization of adipose tissue derived mesenchymal stem cells: MSCs were characterized with alkaline phosphatase; different molecular markers like CD29, CD44, CD166, CD99, and CD34 were expressed whereas no expression was observed for CD45 and CD71 markers in cattle adipose tissue-derived MSCs. The cultured MSCs were aseptically isolated from the culture flask and injected around 107cells/animal in the site of the wound.

Treatment of mastitis and metritis with mesenchymal stem cells: We have assessed the efficiency of MSCs for treatment of mastitis cattle. First the identification and classification of animals were performed as per the veterinary clinical recodes. Mastitis cases were identified based on SSC and CMT score (Hand *et al.*, 2012). Metritis cases were identified based on polymorphonuclear cells (PMN) count of vaginal secretion. Treatment of mastitis and metritis was performed with allogenic mesenchymal stem cells @ dose 10^7- 10^8 cells/injection at the route of delivery Local and Intravenous two injections at an interval of 15 days.

Confirmation for presence of MSCs on wound areas of treated animals: Tissue biopsy was taken to collect the cells of wound areas of animals to differentiate the injected allogenic MSCs and its own cells of animals. For this experiment, genomic DNA was isolated from biopsy cells and PCR was performed using DRB II gene primer to differentiate autologous and allogenic cells after confirmation of healing of wound. Genomic DNA from blood of allogenic treated animal was also analyzed with this primer to see the allogenic MSCs will be reached in blood or not.

We have determined the efficacy of MSCs to treating mastitis and metritis in bovine by intravenous application of MSCs. Clinical improvement was assessed by sign and symptom of healing one every progressive day. Immunomodulatory molecular profile was also performed as blood cells were collected weekly to see the secretion of Anti-inflammatory cytokine IL10, TNF alpha, angiogenic vascular endothelial growth factor (VEGF) and angiopoietin (ANG) and Antimicrobial cytokine: IL37

Results after treatment of mastitis, metritis and hoof wound with MSCs: We have already treated the mastitis, metritis and hoof wound more than 500 cattle, 35 bulls and 55 buffaloes with MSCs at NDRI cattle yard, Gausala Hospital, Shrikishna Gosala and Radhakrishna Gosala at Karnal and all the patients were cured completely one an average 30 days. Tibial bone fractures of mice were treated with allogenic and xenogeneic MSCs cured in 35 days. The photographs and videos were taken of all the treated animals for the records. The videos and photographs were arranged and edited for presentation.

We have observed that all the cytokine IL10, TNF alpha, angiogenic vascular endothelial growth factor (VEGF), angiopoietin (ANG) and Antimicrobial cytokine: IL37 genes were expressed significantly more than the control animals in Real-time PCR.

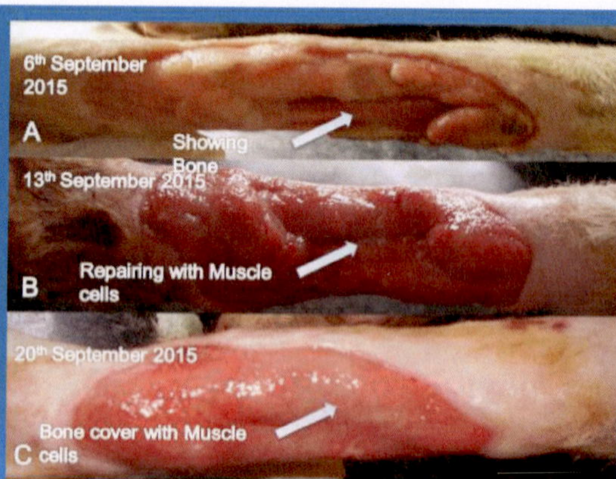

Treatment of Dog with Mesenchymal Stem cell: The Ulna bone was showing (A). Repair of wound with muscle cells after 7 days (B). The wound was healed with mesenchymal stem cells after 14 days (C).

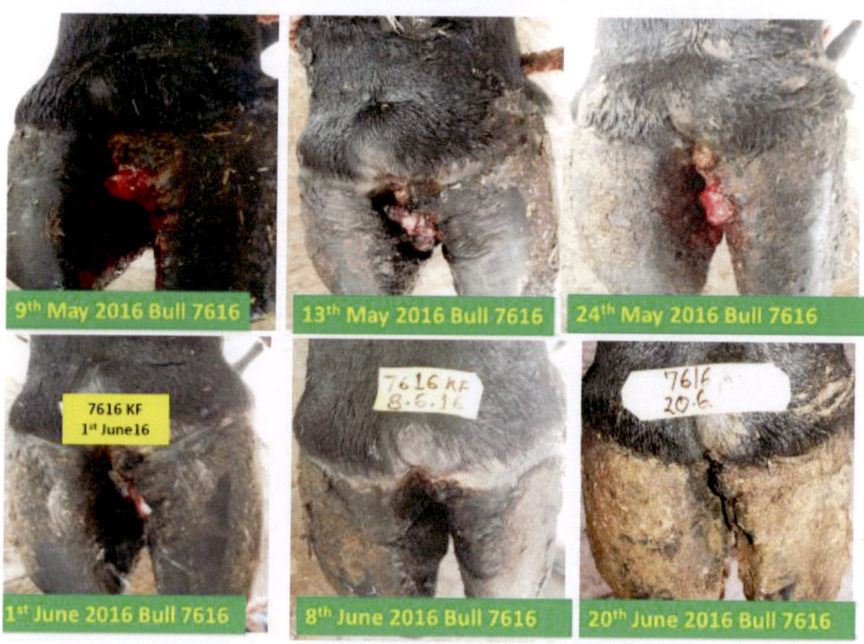

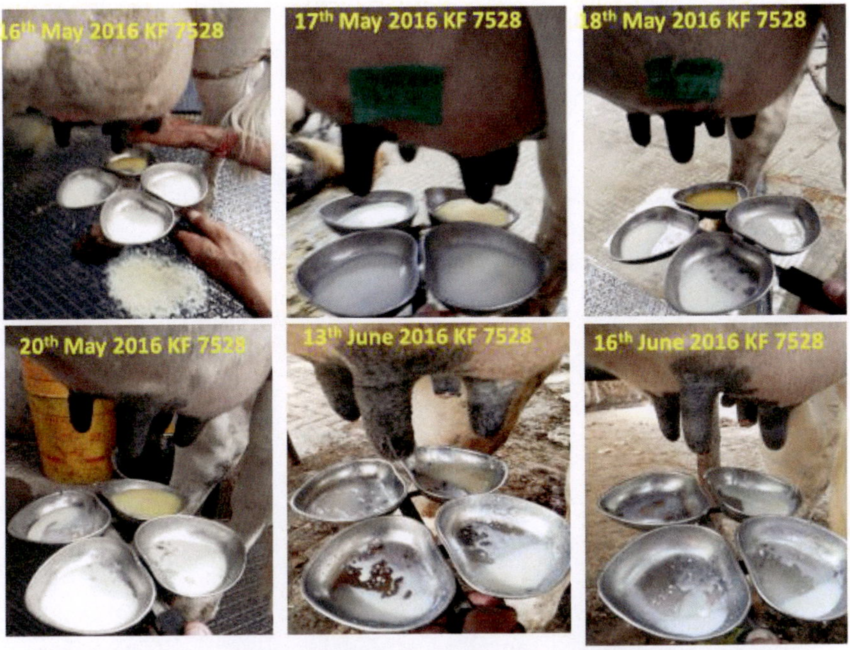

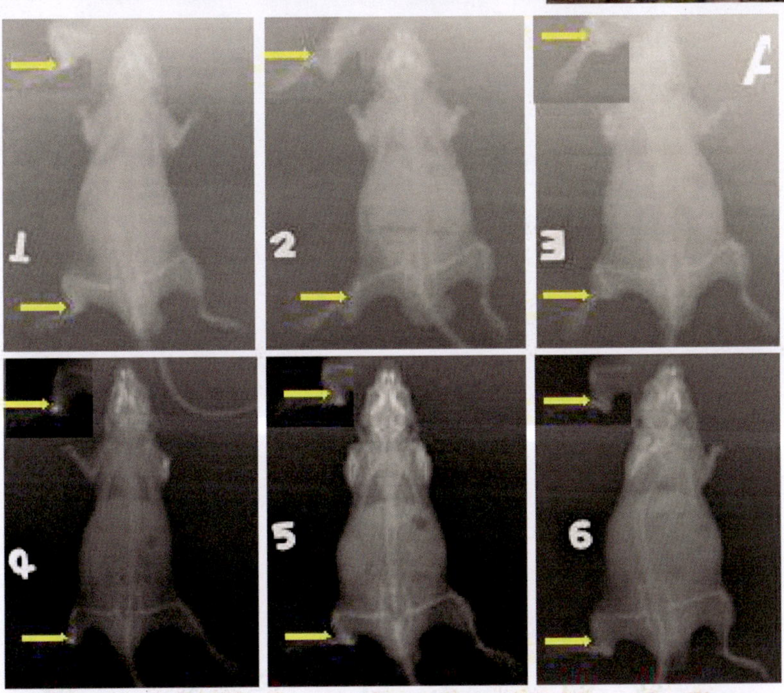

Allogeneic MSCs Group: Complete tibial bone healing was observed on day 35th

Significant achievements

- Isolation, characterization, culture and cryopreservation of adipose tissue derived mesenchymal stem cells from cattle, buffaloes and mice.
- The mesenchymal stem cells were isolated from adipose tissue derived from fat pad of rump site tail head region and slaughterhouse of cattle and buffalo.
- The MSCs were formed monolayer within 10 days of in vitro culture condition.
- Expression profile of different markers like CD29, CD44, CD166, CD99 and CD34 are expressed whereas no expression was observed for CD45 and CD71 markers in cattle adipose tissue-derived MSCs.
- Cryopreservation of MSCs into LN2 for further used in future.
- The MSCs were used for treatment of mastitis, metritis and hoof wound of cattle and buffaloes at NDRI cattle yard, Gausala Hospital, Shrikishna Gausala and Radhakrishna Gausala at Karnal.
- Treatment of mastitis, metritis and hoof wound of cattle and buffaloes and tibial fracture of mice with MSCs and cured the animals completely and permanently within an average 30 days period of time.

Significant Technologies developed: We have already treated the mastitis, metritis and hoof wound more than 500 cattle, 35 bulls and 55 buffaloes with MSCs at NDRI cattle yard, Gausala Hospital, Shrikishna Gosala and Radhakrishna Gosala at Karnal and all the patients were cured completely one an average 30 days. Tibial bone fractures of mice were cured in 35 days treated with allogenic and xenogeneic MSCs. So the economic condition of the farmers of our country will be definitely improved from this study as the health condition of their animals will be improved a lot after treatment of mesenchymal stem cells.

CONCLUSION

In the present day, adult mesenchymal stem cells therapy could be possible to cure the mastitis, metritis and hoof wound of cattle and buffaloes permanently within a short period of time. As the available reports, these diseases cause huge economic loss in our country. We have already treated mastitis, metritis and hoof wound of cattle and buffaloes with mesenchymal stem cells and cured all the animals completely and permanently. All the diseases suffering animals were cured with this treatment of MSCs and drastically reducing the present treatment cost on antibiotics and other therapies to the suffering animals. The farmers of our country will be highly benefited as their animals will be given enhancing rate of milk production, increase conception rate and improve the health of animals. Hence the farmers of our country will gain double income from their animals and remove their agony due to their suffering animals. To best of our knowledge, we can treat the other diseases like diabetes, cancer, and fracture with MSCs that has not been treated with MSCs in livestock in our country even in the world.

REFERENCES

1. Balasundaran B, Gupta AK, Dongre VB, Mohanty TK, Sharma PC, Khat K, Singh RK (2011). Influence of genetic and non-genetic factors on incidence of post-partum utero-vaginal complications in Karan fries cows. *Indian J Anim Res.*; 45:192–7.

2. Bogni, C., Odierno, L., Raspanti, C., Giraudo, J., Larriestra, A., Reinoso, E., Lasagno, M., Ferrari, M., Ducrós, E., Frigerio, C., Bettera S., Pellegrino, M., Frola, I., Dieser, S., and Vissio, C. (2011). War against mastitis: Current concepts on controlling bovine mastitis pathogens. Science against microbial pathogens: communicating current research and technological advances. A. Méndez-Vilas (Ed.). http://www.formatex.info/microbiolog y3/book/483-494.

3. Bretzlaff, K. (1987). Rationale for treatment of endometritis in the dairy cow. *Vet Clin North Am Food Anim Pract.* 3(3):593-607

4. Chaudhari, R.J., Gulavane, S.U., Rangnekar, M.N., Gatne, M.M., Ingole S.D. and Jawale, R.S. (2017). Effect of Endometritis Diagnosed by Cytology on Reproductive Parameter in Postpartum Crossbred Cows. *International Journal of Current Microbiology and Applied Sciences* 6(8):2585-2590.

5. Deori, S. and Phookan, A. (2015). Bovine Postpartum Metritis and its Therapeutics: A Review. *Indian Journal of Science and Technology*, 8(23), DOI: 10.17485/ijst/2015/v8i23/5238

6. Dubuc J, Duffield TF, Leslie KE, Walton JS, LeBlanc SJ (2010). Risk factors for postpartum uterine diseases in dairy cows. *J Dairy Sci.* 93:5764–71.

7. Duschera D., Barreraa J., Wong VW., Maana ZN., Whittama AJ., Januszyka M., Gurtner G C. (2016). Stem Cells in Wound Healing: The Future of Regenerative Medicine? *Gerontology*, 62:216–225.

8. Gilbert R.O. (2011). The effects of endometritis on the establishment of pregnancy in cattle. *Reprod Fertil Dev.*; 24(1):252-7.

9. Halasa, T., K. Huijps, O. Østerås, and H. Hogeveen. (2007). Economic effects of bovine mastitis and mastitis management: A review. *Vet. Quart.* 29(1):18-31.

10. Hand, K., A. Godkin, and D. Kelton. (2012). Milk production and somatic cell counts: A cow-level analysis. *J. Dairy Sci.* 95(3):1358-1362.

11. Hand, K., A. Godkin, and D. Kelton. (2012). Milk production and somatic cell counts: A cow-level analysis. *J. Dairy Sci.* 95(3):1358-1362.

12. Harmon, R. (1994). Physiology of mastitis and factors affecting somatic cell counts. *J. Dairy Sci.* 77(7):2103-2112.

13. Hulkower, K.I. and Herber, R.L. (2011). Cell Migration and Invasion Assays as Tools for Drug Discovery. *Pharmaceutics* 2011, 3, 107-124

14. Liang, D. (2013). Estimating the economic losses from diseases and extended days open with a farm-level stochastic model. University of Kentucky, USA

15. Overton, M. and J. Fetrow. (2008). Economics of postpartum uterine health. Proc. Dairy Cattle Reprod. Council Conv., Omaha, NE: 39-43.

16. PDADMAS (2011). PD_ADMAS News. Project Directorate on Animal Disease Monitoring and Surveillance, January - June, 1(1): 8.

17. Philpot, W. N. and S. C. Nickerson. (2000). Winning the fight against Mastitis. Westfalia Surge, Incorporated, 2000 - Dairy cattle - 192 pages

18. Singh, D., Kumar, S., Singh, B. and Bardhan, D. (2014) Economic losses due to important diseases of bovines in central India, 7(8): 579-585.

19. Singh, R., Prasad. S. and Raghuvanshi, S. (2008). Economic evaluation of Mastitis in Bovine: Proceedings of 4th Convocation of Indian Society of Veterinary Surgery, at Uttar Pradesh Pandit Deen Dayal padhyaya PashuChikitsa Vigyan Vishwavidyalaya evam Go-Anusandhan Sansthan (DUVASU), Mathura.

20. Sinha MK and Thombare NN (2013). Incidence and impacts of clinical mastitis in dairy cattle farms: case of Maharastra Farmers. http://mpra.ub.uni-muenchen.de/54155.

21. Wilson, D.J., González, R.N., Hertl, J., Schulte, H.F, Bennett, G.J., Schukken, Y.H and Gröhn, Y.T. (2004). Effect of Clinical Mastitis on the Lactation Curve: A Mixed Model Estimation Using Daily Milk Weights. *J. Dairy Sci.,* 87:2073 - 2084.

22. Xin Wei, Xue Yang, Zhi-peng Han, Fang-fang Qu, Li Shao, Yu-fang Shi, (2013). Mesenchymal stem cells: a new trend for cell therapy. *Acta Pharmacologica Sinica* 34: 747–754.

Current Status of Researches in Biosciences (2020) : 181-191
Editor : Prof. P.C. Joshi, N. Joshi, Reshman Yasmin, and D. K. Mansotra
Today & Tomorrow's Printers and Publishers, New Delhi - 110 002

MODELING AND DOCKING STUDIES OF ANTI-BACTERIAL PEPTIDE CERCOPIN1 IN *ASCARIS SUUM*

Jayasimha Rayalu Daddam, Madhusudhan Reddy and Nagaraja Rao Poduri

Department of Zoology, Osmania University, Hyderabad

Email: nagarajaraop@yahoo.com

ABSTRACT

Cecropins are linear cationic antibacterial peptides that have potent activities against microorganisms. A three dimensional model of Cecropin1 was generated using MODELLER9V7 software using 1F0D as template. With the aid of the molecular mechanics and molecular dynamics methods, the final model is obtained and further assessed by PROCHECK and Verify 3D graph programs, which showed that the final refined model is reliable. The results indicated that ARG216, LYS234, ALG278, THR279 in this Cecropin1 are important determinant residues in binding process as they have strong hydrogen bonding with these compounds. These hydrogen bonding interactions play an important role for stability of the complex. Finally this peptide was docked to Sortase, membrane protein of *Propioni bacterium*.

KEYWORDS: *Propioni bacterium*, Homology Modelling, Cecropin1, Sortase.

INTRODUCTION

Insects are continually presented to possibly destructive pathogens through contact, ingestion and inward breath [1,2]. The survival of bugs in an organism flourishing environment relies on upon humoral and cell reactions to successfully execute the attacking microorganisms since creepy crawlies do not have the versatile invulnerability equipped for delivering antibodies [3–5]. An investigation of the humoral invulnerable reaction inside the structure of a septic harm in life forms has prompted to the present worldview in which two particular intracellular flag transduction pathways, the safe insufficiency (IMD)2 and Toll pathways, manage the translation of several qualities by controlling the atomic take-up of the NF-jB interpretation variables Relish, Dorsal and Dorsal-related insusceptibility consider (DIF) [1,5–9]. The traditional effector particles in the humoral reactions are antimicrobial peptides (AMPs), which are constitutively communicated or actuated in fat bodies and hemocytes and discharged into the hemolymph to execute attacking microorganisms viably [2,4,9–11]. Most AMPs are integrated not long after microbial disease and act quickly to kill a wide scope of microorganisms. The AMPs found to date have been isolated into a few gatherings in light of their length, their optional and tertiary structure, and the nearness or nonattendance of disulfide scaffolds [12–14]. These AMPs show bactericidal, fungicidal, virucidal and tumoricidal properties, and the way that they can possibly defeat bacterial resistance makes them promising contender for remedial medications [12–15]. In *Drosophila melanogaster*, drosomycin shows fungicidal movement at micromolar fixations and is dynamic for the most part on filamentous growths. Different AMPs, for example, cecropins, attacins, drosocins, and diptericins, are dynamic basically against Gram-negative microscopic organisms, while defensin is powerful against Gram-positive microbes [4–6,10,11]. Cecropins are a group of essential antibacterial peptides created by bugs [16,17]. They have low atomic weights, are water soluble, and have an expansive range of antibacterial exercises, yet they can't lyse eukaryotic cells [17–19]. Cecropin was initially detached from *Hyalophora cecropia* [17,20], and from that point forward, numerous cecropin peptides and qualities have been accounted for in lepidopteran and dipteran species [2,18,21–23]. Cecropins show movement against Gram-positive and Gram-negative microscopic organisms by shaping ionic pores to pulverize the ionic adjust of the bacterial film [16]. In addition, concentrates identified with AMPs, for example, cecropins may give knowledge into the intrinsic invulnerability of spineless creatures, and these AMPs might be utilized to plan novel expansive range anti-toxins for use in people.

MATERIALS AND METHODS

Homology Modelling of Cecropin1: The sequence from *Propioni bacterium* was submitted to SBASE server for domain prediction. The domain was submitted to Interproscan to identify the function and was searched to find out the related protein structure to be used as a template by the BLAST (Basic Local Alignment Search Tool) program against PDB (Protein Databank). Sequence that showed maximum identity with high score and less E-value was used as a reference structure to build a three dimensional model for Cecropin1.

Alignment: The template sequence collected from Uniprot database was aligned with Cecropin1using CLUSTALX software. The structurally conserved regions (SCRs) of both protein and template were determined by multiple sequence alignment.

3D Model building: The initial model of Cecropin1was built by using homology-modeling methods and the MODELLER9V7 software. It is a program for comparative protein structure modeling optimally satisfying spatial restraints derived from the alignment and expressed as probability density functions (pdfs) for the features restrained. The pdfs restrain Cα-Cα distances, main-chain N-O distances, main-chain and side-chain dihedral angles.

Variable regions (VRs) are defined as the sections of the proteins that exist between the SCRs in the sequence. To construct the structural variable regions, a loop-searching algorithm over the data bank of known crystal structure was used. The 3D model of a protein is obtained by optimization of the molecular pdf such that the model violates the input restraints as little as possible. The molecular pdf is derived as a combination of pdfs restraining individual spatial features of the whole molecule. The optimization procedure is a variable target function method that applies the conjugate gradients algorithm to positions of all non-hydrogen atoms.

Molecular Dynamics studies of Cecropin1: The initial model was improved by Energy Minimization. After 1000 steps of Conjugate Gradient (CG) minimization performed, Molecular Dynamics (MD) simulation was carried out to examine the quality of the model structures by checking their stability via performing 500 ps simulations under 1 atm constant pressure and 310 K constant temperature (NPT). An explicit solvent model TIP3P water was used. The homology model was constructed with a 10 Å water cap from the centre of mass of Cecropin1. No periodic boundary conditions were included in this study. A cutoff of 12 Å (switching function starting at 10) for van der Waals interactions was assumed. An integration time step of 2 fs was used, permitting a multiple time-stepping algorithm to be employed in which interactions involving covalent bonds were computed every one time step and short-range non bonded interactions were also computed in every two time steps, and long-range electrostatic forces were computed every four time steps. The pair list of the non bonded interaction was recalculated in every ten time steps with a pair list distance of 13.5 . The short-range non bonded interactions were defined as van der Waals and electrostatic interactions between particles within 12 . A smoothing function was employed for the van der Waals interactions at a distance of 10 . All calculations mentioned above used the NAMD 2.5 software using CHARMM27 force field for lipids and proteins. CHARMM27 force-field parameters were used in all simulations in this study. Finally, the structure having the least energy with low RMSD (Root Mean Square Deviation) was used for further studies. In this step, the quality of the initial model was improved.

Structure Validation: After the optimization procedure, the final structure obtained was checked by Ramachandran's map using PROCHECK (Programs to check the Stereo chemical Quality of Protein Structures) (Brunger, 1992) and environment profile using ERRAT graph (Structure Evaluation server).

Docking studies: Docking of Cecropin1 with Sortase from ***Propioni bacterium*** was carried out using GOLD (Genetic Optimization of Ligand Docking) software which was based on genetic algorithm (G). This process is based on protein partial flexibility and ligand full flexibility. In this work, Cecropin1 was docked to Sortase active site. The binding of Cecropin1 to the amino acids in the active site were studied using molecular mechanics method. The options for GA run were taken such as population size (100), pressure (1.1), number of operations (10,000), number of island (1) and niche size (2). Operator options for crossover, mutation, and migration were set to 100, 100 and 10 respectively. Default parameters of 6.0 for van der Waals forces and 3.0 (dH-X) for hydrogen bonds were selected. During docking, the default algorithm speed was selected and the compound binding site in the Sortase was defined within a 10 radius with the centroid as CE atom of ASP215. The number of poses for compound was set 10, and termination was assigned if the top three bound conformations of Sortase were within 1.5 RMSD. After docking, the individual binding poses of Sortase were studied and also their interactions. The best and most favorable conformation of Sortase was selected.

Gold Score fitness: Gold Score performs a force field based scoring function and was made up of four components: 1. Protein-ligand bonding (external H-bond); 2. Protein-ligand van der Waals energy (external vdw); 3. ligand internal van der Waals energy (internal vdw); 4. ligand intramolecular hydrogen bonding (internal- H- bond). The external vdw score was multiplied with 1.375 factor when the total fitness score was calculated. This was an correction to develop protein-ligand hydrophobic interaction. The fitness function has been minimized for the prediction of ligand binding positions.

GoldScore = S (hb_ext) + S (vdw_ext) + S (hb_int) + S (vdw_int)

RESULTS AND DISCUSSION

Homology Modeling of Anti microbial peptides: A high level of sequence identity should guarantee more accurate alignment between the target sequence and template structure. The sequences of proteins were aligned to those of the templates in order to identify the conserved amino acid residues.

In the results of BLAST search against PDB, only the 1F0D,was the selected template protein for homology modelling, which has a high level of sequence identity with the Cecropin1 domain. Structurally conserved regions (SCRs) for the model and the template were determined by superimposition of the two structures and multiple sequence alignment was performed with the online version of CLUSTALW program to identify the set of conserved residues in the alignment (Figure 1).

```
MRYIRLCIISLLATLPLAVHASPQPLEQIKLSESQLSGRVGMIEMDLASGRTLTAWRADE 60
---------------------SPQPLEQIKLSESQLSGRVGMIEMDLASGRTLTAWRADE 39
                     ***************************************

RFPMMSTFKVVLCGAVLARVDAGDEQLERKIHYRQQDLVDYSPVSEKHLADGMTVGELCA 120
RFPMMSTFKVVLCGAVLARVDAGDEQLERKIHYRQQDLVDYSPVSEKHLADGMTVGELCA 99
************************************************************

AAITMSDNSAANLLLATVGGPAGLTAFLRQIGDNVTRLDRWETELNEALPGDARGTTTPA 180
AAITMSDNSAANLLLATVGGPAGLTAFLRQIGDNVTRLDRWETELNEALPGDARDTTTPA 159
******************************************************.*****

SMAATLRKLLTSQRLSARSQRQLLQWMVDDRVAGPLIRSVLPAGWFIADKTGAGERGARG 240
SMAATLRKLLTSQRLSARSQRQLLQWMVDDRVAGPLIRSVLPAGWFIADKTGAGERGARG 219
************************************************************

IVALLGPNNKAERIVVIYLRDTPASMAERNQQIAGIGAALIEHWQR 286
IVALLGPNNKAERIVVIYLRDTPASMAERNQQIAGIGAALIEHWQR 265
**********************************************
```

Fig 1: Alignment of Cecropin1 with template

In the following study, we have chosen 1F0D as a reference structure for modeling Cecropin1 domain. Coordinates from the reference protein (1F0D) to the SCRs, structurally variable regions (SVRs), N-termini and C-termini were assigned to the target sequence based on the satisfaction of spatial restraints.

The overall sequence identity between Cecropin1 and the 1F0D amounts to 31%. As stated elsewhere, 30% of sequence identity may be sufficient for creating an acceptable description of the binding site. The mere percentage of sequence identity between target and template, however, is not the only criterion for the final quality of the model. In the modeller we will get a 20 PDB out of which we select a least energy.The energy unit will be in kilo joule. All side chains of the model protein were set by rotamers. The final stable structure of the Cecropin1 protein obtained by the help of SPDBV and it is evident that this domain has 10 helices and 11sheets as shown in the Figure 2.

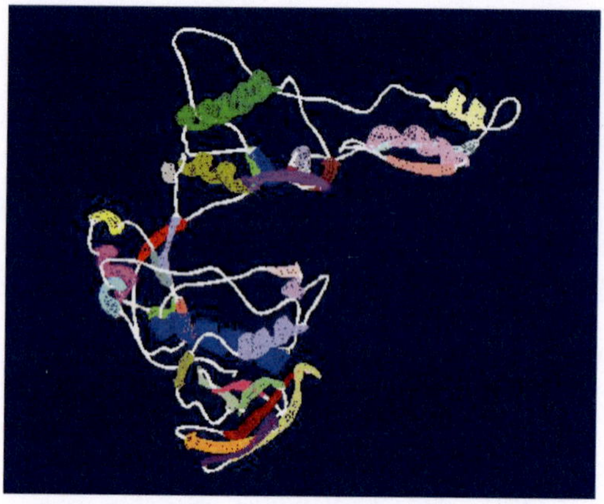

Fig 2: Cecropin1 structure

Computed 3D structure of homology model obtained from SPDBV - Cecropin1 that contains 10 helices and 11 -sheets

The structure having the least energy with low RMSD (Root Mean Square Deviation) which was obtained by the NAMD (Fig 3).

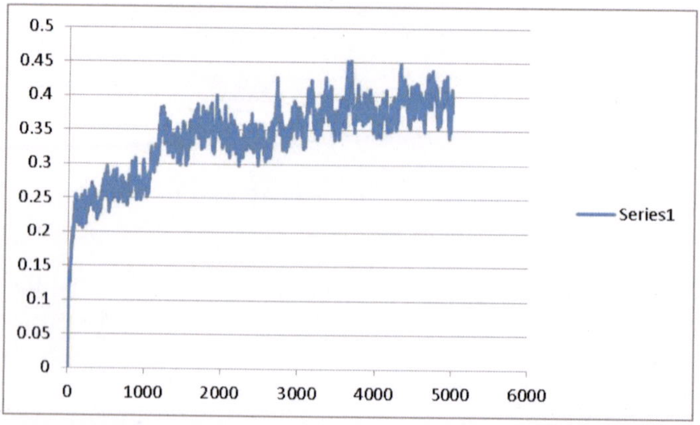

Fig 3: Calculated RMSD graph of molecular dynamics simulations of Cecropin1 protein using NAMD software. Time (Ps) was taken in X-axis and RMSD in Y-axis

The final structure was further checked by verify 3D graph and the results have been shown in Figure . The overall scores indicates acceptable protein environment (Fig 4).

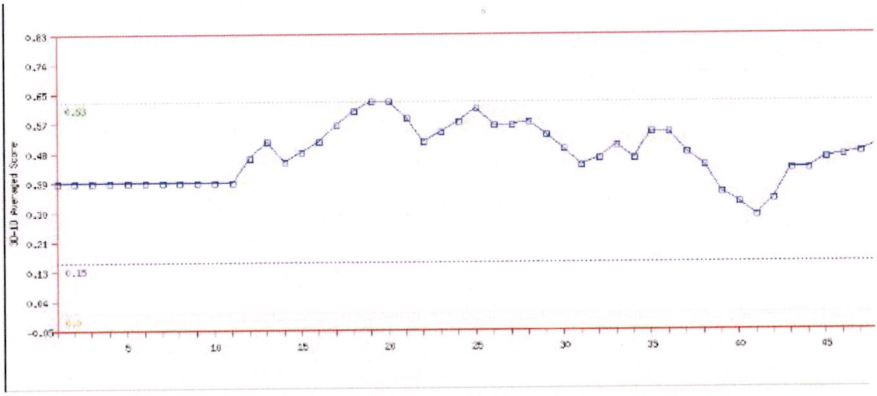

Figure 4: The 3D profiles verified results of Cecropin1 model; overall quality score indicates residues are reasonably folded.

Validation of Cecropin1 Domain: The formal validation of the created models was done with PROCHECK (Laskowski *et.al* 1993, Morris et.al 1992). After the refinement process, validation of the model was carried out using Ramachandran plot calculations computed with the PROCHECK (Programs to check the Stereo chemical Quality of Protein Structures). The ψ and φ distributions of the Ramachandran plots of non-glycine, non-proline residues are summarized in(Fig-5). The RMSD (Root Mean Square deviation) for covalent bonds and covalent angles relative to the standard dictionary of Cecropin1 was -2.27 and -0.35 Å. Altogether 99.6% of the residues of Cecropin1 was in favored and allowed regions. RMSD values of the new model were the same as for the previous model. These values indicate that the model predicted with our method well matches the crystal structures and the overall PROCHECK G-factor of Cecropin1 was 0.04 and verify 3D environment profile was good.

Table 1: % of residue falling in the core region of the Ramachandran's plot

% of residue in most favored regions	94.0
% of residue in the additionally allowed zones	5.6
% of residue in the generously regions	0.4
% of residue in disallowed regions	0.0
% of non-glycine and non-proline residues	100.0

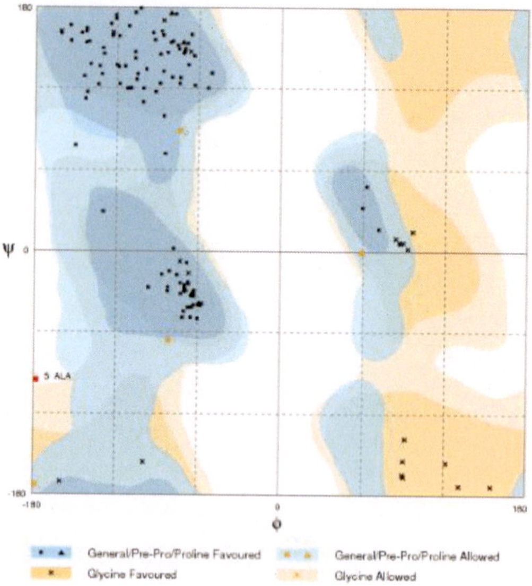

Fig 5: Ramachandran plot by PROCHECK analysis

The Z-score computed by SAVES (Structural Analysis and Verification Server) indicates the overall model quality, and its value for the respective model is related to the Z-scores of all experimentally determined protein structures in the current database indicating whether the Z-score of the particular structure is within the range of reasonable scores. This plot shows the local model quality by plotting energies as a function of amino acid sequence position. As a result, the Cecropin1 shows a Z-score of 0.01. These values indicate that the structural average for the second generation quality control values is within a normal range. Therefore all final refined models did pass the formal evaluation and need to be considered as suitable for further studies.

Identification of Active site: The final structure obtained from Molecular simulation studies was used for the active site identification with the help of CASTp server. The possible active residues of Sortase from *Propioni bacterium* was identified by comparing with template and also using CASTp server (Fig. 6). It was identified that secondary structures were highly conserved and ASP-120, CYS-128, CYS-138, TYR-220, were involved in active site of Sortase.

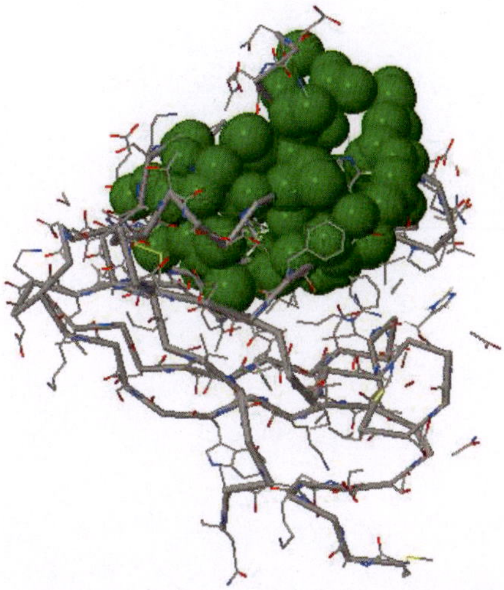

Fig. 6: Active site of Sortasepeptide

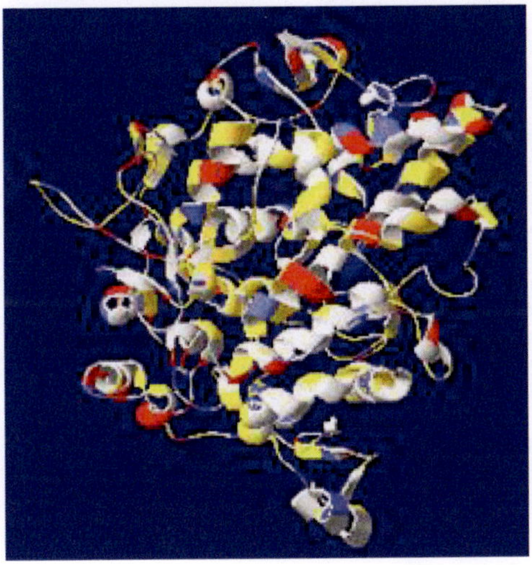

Fig. 7: Structure of Sortase

Docking studies of Cecropin1: Studies of Cecropin1 was performed with Sortase, a membrane protein of ***Propioni bacterium***. The structure of this compound shown in Fig. 8 was constructed and optimized using SPDBV software. The selected docked conformations of Cecropin1 into the Sortasebinding site are shown in Fig.9. The docked results (Table 2) revealed that the Sortase the inhibitory compound of Sortase of *Propioni bacterium*..

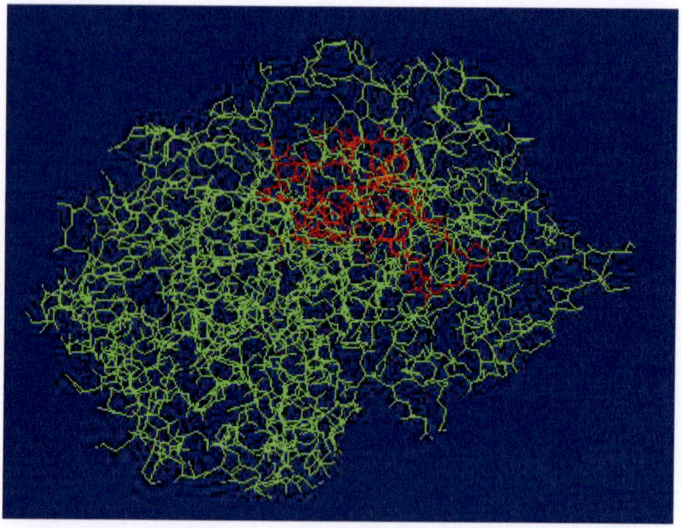

Fig 8: Docking results of Cecropin1 with the active site of Sortase A.

Table 2: Docking results

Docking Score (K.Cal/mol)	Molecule
-23.48	Cecropin1

CONCLUSION

In the present study, docking results revealed the binding orientation of Cecropin1 in the Sortase binding pocket surrounding the active site, which resulted in inhibition of bacteial activity. Based on this results, the bacterial membrane protein Sortase in *Propioni bacterium* plays an important role in anti-bacterial activity. However, further in vitro studies are required to confirm the above result.

REFERENCES

1. Meister, M., Lemaitre, B., & Hoffmann, J. A. (1997). Antimicrobial peptide defense in Drosophila. *Bioessays*, 19(11), 1019-1026.

2. Ferrandon, D., Imler, J. L., Hetru, C., & Hoffmann, J. A. (2007). The Drosophila systemic immune response: sensing and signalling during bacterial and fungal infections. *Nature reviews immunology*, 7(11), 862-874.

3. Lemaitre, B., & Hoffmann, J. (2007). The host defense of Drosophila melanogaster. *Annu. Rev. Immunol.*, 25, 697-743.

4. Lemaitre, B., Reichhart, J. M., & Hoffmann, J. A. (1997). Drosophila host defense:

differential induction of antimicrobial peptide genes after infection by various classes of microorganisms. *Proceedings of the National Academy of Sciences*, 94(26), 14614-14619.

5. Hultmark, D. (2003). Drosophila immunity: paths and patterns. *Current opinion in immunology*, 15(1), 12-19.

6. Tamang, D. G., & Saier Jr, M. H. (2006). The cecropin superfamily of toxic peptides. *Journal of molecular microbiology and biotechnology*, 11(1-2), 94-103.

7. Vallet-Gely, I., Lemaitre, B., & Boccard, F. (2008). Bacterial strategies to overcome insect defences. *Nature Reviews Microbiology*, 6(4), 302-313.

8. Hoffmann, J. A. (2003). The immune response of Drosophila. *Nature*, 426(6962), 33-38.

9. Imler, J. L., & Hoffmann, J. A. (2000). Signaling mechanisms in the antimicrobial host defense of Drosophila. *Current opinion in microbiology*, 3(1), 16-22.

10. Reddy, K. V. R., Yedery, R. D., & Aranha, C. (2004). Antimicrobial peptides: premises and promises. *International journal of antimicrobial agents*, 24(6), 536-547.

11. Otvos, Jr, L. (2000). Antibacterial peptides isolated from insects. *Journal of peptide science: an official publication of the European Peptide Society*, 6(10), 497-511.

12. Gottar, M., Gobert, V., Matskevich, A. A., Reichhart, J. M., Wang, C., Butt, T. M., & Ferrandon, D. (2006). Dual detection of fungal infections in Drosophila via recognition of glucans and sensing of virulence factors. *Cell*, 127(7), 1425-1437.

13. Bulet, P., Hetru, C., Dimarcq, J. L., & Hoffmann, D. (1999). Antimicrobial peptides in insects; structure and function. *Developmental & Comparative Immunology*, 23(4-5), 329-344.

14. Rotem, S., & Mor, A. (2009). Antimicrobial peptide mimics for improved therapeutic properties. *Biochimica et Biophysica Acta (BBA)-Biomembranes,* 1788(8), 1582-1592.

15. Ganz, T. (2001). The role of antimicrobial peptides in innate immunity. *American Zoologist*, 41, 1449.

16. Gobert, V., Gottar, M., Matskevich, A. A., Rutschmann, S., Royet, J., Belvin, M., & Ferrandon, D. (2003). Dual activation of the Drosophila toll pathway by two pattern recognition receptors. Science, 302(5653), 2126-2130.

Current Status of Researches in Biosciences (2020) : 193-200
Editor : Prof. P.C. Joshi, N. Joshi, Reshman Yasmin, and D. K. Mansotra
Today & Tomorrow's Printers and Publishers, New Delhi - 110 002

UTILITY OF PAPER MICROSCOPE – "FOLDSCOPE" IN BIOMEDICAL RESEARCH

Kelevikho Neikha, Bendangtula Walling, Abuno Thepa, Nukshimenla Jamir, and Sarat Chandra Yenisetti*

Drosophila Neurobiology Lab, Department of Zoology, Nagaland University (Central), Lumami 798627 Nagaland, India.

Email: yschandrays@rediffmail.com

ABSTRACT

Foldscope is a facile tool constructed using the principle of "origami" which has a magnification of 2000x and sub micron resolution. Punched only on a flat sheet of paper but robust enough for field study and able to reach in resource constraint settings. Since, many individuals live in rural or underserviced location with limited access to care, and where infection such as malaria, Schistosomiasis and soil transmitted helminths are rife, this novel device may be a stepping stone for future portable diagnosis, as it is a tool with less intricacy and anyone with minimal handling experience can readily use it. We have optimized this paper microscope to study *Drosophila* sex organs, understanding of which will help to study the process of reproduction in *Drosophila* model; further I will be presenting the data relating to study of barr body in buccal smear and basophils from the blood smear. These studies will be of good help for biomedical researchers in their field survey/studies.

Keywords: Foldscope , Barr Body , Buccal smear, Basophil

INTRODUCTION

Foldscope is an origami-based optical microscope that can be assembled from a flat sheet of paper and, a lens that can provide over 2000x magnification with sub micron resolution. Easy portability, flexibility, light weight (8.8g) and minimalistic accommodation (enough to fit in a pocket $70\times20\times2$ mm^3) has made it a robust tool in resource constrained settings and catastrophic situation [1,2]. Traditional microscope are bulky, costly which becomes an obstacle in resource constrained settings, frequently results in under treatment of infectious diseases [3,4]. This expensiveness and fragility of the microscopes inspired Dr Manu Prakash from Stanford University, to create an affordable and sturdy microscope which will reach the people with its frugality. The story started way back in 2011 when he was at a field station in Thailand. He saw an expensive microscope but no one dare to touch it as they worth five times their salary [5]. To overcome these limitations of the conventional microscope, Dr Prakash and team came up with the idea of making affordable and easy to use microscope, by using folded papers and hence named "Foldscope". They developed the first archetype in 2014 as an easily accessible diagnostic tool in the field of research [6]. This innovative tool is apprehending the mind of young children, students, hobbyists, teachers and research workers [7] opening a new window to our unknown micro world and ultimately seeks to make science available to everyone and show "deep connection between science education and global health"[5]. A website community called „Microcosmos is available where foldscopist can share their work and update the latest with other fellows.

The adroitness of this tool can have an immense impact on Medical camp because most of the time poor road condition and electricity becomes a hindrance in developing countries since, any kind of test needs a good microscope along with accessories which is not easily transportable. So, in this kind of campaign patients mostly living in developing and under developed countries suffering from treatable neglected tropical diseases (NTD) [8] went home getting wrong diagnosis or being ask to get blood test which sometimes it s hard to access, in this kind of situations foldscope can be a feasible tool to come in to play.

Foldscope has been used in various field such as study of planktons, plant tissues, microbes, microarthropods, insect, urine crystals [9], histopathology[2] and in diagnosis of Schistomiasis[10]. We have optimized this paper microscope to study *Drosophila* sex organs, understanding of which will help to study the process of reproduction in *Drosophila* model; further we made efforts to study barr body in buccal smear and basophils from the blood smear, knowledge of which will further the use of simple paper microscope in biomedical research. These efforts will be discussed in detail in the present paper.

MATERIALS AND METHODS

Sample preparation

Processing of Drosophila reproductive organs: Using *Drosophila melanogaster* as my sample study, their reproductive organ viz; ovary and testes was dissected placing on a clean slide, added a drop of phosphate buffer saline to avoid any tissues attached, gently transfer them onto a cavity slides and observe under the Foldscope. The entire slides was evaluated for a minimum of 3-5 mins and finally mounted on Samsung Galaxy S8 + to be imaged.

Preparation of polytene chromosome: The salivary glands of the well fed 3rd instar larvae was dissected out in physiological saline (0.7% Nacl) and pass through 1N HCl (fixative) for 1-2 minutes before it was transferred to a drop of 2% Lacto aceto orcein and was left for 30-45 minutes. Then the stained glands will be placed in a drop of 45% acetic acid on a clean slide, a coverslip was then placed and squashed by applying pressure over the cover slip and was observed under Foldscope and imaged on Samsung Galaxy S8+.

Preparation of Barr body from buccal mucosal cells and basophils from blood smear: Buccal mucosal scrapes was taken from mouth sample and stained with alkaline methylene blue for 3 mins, rinsed the stained sample with distilled water and observed under foldscope to be imaged. Similarly, human blood smear was performed on a clean slide and stained using Geimsa stain for 20 mins, washed with distilled water and air dried, finally it was observed under the Foldscope which was mounted and imaged on Samsung Galaxy S8 +. The following slides were observed as follows:

RESULT AND DISCUSSION

1) Demonstration of Ovary and Testes of *Drosophila melanogaster*

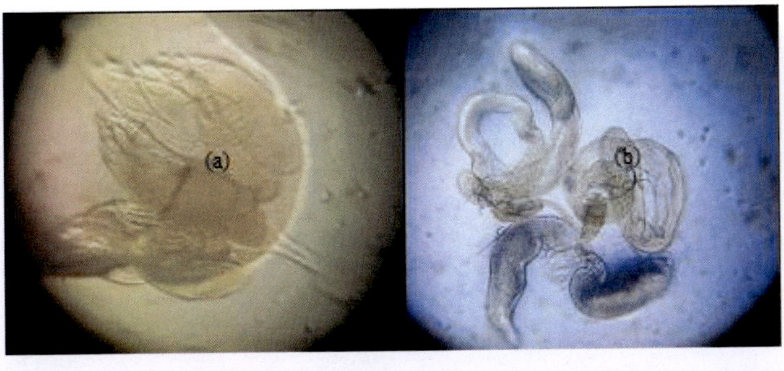

Fig. 1:(a) Ovary and (b)Testes of *Drosophila melanogaster* using Foldscope.

Drosophila is an appealing model for studying adult stems cells. Females have two ovaries and each ovary consist 16-20 tube like structures called ovarioles, which have chains of egg chambers in increasing stages of maturity. In ovarian follicles ecdysteroids, a steroid hormone are found which plays a versatile role in the development of ovary [11]. The ovary of *Drosophila* is also known to be one of the most powerful tools for investigating GSC(Germ line Stem Cells) and niches. Since the ovary consists three types of stem cells which also provides outstanding opportunities for study of stem cells in a given tissue working together to contribute the tissue function and maintenance[12]. The *Drosophila* contains a pair of testes; each is a long blind- ended tube that is coiled around a seminal vesicle. The testes of flies are easily distinguished from other tissues as it is slightly yellowish in colour. At the end of spermatogenesis, the spermatids lose their interconnections and become surrounded by individual plasma membranes. Mature sperm are then released from the open end of the testis into seminal vesicle, where they are stored until needed. Thus, the testis contains a gradient of developmental stages, from stem cells in the niche at the apical end tomature sperm at the basal end. In *Drosophila* testes, ecdysone receptor (EcR with novel composition) is also involved in male fertility, depletion of which causes cell death in male accessory glands which fail to produce seminal fluid proteins leading to sterility of *Drosophila* males [13].

2) Study of *Drosophila* salivary gland and polytene chromosome.

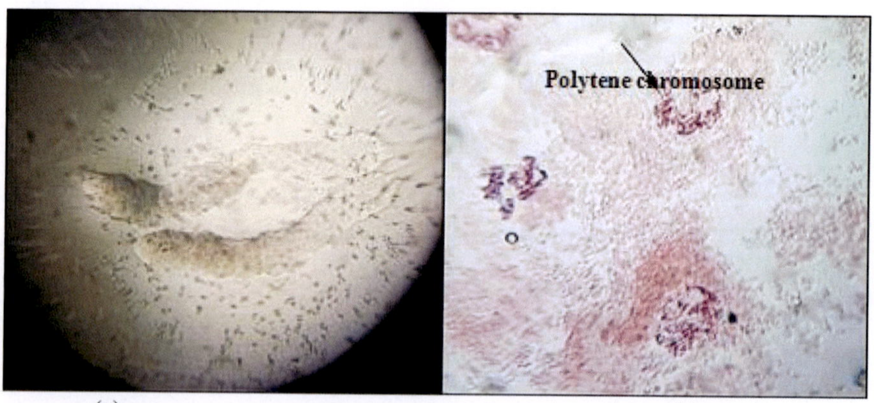

(a) (b)

Fig. 2: a) Salivary glands and b) polytene chromosomes in III instar larvae of *Drosophila melanogaster.*

Drosophila salivary glands consist of two major cell types: secretory cells and duct cells. Secretory cells are columnar epithelial cells that synthesize and secrete high levels of protein. Duct cells are cuboidal epithelial cells that form the simple tubes connecting secretory cells to the larval mouth. In the secretory cells the polytene chromosome are present which is unit to dipterian insects. Polytene chromosomes were first discovered by E.G. Balbiani in 1881. These are giant chromosomes which contain thousands of DNA strands due to "endomitosis". At interphase, distinct thick and thin banding patterns are seen which helps to map chromosomes, identify small chromosome mutations and in taxonomic identification. It is also used to study function of gene and transcription [14].

3) Study of Barr body in Human female buccal mucosal cells

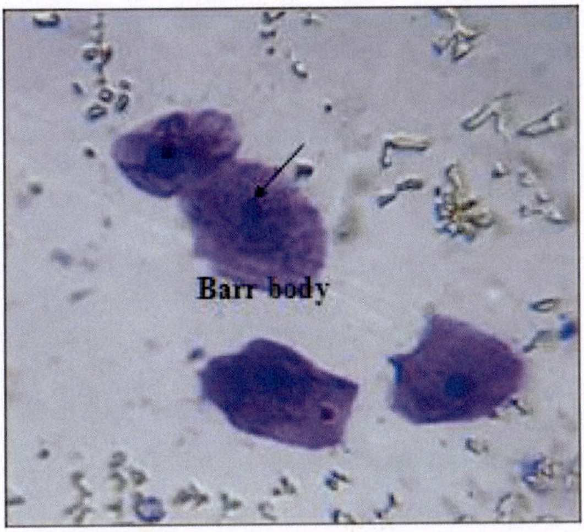

Fig. 3: Demonstration of Human female buccal mucosal cells.

Human buccal mucosal cells with barr body resulting due to hetero-chromatinization of one of the x chromosomes in human females. It was first observed by Murray Barr in 1949 and is found only in female cells as 1X chromosome is sufficient for metabolic activity whereas absent in male somatic cells as only 1 X is present [15]. It has been proven that buccal mucosal scrapes forms the simplest, quick and easiest method of sex determination when information relating to deceased is unavailable. Sex determination becomes the first priority in the process of identification of a person by a forensic investigator in the case of mishaps, chemical and nuclear bomb explosions, natural disasters crime investigation and ethnic studies.

4) Demonstration of Human blood smear showing basophil

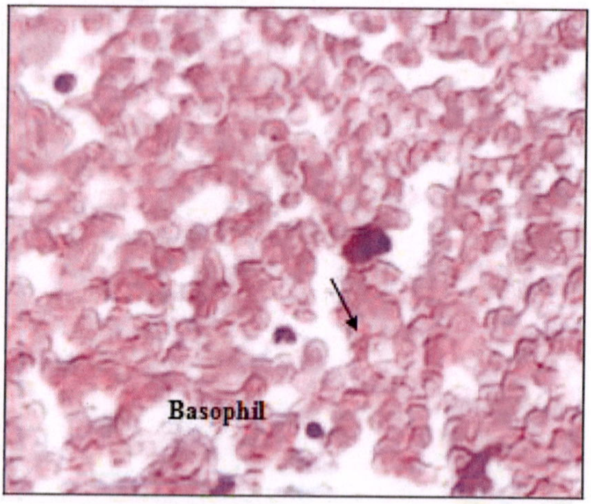

Fig. 4: Human blood smear showing basophil

Human blood consist of two special cells erythrocytes and leukocytes. Human blood smear provides information on the number and shape of cells, which can help doctors diagnose certain blood disorders or other medical conditions. With the help of Foldscope we have observed basophil in the above slide (Fig. 4). Basophil are rarest leukocytes less than 1% with 9-10 μm in diameter [16]. It secretes anti-coagulant and vasodilatory substances as histamines and serotonin which mainly function to mediate hypersensitivity reaction.

CONCLUSION

Using Foldscope, we have demonstrated *Drosophila melanogaster* sex organs, understanding of which will help to study the process of reproduction in *Drosophila* model; further we have demonstrated barr body in buccal smear of human females and basophils from the blood smear, knowledge of which will further the use of simple paper microscope in biomedical research. The future of Foldscope-assisted diagnostic technology is very promising and can be a stepping stone towards treating diseases. Along with smartphone it takes us to a different dimension where the desired samples can be processed, clicked and shared with expertise. As it is a tool with less intricacy; little adjustment is needed on spherical and chromatic aberration [17] and anyone with minimal handling experience can readily use it in every nook and corner of the world.

ACKNOWLEDGEMENTS

This work is supported by the DBT (Department of Biotehnology, India) grant No. BT/IN/Indo-US/Foldscope/39/2015, dated 20-3-2018 awarded to corresponding author Dr. Sarat C. Yenisetti, Department of Zoology, Nagaland University, Lumami.

REFERENCES

1. Cybulski, J. S., Clements, J., & Prakash, M. (2014). Foldscope: origami-based paper microscope. *PloS one*, 9(6), e98781.

2. ASM Waliullah (2018). Application of mobile phone-based portable microscopy in clinical histopathology : A feasibility study. *Int.j.clin.biomed.res.*, 4(2) :16-20.

3. Vasudevan M. (2014). A disposable microscope for a little as $ 1.

4. Bogoch, I. I., Koydemir, H. C., Tseng, D., Ephraim, R. K., Duah, E., Tee, J., ... & Ozcan, A. (2017). Evaluation of a mobile phone-based microscope for screening of *Schistosoma haematobium* infection in rural Ghana. *The American journal of tropical medicine and hygiene,* 96(6), 1468-1471.

5. Suara A. The Foldscope : Making discoveries through experience. http://rightquestion.org/resources/foldscope-making-discoveries-through-experience.

6. "A Microscope to save the world" (2015).*The New Yorker.* Retrieved.

7. Walker D. The Foldscope: Exploring the construction of a beta trial kit, the lenses and their optical performance.

8. Oscar H.,*et al.* (2017). Point-of-care mobile digital microscopy and deep learning for the detection of soil transmitted helminths and *Schistosoma haematobium*.10.1080/16549716 :1337325.

9. Calder R.,*et al.* (2018). Preliminary Studies in the use of the Foldscope paper microscope for diagnostic analysis of of crystals in urine: Issues in the analysis of liquid sample and potential applications in low budget/low tech regions of the world.

10. Ephraim, R. K., Duah, E., Cybulski, J. S., Prakash, M., D'Ambrosio, M. V., Fletcher, D. A., ... & Bogoch, I. I. (2015). Diagnosis of *Schistosoma haematobium* infection with a mobile phone-mounted Foldscope and a reversed-lens CellScope in Ghana. *The American journal of tropical medicine and hygiene*, 92(6), 1253-1256.

11. Belles, X., & Piulachs, M. D. (2015). Ecdysone signalling and ovarian development in insects: from stem cells to ovarian follicle formation. *Biochimica et Biophysica Acta (BBA)-Gene Regulatory Mechanisms*, 1849(2), 181-186.

12. Kirilly, D., & Xie, T. (2007). The Drosophila ovary: an active stem cell community. *Cell research*, 17(1), 15-25.

13. Sharma, V., Pandey, A. K., Kumar, A., Misra, S., Gupta, H. P., Gupta, S., & Ram, K. R. (2017). Functional male accessory glands and fertility in *Drosophila* require novel ecdysone receptor. *PLoS genetics*, 13(5), e1006788.

14. ALANEN, M. (1986). Whole mount electron microscopic maps of *Drosophila melanogaster* polytene chromosomes. *Hereditas*, 104(1), 91-101.

15. https:// biocyclopedia.com/ biotechnology _methods/cell_biology _and _genetics/ buccal epithelial smear and barr body.

16. Wheater, P. R., Burkitt, H. G., & Daniels, V. G. (1987). *Functional Histology A Text and Colour Atlas*. Hong Kong. English Language Book Society/Churchill Livingstone. Second Edition: Longman Group.

17. Denaro, F., Gabriel, M., Noe, A., & Nyaga, S. (2018). Developing Tools for STEM Education: The Foldscope, a Very Inexpensive Monocular Microscope for Biological Research. *Microscopy and Microanalysis*, 24(S1), 1374-1375.

Current Status of Researches in Biosciences (2020) : 201-210
Editor : Prof. P.C. Joshi, N. Joshi, Reshman Yasmin, and D. K. Mansotra
Today & Tomorrow's Printers and Publishers, New Delhi - 110 002

TECHNOLOGICAL INTERVENTION IN DEVELOPING NUTRICEREALS BASED RTE FOOD – A FOOD FOR NUTRITIONAL SECURITY

Harini Venugopal*, H.M. Jayaprakasha and Arun Kumar, H

Dairy Science College, Hebbal, Bengaluru-560024, India

Email: harinivenugopal.sj@gmail.com

ABSTRACT

This investigation was undertaken to formulate RTE foods utilizing malted flours of nutricereals namely finger millet, foxtail millet and little millet along with corn and wheat. It is also endeavor in this study to supplement the product with WPC and Horse gram to enhance protein profile. Composite flour technology makes it possible to blend, mix or fortify one food material with others. So that the resulting fortified mix has not only better nutritional quality but also the necessary attributes for consumer acceptance. With this unique concept the investigation involved in blending of composite flour through replacement method, in the ratios of nutricereals (finger millet, foxtail millet and little millet; 1:1:1) and corn-wheat with 0: 100, 25: 75, 50: 50, 75: 25 and 100: 0 proportions. The RTE food was formulated with Nutricereals and Corn-wheat (75: 25), 20 per cent sugar, 10 per cent horse gram and 8.0 per cent WPC. It was further subjected to extrusion process and oil fried to develop RTE food .The developed millet based composite flour RTE product was found to be highly acceptable without sacrificing the sensory quality The functional aspects of composite flour based RTE were analyzed regarding ER (Expansion ratio), WAI (water absorption ratio), WSI (water solubility Index) and OAC (oil absorption capacity) .The result revealed that RTE food developed with 75:25 nutricereal : corn and wheat flour combination had 2.74 ER, 5.40 WAI, 5.78 WSI, 218.75 WHC, and 61.25 OAC with 3.9 % moisture, 5.5 % fat, 10.07 % protein, 5.32 % crude fiber, 2.48 % minerals and 72.73 % carbohydrate.

KEYWORDS: Minor millets, Whey protein concentrate, Horse gram, RTE food, functional properties.

INTRODUCTION

New product development and continuous innovation in food industry has been partially a reflection of the changing needs of consumer for health, nutrition and convenience. There has been a consistent effort to increase functionality of cereals-based foods by enriching them with higher protein, phenolics, dietary fibre and mineral content so to qualify them as functional foods. Nutritional well-being is a sustainable force for health, development and maximization of human genetic potential. The nutritional status of a community has therefore been recognized as an important indicator of national development.

There is a global concern to collect, introduce, evaluate and utilize vast array of under-utilized crops and legumes such as Minor millets and Horse Gram. In India, they form an important component of the traditional cropping systems and contribute significantly to the regional food and nutritional security and diversity in the national food basket. The small millet grains have longer storage life and can be termed as famine reserve and farmers friendly. The resilience exhibited by them may prove good for their adjustment to different eco systems and make them potential crops. The growing public awareness about nutrition and health care has strengthened research. So that there is need to identify newer sources of nutritional materials with the desirable functional characteristics. Use of underutilized millets and legumes in development of extruded snacks in the form of RTE food is clearly a consequence of the recent trend to satiate the growing needs of modern consumer. The special nutritional. Millets, comprising the small-seeded group of the *Poaceae* family, represent one of the major food- and feed-crops in the semi-arid tropical regions of Africa and Asia. Millets are classified in to major millet such as sorghum and pearl millet. A minor millet crop includes Finger millet, (*Eleusine coracana*), Pearl millet (*Pennisetum glaucum*), Foxtail millet (*Setaria italica*), Kodo millet (*Paspalum setaceum*), Little millet (*Panicum sumatrense*), Proso millet (*Panicumm iliaceum*) and Barnyard millet (*Echinochola utilis*). Minor millets contain 60-70 per cent carbohydrates, 7-11 per cent proteins, 1.5-5 per cent fat, and 2-7 per cent crude fiber and are also rich in vitamins and minerals. They are excellent source of vitamin B, magnesium, and antioxidants. They are also a good source of other dietary minerals like manganese, phosphorus and iron. Millet proteins are good source of essential amino acids except lysine and threonine but are relatively high in Sulphur containing amino acids methionine and cysteine (IIMR, 2016). Several potential health benefits such as preventing cancer and cardiovascular diseases, reducing tumor incidence, lowering blood pressure, risk of heart disease, lowering LDL cholesterol, and rate of fat absorption, delaying gastric emptying, and supplying gastrointestinal bulk were reported for minor millets. Recently, the U.S. Dept. of Agriculture's nutritional guidelines put grains and grain products at the base of the food guide pyramid to emphasize grains or grain product consumption as part of a normal diet for optimal health. The minor millets are therefore recognized as NUTRI – CEREALS (GOI, 2013).

Legumes as good sources of proteins, carbohydrates, several water-soluble vitamins, and minerals, in general they make a major contribution to human nutrition. Horse gram (*Macrotyloma uniflorum*) has been recognised as potential sources of protein and other nutrients. Horse gram is largely cultivated, especially in dry areas of Australia, Burma, India and Srilanka, mainly for animal feed. It is also used as a vegetable in India and is known as the poor man's pulse crop in southern India. Recently, health-promoting and disease-preventing properties have been attributed to these phytochemical with antinutrient effects, thus attracting more and more interest from both researchers and food manufacturer's. Horse gram has been reported to have a lot of medicinal value. The rich fibre content of horse gram helps in reducing the body fat in fast mode. It is believed that consuming horse gram makes our body strong and is also good in treating kidney stones, menstrual problems, obesity, and curing cough and cold.

Whey proteins provide good functional and nutritional properties for formulation novel products which have potential to improve the quality of food products. WPC also have high content of Sulphur containing amino acids which supports antioxidant functions and can be exploited in many food formulations which exhibits multi-dimensional functionality (Jayaprakasha and Bruecker, 1999).

PROCESS OPTIMIZATION FOR THE PREPARATION OF COMPOSITE FLOUR SWEETENED RTE FOOD

1. Formulation of Nutricereal based composite flour sweetened RTE food: Malted finger millet, foxtail millet, little millet was blended at 1:1:1 ratio with 20 per cent sugar. The blended nutricereal based flour was used to replace corn-wheat flour at the rate of 25, 50, 75 and 100 per cent levels. The composite flour was preconditioned to prepare dough as per standard moisture per cent and temperature dough was extruded in a single screw extruder followed by frying at standard temperature .The developed product was subjected to Physico-chemical, Functional, Rheological and Sensory tests to adjudge the best combination of blending to prepare composite flour sweetened RTE food.

2. Supplementation of Horse gram to composite flour sweetened RTE food: For the standardized blend of composite nutricereal namely malted finger millet, foxtail millet, little millet and corn-wheat flour, malted horse gram flour was supplemented at the rate of 10.20, 30 and 40 per cent levels. The supplemented flour was preconditioned to prepare dough as per standard moisture per cent and temperature. The dough was extruded in a single screw extruder followed by frying at standard temperature. The developed product was subjected to Physico-chemical, Functional, Rheological and Sensory tests to adjudge the right proportion of horse gram.

3. Enrichment of WPC to Composite flour sweetened RTE food: The standardized blend of composite nutricereal namely malted finger millet, foxtail millet, little millet, corn-wheat and horse gram flour it was further enriched with WPC at the rate of 4, 6, 8 and 10 per cent levels. The enriched flour was preconditioned to prepare dough as per standard moisture per cent and temperature. The dough was extruded in a single screw extruder followed by frying at standard temperature. The developed product was subjected to Physico-chemical, Functional, Rheological and Sensory tests to adjudge the right proportion of WPC.

The RTE food was prepared by replacing corn-wheat with nutricereals namely finger millet, foxtail millet and little millet (1:1:1).The composite flour was replaced with corn and wheat at 25, 50, 75 and 100 per cent levels, the optimized levels were 75:25, this was further preconditioned with 22 per cent moisture at 90 °C followed by addition of 20 per cent sugar, The selected proportions were supplemented with horse gram at 20 % The optimized level of horse gram and corn-wheat was further enriched with WPC at 8 %. The dough prepared was extruded and fried in refined oil at 110 °C. The results pertaining to functional attributes of RTE food are show cased here under.

Plate 1. Millet based Composite Flour Sweetened RTE Food

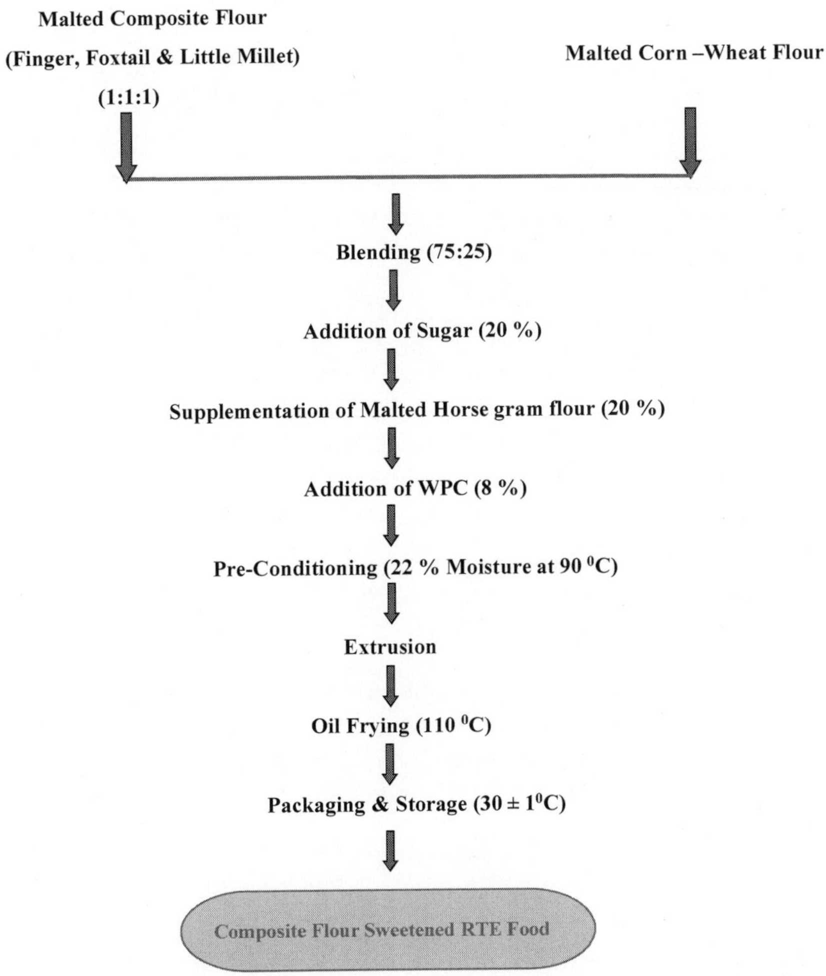

Figure 1. Process flow chart for the production of millet based composite flour sweetened RTE food

4. Effect of millet based composite flour on the chemical composition of composite flour sweetened RTE food: The effect of millet based composite flour on the chemical composition of RTE food is delineated here under (Table.1).The proportions of composite flour had non-significant effect on moisture, fat and mineral contents. The moisture per cent ranged from 3.85 to 3.91 per cent for 25 to 100 per cent replacement of composite flour where control sample had 3.82 per cent moisture. Similar pattern was also observed for fat content. The fat content of control was 5.21 and the composite flour based RTE food had fat content of 5.30 to 5.31 per cent for 25 to 100 per cent level of composite flour indicating that the composite flour incorporation had non-significant effect on fat per cent.

The mineral content also indicated non-significant effect for different proportions of composite flour. The control had 2.32 per cent mineral content whereas 25, 50, 75 and 100 per cent levels of composite flour had 2.37, 2.42, 2.48 and 2.52 per cent of mineral content.

The protein and carbohydrate had significant effect on RTE food. Protein content decreased with increase in the composite flour. The protein content was 11.39, 10.72, 10.07 and 9.41 per cent for 25, 50, 75 and 100 per cent, respectively in the RTE food incorporated with composite flour.

The carbohydrate content significantly decreased with increase in composite flour level. The carbohydrate content of control was 74.61 per cent levels and the formulated RTE food possessed 73.98, 73.35, 72.73 and 72.16 per cent carbohydrate.

The moisture content of composite flour RTE food varied between 3.85 to 3.91. The highest moisture content was observed for 100 per cent levels (3.91) followed by 75 per cent (3.90), 50 per cent (3.87) and 25 per cent (3.85) replacement of nutricereals.

The mineral and fat content exhibited non-significant effect as proportions of nutricereals increased. The fat content varied between 5.30 to 5.59 per cent and mineral content varied between 2.37 to 2.52 per cent for 25 to 100 per cent levels of composite flour respectively as the nutricereals contains low fat profile.

As observed from the result it is acceptable fact that as the levels of composite flour increased protein and crude fiber since the nutricereals are rich in proteins and crude fiber. On the other hand, it was observed that the carbohydrate content of composite flour RTE food had lower per cent when compared to control executing the low profile of carbohydrate in the millets. This is agreeable with Noorfarhzilah *et al.* (2014), Thejeshwini and Rao (2014); Swapnil *et al.* (2015).

Table 1. Effect of millet based composite flour on the chemical composition of composite flour sweetened RTE food

Proportions of Composite flour	Moisture	Fat	Protein	Crude fiber	Mineral	Carbohydrate
				Per cent		
Control	3.82[a]	5.21[a]	12.04[a]	2.00[a]	2.32[a]	74.61[a]
25:75	3.85[a]	5.30[a]	11.39[b]	3.11[b]	2.37[a]	73.98[b]
50:50	3.87[a]	5.41[a]	10.72[c]	4.23[c]	2.42[a]	73.35[c]

75:25	3.90^{a}	5.50^{a}	10.07^{d}	5.32^{d}	2.48^{a}	72.73^{d}
100:0	3.91^{a}	5.59^{a}	9.41^{e}	6.41^{e}	2.52^{a}	72.16^{e}
CD (*P* = .05)	0.13	0.42	0.48	0.50	0.27	0.41

Note:* All values are average of three trails

Similar super scripts indicate non-significance at the corresponding critical difference (CD)

5. Effect of millet based composite flour on functional properties of composite flour sweetened RTE food: The effect of millet based composite flour on functional properties of composite flour sweetened RTE food is depicted in Table. 2. As evident from the response, the ER has non-significant effect on various proportions of composite flour. The control had 3.64 ER whereas the RTE food had 3.34, 3.05, 2.74 and 2.43 for 25, 50, 75 and 100 per cent levels of composite flour, respectively.

The WAI has a significant effect on different levels of composite flour. The control had 8.05 g/100 g whereas treated sample as 7.13, 6.27, 5.40 and 4.51 g/100 g as WAI at 25, 50, 75 and 100 per cent levels of composite flour.

The WSI of control was 3.12 g/100 g for RTE food and it was observed 3.12, 4.45, 5.78 and 7.10 g/100 g for 25, 50, 75 and 100 per cent, respectively with composite flour.

The WHC capacity of RTE food had significant influence as seen in plotted response. The control had WHC of 116.90 g/100 g whereas formulated food with composite flour had 150.20, 184.90, 218.75 and 252.23 g/100 g for 25, 50, 75 and 100 per cent, respectively.

The OAC of RTE food decreased with increase in the proportions of composite flour. The OAC of control was 89.10 g/100 g whereas composite flour based RTE food had OAC of 79.23, 70.50, 61.25 and 51.91 g/100g for 25, 50, 75 and 100 per cent levels respectively. Thereby the results indicate the positive linear effect of proportions of composite flour on the functional properties of formulated RTE food.

The independent variables of formulation i.e., various proportions of composite flour had a positive linear effect on the dependent variable. This can be justified with result.

The ER though exhibits non-significant effect it can be observed from result that there was slight decrease. In our study the use of composite flour increased the protein levels and decreased the starch level in the blend which may have impended full starch gelatization, which indicates that the structure of extruded products depends on starch gelatinization. The findings are in conformity with Omwamba and Mahungu (2014) who developed protein rich RTE extruded snacks from a composite blend rice, sorghum and soy flour. The results are also comparable with Sawanth *et al.* (2013) who blended finger millet, maize flour, rice flour and soy flour in the ratio of 20: 50: 20: 10.

The WAI was found to be decreased this can be postulated that increasing temperature can be assigned to high dextrinization. The relevant findings were also reported by Sawanth *et al.* (2013) and Patil *et al.* (2017).

Table 2. Effect of millet based composite flour on functional properties of composite flour sweetened RTE food

Proportions of Composite flour	ER	WAI	WSI	WHC	OAC
			(g/100g)		
Control	3.64^{a}	8.05^{a}	1.80^{a}	116.90^{a}	89.10^{a}
25:75	3.34^{a}	7.13^{b}	3.12^{b}	150.20^{b}	79.23^{b}
50:50	3.05^{a}	6.27^{c}	4.45^{c}	184.90^{c}	70.50^{c}
75:25	2.74a	5.40^{d}	5.78^{d}	218.75^{d}	61.25^{d}
100:0	2.43^{a}	4.51^{e}	7.10^{e}	252.23^{e}	51.91^{e}
CD (P = .05)	1.27	0.62	0.58	18.0	1.50

The WSI of composite flour RTE food increased with the increase in the proportions of composite flours. Highest WSI was observed for 100 per cent levels of composite flour (7.10 g/100 g). This could be due to depolymerization and degradation of macro-molecular to small molecules imparting high solubility. Similar results were documented by Patil *et al.* (2017) also optimized extrusion process using composite blend of finger millet, horse gram and rice in the proportions of 50: 40: 0 similarly Siddharth (2014) also determined the characteristics of RTE snack by blending little millet, sorghum, foxtail millet, black gram and chickpea flours in the ratios of 40: 20: 20: 10: 0 respectively.

The WHC was maximum for the composite flour RTE food when compared with control. This could be due to high crude fiber content in the composite flour which help in highest moisture retention. The results are in conformity with Deshpande and Poshadri (2011) and Sudha *et al.* (2015) who studied the influence of composite flour pasta using wheat, little millet, composite flour and soybean flour in the ratio of 40:50:5:5. Similarly,

Omwamba and Mahungu (2014) investigated on RTE snack from rice, sorgam and soya in the ratio of 65:20:15.The OAC decreased with increase in the proportions of the composite flour, which could be attributed to hydrophobicity of proteins and also the processing steps undergone by millets before blending.

CONCLUSION

The underutilized and ignored minor millets are now receiving attention all over due to the fact that nutritional potential of minor millets in terms of protein, crude fiber, mineral and low fat are un comparable. Thereby minor millets are coined as **"Nutricereals".** RTE foods are increasingly gaining global attention and acceptance which add variety to diet with convenience. RTE food in association with extrusion technology has numerous advantages over traditional and conventional methods, is one of the most versatile and energy efficient processes currently contributing solutions to world hunger and nutritional problems. The development of food products using composite flour has increased and is attracting much attention from researcher. The innovative trend to blend composite flour is a good new approach to utilize minor millets in the formulation of RTE food enforcing more on food security.

REFERENCES

1. Deshpande, H.W. and Poshadri, A. (2011). Physical and sensory characteristics of extruded snacks prepared from Foxtail millet based composite flours *International Food Research Journal.,* **18:** 751-756.

2. GOI, (2013). State of Indian Agriculture: Ministry of Agriculture, Dept.of Agriculture and Co-operation, Directorate of Economics and Statistics, New Delhi.

3. IIMR: Indian Institute of Millet Research, (2016). Nutritional and Health benefits of Millets. 1- 90.

4. Jayaprakasha, H.M. and Brueckner, H. (1999). Whey protein concentrate: A potential functional ingredient for food industry. *J. Food Sci.Technol.*, **36** (3): 189-204.

5. Noorfarahzilah, M., Lee, J.S., Sharifudin, M.S., Mohd Fadzelly, A.B. and Hasmadi, M. 2014. Applications of composite flour in development of food products. *International Food Research Journal.*, **21**(6): 2061-2074.

6. Omwamba and Mahungu, (2014). Development of a Protein-Rich Ready-to-Eat Extruded Snack from a Composite Blend of Rice, Sorghum and Soybean Flour. *Food and Nutrition Sciences.,* **5**: 1309-1317

7. Patil, Charanjit Kaurand Varghese, (2017). Optimization of extrusion process for development of antioxidant rich extrudates from finger-millet and horse-gram in rice matrix. *International Journal of Chemical Studies.* **5** (6): 981-990.

8. Swapnil, Hiregoudar, Beladhadi, Hosamani, Rramachandra and Nidoni, (2015). Optimization of extrusion process parameters for the development of foxtail millet based extruded snacks. *Karnataka J. Agric. Sci.*, **28** (2): 301-303.

9. Swapnil, S., Margaret, A., Brennan, Susan, L., Mason and Charles, S. Brennan, 2016. The Effects of Fortification of Legumes and Extrusion on the Protein Digestibility of Wheat Based Snack. *Foods.* , **5**:2-18

10. Sawant, A., Thakor, S.B., Swami, A.D. and Divate, (2013). Physical and sensory characteristics of Ready-To-Eat food prepared from finger millet based composite mixer by extrusion. *Agric. Eng. In., CIGR Journal.* **15** (1): 80-85.

11. Sawant, A. A., Thakor, N.J. and Dhekale, J.S. (2015). Application of horse gram flour in extruded products. *International Journal of Development Research.,* **5**(10): 5668-5673.

12. Siddarth, M. (2014). Determination of physical characteristics of Extruded snack food prepared using little millet based composite flours. *International Journal for Research in Applied Science and Engineering Technology.,* **2** (7):213-218.

13. Sudha, Tamilselvi, Preethivardhiniand Karthika, (2015). Studies on the optimization of little millet pasta. *International Journal of Tropical Agriculture.,* **33** (2) :1865-1870.

14. Thejeswini, M.L. and Rao, H.G. (2014). Development of milk and cereal based extruded products. *International Journal of Science, Environment and Technology.,* **3** (5): 1797-1802.

Current Status of Researches in Biosciences (2020) : 211-217
Editor : Prof. P.C. Joshi, N. Joshi, Reshman Yasmin, and D. K. Mansotra
Today & Tomorrow's Printers and Publishers, New Delhi - 110 002

EFFECT OF DIETARY SUPPLEMENTATION OF GRADED LEVELS OF SEA BUCKTHORN LEAF MEAL ON THE GROWTH, IMMUNOCOMPETENCE TRAITS AND BLOOD BIOCHEMICAL ATTRIBUTES OF COLOURED CHICKEN

Vaishali Gupta, Amitav Bhattacharyya and Pankaj Kumar Shukla

College of Veterinary Science and Animal Husbandry, DUVASU, Mathura

Email: amitav6@gmail.com

ABSTRACT

One hundred and twenty day old straight run coloured chicken were distributed into four dietary treatments: T1-basal diet, T2-T1+0.5% Sea buckthorn leaf meal (SBTLM), T3- T1+1% SBTLM, T4-T1+1.5% SBTLM, having three replicates each. There was no significant difference in the growth and biochemical attributes among the treatment groups. However, humoral immune response to 1% SRBC (log_2 titre) and cell mediated immune response to PHA-P was significantly better ($P<0.05$) in T2 compared to other treatment groups. Thus, it may be concluded that dietary supplementation of 0.5% SBTLM elicited better immune response in coloured chicken.

KEYWORDS: Sea buckthorn, Chicken, Feed, Growth, Immunity

INTRODUCTION

India lives in the heart of villages constituting 67.6% of the total population of our country (Anonymous, 2014) with prime occupation as agriculture. In fact, the soul of India is in the toil of the rural areas. Thus, the growth and welfare of India depends on the development and prosperity of villages through improving or uplifting the living conditions of the rural people. Though the agriculture is main activity to be carried out at rural level, animal husbandry sector plays important role to stream line the agriculture activity. Among Animal husbandry poultry occupies leading position to support the rural economy. The total poultry population of India is 729.2 million numbers (Livestock Census, 2012). About 95% of the poultry of India comprises of chicken followed by 3% ducks and 2% other species. Out of the total poultry population of India, the backyard poultry population in rural and urban areas is 67.89 million and 4.3 million respectively. Though Indian poultry industry recorded phenomenal growth, it has been plagued by a series of problems due to high ambient temperature in the tropics accompanied by high relative humidity. Thus, backyard poultry farming is being promoted for sustainability of poultry production and up liftment of socio-economic condition of Indian farmers. Different Central Government agencies have developed several strains of poultry birds for backyard farming. Chabro is a breed, developed by Central Poultry Development Organization especially for the farmers of our country as it is comparatively more adaptive to climatic fluctuations and variable zoo sanitary conditions of the farm as compared to their broiler counterparts. Over the past few decades, antibiotics were used to increase the productivity of the birds. However, the ban on the sub therapeutic level of antibiotics in poultry feeding has led scientists to rethink for a viable alternative. In the recent past, efforts have been made to counteract the adverse effects of various levels of stress and augment the production potential in chicken by using herbs possessing therapeutic potential. Sea buckthorn *(Hippophae rhamnoides)* is a naturally growing shrub that is native to Eastern Europe and Asia and is found in the high altitudes of Indian Himalayan region. Few studies have been undertaken on supplementation of SBT leaf meal and giloe in broiler (Hu, 2000; Hu and Guo, 2006; Chen *et al.,* 2011; Kulkarni *et al.,* 2011; Singh *et al.,* 2014). However, detail studies are necessary to assess the effect of feeding sea buckthorn leaf meal (SBTLM) at various levels in coloured chicken. Hence, a study was designed to assess the effect of SBTLM at graded levels on the production performance and immunity of coloured chicken.

MATERIALS AND METHODS

Raw *Hippophae rhamnoides* leaves were procured from District Lahul-Spiti in Himachal Pradesh. Fresh leaves were sun dried in a clean, dust free environment and ground to obtain fine powder. The powder formed was packed in an airtight container.

Birds and Feed: 120 day old straight coloured chicken (Chabro) were distributed into four dietary treatments: T1-basal diet (Broiler starter diet till 4 weeks of age and there after broiler finisher diet till 8 weeks of age; BIS, 2007), T2- T1+0.5% SBTLM, T3- T1+1% SBTLM, T4- T1+1.5% SBTLM, having three replicates each. The birds were housed in deep litter system. Water was offered *ad lib.*

Body Weight gain and Feed Conversion Ratio: Weekly body weight and feed consumption were recorded and thereafter phase wise body weight gain and feed conversion ratio were determined.

Blood Biochemical Attributes: Blood was collected from six birds of each group at the end of the experiment from the wing vein with the help of heparinized syringes and poured into sterile tubes. The blood samples were centrifuged for the 10-15 min at 2500 rpm. Plasma was separated and stored in refrigerator (-20^{0}C) until analyzed. Total plasma protein, total plasma cholesterol, plasma uric acid, plasma GOT (Glutamate Oxaloacetate Transaminase), plasma GPT (Glutamate Pyruvate Transaminase) and plasma ALP (Alkaline Phosphatase) was determined by using commercial kits of Arkray Healthcare Pvt. Ltd., India, according to the manufacturer's instructions.

Immunocompetence Traits: The general innate immune-competence status of chabro birds was assayed by measuring two important immunocompetence traits as antibody response to SRBC and cell mediated immune response to PHA-P at 8 week of age.

Antibody Response to Sheep Red Blood Cells (SRBC): The microtitre plate haemagglutination procedure as described by (Siegel and Gross 1980) with slight modifications was followed to measure total HA antibody titres in chabro birds on day zero and day 5 post injection.

Preparation of Sheep Red Blood Cells (SRBC) Suspension: Blood from jugular vein was collected from healthy sheep in Alsever's solution. The red blood cells were washed thrice in PBS (phosphate buffer saline, pH 7.2). Finally 1% suspenstion of SRBC in PBS (V/V) was prepared.

Immunization and Harvesting of Immune Serum: 1 ml of 1% (V/V) of SRBC suspension was injected to 6 birds of each treatment group. About 3 ml of blood on 0 and 5th day post immunization (dpi) were collected from jugular vein. The blood was endorsed to clot in an incubator having temperature of 37^{0}C for 1 hour. The blood was endorsed to retract after detaching it from sides of its container and left at 4°C. Centrifugation of blood was carried out at 2000rpm for 5-10 minutes as it facilitated rapid collection of serum. The antibody titer was determined by HA methods (Vander Zijpp 1983; Siegel and Gross, 1980). Antibodies were determined by means of a mercaptoethanol (ME) HA test as per the method described by Martin *et al.* (1989) with slight modification.

In vivo Cell Mediated Immune Response: The cellular immune response was assessed by cutaneous basophilic hypersensitivity test *in vivo* by using PHA-P (Phytohaemagglutinin, lectin from *Phaseolus vulgaris*) (Corrier and De Loach, 1990). Coloured chicken were injected intra-dermally between 3rd and 4th toe of the right foot or on the wattle with 0.1 mg PHA-P in 0.1 ml of PBS (1 mg PHA-P/ml of PBS). The right foot received 0.1 ml of PBS and served as control. The thickness of inter-digital skin was measured using micrometer (AMES) at 0 and 24 hr after injection. The skin swelling was calculated by subtracting the skin thickness at 0 hr from that of after 24 hours of injection. The foot web index (FWI) was determined as the difference between interdigital swelling values of PHA-P injected and control foot.

STATISTICAL ANALYSIS

Data were subjected to one-way analysis of variance in a completely randomized design (Snedecor and Cochran 1994). Homogenous subsets were separated using multiple range test described by Duncan (1955).

RESULTS AND DISCUSSION

Phase wise body weight gain: The phase wise body weight gain has been presented in Table 1. There was no significant difference in body weight gain among the treatment groups during 0-4 weeks, 4-8 weeks and 0-8 weeks of age. However, Sharma *et al.* (2017) reported that dietary supplementation of 0.5% SBTLM in turkey poults resulted in significantly higher ($P<0.05$) body weight gain during 4-8 weeks and 0-8 weeks of age.

Table 1: Effect of dietary supplementation of Sea buckthorn leaf meal on the body weight gain of coloured chicken at different phases of growth.

Treatment	0 to 4 Weeks	4 to 8 weeks	0 to 8 weeks
T1	614.87	1034.00	1648.87
T2	592.73	996.07	1588.80
T3	581.87	968.80	1550.67
T4	539.93	1008.77	1548.70

Pooled SEM	14.89	12.91	23.37
Sig Level	NS	NS	NS

NS: Not Significant (P>0.05) SEM: Standard Error of Means

Phase wise feed conversion ratio: The pase feed conversion ratio has been presented in Table 2. There was no significant difference in FCR among the treatment groups during 0-4 weeks, 4-8 weeks and 0-8 weeks of age. However, Sharma *et al.* (2017) reported that dietary supplementation of 0.5% SBTLM in turkey poults resulted in significantly better (P<0.01) feed conversion ratio during 4-8 weeks and 0-8 weeks of age.

Blood Biochemical Parameters: Effect of SBTLM supplementation in feed on total plasma protein, total plasma cholesterol, plasma uric acid, plasma GOT, plasma GPT and plasma ALP have been presented in Table 3. There was no significant difference in the biochemical attributes among the treatment groups.

Table 2: Effect of dietary supplementation of Sea buckthorn leaf meal on the FCR of coloured chicken at different phases of growth.

Treatment	0 to 4 week	4 to 8 week	0 to 8 week
T1	1.92	2.69	2.41
T2	1.97	2.79	2.48
T3	1.96	2.86	2.52
T4	2.17	2.75	2.55
Pooled SEM	0.06	0.06	0.04
Sig Level	NS	NS	NS

NS: Not Significant (P>0.05) SEM: Standard Error of Means

Table 3: Effect of dietary supplementation of Sea buckthorn leaf meal on blood biochemicals (protein, uric acid, Cholesterol, alkaline phosphatase, alanine aminotransferase and aspartate transaminase) of coloured chicken at 8 weeks of age.

Treatment	Protein (g/dL)	Uric acid (mg/dL)	ALP (IU/L)	ALT (IU/L)	AST (IU/L)
T1	4.84	6.76	336.97	22.76	3.09
T2	4.44	6.72	456.97	28.73	3.98
T3	4.92	6.80	314.25	23.21	4.64
T4	4.87	5.11	439.68	21.66	6.41
Pooled SEM	0.10	0.33	53.91	1.57	0.99
Sig Level	NS	NS	NS	NS	NS

NS: Not Significant (P>0.05) SEM: Standard Error of Means

Humoral Immune Response: The humoral immune response (response to 1% SRBC HA titre) of the treatment groups have been compiled in Table 4. T2 birds had a significantly higher (P<0.01) humoral immune to 1% SRBC (log_2 titre) than other treatment groups. Similarly, T2 birds had a significantly higher (P<0.05) IgM response to 1% SRBC (log_2 titre) than T1 and T4 and comparatively higher than T3. However, Sharma (2017) reported no significant difference in the humoral immune response after dietary supplementation of 0.5% SBTLM in turkey poults.

Cell Mediated Immune Response: Effect of SBTLM feeding on cell mediated immune response to PHA-P was determined as FWI (Foot Web Index). T3 group birds had a significantly higher (P<0.05) FWI compared to T1 and T3 and comparatively higher than T4 (Table 4). However, Sharma (2017) reported no significant difference in the cell mediated immune response after dietary supplementation of 0.5% SBTLM in turkey poults.

Table 4: Effect of dietary supplementation of Sea buckthorn leaf meal on the humoral immune responses [antibody titer (log_2) values] to 1% SRBC and cell mediated immune response measured as Foot Web Index (FWI) to PHAP in coloured chicken at 8 weeks of age.

Treatment	HA	IgG	IgM	FWI
T1	7.33^{a}	3.83	3.50^{a}	0.79^{a}
T2	11.17^{b}	3.67	7.50^{b}	1.30^{b}
T3	8.67^{a}	3.00	5.67^{ab}	0.94^{a}
T4	8.67^{a}	3.83	4.83^{a}	1.00^{ab}
Pooled SEM	0.38	0.30	0.50	0.06
Sig Level	P<0.01	NS	P<0.05	P<0.05

Means bearing different superscripts within a column differ significantly (P<0.05)

NS: Not Significant (P>0.05) SEM: Standard Error of Means

CONCLUSION

Thus, it may be concluded that dietary supplementation of 0.5% SBTLM elicited better immune response in coloured chicken.

REFERENCES

1. Anonymous, (2014). http://www.tradingeconomics.com/india/rural-population-percent-of-total-population-wb-data.html

2. Chen, X., Zhao, W., Liu, H. N., Su, J., Zhang, Z. H. and Li, Y. (2011). Effect of sea buckthorn leaves on growth performance and calcium metabolism in Arbor Acres broilers. *Dongbei Nongye Daxue Xuebao,* 42: 19-24.

3. Corrier, D.E. and Deloach, J.R. (1990). Evaluation of cell mediated, cutaneous basophil hypersensitivity in young chickens by interdigital skin test. *Poultry Science*, 69: 403-408.

4. Hu, J. Z. (2000). Eco-economic values & comprehensive development techniques of sea buckthorn. Zhengzhou: *The Yellow River Water Conservancy Press* (in Chinese).

5. Hu, J. Z. and Guo, X. F. (2006). Evaluation of nutrient value of sea buckthorn in north China. *Forestry Studies in China*, 8: 50-52.

6. Kulkarni, R. C., Mandal, A. B., Munj, C. P., Dan, A., Saxena, A. and Tyagi, P. K. (2011). Response of coloured broilers to dietary addition of geloi (*Tinospora cordifolia*) during extreme summer. *Indian Journal of Poultry Science,* 46(1): 70-74.

7. Livestock Census, (2012). All India Report. Ministry of Agriculture, Department of Animal Husbandry, Dairying and Fisheries, Krishi Bhawan, New Delhi

8. Martin, A., Gross, W.B. and Siegel, P.B. (1989). IgG and IgM responses in high and low antibody selected lines of chickens. Journal of Heredity, 80: 249-252.

9. Sharma, A., Shukla, P.K., Bhattacharyya, A., Kumar, U., Roy, D., Yadav, B. and Prakash, A. (2018). Effect of dietary supplementation of sea buckthorn and giloe leaf meal on the body weight gain, feed conversion ratio, biochemical attributes and meat composition of turkey poults, Veterinary World, 11(1): 93-98.

10. Sharma, A. (2017). Effect of feeding sea buckthorn and giloe leaf meal on the performance of turkey poults. M.V.Sc. thesis, DUVASU, Mathura, India

11. Siegel, P. B. and Gross, W. B. (1980). Production and persistency of antibodies in chickens to sheep erythrocytes. 1. Directional selection. *Poultry Science,* 59: 1-5.

12. Singh, A., Kaushik, P. K., Yadav, P. K. and Yadav, P. (2014). Effect of bael (*Aegle Marmelos*) and giloy (*Tinospora Cordifolia*) alone and in combination on growth and feed conversion of broiler chicks. *Global Journal of Research Analysis*, 3: 96-99.

13. Snedecor, G.W and Cochran, W.G. (1994). Statistical Methods. 9th ed. The Iowa, State University Press, Ames, Iowa.

14. Van der zijpp, A. J. (1983). The effect of genetic origin, source of antigen and dose of antigen on the immune response of cockerels. *Poultry Sci*ence, 62: 205.

Current Status of Researches in Biosciences (2020) : 219-223
Editor : Prof. P.C. Joshi, N. Joshi, Reshman Yasmin, and D. K. Mansotra
Today & Tomorrow's Printers and Publishers, New Delhi - 110 002

TOXIC HEAVY METALS IN MILK IN THE POLLUTION IMPACT AREAS OF ERNAKULAM AND CHALAKUDY AT KERALA- A CASE STUDY AND THE NEED TO DEVELOP SAFETY STANDARDS

Indu B.

Dept.of Dairy Chemistry, College of Dairy Science and Technology, Kerala Veterinary and Animal Sciences University)

Email: indubthayyil@gmail.com

ABSTRACT

Heavy metals are naturally occurring elements that have a high atomic weight and a density at least five times greater than that of water. Their industrial, domestic, agricultural, medical and technological use have led to their wide distribution raising concerns over their potential effects on human health and the environment. Among the heavy metals, lead, arsenic, cadmium, chromium and mercury are of great public health importance. They are also classified as human carcinogens (known or probable) according to the U.S. Environmental Protection Agency, and the International Agency for Research on Cancer. So a study was under taken to quantify heavy metals in milk in the pollution impact areas of Ernakulam and Chalakudy of Kerala state. A total of one hundred and two samples of fresh cow milk were analyzed for the presence of heavy metals. The levels of heavy metals were quantified by atomic absorption spectroscopy. The results indicated the presence of heavy metal, lead in the range of 0.001- 0.09, and <0.1 and <0.05 for Cadmium, and Arsenic respectively. The result showed remarkable presence of lead in many samples and it point to a fact that there should be more meaningful rows from Food Safety and Standard Authority of India (FSSAI), the statutory body for laying down science based standards for articles of food, to precisely specify the level of heavy metals in milk to ensure the safe milk for human consumption.

KEYWORDS: Raw cow milk, Heavy metal, Atomic Absorption Spectroscopy.

INTRODUCTION

The toxicity of heavy metals depends on several factors including the dose, route of exposure, and chemical species of the metal as well as the age, gender, genetics, and nutritional status of exposed individuals. Because of their high degree of toxicity, lead, Cadmium, Arsenic are of high public health significance. These metallic elements are considered systemic toxicants that are known to induce multiple organ damage, even at lower levels of exposure. With the assumption that heaviness and toxicity are inter-related, heavy metals also include metalloids, such as arsenic, that are able to induce toxicity at low level of exposure (Bradl, H, 2005). In recent years, there has been an increasing ecological and global public health concern associated with environmental contamination by these metals. Reported sources of heavy metals in the environment include industrial, agricultural, pharmaceutical and domestic effluents (He ZL *et al*., 2005). The presence of toxic elements in the food chain is the result of the environmental pollution and their concentration need to be controlled constantly.

The food chain is an important source of Cd and Pb accumulation, especially for plants grown on polluted soils. Significant amounts of Cd and Pb can be transferred from contaminated soil to plants and grass, causing accumulation of these potentially toxic metals in grazing ruminants, particularly in cattle (López Alonso *et al.,* 2003). Accumulation of Cd and Pb in ruminants causes toxic effects in cattle, but also in humans consuming meat and milk contaminated with toxic metals.

Cd and Pb are amongst the elements that have caused the most concern in terms of adverse effects on human health. This is because they are readily transferred through food chains and are not known to serve any essential biological function. Lead is a pervasive and widely distributed environmental pollutant with no beneficial biological roles. The poisoning is more common in farm ruminants, which are considered most susceptible to the toxic effects of lead. For that reason, the concentration of Cd and Pb in cow's milk should be monitored to ensure the consumers' health.

Accumulation of Lead and Cadmium in ruminants causes toxic effects in cattle and also in humans consuming meat and milk contaminated with toxic metals. Lead is the most systemic toxicant that affects several organs in the body including the kidneys, liver, central nervous system, hematopoietic system, endocrine system, and reproductive system. Arsenic exposure affects virtually all organ systems including the cardiovascular, dermatologic, nervous, hepatobilliary, renal, gastro-intestinal, and respiratory systems (González -Weller *et al.*, 2006).

MATERIALS AND METHODS

A total of one hundred and two samples of fresh cow milk were randomly selected and analyzed for determination of essential and toxic heavy metals such as N, Pb, Cd, and As. The samples were procured from the houses and milk societies of Ernakulam and Chalkkudy, sites located in places with different pollution impact (Industrial plants and effluents). Individual samples of 30ml of liquid cow milk were taken in sterile plastic containers and then were labelled, sealed, and stored at suitable cold temperature (0–5°C).

The sample treatment involved digestion and preparation using Perkin Elmer Titan MPS equipment. The HNO_3/H_2O_2 based digestion process was adopted for the identification of trace elements by Perkin Elmer Atomic Absorption Spectrophotometry (AAS) with graphite furnace and flame. Initially 6mL of fresh cow milk was poured into plastic vessel within the pressure jacket in the digestion flask followed by addition with 12mL of 70% Nitric acid and 2mL of 30% Hydrogen peroxide. These solutions were digested using microwave digestion apparatus for 50 minutes and cooled it for 30 minutes. Then it is transferred to a 50mL standard flask and made up to the mark using double distilled water. The digested sample was stored in properly labelled and sealed Duropack PET bottles. It is then used for AAS analysis. The temperature-time-power combinations were optimized for each element, and the accuracy, precision, selectivity, and sensitivity were verified by using Atomic Absorption Spectrophotometer with reference sample. The blanks were made in the same way without using any sample.

Standards were prepared with concentrations of 0.5 ppm, 1.0 ppm, 2.0 ppm, 4.0 ppm, 8.0ppm, and 10.0ppm. The same procedure was followed for the preparation of each element under the study. Using Beer-Lamberts law, calibration curves for each element was drawn from which the concentrations of heavy metals in the samples were determined.

RESULTS AND DISCUSSION

Table 2 Toxic Heavy metal concentrations (Lead, Cadmium and Arsenic) in raw cow's milk

Toxic Heavy metal	Normal Value as per FSSAI* for milk	Obtained Range	Mean ±SE
Pb (ppm)	Nil *	0.001-0.09	0.01386± 0.01217
Cd (ppm)	Nil *	0-0.001	0
As (ppm)	<0.1 (in milk)	0-0.003	0

- **FSSAI specifies standards only for Infant milk substitute and Infant foods and not for milk.**
- *** Food Safety and Standard Authority of India.**

The mean values of Cadmium and Arsenic are almost zero, but the toxic element lead is present in the range of 0.001-0.09.

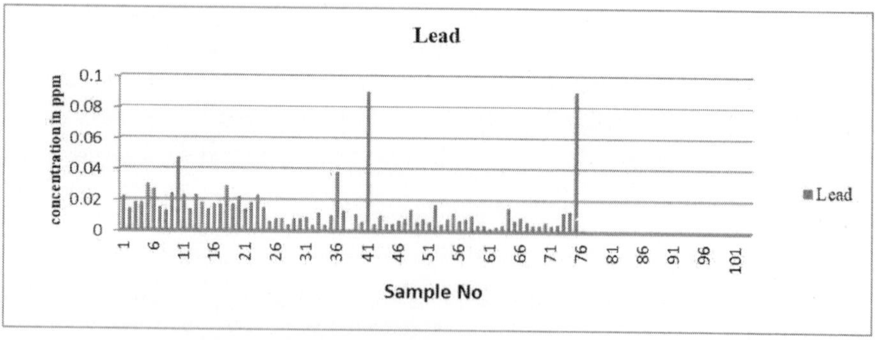

Figure 1:-Concentration of toxic heavy metal, lead in raw milk.

CONCLUSION

The toxic heavy metals such as Lead, Cadmium, and Arsenic are found to be present in the samples, but it is strange to see that the FSSAI, the single reference point for articles of food and to regulate their manufacture, storage, distribution, sale and import, has not set the limits for pure milk. The main apprehension is in the case of lead. The National Institute of Environmental Health Sciences reports that no amount of lead in the blood is safe.

The FSSAI enforces to have a certain limit for lead content in," Infant Milk Substitutes and Infant Foods but not in pure milk. There are two contradictions in this case; 1) Milk is consumed by all groups of children and not only limited to" infants" where even the low lead exposure can cause permanent intellectual disability. 2) The lead poisoning is building up slowly and may be harmful to the adults too , especially by contamination through milk, a regular diet component. So, the FSSAI* should clearly specify the permissible limit of lead and other heavy metals in milk to safeguard this.

*** Food Safety and Standard Authority of India.**

ACKNOWLEDGEMENTS

The work on "Antibiotic Residues in Milk' was supported by College of Dairy Science & Technology, Mannuthy, Kerala Veterinary and Animal Sciences University, Pookode,Wayanad under the State Plan Project.

REFERENCES

1. Ayar, A., Sert, D., & Akýn, N. (2009). The trace metal levels in milk and dairy products consumed in middle Anatolia—Turkey. *Environmental monitoring and assessment, 152*(1-4), 1-12.

2. Bradl, H. (Ed.). (2005). *Heavy metals in the environment: origin, interaction and remediation* (Vol. 6). Academic Press.

3. Duffus, J. H. (2002). "Heavy metals" a meaningless term? (IUPAC Technical Report). *Pure and applied chemistry, 74*(5), 793-807.

4. Fergusson, J. E. (1990). *Heavy elements: chemistry, environmental impact and health effects.* Pergamon.

5. F.No.2-15015/30/2010 of FSSAI, 2011.

6. González-Montaña, J. R., Senís, E., Gutiérrez, A., & Prieto, F. (2012). Cadmium and lead in bovine milk in the mining area of the Caudal River (Spain). *Environmental monitoring and assessment, 184*(7), 4029-4034.

7. He, Z. L., Yang, X. E., & Stoffella, P. J. (2005). Trace elements in agroecosystems and impacts on the environment. *Journal of Trace Elements in Medicine and Biology, 19*(2), 125-140.

8. López, A. M., Prieto, M. F., Miranda, M., Castillo, C., Hernandez, J., & Benedito, J. L. (2003). Cadmium and lead accumulation in cattle in NW Spain. *Veterinary and human toxicology, 45*(3), 128-130.

9. Pavlović, I., Sikirić, M., Lukač-Havranek, J., Plavljanić, N., & Brajenović, N. (2004). Lead and cadmium levels in raw cow's milk from an industrialised Croatian region determined by electrothermal atomic absorption spectrometry. *Czech Journal of Animal Science, 49*(4), 164-168.

10. Rao, A. (2005). Trace element estimation–Methods & clinical context. *Online Journal of Health and Allied Sciences, 4*(1).

Current Status of Researches in Biosciences (2020) : 225-239
Editor : Prof. P.C. Joshi, N. Joshi, Reshman Yasmin, and D. K. Mansotra
Today & Tomorrow's Printers and Publishers, New Delhi - 110 002

HYPERSENSITIVITY REACTIONS IN THE LIVER INDUCED BY DIFFERENT DOSES OF EGG ANTIGEN OF *ASCARIDIA GALLI* IN *GALLUS DOMESTICUS*

Juhie Agarwal[1], Savita Rani[2], and Dushyant Kr. Chauhan[3]

[1]Department of Zoology, Vardhaman College, Bijnor

[2]Department of Zoology, D.N. College, Bijnor

[3]Department of Zoology, C.C.S University campus Meerut

Email Id: juhie07@gmail.com

ABSTRACT

During present investigation hypersensitivity reactions have been studied in the liver of *Gallus domesticus* against *A.galli* which triggers immune responses inside the host. The immune responses are also capable of causing injuries. The injury is referred as sensitivity in the host. Sometimes secondary immune response are excessive inducing various changes in the host. Sometimes these responses are severe. This inappropriate response called hypersensitivity reaction. A high dose (2000 eggs/ chicks) and low dose (1000 eggs /chicks) of egg antigens have been used to stimulate the sensitivity inside host on 15 days old white leg horn chicks. After 10days high dose and low dose of egg antigens have also be used to hyper sensitized the chicks. Liver tissues have been taken out from the sacrificed host at the 25th, 35th and 45th day of the experiment and processed to observe immunopathological changes as a part of hypersensitivity reactions. During present investigation changes in the hepatocytes, sinusoids, central veins were observed in the sensitized and hyper sensitized chicks with the low and high doses of the embryonated eggs of *A. galli*. Changes observed were cloudy swelling, focal collection of lymphocyte. edema, pyknosis, karyolysis and karyokinesis.

KEYWORDS-Immune reactions, *Ascaridia galli,* Egg anitens, *Gallus domesticus*, Hypersensitivity reactions, Edema, Pycknosis, Karyokinesis.

INTRODUCTION

Most of the micro parasitic and macro parasitic infections and infectious disease are flourishing successfully in present day era. These parasites induce primary and secondary immune response. Adaptive immunity plays important functions of host defense. The host immunity can cause a variety of structural, biochemical, haematological and immunological changes in micro and macroparasites. The immune responses are also capable of causing injuries. The injury is referred as sensitivity.in the host. Sometimes secondary immune response are excessive inducing various changes in the host. Sometimes these responses are severe. This inappropriate response called hypersensitivity reaction.

These response are called as type1 hypersensitivity, type2 hypersensitivity, type3 hypersensitivity, type4 hypersensitivity, type 5 hypersensitivity reactions. All above hypersensitivity reactions cause excessive, undesirable, damaging or discomfort (uneasiness). The hypersensitivity reactions are inappropriate reactions. These reactions refer to immunopathological changes. The hypersensitivity reaction is associated with various clinical and subclinical manifestations and inflammatory reactions. The inappropriate immune responses (hypersensitivity) are also termed as allergic reactions or inflammatory reactions. The inflammatory responses induce significant tissue damage and even death sometimes.

Parasitic worms (helminthes) are highly prevalent in animals and in humans in developing countries. Parasitic worms can be found in the body either in tissues as larval forms or within the gastrointestinal or respiratory tract as adults. These worms, which live in the intestine and stomach walls, secrete antigens that act as allergens. For larger worm –as for gastrointestinal nematode, the host will often develop inflammation and hypersensitivity.

Hartmann (2003) studied the modulation of host immune responses by nematodes. He revealed that parasitic nematodes possess multiple specific capacities for immune-modulation. Parasitic infection account for hundreds of millions of dollars annual losses for livestock and poultry industry throughout the world. The most costly parasite in terms of production losses are the gastrointestinal nematodes in ruminants and poultry (Gamble and Zarlenga, 1986).The parasites change or modulate the immune system, biochemical and haematological parameters of the hosts and also lead to malnutrition in the chicks which results in the decreased return of products derived from poultry (WHO, 1967). Of the many parasitic infection ascaridiasis due to *A.galli* is a debilitating and commonly fatal discease of domestic fowl which often sets back the progress of the poultry industry economic loss (Dubinsky *et al.*, 1976; Matta, 1981; Dabur, 1996). Ascaridiasis is a common gastrointestinal disease and is caused by an enteric nematode parasite, *Ascaridia galli* (Shrank, 1788 and Freeborn, 1923) Infections with the fowl nematode *Ascaridia galli* (Shrank 1788) are widespread among domesticated and wild birds throughout the world (Souls by 1982). The life cycle of *A.galli* is direct, involving two principal populations; the sexually mature parasite in the gastrointestinal tract and the infective stage (L3) the in form of a resistant egg in the environment (Ackert, 1931; Todd & Crowdus, 1952; Araujo & Bressan, 1977). The unsegmented eggs passes with the faeces of the host into soil, where they develop into infective stage in 8-16 day. Infective eggs are ingested by birds and hatch in proventricullus depending upon several factors and hatching may occur at different periods of time. After attaining sexual maturity and after copulation the female lays large number of eggs. The eggs are passed out in soil along with faeces of the host for infection of new host. The eggs develop in the open, reaching the infective stage (L3) in 10 to 20 days or longer depending on temperature and relative humidity (Ackert, 1931; Riedel, 1947; Reid, 1958; Reid, 1960). The parasitic infection tend to be long and chronic and is associated with in appropriate immune response, immunodepression to superimposed infection and immunopathological damage.

The immune status of host plays a decisive role in resistance to parasites. There are many reports confirming the direct immune-suppressive influence of helminthes of their products on the host (Terry and Hudson, 1982; Price and Turner, 1986).

MATERIAL, METHOD AND EXPERIMENTAL DESIGN-

Various aspects of hypersensitivity reactions consist of the following studies-

Experimental host- Male white leg horn chicks

Experimental parasite- *Ascaridia galli*

Dose of infection -Low dose – 1000 embryonated pure eggs/chick.

High dose – 2000 embryonated pure eggs/chick.

Culturing of the eggs- Adult worms were isolated from the intestine of fowl from the local abattoir. *A .galli* eggs were cultured upto infective stage. The worms were gently squeezed with uniform pressure until the entire reproductive organs passed out into the culture medium in the petridish. Now fertilized eggs have been separated from the uteri. The eggs were kept in sterile solution at 32°C for embryonation for 2-3 weeks. Now these embryonated eggs per dose (estimated by dilution technique) have been given orally for sensitization and hypersensitization of the chicks.

Infection to experimental host- White leg horn chicks were grouped and labeled properly. The inocula with desired number of embryonated infective eggs (1000 eggs as low dose and 2000 eggs as high dose) were administered orally to 15 days old WLH chicks. After infection, the chicks were kept separately in spacious wooden cages in the animal house.

Maintenance of experimental host and Experimental design- During present investigation, one day old white leg horn chicks (*Gallus domesticus*) were purchased from the local hatchery and kept at normal room temperature. After 15 days, chicks were divided into three groups: Group (A)-control, Group (B) -sensitized and hyper sensitized with the low dose and Group(C)-sensitized and hyper sensitized with the high dose. Each group comprises of 10 chicks.15 days old white leg horn chicks were sensitized with the low and high doses.2 Chicks from each group were autopcised after 10 days of the sensitization (25th day) and on the same day, rest of the chicks (4 sensitized chicks) from Group (B) and Group(C) were hyper sensitized with the low dose and high dose subsequently. Then 2 chicks from each group were autopcised at 35th and 45th days of the experimental design. For the study of inflammatory changes. Intestinal tissues were taken out at the time of post-mortem examination and these tissues had been fixed and then processed for microscopic examinations and to study the hypersensitivity reactions.

Tissue processing for Immunopathological studies- The immunopathological changes were studied in liver of both control and experimental groups. The organ was cut and collected immediately from the chicks sacrified freshly in the laboratory. The tissues were fixed in 10% formalin. The fixed tissues were further processed for microtomy. Paraffin blocks were prepared and sections were cut at a thickness of 5-6 u using a rotatory microtome. The sections were stained with haematoxylins and eosin and permanent slides were prepared using mounting media Entellon.

Immunopathological changes in the liver of W.L.H. chicks -

Group (A) - Nonsensitized group (control) at 25th, 35th and 45th days as per experimental design – (Fig- 1, 2, 3)

Liver of the nonsensitized chicks showed following structure-The hepatocytes appeared roughly triangular or polygonal shape. The hepatocyte impinged upon a sinusoid and apically upon a bile canaliculus and closely to its neighbours by distinct terminal bar. The bile canaliculi between the two layers of cells form a network of ducts eventually into a canal of Hering which entered the portal canal and merged with the fine branches of the bile duct. The cytoplasm of the hepatocytes were slightly granular in appearance and nucleus was centrally located and was rounded in shape having 1 to 2 nucleoli. The walls of the sinusoids were composed of endothelial cells and scattered phagocytic kupffer cells. The portal tract consisted of one portal vein, hepatic artery and the bile duct. The histological structure of the liver of chick revealed that it was composed of solid mass of hepatic cells around central vein (CV) or hepatic plate throughout the section. The hepatic cells were irregularly surrounded by more or less cylindrical different types of sinusoids. The surface of the liver was covered by peritoneal layer of mesothelium. This layer was underlined by dense connective tissue layer called Glisson's capsule. This connective tissue layer was carried inside the substance of the liver by the portal and hepatic canals. Elastic fibers were found in small number only throughout the connective tissue of the liver.

Group (Ba_1) - Sensitized group with low dose at 25th day as per experimental design- (Fig-4)

After 25th day of the sensitization with the low dose of the embryonated eggs of *A.galli*, several changes have been observed in liver .Central vein was observed slightly dilated.

Group (Ca_1) - Sensitized group with high dose at 25th day as per experimental design- (Fig-5)

After 25th day of the sensitization with high dose of the embryonated eggs of the *A.galli*, the degenerative changes were very severe. Central vein become dilated, filled with red blood cells. Thick band of fibrotic zone was also evident around the central vein. Dilations in sinusoids were quite marked.

Group (Ba_2) - Sensitized group with low dose at 35th day as per experimental design- (Fig-6)

After the 35th day of sensitization following changes have been observed-

The central vein with thick band of fibrotic layer, was observed to be dilated. The hepatocytes revealed marked degenerative changes. The focal collections of lymphocytes revealing chronic inflammatory cells were very clearly observed. Kupffer cells were present in the sinusoidal area.

Group (Bb_1) - Hyper sensitized group with low dose at 35th day as per experimental design

After the 10 days of the hypersensitization of the sensitized chicks with low dose, severe changes have been observed. There was dilation and hemorrhages in the central vein. A thick band of fibrotic zone was observed all around the portal region. Focal collection of lymphocytes was evident around the central vein. The hepatocytes showed swelling with mild edema. Prominent kupffer cells were clearly observed.

Group (Ca_2) - Sensitized group with high dose at 35th day as per experimental design – (Fig-7)

At the 35th day in the sensitized chicks with the high dose severe changes have been observed. A thick band of fibrotic layer was also observed all around the central vein. Necrosis at the side of the central vein was clearly observed. The hepatocytes rupturing in their wall and degenerative changes were very severe showing swelling and distinct edema. Prominent kupffer cells were clearly seen.

Group (Cb_1)-Hyper sensitized group with high dose at 35th day as per experimental design – (Fig-8)

The following immunopathological alterations have been observed in the chicks after 35th day of the hypersensitization. The wall of the central vein was ruptured but showed thick band of fibrotic layer. Focal collection of lymphocytes was clearly seen around the portal tract. Lack of interstitial spaces between the sinosoids was also observed.

Group (Ba_3) - Sensitized group with low dose at 45th day as per experimental design- (Fig- 9)

In this group the pathological changes were more severe. Liver showed degenerative changes. Severe congestion in sinosoids were clearly observed. Central vein was highly dilated.

Group (Bb_2) - Hyper sensitized group with the low dose at 45th day as per experimental design – (Fig-10)

At the 45th day several changes have been observed in the hyper sensitized chicks with the low dose. Central vein was observed to be highly dilated and ruptured. Focal collection of lymphocytes was clearly observed.

Group (Ca_3) - Sensitized group with high dose at 45th day as per experimental design-– (Fig- 11)

At 45th day severe changes have been observed in the sensitized chicks with the high dose. Highly dilated central vein with thick band of fibrotic layer was clearly observed. Kupffer cells were also seen. Hemorrhages near the central vein also have been observed.

Group (Cb_2) - Hyper sensitized group with high dose at 45th day as per experimental design – (Fig-12)

At 45th day hyper sensitized chicks with the high dose showed following changes. The central vein revealed formation of two dilated horn like process and also infiltration of blood cells. Chronic inflammatory cells were observed all around the central vein. The wall of the portal canal was observed to be ruptured .A thick band of fibrotic zone was observed all around the portal region. Focal collection of lymphocytes was evident around the central vein. The hepatocytes showed swelling with mild edema.

During present investigation changes in the hepatocytes, sinusoids, central veins were observed in the sensitized and hyper sensitized chicks with the low and high doses of the embryonated eggs of *A. galli*. Changes observed were cloudy swelling, focal collection of lymphocyte. edema, pyknosis, karyolysis and karyokinesis.

DISCUSSION

During present investigation changes in the hepatocytes, sinusoids, central veins were observed in the sensitized and hyper sensitized chicks with the low and high doses of the embryonated eggs of *A. galli*. Changes observed were cloudy swelling, focal collection of lymphocyte. edema.. Changes in hepatocyte have been reported by Mossoud (1981) in the sheep naturally infected with *Dicrocoelium dendriticum*. Klei and Crowell (1981) studied *Brugia Pahangi* infected birds and found that infected liver showed large granulometa composed of lymphocyte, macrophages, eosinophils, and giant cells. Gradinarski and Svilennov (1982) found that the histological changes in the liver of lamb with experimental or natural *Cysticercus tenuicollis* infection.

Changes in liver have been attributed due to disturbance of lipid metabolism, and protein metabolism (Prodan, 1933, Huges *et al.*, 1944 and Karmaker, 1989). The various immunopathological changes observed as a result of experimental infection of *Ascaridia galli* have been studied by

Dushyant *et al.,* (2002). The present author is of opinion that *Ascaridia galli* being inter intestinal parasite induce the immunopathological changes in the liver. Cox (1982) been given an account regarding parasite induce pathology. In schistosomiasis egg antigen stimulates production of collagen resulting into silicosis in which silicon compound stimulate macrophages to induce factor that active fibroblast for producing collagen and eventually fibrosis. In the present study fibrosis in the central vein have also been observed. In fascioliasis it has been observed that excretory product of adult fluke have a duct deleterious effect on the liver cells of host by blocking mitochondrial ATP synthesis (Cox 1980).The excretory and secretory metabolites, exotoxins and endotoxins produced by intra-intestinal parasities might be responsible for causing various immunopathological alterations in other organs also as in liver, kidney spleen besides intestine (Dushyant *et al.,* (2002).

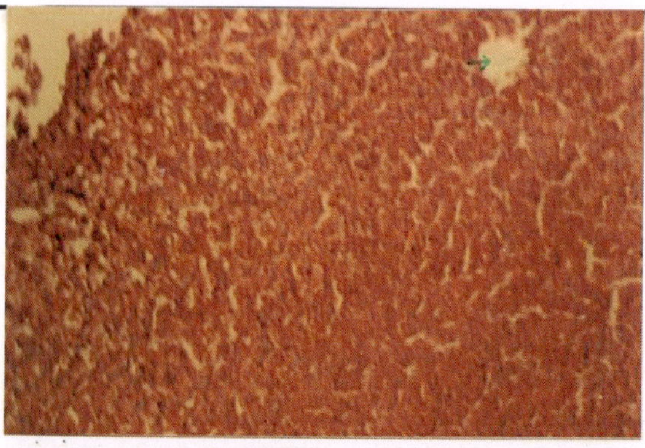

Fig-1 Liver –nonsensitized at 25th day of the infection

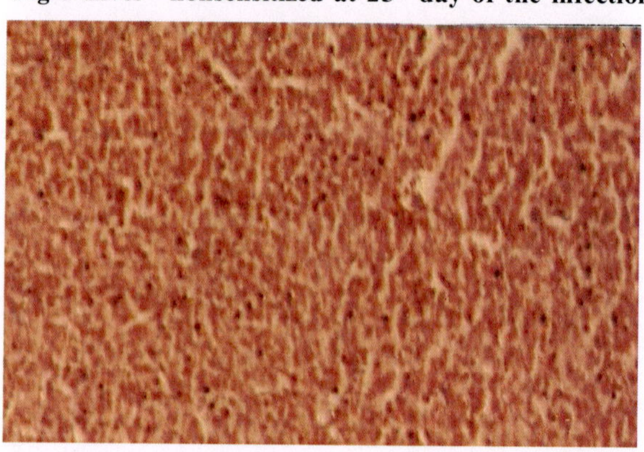

Fig-2 Liver –nonsensitized at 35th day of the infection

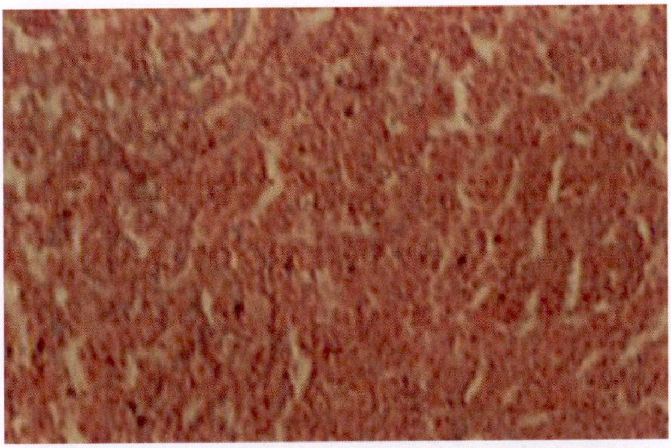

Fig-3 Non-sensitized at 45th day of the infection

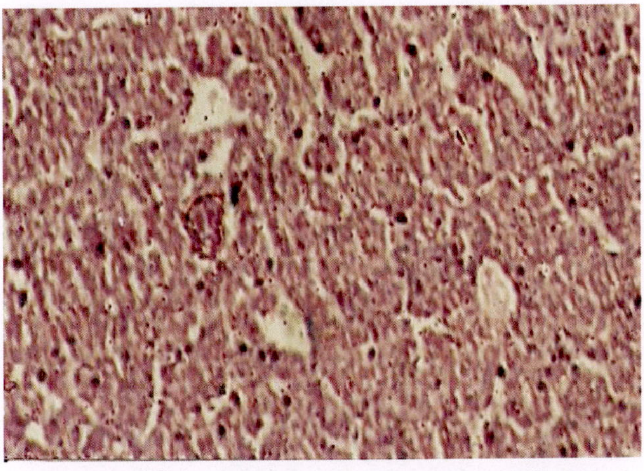

Fig-4 Liver –sensitized with low dose at 25th day

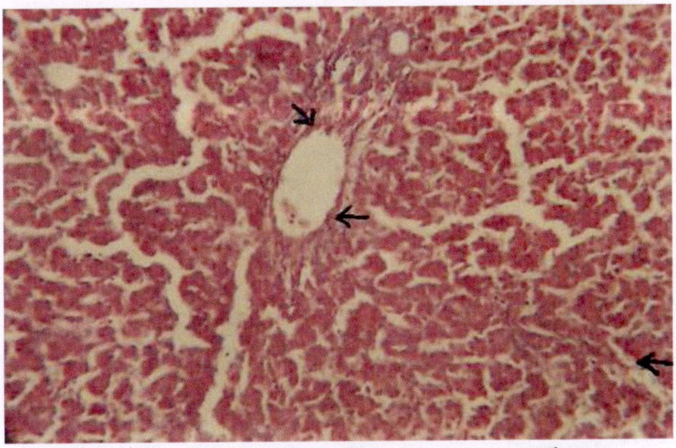

Fig-5 Liver-sensitized with high dose at 25th day

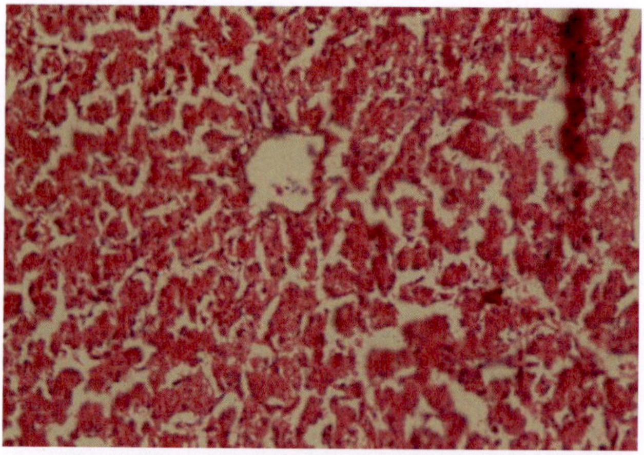

Fig- 6 Liver –sensitized with low dose at 35th day

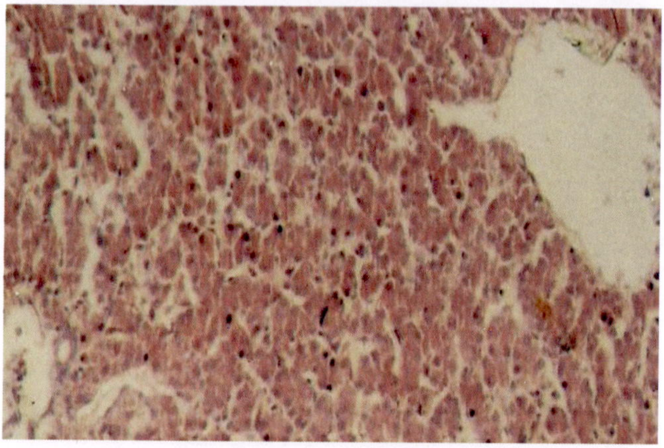

Fig-7 Sensitized with high dose at 35th day

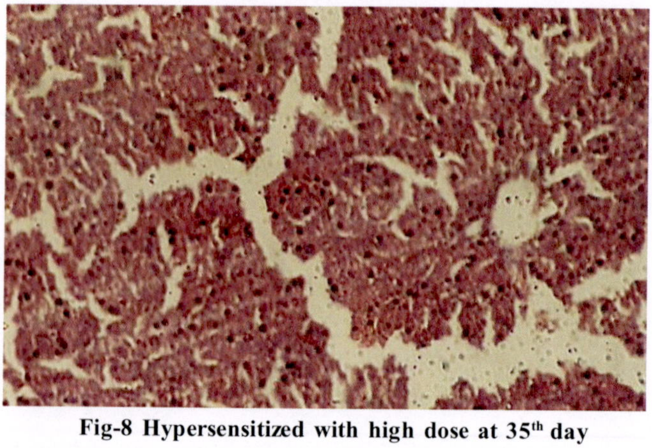

Fig-8 Hypersensitized with high dose at 35th day

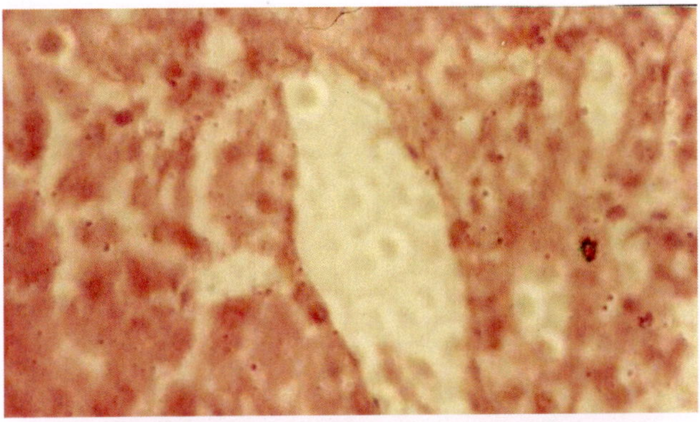

Fig-9 Sensitized with low dose at 45th Day

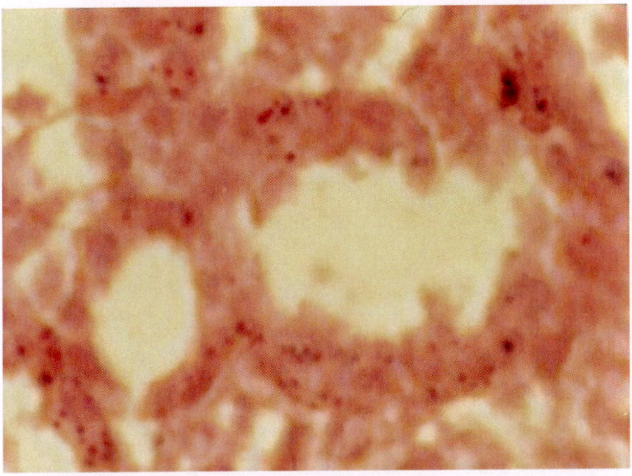

Fig-10 Hypersensitized with low dose at 45th day

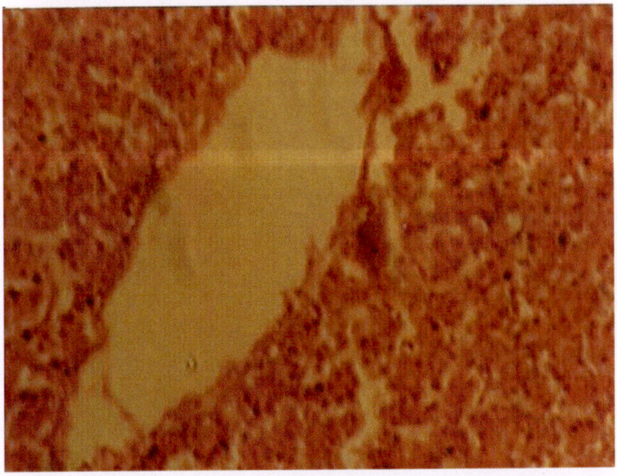

Fig-11 Sensitized with high dose at 45th day

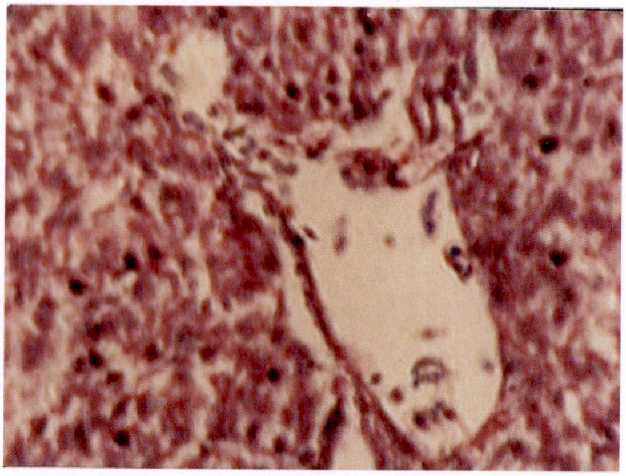

Fig-12 Hyper sensitized with high dose at 45th day

REFERENCES

1. Ackert J.E. (1938). On the nutrition of the fowl nematode *Ascaridia lineata* (Schneider). *Transactions of the American Microscopical Society* 2: 218-222.

2. Ackert J.E., Graham G.L., Nolf L.O. & Porter D.A. (1931). Quantative studies on the administration of viable numbers of nematode eggs (*Ascaridia lineata*) to chickens. *Transactions of the American Microscopical Society* 3: 206-214.

3. Ackert J.E. (1931). The morphology and life history of the fowl nematode *Ascaridia lineata* (Schneider). *Parasitology* 23: 360-379.

4. Ackert IE. and Herrick, C.A. (1928). "Effect of nematode, *Ascaridia lineate* (Schneider) on growing chickens". *J. Parasitol.* 15, 1-13.

5. Ackert J.E. & Nolf L.O. (1931). Resistance of chickens to parasitism affected by Vitamin B. *American Journal of Hygiene* 13: 337-344.

6. Alimora, V.D (1972) "Decreased meat production in chicken infected with *Ascaridia*". Atlas of helminthes of domestic and wild ruminant of Kazabhstam pg –22-2 83(6): 614-7.

7. Anderson, J.R. (1985). Muir's Text Book of Pathology.12th edition English language book society. Edward Arnold.

8. Anita H.J., Biggelaar V., Yazdanbakhsh M., (2004)."The induction of immunoregulation prevents the development of immunopathology in chronic helminth infections and allergy. Allergic disease and the environment. 53:97-116.

9. Apte, S.K. (1980). "Immune response and serum protein pattern in chickens experimentally infected with *Ancylostoma caninum".* Ph.D. thesis, Vikram University, Ujjain, India.

10. Askenase, P.W. (1977). "Immune inflammatory response to parasites: the role of basophils, mast cells and vasoactive amines"Am. J. Tropic Med. Hyg 26, 96-103.

11. Auriault, *C,* Pancre, *V.,* Welowczuk, I., Asseman and I. Veroeaerde (1996) "Cellular

immune response and pathology in schistosomiasis". *Parasitol.* 3.199-208.

12. Awasthi, B.L., Hafeez, H.W., Kathiria and Hasnani, J.J. (1986). "Agar gel electrophoretic pattern of serum proteins of rabbits hyperimmunized with *Gigantocotyle explanatum paraamphistomatidae* antigen". *Ind. J. Anim. Sci. 56:* 339-340.

13. Babu, S., Klei, T.R. and Ranjan, T.V. (1999). "Immunity in experimental murine filariasis, role of T and B cells revisited". *Infection and Immunity.* 67(6), *3*166-*3*167.

14. Bancrof A J., Grncis, R.K. (1999). "Th-1 and Th-2 and immunity to intestinal helminth". *Chemical immunology.* 717.

15. Barriga, O.O. (1984). "Immunomodulation by nematodes".A review.*Veterinary parasitology.* 14(314), 299-320.

16. Befus, A.*D.* and Bienenstock, J. (1982). "Factors involved in symbiosis and has resistance at the mucosa-parasite interface. *Progress in allergy.*31, 76-117.

17. Bradley L.M. (2003). Migration and T-lymphocyte effector function." *Current opinion in immunology* 15:343-348.

18. Cheng, T.C. (1973). "General parasitology". *Academic press, New York* and London.

19. Chitwood, B.G. and Chitwood, M.B. (1950). "Introduction to nematology". *University* Park *Press.* Baltimore.

20. Coombs and Gel- See essential immunology (1994). *Blackwell scientific publications,* U.K.

21. Grancis, R.K. (1997). "Enteric helminth infection, Immunopathology and resistance during intestinal nematode infection". *Chemical immunology. 66,*41-46.

22. Grzych, J.M., GrezeL D., XU C.B. (1993). "IgA antibodies to a protective antigen in human *Schistosoma mansoni*". *Journal of Immunology* 150, 527-535.

23. Idi, A., Permin A. and Murreli, K.D. (2004). "Host age only partially affects resistance to primary and secondary infections with *Ascaridia galli* (Schrank, 1788) in chickens". *Veterinary parasitology.* 1.*22*,221-231.

24. Ikeme M.M. 1971c. Weight changes in chickens placed on different levels of nutrition and varying degrees of repeated dosage with *Ascaridia galli* eggs. *Parasitology* 63: 251 Ina-36 (5): 272-274.

25. Ishaque, S. (1989). "Experimental ascaridiasis in WLH chickens: adaptive immune responses in relation to serum protein profiles by gel electrophoresis". *Ph.D. Thesis. Vikram University, Ujjain, India.*

26. Jacyson, J.F., Abrahamsohn, I.A. and Macedo, M.S. (2001). "Modulation of delayed type hypersensitivity during the time course of immune response to a protein antigen". *Immunology.* 102(3): 373-379.

27. Joshi. S.C. and Johri, G.N. (1978). "Electrophoretic studies on serum protein of fowl in *Ascaridia galli.* Infection I. Serum protein changes when infected with single dose". *Bioresearch.* 2, 1-5.

28. Klei, T.R. (1997). "Immunology control of gastrointestinal nematode infection". *Vet. Parasitology*. 72 (314):507-523.

29. Klei, T.R. and Crowell,W.A.(1981). Pathological changes in kidneys ,liver and spleens of *Brugia pathangi* infected birds. *Journal of Helminthol*. 55(2), 123-131.

30. Kolte, S.W., Narsapur V.S. and Game, M.L. (1995). "Studies on immune response in experimentally induced visceral larva migrans". *International Jr. Animal Science*. 10(2), 261-265.

31. Madan, J. (1992). "Immunobiochemical and immunization studies in experimental ascaridiasis". Ph.D. Thesis, Meerut University, Meerut India.

32. Maizels, R.M., Holland, MJ. (1998). "Parasite immunity pathways for expelling intestinal heiminthes". *Current biology*. 8(20). R711-R714.

33. Malviya HC, Dwivedi P, Varma TK. (1988) "Effect of irradiated *Ascaridia galli* eggs on the growth and cell-mediated immune response in chickens". Vet Parasitol. Apr; 28(1-2): 137-41.

34. Margaret, A. and Wissman (2001). Exotic Pet, Vet. Net Avian Practice. *A. summary*.

35. Massoud, J.1981:Histopathology of liver in Iranian sheep naturally infected with *Dicrocodium dendriticium.Annuals of tropical Medicine and Parasitol*.75(3):293-298

36. Mcclure S.J. and Emery D.L. (1994). "Cell mediated response against gastrointestinal nematode parasites of ruminants'". *Cell mediated immunity in ruminants*. 213-227'.

37. Miller H.R.P. and W.F.H. Jarret (1978). "Immune reaction in mucous membranes, intestinal mast cell response during helminth expulsion on the rat". *Immunology*. 6, 167-259.

38. Miller, H.R.P. (1984). "The protective mucosal response against gastrointestinal nemataodes in ruminants and laboratory animals". *Veterinary immunology and immunopathology*. 65, 167-259.

39. Moiscenko, K.G. (1981). "Serum protein in sheep with mixed infection". Alma Ata. USSR. *Vastochnoe aldelenic vaskhnil*. 94-96.

40. Moll, H. (1996). "Immune responses to parasites, the art of distinguishing the good from the bad". *Immunology today*. 17(12), 551-552.

41. Rajkhowa, C, Gogoi, A.R., Borkakoty *and* M.R. Sharma {1996). "Changes in serum protein fraction in *cattle* experimentally infected with *Schistosoma indicum*". *J of Vetrinary parasitology*. 10(2), 185-189.

42. Riedel B. 1947. New technique on culturing and feeding Ascarid eggs. Transactions of the American Microscopical Society 66: 396-397.

43. Robert, J.A. (1968). "Adoptive transfers of immunity to Babesia rodhani by spleen cells from immune rats". *Aust. J. exp Bio. Used Sci*. 46: 807.

44. Rogers,,W.P. and Summ, Erville, P.I.(1975). "The infection stage of nematode parasite and its significance in parasitism". *Adv Parasitol*. 1: 107-109.

45. Roth, R.L. and Levy, D.A. (1980). *"Nippostrongylus brasiliensis* peripheral leucocyte responses and correlation of basophils with blood histamine concentration duringinfection in rats". *Experimental parasitology.* 50, 331 -341.

46. Rothman K.J. (1986). Modern Epidemiology. 358pp. Little, Brown and Company,

47. Rothwell, T.L.W. (1989). "Immune expulsion of parasitic nematodes from the alimentary tract". *Int. Journal of Parasitol.* 19: 139-168.

48. Singh, S., Banerjee, D.P. and Chaudhri, S.S. (1988). "Immunization of sheep with irradiated infective larval of *Haemonchus contortus". Ind. J. Am. Sci.* 53, 303.

49. Soltys, J., Goyal, P.K. Wakelin, D. (1999). "Cellular immune responses in mice infected with the intestinal nematode *Trichuris muris". Exp. Parasitological.* 92(1), 40-47.

50. Soulsby E.J.L (1982) "Helminth Artropods and Protozoa of Domesticated animals". *Bailleri Tindall London.*

51. Stefan. M.G., Cristiano, L.M., Jeffrey, B. and Radrigo, CO. (2002). "Cellular response and cytokine profiles in *Ascaris lumbricoides* and *trichuris trichura* infected parasites". *Parasite immunology.* 24, 499-509.

52. Stephenson, L.S., Pond,W.G., Nitsheim,M.C.Krook, L.P.andCrompton, D.W.T. (1980). "*Ascaris summ*:Nutreint absorption growth and intestinal pathology in youngs pigs experimentally infected with 15 days old larvae." *Exp. Prasitology* 49:15-25.

53. Vibe-Peterson, G. and Nielson. K. (1979). "Vermious enteritis and thromboembolic celic in the horse". *Nordisk Veterinaer medicine.* 31, 385-191.

54. Vincent. A.L. and Ash, L.R. (1978). "The lymphatic pathology of *Brugia pahangi* in the Mongolian jird". J. Parasitol. 66: 613-620.

55. Vonbrand, T. (1973). "Biochemistry of parasites". *Academic press. New York,* USA.

56. Wakelin,D.(1997)."Parasites and the immune system: conflict or compromise." *Bioscience.* 47 (1):32-40 (En.38 ref.).

57. Wang Kaigong and XU Leren (1996). "Effect of F. hepatica of infection on the number of mucosal mast cells in the bile ducts of ruminants". *Acta Veternaria et. zootechinca sinica.* 27(3), 265-268.

58. Warmer, B. and Polley, L. (1999). *"Ascaris suunr.* Seasonal egg development rates in Saskatchewan pig barn". *Vetrinary parasitology.* 85(1), 71-78.

59. Weatherly, N.F. (1983). "Anatomical pathology. In *Trichinelia* and *Trichinosis..* Plenum Press". New York. 173-205.

Current Status of Researches in Biosciences (2020) : 241-245
Editor : Prof. P.C. Joshi, N. Joshi, Reshman Yasmin, and D. K. Mansotra
Today & Tomorrow's Printers and Publishers, New Delhi - 110 002

NUTRITIONAL EVALUATION OF *MELIA DUBIA* TREE LEAVES BY *IN VITRO* AND *IN VIVO* METHODS

Farha Yasmeen, Prabhu, T. M., Giridhar, K. S. and Nagabhushan, V.

Department of Animal Nutrition, Veterinary College, Hebbal, Bengaluru-24 Karnataka, India.

ABSTARCT

Melia dubia is fast growing multipurpose tree species. Tree fodders contain higher protein and minerals during growth than do grasses. *Melia dubia* leaves were selected for chemical composition evaluation and *in vivo* feeding trial. Fresh leaves from various places were hand plucked, the CP was more in *Melia dubia* leaves collected from Hassan *i.e.,* 17.62 %. NDF was higher in *Melia dubia* leaves collected from Hassan *i.e.,* 37.94 % and ADF was found to be more for leaves from Sira *i.e.,* 25.24 % . *In vivo* feeding trial of 1 month duration was conducted using twelve male sheep (rams) randomly distributed to two groups of six each. T1 (control) was fed with finger millet straw (FMS) and Concentrate Feed Mixture (CFM) while in other group (T2), 25% of FMS was replaced with *M. dubia* tree leaves. DMI (g/d) for T1 and T2 groups was 712.79 and 738.01, respectively. DOMDM (%) for T1 and T2 was 54.32 and 59.71, respectively. Based on these results it was concluded that the *Melia dubia* tree leaves can be considered as supplementary source of nutrients for sustainable sheep production.

Keywords: *Melia dubia,* Chemical composition, Feeding trial, Sheep, Intake, Digestibility.

INTRODUCTION

Tree fodders can be used as potential source for feeding livestock because they are generally rich in protein and minerals and thus can be supplemented to low quality diets and significantly improve feed intake as well as performance of small ruminants. The grazing ruminants usually feed on the young foliage of shrubs on the rangelands. The leaves serve as a concentrate source to goat and sheep population at a very low cost. The parts of tree fodders commonly used as feed is the tender shoots, twigs and leaves of shrubs and woody plants and also fruits and pods. Selection of any trees for fodder is based on, (i) high palatability with minimum of toxic element (ii) high nutritious value (iii) hardy (tolerate to drought and cold) and fast growing; iv) easy in establishment either by seeds, cuttings, root suckers, etc; (v) good coppicing ability so as to tolerate lopping injury with quick growing coppicing shoots; (Rai *et al.,* 2007). The tree leaves are low in phosphorus and high in calcium, resulting in wide calcium: phosphorus ratio. The higher calcium content of the tree leaves is of no value without phosphorus supplementation when feeding to sheep and goats. Rajaguru (1990) has reported that tree leaves are a good source of minerals. *Melia dubia* (*Melia azedarach* Linn) belongs to the family *Meliacea*. Barman *et al.* (2003) evaluated influence of *Melia azedarach* leaves in goats. Based on the chemical composition, intake and digestibility these leaves are considered good source of fodder to meet the maintenance requirement of small ruminants.

MATERIALS AND METHODS

The present study was done to evaluate nutritive value of *Melia Dubia* tree leaves based on their chemical composition and *in vivo* feeding trial in sheep. Fresh tree leaves *of Melia dubia* (Hebbevu, Kan.), selected for evaluation from various places namely GKVK campus Bangalore, Sira and Hassan were hand plucked, oven dried at 55°C for 48h and ground through 1mm sieve for further analysis. Finger millet straw (FMS), Compounded Feed Mixture (CFM) and tree leaves of *Melia dubia* were used in preparing experimental diets. These ingredients were ground to 1mm particle size in a Willey mill and stored in air tight polyethylene containers for analysis of DM, CP, OM, EE, TA and fibre fractions.

Analysis of tree leaves: The DM content of tree leaves was analyzed by drying the leaves in hot air oven. The ash content in the samples was estimated as residue obtained after incineration of samples. Crude protein was analyzed with macro kjeldahl standards (AOAC, 2005). The ether extract (EE) ether using the procedure of AOAC (2005). The NDF and ADF were determined according to the methods described by Van Soest *et al*. (1991).

Feeding trial: The feeding trial was conducted at Veterinary College, Shivamogga. The place of trial is a tropical wet and dry region. Twelve nondescript male rams of comparable age (9-11 months) and average body weight of 27.52 ± 0.05 kg were purchased from Sira. The adjustment period of 24 days was provided. Following adjustment period, rams were randomly allotted to two groups of six each for conducting feeding trial of one month duration. In the control group, rams were offered with FMS and CFM and for other group, 25 % of FMS of control was replaced with the *Melia dubia* leaf hay and CFM. Daily intake of straw, leaf hay and CFM was recorded.

Digestion trial: Six days digestion trial was conducted in metabolic cages using all the sheep. The metabolic cage had the provision for separate collection of faeces. The sample of the diets offered, ort and the faecal voidings was analyzed for chemical composition as per AOAC (2005) and forage fiber fractions as per Van Soest *et al.* (1991) to obtain digestibility (%).

Dung Sampling, analysis and calculation: Dung voided by each sheep over 24 h was weighed. After proper mixing, 1/10th of total voided faeces from each animal was sub sampled for dry matter estimation. The samples collected over six days were pooled, ground and stored in an air tight containers for further chemical analysis. Sampling for nitrogen estimation was done separately, 1/100th of total voided faeces was sub sampled in an air tight container of known weight with 5 ml of 25 per cent sulphuric acid (H_2SO_4) added every day as a preservative. The pooled and ground dung samples of individual sheep were subjected to proximate analysis (AOAC, 2005). The fiber fractions viz., neutral detergent fiber (NDF) and acid detergent fiber (ADF) were determined according to Van Soest *et al.* (1991). Nitrogen in the dung sample was determined by macro kjeldhal method as per the procedure of AOAC (2005).

RESULTS AND DISCUSSIONS

The chemical composition of sun dried leaves of *Melia dubia* collected from three different places namely Hassan, Sira and Bengaluru are presented in Table 1. The crude protein content of tree leaves varied from 18.3 to 30.4 % (Dzowela *et al., 1*995); 11.8 to 14.2 % (Jamuna *et al.*, 2012). The OM content of *Melia dubia* tree leaves was higher (91.28 %) indicating higher energy potential. The differences in the composition of tree fodder might be due to differences in the genotype, stage of maturity, environment and stage of harvesting (Elghandour *et al.*, 2014; Ramachandran *et al.*, 2015).

The overall mean dry matter intake (DMI) (as g / d) from total diet for T1and T2 groups was 712.79 and 738.01, respectively. The overall mean DMI from total diet as a per cent of body weight was 2.47 and 2.62, respectively. The overall mean organic matter intake (OMI) (as g/d) for T1 and T2 groups was 657.38 and 678.98, respectively. The overall mean intake of OM as a per cent of body weight was 2.47 and 2.62, respectively. The overall mean CPI as g per day and per cent of body weight for T1 and T2 groups was 71.15 & 81. 63 and 0.24 & 0.29, respectively. The overall NDF intake (as g/d and % BW) for T1 and T2 groups was 408.76 & 391.57 and 1.42 & 1.39, respectively. Overall ADF intake (as g/d and % BW) for T1 and T2 groups was 214.51 & 211.58 and 0.74 & 0.75, respectively. The digestible organic matter in dry matter (DOMDM, %) was 54.32 and 59.71 for T1 and T2 groups, respectively. The mean apparent digestibility (%) values for T1 and T2 groups with respect to DM, OM and CP were 57.35 & 63.55; 58.98 & 64.97 and 40.75 & 57,83, respectively. The apparent digestibility for DM, OM and CP between the two groups was significantly ($P \leq 0.05$) different. The mean apparent digestibility (%) values for T1 and T2 groups with respect to NDF and ADF were 55.79 & 59.96 and 33.16 & 41.52, respectively. The apparent digestibility for NDF and ADF between the two groups was non-significant ($P \geq 0.05$).

Table 1: Chemical composition of *Melia dubia* tree leaves

Parameter	***Melia dubia* tree leaves**		
	Chemical composition (%, DMB)		
	Hassan	Sira	Bengaluru
Dry matter	91.79	90.59	87.99
Organic matter	91.22	90.61	91.28
Crude protein	17.62	16.97	16.36
Ether extract	2.22	2.82	2.52
Total ash	8.78	9.39	8.72
AIA	0.16	0.22	0.17
Neutral detergent fiber	37.94	37.82	32.92
Acid detergent fiber	23.81	25.24	21.91

CONCLUSION

On the basis of chemical composition analysis and *in vivo* feeding trial using *Melia dubia*, it was concluded that, the tree leaves could be incorporated to substitute 25 per cent of basal dry roughage to sustain small ruminant production in India.

REFERENCES

1. A.O.A.C., (2005). *Official methods of analysis,* Association of official analytical chemists, 18th Ed. Arlington, VA.

2. Barman, K. C., Das, A. K., Konwar, B. N. and Sarma, S. (2003). Nutritive value of ghora neem (*Melia Azedarach* Linn) leaves for crossbred goats. *Indian J. Anim. Nutr.* 20(2): 221-223.

3. Dzowela, B.H., Hove, L., Topps, J.H. and Mafongoya, P.I., (1995). Nutritional and anti-nutritional characters and rumen degradability of dry matter and nitrogen for some multipurpose tree species with potential for agroforestry in Zimbabwe. *Anim. Feed Sci. Technol.* 55: 207-214.

4. Elghandour, M. M. Y, Vazquez, N. J. C, Salem Azm, Kholif A. E, Martinez Castane J. S, Camacho L. M, and Cerrillo-Soto M. A., (2014). Effects of Saccharomyces *cerevisiae* at direct addition or pre-incubation on *in vitro* gas production kinetics and degradability of four fibrous feeds. *Ital. J. Anim. Sci.,* 13:295-301.

5. Jamuna, K. V., Remadevi, O. K., Prabhu, T. M and Chandrapal Singh, K. (2012). Nutritional evaluation of fodder tree leaves. *Indian Vet. J.,* 89(12): 35-38.

6. Rai, P., Ajit and Samanta, A.K. (2007). Tree leaves, their production and nutritive value for ruminants: a review. *Animal Nutri. Feed Technol.,* 7: 135-159.

7. Rajaguru, A.S.E., (1990). Availability and use of shrubs and tree fodders in Sri Lanka. In Devendra, C. ed., *Shrubs and tree fodders for farm animals*: Proceedings of a workshop in Denpasar, Indonesia, 24-29, July 1989. International Development Research Centre, Ottawa, Ont, Canada. 237-243.

8. Ramachandran, M., Bharathidhasan, A and Balakrishnan, V. (2015). Nutrient composition, *in vitro* true digestibility (IVTD) and methane production potential of fodder tree leaves *Indian J. Anim. Sci.,* 85 (5): 494-497.

9. Van Soest, P. J., Robertson, J. B. and Lewis, B. A., (1991). Methods for dietary, neutral detergent fiber, and non-starch polysaccharides in relation to animal nutrition. *J. Dairy Sci.*, 74:358-359.

Current Status of Researches in Biosciences (2020) : 247-255
Editor : Prof. P.C. Joshi, N. Joshi, Reshman Yasmin, and D. K. Mansotra
Today & Tomorrow's Printers and Publishers, New Delhi - 110 002

STUDY OF THE AWARENESS ABOUT AIDS/HIV AMONG THE STUDENTS OF R. N. COLLEGE HAJIPUR, BIHAR, INDIA

Rakesh Mohan[1], Sushma Kumari[2], Vijay Kumar[3] and Suchitra Kumari[4]

[1, 2, &3] *Department of Zoology, R N. College, Hajipur)*

[4]*B.R.A.B.Univ. Muzaffarpur, Bihar*

Email: ranjan2270@rediffmail.com

ABSTRACT

In this research paper it was try to know the awareness of HIV /AIDS among the students of R. N. College, Hajipur by the method of questionnaire and interview. The research works were carried during the month of August 2019. A set of 15 questions were framed in the form of questionnaire, as adopted from the Research paper – "Mohan Rakesh, Study of the awareness about AIDS/HIV among the selected categories of the students of the locality Agarwa & Bluatal of Motihari town, East Champaran, Bihar, Res Expo Int Multi Res Journal ISSN: 2250-1630, March-2014, p27" to know the awareness of the HIV/ AIDS among our students. A total of 900 students of different stream of Bachelor courses were interviewed with the help of the students of M.Sc. (Zoology)

KEYWORDS: Hajipur, AIDS awareness HIV, Monogamy, Unsafe sex.

INTRODUCTION

Acquired immune deficiency syndrome (AIDS) is a fatal disease of the human immune system, caused by the human immunodeficiency virus (HIV) .This condition progressively reduces the effectiveness of the immune system and leaves individuals susceptible to opportunistic infection and tumour. HIV is transmitted through direct contact of a mucous membrane or blood stream with a body fluid containing HIV. It is also transmitted through reuse of contaminated syringe, via transfusion of a blood & blood products derived from on infected person's blood. Till date HIV /AIDS is not curable disease so it is better to check spread of the disease to aware the students about the HIV

MATERIALS AND METHODS

Study sample:

Our targeted students were students of R. N. College, Hajipur. Among them some were local students, remaining were migrated from the rural areas of District Vaishali. They came Hajipur to get admission in the college in various courses.

Total Number of students interviewed - 900

Age limit (in years) of students interviewed – 18-28

Education status of students interviewed -

1. B.Sc. 1st yr- -150
2. B.A. 1st yr - 150
3. B.Sc. 2ndt yr- -150
4. B.A. 2ndt yr - 150
5. B.Sc. 3rd yr- -150
6. B.A. 3rd yr - 150

Research method : Survey of 900 students with the questionnaire and interview method was conducted. We had prepared a questionnaire of a set of 15 questions.

OBSERVATIONS AND DISCUSSION

In the response of question number 1, 74% students admitted that a person should not have many sex partners, only 13% students were in view to have many partner for sex relationship. Such type of answer may be due to culture and religious prejudice as monogamy is a basic principle of our ethics.

In the response of question number 2, 83% students responded in positive, according to them unsafe sex could induce HIV/AIDS. In the response of question number 3, about 60% students responded in positive.

8% students answered in negative and rest did not aware about this fact. Here it was noticed that maximum person were aware about the danger to give birth a child by HIV infected lady.

In the response of question number 4, 54% person admitted a child could be infected with HIV/AIDS, by breast feeding with infected mother. 13% person answered in negative. Scientifically it is found that only 14% child get infection of HIV/AIDS through breast feeding.

In the response of question number 5 &6 about half of the students had view that AIDS are transmitted via kiss. They did not differentiate between kiss with saliva and kiss without saliva. After the analysis the answer of the question number 5 and 6 of our questionnaire we found that 50% person were aware about the spread of HIV/AIDS through kiss and 50% students did not aware about the spread of HIV/AIDS through kiss.

In the response of question number 7, 75% students admitted that a person could be infected by using injection syringe/needles, which had been used previously by other infected person. 16% person had view that there was no risk to use a syringe/needle which was used by infected person.

In the response of question number 8, 78% students knew about the risk of blood transfusion without HIV test to a patient as they response this question in positive. This was the good sign of HIV/AIDS awareness. Again near about 22% students did not aware about the risk of blood transfusion without serological test.

In the response of question number 9, it was indicated that 47% students had wrong information about the transmission of HIV virus through mosquito bite. Only 22% students knew the fact that HIV virus could not be transmitted through the mosquito bite. So it was needed to remove such type of miss concept from the students of our study area. Here it was found that maximum students of science stream gave right answer and students of B. A. courses were found to be unaware about the fact.

In the response of question number 10, only 50% students admitted that a person could not be infected with HIV, after wearing cloth of AIDS patient. Thus only about half of the students had right information about this fact.

In the response of question number 11; only 40% person knew HIV virus could not be transmitted through the shaking hand with infected person but 24% students had miss concept that a person could get infection through hand shaking with infected person.

In the response of question number 12, it is concluded that only 30% students knew that HIV/AIDS cannot be transmitted with living together and sharing food with infected person. Here also large number of students had no or wrong information regarding this fact.

In the responses of question number 13, 17% students had miss concept that HIV/AIDS could be transmitted with using common lavatory and bathroom.

In the response of question number 14, it was clear that a good number of students (i.e. 69%) used condom but use of condom was not only for the prevention of HIV/AIDS, but also for preventing sex transmitted disease (such as STD) and for birth control.

In the responses of question number 15, 94% students want to provide sex education about HIV/AIDS in school/collage. A few did not want to provide sex education as they had concept that sex education could divert the concentration of the students from education to sex.

CONCLUSION AND SUGGESTIONS

During the study, it was found that students of science stream especially of biology students were more aware than that of Arts students.

In the response of question number: - 1,2,3,4,5,6 it was found that about 40% students had miss concept or no idea about this fact. This was not very good indication. The wrong concept about the facts raised in these questions can induce a type of untouchability with AIDS patient. HIV/ AIDS patient cannot live in society with dignity.

Therefore we have to remove such type of miss concept among the students. For this there are very important role of government and various NGO society and others. Government should also include these facts in school/college syllabus.

After the analysis of the response for the question number 7, 8, 9, 10 and 11, it was found that maximum peoples were aware about the facts raised in these questions. This was the good indication. But again we have to try 100% students be aware about the transmission of HIV/AIDS virus. For this following measures should be taken.-

1. During study it was found that students had miss concept about the HIV/AIDS. To remove the wrong information among students help of NGO, educational institute, social workers, is needed.

2. Red ribbon and eco club in each collage should be established
3. We should provide moral sex education among students
4. All types of untouchability with AIDS patient in society must be removed
5. It was found that awareness program among girls, boys, women can help better to reduce transmission of virus and also to reduce untouchability with aids patient.

REFERENCES

1. Annonymous. (1997) World Health Organization, WHO, Geneva, Switzerland.
2. En.wikipedia.org /wiki/HIV
3. AIDS.gov/his AIDS basic
4. hivinsite.ucsf. edu.
5. WWW.nim.nih.Gor/1000602.htm
6. www.census2011.co.in census city.
7. Chandler J *et al.,* (2000). The place of gold in rapid test, IVD technology.
8. Mohan Rakesh, Study of the awareness about AIDS/HIV among the selected categories of the people of the locality Agarwa & Bluatal of Motihari town ,East Champaran, Bihar, *Res Expo Int Multi Res Journal* ISSN: 2250-1630,

QUESTIONNAIRE

Name:-

Address:-

Sex:-

Age:-

Income:-

Qualification:-

Serial No	Questions	Responses		
		Yes	No	Do not know
01	Should a person (male or female) continue sex relationship with several partners?			
02	Can AIDS/HIV infection be transmitted via unsafe sex i.e copulation without protection?			
03	Can AIDS/HIV infection be transmitted in foetus from infected pregnant women?			
04	Can an AIDS/HIV infected lactating women transmit HIV virus in her child during breast feeding?			
05	Can a person get HIV infection if he/she kiss (with saliva) to a partner having AIDS/HIV infection?			
06	Can a person get HIV infection if he/she kiss (without saliva) to a partner having AIDS/HIV infection?			
07	Can a person get HIV infection by using syringe/needle which has been used previously by the infected person?			
08	Can a person get HIV infection if blood, collected from the HIV infected person, is transfused in a healthy person unknowingly?			
09	Does transmission of HIV virus occur through mosquito bite?			
10	Can a person get infection of HIV virus after wearing cloth of AIDS patient?			
11	Can a person get HIV infection by shaking hand with infected person?			

12	Can HIV/AIDS be transmitted with living together and sharing food with infected person?			
13	Can HIV/AIDS be transmitted with using common lavatory and bathroom?			
14	Do you use condom?			
15	Should sex education with information about AIDS/HIV be given to school and college students?			

Questionnaire is adopted from Research paper "Study of the awareness about AIDS/HIV among the selected categories of the students of the locality Agarwa & Bluatal of Motihari town ,East Champaran, Bihar", Res Expo Int Multi Res Journal ISSN: 2250-1630,

Serial No.	Question	TABLE- Answer of the Targeted students in the response of questionnaire							
		Number of person who answer the question in affirmative i.e. ***YES***		Number of person who answer the question in negative i.e. ***NO***		Number of person who answer the question ***"Don't know"***		Total targeted students	
1	Should a person (male or female) continue sex relationship with several partners?	120	13%	666	74%	114	13%	900	100%
2	Can AIDS/HIV infection be transmitted via unsafe sex i.e. copulation without protection?	750	83%	63	7%	87	10%	900	100%
3	Can AIDS/HIV infection be transmitted in foetus from infected pregnant women?	540	60%	75	8%	285	32%	900	100%
4	Can an AIDS/HIV infected lactating women transmit HIV virus in her child during breast feeding?	489	54%	120	13%	291	32%	900	100%
5	Can a person get HIV infection if he/she kiss (with saliva) to a partner having AIDS/HIV infection?	423	47%	267	30%	210	23%	900	100%
6	Can a person get HIV infection if he/she kiss (without saliva) to a partner having AIDS/HIV infection?	435	48%	270	30%	195	22%	900	100%

7	Can a person get HIV infection by using syringe/ needle which has been used previously by the infected person?	678	75%	144	16%	78	9%	900	100%
8	Can a person get HIV infection if blood, collected from the HIV infected person, is transfused in a healthy person unknowingly?	687	76%	153	17%	60	7%	900	100%
9	Does transmission of HIV virus occur through mosquito bite?	420	47%	195	22%	285	32%	900	100%
10	Can a person get infection of HIV virus after wearing cloth of AIDS patient?	225	25%	450	50%	225	25%	900	100%
11	Can a person get HIV infection by shaking hand with infected person?	216	24%	357	40%	327	36%	900	100%
12	Can HIV/AIDS be transmitted with living together and sharing food with infected person?	267	30%	276	31%	357	40%	900	100%
13	Can HIV/AIDS be transmitted with using common lavatory and bathroom?	150	17%	477	53%	273	30%	900	100%
14	Do you use condom?	510	57%	183	20%	207	23%	900	100%
15	Should sex education with information about AIDS/HIV be given to school and college students?	849	94%	45	5%	6	1%	900	100%

Current Status of Researches in Biosciences (2020) : 257-265
Editor : Prof. P.C. Joshi, N. Joshi, Reshman Yasmin, and D. K. Mansotra
Today & Tomorrow's Printers and Publishers, New Delhi - 110 002

ANTIBACTERIAL ACTIVITY OF PLANT SPECIES- CLOVES, CINNAMON AND BLACK PEPPER ON BACTERIAL STRAIN ISOLATED FROM SOIL

Kamal Jaiswal, Suman Mishra, Amit Singh and Saumya Sharma

Department of Zoology, Babasaheb Bhimrao Ambedkar University, Lucknow-226025

Email id- drkamaljaiswal@gmail.com

ABSTRACT

Unani and Ayurvedic medications have been practiced for hundreds of years by a considerable fraction of the population in the treatment of various human afflictions. Now –a- days, a substantial number of pathogens have developed resistance to multiple antibiotics in either narrow or broad spectrum (Multiple Drug Resistance), which is quite alarming. It does not allow developing complete immunity against all antimicrobial agents. In past years, many studies have been conducted in finding the solutions for the menace. Reports have documented about the existence of microbial treatments which are believed to be permanent in treating the ailments caused by bacteria as herbal products are not toxic. Three herbal plants were selected based on their traditional usage in India to assess their antibacterial properties- *Syzygium aromaticum* (Cloves), *Cinnamomum verum* (Cinnamon), and *Piper nigrum* (Black pepper). The bacteria were isolated from soil and colonies were purified by sub culturing using streaking plate method, further gram staining was also conducted. Bacteria showed purple stains thus indicating bacteria to be Gram positive (+). Colonies of *Staphylococcus aureus* were cultured. Antibacterial activity of methanolic extracts of herbal pants was determined at 50% and 100% concentration solution by disc diffusion assay on gram positive soil bacteria. Results concluded that, methanolic extracts of herbs at 100% concentrated solution have more inhibitory potency against bacterial colonies where as 50% concentrated solution of plant extract showed less inhibitory potency against bacterial colonies. Of the three herbal plants, *Cinnamomum verum* (Cinnamon), showed maximum inhibitory effect on the growth of *Staphylococcus aureus*.

Keywords: Antibiotics, Multiple drug resistance, Disc diffusion assay, Antibiotic activity.

INTRODUCTION

Plants are proved to be noteworthy source of natural products, since time immemorial in maintaining the health of human beings, through natural therapies (Gislene *et al.*, 2000). Herbs like spices have been used for so long not only for the purpose of varieties of flavour or a distinctive aroma of the food but also due to their antimicrobial properties (Nanasombat *et al.*, 2002). Spices contribute in generating a sharp tanginess in beverages and food (Perween *et al.*, 2006). Moreover, among these spices, there are some herbs that are most commonly used as natural antimicrobial agents to treat various microbial as well as bacterial infections. Some natural compounds found in various spices show high antimicrobial potency (Indu, 2006). Obtaining biologically active compounds from natural sources and their examination in the scientific labs have proved to be of great interest to scientists who are looking for new drug sources useful in treating infections. In recent years many studies have been reported, dealing with antibacterial testing of medicinal plant extracts for their biological, chemical composition and therapeutic activities. An increased number of pathogens have also developed resistance to multiple antibiotics in either narrow or broad spectrum (Multiple Drug Resistance), threatening to develop complete immunity against all antimicrobial agents and, therefore, are untreatable (Cohen, 1992). Thus, it is vital to perform the actions that can control the problem of multiple drug resistance by using the plant extracts that contain phytochemicals having antibacterial properties (Agaoglu *et al.,* 2007). Keeping the same in view, the present study was conducted to find out the antibacterial potency of three herbal spices including *Syzygium aromaticum* (Cloves), *Cinnamomum verum* (Dalchini), and *Piper nigrum* (Black pepper) against pathogenic bacteria, since spices have been signified for their property of preserving foods and imparting medicinal values due to the presence of bioactive antibacterial compounds (Shelef 1983; Papp *et al,*. 2007).

MATERIALS AND METHOD

Study Area: Soil sample was collected from BBAU campus Lucknow.

The sample was then diluted at different concentrations. This sampling point was chosen so as to get abundant microflora. 2gm of soil was mixed in 10 ml distilled water to make 1/10 concentrated dilution. 1ml of 1/10 was taken and mixed with 9 ml to make 1/100 and 1ml of 1/100 was taken and added with another 9ml of water to make 1/1000 solutions (Fig-1).

Type of Agar: Luria Bertini Agar was used in the study.

Composition of LB agar: Casein enzymatic hydrosylate 10.00 g/L, Yeast extract 5.00 g/L, Sodium chloride 10.00 g/L

Method of preparation: 40 gm of LB agar was weighed in two turns and was suspended in a 500 ml conical flask. The amount of distilled water added was just enough to dissolve the LB agar powder by constant stirring. Volume was made up to 250ml so as to yield 1L of media. The mouth of conical flask was tightly sealed by a cotton plug and covered by a butter paper. It was then autoclaved at 121 degree C for 45minutes. The flask was kept empty more than half of its capacity to avoid spillage of media during autoclave procedure.

Maintenance of Culture plates: Laminar air flow cabinet was prepared for work by wiping down the area with alcohol.

- The media was allowed to cool and was poured in autoclaved petri plates (when it was warm but not hot) to avoid vapour formation in plates.
- The cotton plug was removed from the flask and the lid of the plates was partially lifted just enough to pour the media.
- Pouring was done near the flame of burner to avoid any sort of contamination by microbes present in surrounding environment.
- The plates were sealed using a parafilm.
- After three or four uses the neck of conical flask containing media was heat sterilized by passing it through the flames of burner.
- The plates were allowed to stand at room temperature for some time and then kept in refrigerator to maintain 4 degrees temperature to avoid contamination before being used (Fig-2).

Isolation solution of Bacteria from Soil solution sample: 100μl of water sample was spread on LB agar plates. The plates were incubated at 37°C. The colonies were enumerated, characterized and readings were recorded after 24 hours (Fig-3).

Sub culturing of colonies: Colonies were purified by sub culturing using the streaking plate method. Platinum loops were first sterilized by flaming the wire loop in the light blue area of Bunsen burner flame until it was red hot. Once sterilized, the loop was allowed to cool. The lid of the agar plate was opened just sufficiently enough to streak the plate with the inoculation loop (Fig-4).

Gram Staining:

Preparation of slides: The slides were wiped with spirit or alcohol. After cleaning the slides were air dried.

Preparation of the smear: With a sterile cooled loop, a drop of sterile water was placed on the slide. The loop was sterilized and cooled again and a very small sample of bacterial colony of *Staphylococcus aureus*, was picked and gently stirred into the drop of water on the slide.

Heat fixing: Heat fixing kills the bacteria in the smear, firmly adheres the smear to the slide, and allows the sample to more readily take up stains. The smear was allowed to air dry. After the smear was air dried, the slide was held at one end and the entire slide was passed through the flame of a spirit lamp two to three times with the smear side up.

Gram staining: Crystal violet was added to the smear and was left to stand for 1min. The slide was rinsed gently with tap or distilled water. Gram's iodine was added to the smear and was left to stand for 1minute. The slide was rinsed gently with tap water or distilled water. The smear appeared as a purple circle on the slide. The slide was decolorized using 95% ethyl alcohol. The slide was tilted slightly and alcohol was added drop by drop for 5 to 10seconds. Further, the slide was again rinsed with water. Then, safranin was added as a counterstain and was left to stand for 45 seconds. The slide was titled slightly and gently rinsed with the distilled water. The smear was observed under the microscope (Fig-5).

Collection of plant material: 10 gm each of the three plant species namely cinnamon, cloves and black pepper were purchased from the local food store in crude forms.

Preparation of Methanolic Extracts of Plant Samples: Few grams each of the three plant species (cinnamon, cloves and black pepper) was crushed to make a fine powder using mortar and pestle (Fig-6).

Three gram of cinnamon was added to 3ml of methanol, two gram of cloves in 2ml of methanol and four gram of black pepper in 4ml of methanol in a glass reagent bottle. The bottle was shaken and left to stand at room temperature for 5 days with regular shaking. After few days, the suspension was filtered using muslin cloth and the filtrate was collected. The concentration of methanolic extract was 100%.This was considered as stock solution and also tested. For preparation of another extract with 50% concentration, 1ml each of three different plant extracts (100%) was added to 1 ml of water in three different test tubes i.e 1:1 dilution (Fig-7).

Preparation of Filter paper discs: Whatsman filter paper discs were generated, autoclaved, and allowed to be dipped in both 100% and 50% concentrated plant extract solutions for 2.5 hours so that the paper absorbs the sample completely.

Antibacterial Activity testing using Disc Diffusion Method: 100 micro litre of liquid bacterial culture (1/100 dilution) was spread on separate agar plates by spread plate method. The plates were incubated for 1 hour at room temperature. Soaked discs were put at four corners of these plates. The diameter of clear zone of inhibition formed was measured.

RESULTS

After successfully placing the soaked discs on the plates, the plates of all the three plant extracts were allowed to rest for 24h. The results were obtained showed the inhibitory zones produced by cloves, cinnamon and black pepper (Fig 8-10). After 24 hours of incubation, plant extract got diffused into the bacterial colony and showed bactericidal activity. The concentration of the plant extract was highest next to the disc, and started to decrease as the distance from the disc increased. If the compound is effective against bacteria at a certain concentration, no bacterial colonies will grow in that region. The results showed a varying degree of the size of inhibition zones of herbal spices. The size of inhibition zones obtained, were inversely proportional to the increase in dilution of plant extracts. Table-1 shows the clear description of the zone of inhibition of plants extracts.

Table-1: Methanolic extracts of plant species showing zone of inhibition.

Plant extractstested on *Staphylococcus aureus* colony	Time (in hours)	Inhibition zone (I.Z) AT 50% conc. in (mm)	Inhibition zone (I.Z) AT 100% conc. in (mm)
CLOVES	24 HOURS	3 mm	3.2 mm
CINNAMON	24 HOURS	4.2 mm	10 mm
BLACK PEPPER	24 HOURS	2 mm	4 mm

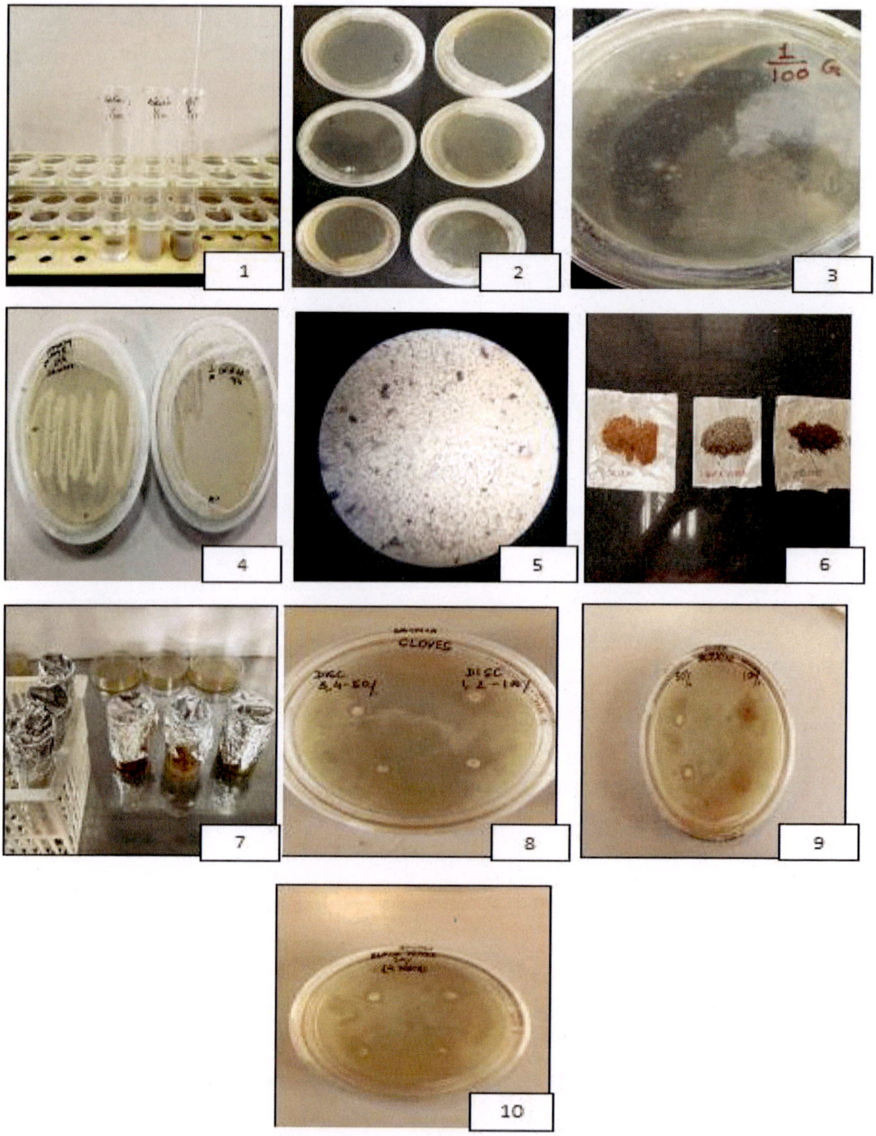

Plate –A: Explanation of the images taken during study is as follows:

1. Soil sample collected and diluted. 2. Clear plates having media. 3. Bacterial colonies after 24 hours of incubation 4. Streaking for the purification of colony of *Staphylococcus aureus.* (subculture). 5. Light microscopic images of Gram positive bacteria *Staphylocuccus aureus.* 6. Powdered samples. 7. 50% and 100% conc. sol of PE. 8. Plate having clove as a plant extract showing zone of inhibition at different conc. after 24 hours of incubation. 9. Plate having cinnamon as a plant extract showing zone of inhibition at 50% and 100% conc. after 24 hours of incubation. 10. Plate having black pepper as a plant extract showing zone of inhibition at different concentration after 24 hours of incubation.

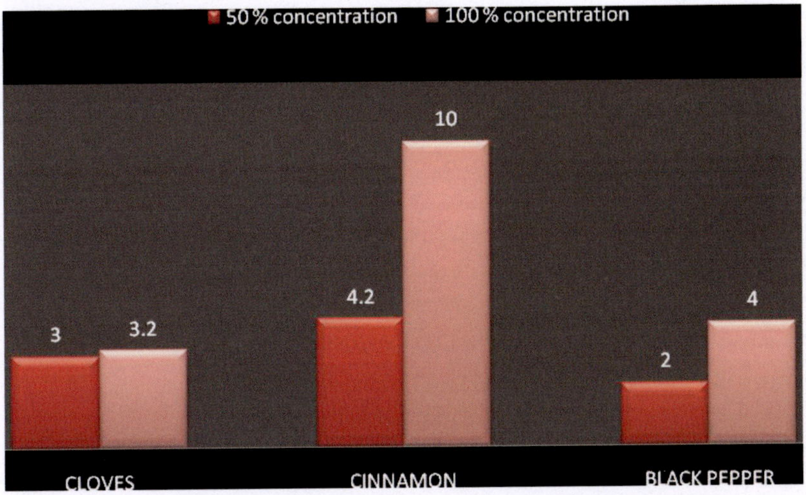

Fig.11 Graphical representation of zone of inhibition (in mm)

Methanolic extract of cloves, cinnamon, and black pepper on the spherical bacteria shown in the table and represented by graph, reveal that methanolic extract of these plants at 100% solution have more inhibitory activity against *Staphylococcus aureus* colonies isolated from soil whereas methanolic extract of same plant species at 50% solution have less inhibitory potency against bacterial colonies isolated from soil in this study. Also it is concluded that, cinnamon (dalchini) has the highest inhibitory effect, of all the three plant extracts. This inhibitory effect of dalchini is due to main constituents like cinnamaldehyde, which showed antibacterial and bactericidal activity (Fig- 11).

DISCUSSION

Staphylococcus aureus is a major pathogenic bacterium that can cause a wide variety of clinical problems. *S. aureus* is found in normal human flora as well as in environment. It is generally found on the skin and mucous membranes of many healthy individuals. It does not cause infection on healthy skin normally; how so ever if comes in the contact of blood or the internal tissues, then these bacteria bring up a variety of severe infections (Rasigade, 2014). In the present study, *Staphylococcus aureus* bacteria islotated from the soil were cultured in agar plates. These were the round-yellow colonies clearly visible after the streaking plate method. Gram staining revealed that *S. aureus* is a Gram-positive bacteria (stained purple or violet by Gram stain) that are cocci-shaped (spherical) arranged in the form of clusters which resembles to a cluster of ripen grapes. According to the reports of Chamber, (2009), *Staphylococcus aureus* is resistant to almost every antibiotic that has ever been introduced in the market. However, resistance is obtained by horizontal gene transferring from outer sources, although antibiotic selection and chromosomal mutation also play a significant role. This tormenting resistance of *S. aureus* provoked Alexander Fleming to discover penicillin, escorting in "era of antibiotics." Penicillin proved itself as a miracle drug: fatal infections were found to be cured by penicillin. Yet, till the mid-1940s, only after few years of its introduction in to the scientific laboratories, penicillin resistance was experienced in sanatoriums and within next ten years it emerged as the biggest problem in a certain clique. *S. aureus* is exceptional in its property to develop resistance to any antibiotic. According to the studies of Bhawna *et al.* (2014), few common Indian spices showed an inhibitory effect against the growth of certain bacterial pathogens in tissue culture. Her results suggest that turmeric, clove, black pepper, ajwain and dalchini powder, produce significant antibacterial effects. Antibacterial activity by methanolic extract of dalchini was maximim against *Staphylococcus aureus*. In the present study, the bacteria were treated with a methanolic extracts of cloves, cinnamon and black pepper because of their antibacterial properties. The bacterial strains were investigated individually for antibacterial activity by disc diffusion assay. The results showed that cent percent concentration of all plant extracts is inhibitory in nature. Cinnamon showed a remarkable effect, as per with the findings of Bhawna *et al.* (2014) maximum inhibitory effect on the bacterial strain out of the three herbal plant methanolic extracts was evident.

CONCLUSION

Due to less enthusiasm among the youthful age and their tendency to move to urban areas for rewarding occupations, wealth of knowledge about medicinal plants is continuously declining. The high costs of antibiotic medicines, adulteration and increased side effects of synthetic drugs make it tough for people to use them in curing various ailments. Therefore considering the high requirement of these herbal plant spices in the current world in human services framework, it is of utmost importance that natural plants ought to be developed and proliferated. The need of hour is to strap up our traditional knowledge regarding herbal plants and conserve it for the betterment of society as they pursue no side effects and show good results in less span of time. The present study was conducted to check the antibacterial activity of three herbal plants, out of which cinnamon showed the maximum inhibitory effect against soil bacteria.

ACKNOWLEDMENTS

Thanks our Hon'ble Vice-Chancellor Prof. Sanjay Singh and other authorities of the University for Necessary Infrastructural Facilities to carry out the research work.

REFERENCES

1. Agaoglu, S., Dostbil, N. and Alemdar, S. (2007). Antimicrobial activity of some spices used in the meat industry. *Bulletin of the Veterinary Institute in Pulawy.* 55:53-57.

2. Bhawana Pandey , Shabina Khan and Sheetal singh (2014). A Study of antimicrobial activity of some spices. *Int.J.Curr.Microbiol.App.Sci.* 3(3): 643-650.

3. Chambers, H. F., & DeLeo, F. R. (2009). Waves of resistance*: Staphylococcus aureus* in the antibiotic era. *Nature Reviews Microbiology,* 7(9): 629–641.

4. Cohen ML (1992). Epidemiology of drug resistance: implications for a post antimicrobial era. *Science* 257: 1050-1055.

5. Gislene, G.F.N., Juliana, L., Paulo, C.F., and Giuliana, L.S. (2000). Antibacterial Activity of Plant Extracts and Phytochemicals on Antibiotic Resistant Bacteria. *Brazilian J. Microbiology.* 31: 247-256.

6. Indu, M. N., Hatha, A. A. M., Abirosh, C., Harsha, U., & Vivekanandan, G. (2006). Antimicrobial activity of some of the south-Indian spices against serotypes of *Escherichia coli, Salmonella, Listeria monocytogenes* and *Aeromonas hydrophila. Brazilian Journal of Microbiology*, 37(2): 153–158.

7. Nanasombat, S. and Lohasupthawee, P. (2005). Antibacterial Activity of Crude Ethanolic Extracts and Essential Oils of Spices against Salmonellae and other Enterobacteria. *KMITL Science Technology J.* 5(3):527-538.

8. Nanasombat, S., Prasertsin, V., Graisin, K., Shain, H. and Thanaboripat, B. (2002). Efficacy of New Enzyme- Linked Immunosorbent Assay for Rapid Detection of Salmonella in Foods. *Government Pharmaceutical Organization Report, Bangkok.* 51: 53- 57.

9. Papp, L.V., Holmgren, J. Lu. A. and Khanna, K.K. (2007). From Selenium To Selenoproteins: Synthesis, Identity, and Their Role In Human Health. *Pakistan Journal Botany.* 9(7): 775 806.

10. 10 Perween, T. and Nazia, M.A.C. (2006). Bactericidal Activity of Black Pepper, Aniseed and Coriander against Oral Isolates. *Pakistan J. Pharmaceutical Sciences.* 19(3): 214-218.

11. Rasigade JP, Vandenesch F. (2014). *Staphylococcus aureus*: a pathogen with still unresolved issues. *Infect. Genet. Evol.*Jan;21:510-4.

12. Shelef, L.A. (1983). Antimicrobial effects of spices. *J. Food Safety.* 6:29-34.

Current Status of Researches in Biosciences (2020) : 267-274
Editor : Prof. P.C. Joshi, N. Joshi, Reshman Yasmin, and D. K. Mansotra
Today & Tomorrow's Printers and Publishers, New Delhi - 110 002

EVALUATION OF MICROBIAL LOAD AT CERTAIN NON- SPECIFIC IMMUNE SITES OF *MACROBRACHIUM ROSENBERGII* SUPPLEMENTED WITH *CYNODON DACTYLON*

Jasmine Anand

Department of Zoology, T.K.M.M.College Nangiarkulangara, Alappuzha

Email: ja7210@gmail.com

ABSTRACT

The present investigation was carried out to assess the alterations in the microbial load at certain non- specific immune sites (gut, gill and exoskeleton) of *Macrobrachium rosenbergii* supplemented with *Cynodon dactylon.* Each experimental set up consist of 15 prawns per test dose and equal number of control reared in 100 litre tanks. Prawns were fed with the control and medicated diet (250mg/kg) prepared in the laboratory and was fed at 10 % of the body weight. After 4 months the total microbial load at non-specific immune sites of prawns in both control and medicated culture were isolated and identified. In the present work the medicated diet effectively suppressed the bacterial growth in the non-specific immune sites and promisingly enhanced the growth of probiotic bacteria and there by reduced the pathogenecity and helps in immunomodulation.

Keywords: *Macrobrachium rosenbergii, Cynodon dactylon,* non-specific immune sites

INTRODUCTION

Fish contributes substantially to the domestic food security of India. To meet the increasing demand for fish and shellfish, aquaculture practices such as semi-intensive and intensive aquaculture practices have become very common. Intensive fish farming creates a highly stressful environment for the fish with consequent suppression of the immune response resulting in increased susceptibility to diseases. Traditionally, antibiotics have been used as one of the most important chemotherapeutics in aquaculture for the prevention and treatment of bacterial diseases and also to avoid an acute infection. But problems arise with indiscriminate use of these antibiotics. Antibiotic therapy is undesirable as it leads to the development of antibiotic- resistant bacteria, immunosuppression and destabilization of the helpful bacterial populations (Hiam El-Desouky *et. al.,*2012, Mohamed A Farook *et. al.,* 2015)..

Problems with antibiotics, drugs and chemical treatments and to some extent vaccines to prevent diseases in fish set a stage for a new concept in disease prevention using probiotics (Lilley and Stilwell, 1965) and immunostimulants (Anderson, 1992). Introduction of organic acid as an alternative to antibiotics in fish feed may present an interesting area of research. Some promising results using organic acids have already been documented in livestock and poultry diets, although their use in aqua feed has been very limited to date. Antibiotics fed to farm animals as growth promoters have been associated with the spreading of bacterial resistance. This potential risk, combined with consumer demand for a medicinal-product free food chain, necessitates a reduction in the use of antibiotics in food production. Dietary acidification lowers gastric pH, maintains a more desirable microflora and enhances the digestion process. Promising results were found using organic acids in weanling piglets and chicks, where the stomach is not yet fully developed and functional, and also in fishes (Baruah *et al* 2006).

Recently, a trial with a commercial product (Raafres AQ, Guybro Chemicals, Mumbai, India) containing organic acids and their salts along with _-glucan and mannan oligosaccharides and other "antistress" factors has shown a reduction in the pH of the feed and in the gastrointestinal tract of *Macrobrachium rosenbergii* .Further, inhibition of the growth of pathogenic bacteria like *E coli, Vibrio* and *Salmonella* was demonstrated, as was an improvement in the absorption of nutrients through better digestion of complex food molecules and by increasing the size of villi in the intestine.

Bacterial species of the gut can influence the health and robustness of the host. Extreme examples of the influence of the gut flora include the negative effects of pathogenic organisms and, in contrast, the total reliance that ruminants have on their gut flora for the assimilation of organic carbon from the environment (Kennedy *et al.* 1991). More subtle interactions have usually been much less characterized and understood but are, nevertheless, important and have been shown to impact on the health and growth of the host (McBee, 1971; Phillips, 1984; Vitalis *et al.*, 1988). The hard cuticle, a physical barrier that also may contain antimicrobial factors, can be considered as the external defense in crustaceans.

The influence of the gut flora on the host is clearly of great interest in aquaculture, particularly where poor productivity and/or stock losses are widespread (Lavens and Sorgeloos, 2000). Within marine and other aquatic animals, the colonization of the digestive system by micro-organisms is influenced by a number of both host- and non-host-related factors (for a review see Harris 1993). Such factors include the ingestion of the surrounding free-living bacterial community (Austin and Austin 1989), physicochemical aspects of the gut (Hood *et al.,* 1971; Huq *et al.,* 1986).

Comparisons of the bacterial community of the gut in wild and cultured prawns have, thus far, only been studied in *Penaeus japonicus* Bate where no significant difference in gut bacterial composition was reported between wild and cultured varieties (Yasuda and Kitao, 1980).

MATERIALS AND METHODS

The present investigation was carried out to assess the alterations in the probiotic microbial load at certain non-specific immune sites of *Macrobrachium rossenbergii* supplemented with *Cynodon dactylon.*

Experimental Set Up: *M. rosenbergii* (post larvae (PL-30) were purchased from the prawn hatchery of ADAK (Agency for Development of Aquaculture in Kerala) and acclimated in the laboratory for two weeks water quality parameters were maintained everyday and monitored periodically, as per Trivedi and Goel, (1984). Prawns were fed with the control and medicated diet (0.25%) prepared in the laboratory (FAO 1978). Experiments were carried out in glass tanks of 1000L capacity with proper aeration was provided. 50% of water was renewed daily during removal of waste feed and faecal matter. In each sets, 15 prawns were fed at 4 % of the body weight / day. After 2 and 4 months, the total microbial load of the non- specific immune sites of prawns of both control and medicated culture were done (Kannan *et., al* 2002).The isolation and identification of major microbes were also done (Buck *et al.*, 1963; Akagawa-Matsushita *et al.*, 1992; Holt *et al.*, 1994; Webster, 1995).

RESULTS AND DISCUSSION

The results obtained according to the experimental design are presented in Table 1-4. Table 1 represents the physico-chemical characters of culture water. A marginal increase in the acidity of water with a corresponding drop in dissolved oxygen of the culture media is observed. Table.II represents the average microbial load in culture tanks. The values indicate a considerable fall in the total microbial load in the experimental tanks over the period monitored.

Table.III shows the total bacterial count of gill, gut and exoskeleton of *M.rosenbergii* after 2 and 4 months, of *Cynodon dactylon* treatment. Though there was a decrease in the total microbial load at the tissue analysed after 2 and 4 months,a remarkable fall in the values is observed and the gut tissue(CFU/gm)31x10^5 and 25x10 ^{5} in the controls at 2 and 4 months, the same for *Cynodon dactylon* treated prawn tissue was 10x10^5 and 14x10^5.

Table. IV. shows the distribution of main bacterial groups and genera isolated from gill, gut and exoskeleton at the end of the treatment schedule. The major bacteria isolated from the gills were *Arthrobacter type I, Vibrio type II, Pseudomonas type I, Streptococous, Acinitobacter type II.* The major bacteria isolated from the gut were *Arthrobacter type I, Vibrio type II, Streptococous type II, Acinitobacter type I & II* and *enterobacteriacea.* The major bacteria isolated from the exoskeleton were, *Vibrio type II, Pseudomonas type II, Streptococous type II, Acinitobacter type I& II(47)and enterobacteriacea.*

From the above observations its evident that, the microbial flora at the non-specific immune sites such as gill, gut and exoskeleton have changed qualitatively and quantitatively. This might be due to the alteration in the probiotic bacteria pool at the said sites, probably due to the antibacterial active principles present in *Cynodon dactylon.*

Table-I Total water quality parameter during the treatment schedule

Parameter	Water quality parameters of culture tanks	
	Control	Experimental
1.Tempeature (° C)	28±0.13	28±0.15
2.pH	6.9±0.3	6.8±0.02
3.Dissolvedoxygen(mg/L)	6±0.14	2.8±0.01
4.Salinity(ppt)	2.39±0.21	2.43±0.11

Table II. Total microbial load of culture water during the treatment schedule

Sample	Total plate count(CFU/ml)	Total plate count(CFU/ml
	2 months	4 months
Control	100 x 10^6	92x 10^6
Expermental	80 x 10^6	71x 10^6

Table III. Total microbial load during the treatment schedule at gill, gut and exoskeleton of *Macrobrachium rosenbergii*

Month	Sample	Gill(CFU/gm)	Gut(CFU/gm)	Exoskeleton (CFU/gm)
Two Months	Control	46x10^4	31x10^5	38x10^5
	Treated	18x10^4	10x10^2	12x10^4
Four Months	Control	25x10^4	25x10^5	20x10^5
	Treated	16x10^3	18x10^2	10x10^4

Table IV. Dominant bacterial geneus identified at gill a,gut and exoskeleton of *M.rosenbergii* fed with *Cynodon dactylon* extract.

Bacterial isolates/gm	Control	Gill	Control	Gut	Control	Exoske-leton
Pseudomonas I	15	20	-	-	-	-
Pseudomona II	5	-	-	-	5	2
Vibrio type II	10	7	6	2	10	8
Streptococcus II	4	-	8	2	4	1
Strepto coccus I	4	1	-	_	4	_
Acinetobacter I	289	267	370	355	350	343
Acinetobacter II	21	4	10	2	50	47
Arthrobacter I	10	7	5	2	5	-
Enterobacteriaceae	15	-	8	3	3	2

According to Gatessoupe *et al,* (1989) gut flora, which play prominant part in animal nutrition, do not have a major role in digestion in cold water fish but do play a significant role in intestinal fermentation in warm water fishes. Apart from enzymes, the gut bacteria bring in additional nutrients like vitamin B12. Though lactic acid bacteria (LAB) like

Streptococcus, Lactobacillus and *Camobacterium* belong to the normal microflora of the gastrointestinal tract in healthy fish population of LAB in digestive tract is affected by nutritional and environmental factors like polyunsaturated fatty acids, chromic oxide, stress, salinity etc. LAB isolates from fish gut can act as probiotics and prevent colonization of gut by pathogenic bacteria (Ringo and Gatesoupe, 1998), Sujita and Shibuya (1996) A case of 'good vibrio' (which form yellow colonies in TCBS agar) acting as probiotic in fish was reported by Austin *et al.* (1995). Smith and Davey (1993) isolated a fluorescent strain of pseudomonid bacteria that could competitively inhibit the growth of fish pathogen *A. salmonicida* in culture media, probably due to competition for iron.

The pytonutrients present in the medicinal plants act synergistically for the development of probiotic bacteria and aid in antimicrobial activity. Herbal diets prepared from five different herbs such as *Adathoda vasika, Murraya koenigii, Ocimum basilicum, Psoralea corylifolia* and *Qurcus infectoria* were effectively suppressed the pathogens such as *Pseudomonas aeruginosa, Staphylococcus aureus, Aeromonas hydrophila, Vibrio harveyi*, and *V. parahaemolyticus* in the *P. monodon* (Kuttanappilly *et.al.*, 2004, Velmurugan *et. al*; 2010, Shabana *et. al*; 2019) showed a positive effect on immunomodulation and antimicrobial activity. Immunomodulatory effect of *Cynodon dactylon* against white tail disease of giant freshwater prawn, *Macrobrachium rosenbergii* has been studied by Mohamed_A Farook *et. al.,* (2015) which shows similar results.

CONCULSION

In the present work, the *Cynodon dactylon* methanolic extract supplementation had very effectively suppressed the bacterial growth in the culture media as well as on specific immune sites such as gill, gut and exoskeleton of *M. rosenbergii*. This can be substantiated by the works of Hiam El-Desouky *et. al.* (2012), Mohamed A Farook *et al.* (2015), Shabana *et al.* (2019).

REFERENCES

1. Akagawa-Matsushita, M., Matsuo, M., Koga, Y. and Yamasato, K. (1992). *Alteromonas atlantica* sp. nov. and *Alteromonas carrageenovora* sp. nov., bacteria that decompose algal polysaccharides. *International Journal of Systematic Bacteriology* 42, 621–627.

2. Austin, B. and Austin, D.A. (1989). General introduction. In Methods for the Biological Examination of Fish and Shellfish ed. Austin, B. and Austin, D.A. pp. 1–24. Chichester: Ellis Horwood.

3. Baruah K, Pal AK, Sahu NP, Jain KK, Mukherjee SC and Debnath D (2005).

Dietary protein level, microbial phytase, citric acid and their interactions on bone mineralisation of *Labeo rohita* (Hamilton) juveniles. *Aquaculture Research 36*. Pp 803-812

4. Buck, J.D., Meyers, S.P. and Leifson, E. (1963). Pseudomonas (Flavobacterium) piscida Bein comb. nov. *Journal of Bacteriology* 86, 1125–112

5. Gatesoupe, F.J., T.Arakawa and T.Watanalee (1989). The effect of bacterial additives on the production rate and dietary value of rotifers as feed for Japanese flounder, *Paralichithys olivaceus, Aquaculture,* 83 : 39-44.

6. Harris, J.M. (1993). The presence, nature, and role of gut microflora in aquatic invertebrates — A synthesis. *Microbial Ecology* 25, 195–231

7. Hiam El-Desouky, Amel El-Asely, A.A. Shaheen and A. Abbass (2012). Effects of *Zingiber officinalis* and *Cyanodon dactylon* on the Growth Performance and Immune Parameters of *Macrobrachium rosenbergii.World Journal of Fish and Marine Sciences* 4 (3): 301-307

8. Holtt, J.G., Krieg, N.R., Sneath, P.H.A., Staley, J.T. and Williams, S.T. (ed.) (1994). Bergey's Manual of Determinative Bacteriology, 9th edn. Baltimore: Williams & Wilkins.

9. Hood, M.A., Meyers, S.P. and Colmer, A.R. (1971). Bacteria of the digestive tract of the white shrimp Penaeus setiferus. *Bacteriological Proceedings* 71, 48

10. .Huq, A., Huq, S.A., Grimes, D.J., O'Brien, M., Chu, K.H., Capuzzo, J.M. and Colwell, R.R. (1986). Colonization of the gut of the blue crab (Callinectes sapidus) by Vibrio cholerae. *Applied and Environmental Microbiology* 52, 586–588

11. Kennedy, D.G., Young, P.B., McCaughey, W.J., Kennedy, S. and Blanchflower, W.J. (1991). Rumen succinate production may ameliorate the effects of cobalt-vitamin B-12 deficiency on methylmalonyl Co A mutase in sheep. *Journal of Nutrition* 121, 1236–1242.

12. Kuttanapilly V. Lalitha and Poothuvallil K. Surendran, (2004). *Aquaculture Research*; 35:629-635.

13. Lavens, P. and Sorgeloos, P. (2000). Experiences on importance of diet for shrimp postlarval quality. *Aquaculture* 191, 169–176.

14. Lilley, D.M., and Stillwell, R.H. (1965). *Sci.* 147, 747.

15. Maeda, M. and I.C.Liao (1994). Microbial processes in aquaculture environment and their importance for increasing crustacean production. *JARQ* 28(4):283- 288.

16. McBee, R.H. (1971). Significance of intestinal microflora in herbivory. *Annual Review of Ecological Systematics* 2, 165–176.

17. Mohamed A Farook *et al.,* (2015). Immunomodulatory effect of *Cynodon dactylon* against white tail disease of giant freshwater prawn, *Macrobrachium rosenbergii* (de Man, 1879*) Aquaculture Research* 47(11)

18. Phillips, N.W. (1984). Role of different microbes and substrates on the midgut of the fiddler crab Uca pugilator. *Mycologia* 66, 669–675.

19. Ringo, E. and F.J.Gatesoupe (1998). Lactic acid bacteria in fish: a review. *Aquaculture* 160: 177-203.

20. Shabana M. S., M. Karthika and V. Ramasubramanian (2019). Effect of dietary *Citrus sinensis* peel extract on growth performance, digestive enzyme activity, muscle biochemical composition, and metabolic enzyme status of the freshwater fish, *Catla catla. The Journal of Basic and Applied Zoology.* volume 80, Article number: 51

21. Sugita, H., Ueda, S., Ueda, R. and Degushi, Y. (1987). Morphology and brief characteristics of Vibrio sp. isolated from marine Crustacea. Bulletin of the College of Agriculture and Veterinary Medicine Nihon University 44, 160–164

22. Suglta, H. and K.Shibuya (1996). Antibacterial abilities of intestinal bacteria in freshwater cultured fish. *Aquaculture,* 145(1/4) : 195-203

23. Velmurugan S, S.M.J.Punitha, M.Michael Babu, T. Selvaraj, T. Smith, D, and S. Davey, (1993). Evidence of competitive exclusion *of Aeromonas salmonieida* from fish with stress - inducible furunculosis by a fluorescent pseudomonad. *Journal of Fish Disease,* 16 : 521-524.

24. Vitalis, T.Z., Spence, M.J. and Carefoot, T.H. (1988). The possible role of gut bacteria in nutrition and growth of the Sea Hare Aplysia. *The Veliger* 30, 333–341

25. Webster, N.S. (1995). The isolation and identification of bacterial flora from the Redclaw Crayfish, Cherax quadricarinatus. Honours Thesis, Department of Microbiology and Immunology, James Cook University, Townsville, Australia.

26. Xu, Y. J., Wu, X. H., Tan B. K. H., Lai, Y. H., Vittal, J. J. and Imiyabir, Z. (2000). Flavanol- cinnamate cycloadducts and diamide derivatives from Aglaia laxiflora. *Journal of Natural Products* 409, 473-476

27. Yasuda, K. and Kitao, T. (1980). Bacterial flora in the digestive tracts of prawns, Penaeus japonicus Bate. *Aquaculture* 19, 229–234.

Current Status of Researches in Biosciences (2020) : 275-278
Editor : Prof. P.C. Joshi, N. Joshi, Reshman Yasmin, and D. K. Mansotra
Today & Tomorrow's Printers and Publishers, New Delhi - 110 002

DIFFERENTIAL NUTRITIONAL QUALITY OF MILK AMONG DIFFERENT BREEDS OF CATTLE AND BUFFALOES

Pooja Ch, Venkataram K, Sushma A, Jayasri K, Padmaja K, Eswara prasad P and Raman D*

Department of Vety. Biochemistry, College of Vety. Science, SVVU, Tirupati, AP,India

**NKBC, Chintaladeevi, A.P*

Email: kondetibicm6@gmail.com

ABSTRACT

Milk is considered as an important protein source in human diet. Hence, this study was conducted to evaluate composition of milk in samples collected during early lactation period from different breeds of buffaloes and cows located at NKBC, Chintaladeevi, Andhra Pradesh. Jafarabadi milk has lowest fat and lactose concentration while its protein content was highest compared to other buffalo breeds. Whereas SNF content was lowest in Pandharpuri buffalo milk. Gir has highest fat content while the protein, lactose and SNF concentration was lowest among cattle milk. The data suggests that relative proportion of milk composition varies with the breed of animals and hence selection of breed for better milk price is essential.

KEYWORDS: Milk composition, Breed, Cattle, Buffaloes.

INTRODUCTION

Milk is a complex biological fluid produced by mammary glands of all mammals. Milk contains essential nutrients needed for growth and development of human health as well as the calf. Most common economically important milk constituents include fat, protein, SNF and lactose. Milk composition is economically important to milk producers, processors, and nutritionally important to consumers. Milk composition is influenced by many factors that include nutrition, parity and stage of lactation (Gurmessa and Melaku, 2012).

Milk from different cattle breeds holds distinct composition profiles because of genetic background (Yang *et al.*, 2013). In dairy industry, it is of major interest to milk processors and consumers because the concentration of certain components in milk affects the pricing policy of milk in the market thereby directly affect the economy of milk production as well as economic condition of dairy farmers. The effect of genetic factors on the milk constituents of dairy buffaloes is rather scanty under Indian conditions. Hence, the present study was taken up to compare the breed wise nutritional quality of cattle and buffalo milk. Determining differences in milk composition may be of interest due to their impact on human health and manufacturing properties of milk (Riel, R., 1985).

MATERIALS AND METHODS

1. **Collection of milk samples:** Milk samples were collected during early lactation stage from cattle (Ongole, Tharparkar, Kangayam, Deoni, Kankrej and Gir) and buffalo (Murrah, Mehasana, Pandharpuri, Banni and Jafarabadi) breeds. Six samples were collected from each breed maintained by National Kamadhenu Breeding Centre, Chintaladeevi, Andhra Pradesh.
2. **Determination of milk composition:** The milk samples were pre heated to 37°C and then the percent fat, lactose and SNF were measured using automatic analyser, MilkoScan Minor 4.

RESULTS AND DISCUSSION

According to Gurmessa and Melaku (2012) factors such as breed and health of animal, stage of lactation, feeding systems, seasonal changes, milking frequency and milking systems, nutrition and genetics can cause variation in relative proportion of milk composition.

Table 1: Nutritional quality of milk in cattle breeds

Breed	Lactose %	Fat %	SNF %	Protein g %
Deoni	5.28 ± 0.25CD	3.17 ± 0.26A	9.63 ± 0.05B	3.47 ± 0.03AB
Tarparkar	5.22± 0.48BC	4.10 ± 0.21C	9.24 ± 0.02B	3.43 ± 0.03A
Kankrej	5.42± 0.24E	4.49 ± 0.17D	8.76 ± 0.09B	3.72 ± 0.01C
Gir	4.86 ± 0.28A	5.18 ± 0.05E	6.8 ± 1.36A	3.35 ± 0.04A
Ongole	5.11 ± 0.24B	3.62 ± 0.84B	9.24 ± 0.07B	3.35 ± 0.04A

Means with different superscripts (A, B, C, D, E) with in a column differ significantly ($p < 0.01$).

Table 1 shows that among the cattle breeds, highest lactose (5.42± 0.24) and protein content (3.72± 0.01) was observed in Kankrej milk.Gir milk has highest fat content (5.18± 0.05) and lowest SNF% (6.8± 1.36). Whereas Deoni has highest SNF% (9.63± 0.05) and lowest fat% (3.17± 0.26) which shows an inverse relationship between fat and SNF content of milk.

Table 2 shows that among the buffalo breeds, lactose content was highest (5.08± 0.02) in Murrah while it is lowest (3.35± 0.04) in Jafarbadi milk. Protein content was highest (4.84± 0.03) in Jafarbadi whereas lowest (3.08± 0.01) in Pandharpuri. SNF content was highest (9.56 ± 0.14) in Murrah milk whereas lowest (7.89±0.01) in Pandharpuri.Pandharpuri milk has highest fat (9.36± 0.08) content followed by Murrah, Banni, and Jafarabadi (3.72 ±0.01).

Highest fat, SNF, protein and lactose content was found in buffalo milk compared to cow milk (Kapadiya *et al.,* (2016). Similar results were observed in our study. Misra *et al*., (2008) conducted a comparative study on four breeds of Indian buffaloes (Bhadawari, Mehasana, Murrah, Surti) and reported the highest fat percent in Murrah (7.53±0.19) and lowest in Surti (6.17±0.02) buffalo milk. Our results are in agreement with Prasad *et al*. (2018) who reported highest fat % in buffalo milk (6.42%) when compared to cow (4.06%).

Table 2: Nutritional quality of milk in buffaloe breeds

Breed	Lactose %	Fat %	SNF %	Protein g %
Murrah	5.08 ± 0.02E	7.39 ± 0.08 C	9.56 ± 0.14D	3.51 ± 0.10 C
Banni	4.48 ±0.02 D	5.33 ± 0.10 B	8.48 ± 0.17B	4.10 ± 0.08D
Mehasana	4.10 ± 0.03B	5.18 ± 0.06 B	9.06 ± 0.02C	3.27 ± 0.01B
Pandharpuri	4.30 ± 0.01C	9.36 ± 0.08D	7.89 ± 0.01A	3.08 ± 0.01A
Jafarabadi	3.35 ± 0.04A	3.72 ± 0.01A	8.51 ± 0.11B	4.84 ± 0.03E

Means with different superscripts (A, B, C, D, E) with in a column differ significantly ($p < 0.01$).

Woodford *et al*. (1986) reported that between and within breeds, fat varies the most and lactose the least. Similar results were noticed in our study between the buffalo breeds where fat content ranges from 3.72% to 9.36% and lactose ranges from 3.35 to 5.08% and in between cattle breeds fat content ranges from 3.17% to 5.18% and lactose ranges from 4.86% to 5.42%. Our results are in accordance with Gaunt (1980) who reported that cattle contain lowest fat compared to buffaloes. Ladd *et al.,* (2017) reported lactose as a valuable nutrients because it favours intestinal absorption of calcium, Mg and phosphorus. As reported by Singh *et al.,* (1979) gross composition of buffalo milk is affected by the breed.

It was concluded that fat content varies much compared to lactose among breeds of cattle and buffaloes. Our results suggest that fat content is responsible for antioxidant activity of milk.

REFERENCES

1. Kapadiya, D. B., Prajapati, D. B., Jain, A. K., Mehta, B. M., Darji, V. B., & Aparnathi, K. D. (2016). Comparison of Surti goat milk with cow and buffalo milk for gross composition, nitrogen distribution, and selected minerals content. *Veterinary world*, *9*(7), 710.

2. Gurmessa, J., & Melaku, A. (2012). Effect of lactation stage, pregnancy, parity and age on yield and major components of raw milk in bred cross Holstein Friesian cows. *World Journal of Dairy & Food Sciences*, *7*(2), 146-149.

3. Gaunt, S. N. (1980). Genetic variation in the yields and contents of milk constituents. *Bulletin, International Dairy Federation*, (125), 73-82.

4. Lad, S. S., Aparnathi, K. D., Mehta, B., & Velpula, S. (2017). Goat milk in human nutrition and health-A review. *International Journal of Current Microbiology and Applied Sciences*, *6*(5), 1781-1792.

5. Misra, S. S., Sharma, A., Bhattacharya, T. K., Kumar, P., & Roy, S. S. (2008). Association of breed and polymorphism of αs1-and αs2-casein genes with milk quality and daily milk and constituent yield traits of buffaloes (Bubalus bubalis). *Buffalo Bull*, *27*(27), 294-301.

6. Prasad, N., Shukla, S., and Ramteke, P.W. (2018). Physico-chemical properties of milk and dairy products collected from Allahabad city, India. *International Journal of Current Microbiology and Applied Sciences,* 7(7):1662-1666.

7. Riel, R., (1985). Composition and physio chemical structure of milk .In: *Dairy Science and Technology, principles and Application.* Les Presses de L University Laval, Quebec, Canada.

8. Singh, A., Basu , S. B., Bhatta, K. L., (1979). Milk fat and SNF percentages of Murrah buffaloes. *Indian Journal of Dairy Science;* 32:446-449.

9. Woodford, J.A., Jorgensen, N.A., and Barrington, G.P., (1986). Impact of dietary fiber and physical form on performance of lactating dairy cows. *J. Dairy Sci.*69:1035.

10. Yang, Y., Bu, D., Zhao, X., Sun, P., Wang,, J., and Zhou, L., (2013). Proteomic analysis of cow, yak, buffalo, goat and camel milk whey proteins: Quantitative differential expression patterns. Quantitative differential expression patterns. *Journal of Proteome Research,* 12(4):1660-1667.

Current Status of Researches in Biosciences (2020) : 279-283
Editor : Prof. P.C. Joshi, N. Joshi, Reshman Yasmin, and D. K. Mansotra
Today & Tomorrow's Printers and Publishers, New Delhi - 110 002

A PILOT STUDY ON FEEDING PRACTICES FOLLOWED IN LACTATING DAIRY CATTLE UNDER FIELD CONDITIONS OF TAMIL NADU, INDIA

A. Bharathidhasan* and G. Kalaivani

Associate Professor and Head, Veterinary University Training and Research Centre, Tamil Nadu Veterinary and Animal Sciences University, Vellore -632 009.

Email ID: bdhasana@gmail.com

ABSTRACT

Dairying is a main source of income for many rural families in India. Most of the small holding rural dairy farmers feed the cattle with agricultural by products with very limited quantity of concentrate feed. The feeding of paddy straw without concentrate feed leads to decrease the milk production of dairy cattle due to imbalanced feeding. Further the deficiency of critical nutrients also leads to reduce the production and reproduction of dairy animals. Hence, a pilot study was conducted to assess the feeding practices followed in lactating dairy cattle under field conditions in two villages of Tamil Nadu. Data was collected about the feeding practices followed for 50 dairy cattle under field conditions. Based on the milk yield and feeding practices of the surveyed dairy cattle were categorized into four groups. Group I was high yielders fed with roughage and concentrate ratio of 71: 29 with milk production of 18 litre per day and Group II was medium yielders fed with roughage and concentrate ratio of 76: 24 with milk production of 8 litre per day. Group III was average yielders fed with roughage and concentrate ratio of 80: 20 with milk production of 5 litre per day and Group IV was low yielders fed with roughage and concentrate ratio of 88:12 with milk production of < 4 litre per day. The dry matter intake (DMI) of group I, II, III and IV was deficit by

8.5, 24, 25 and 40 %, respectively than the standard values. The given total digestible nutrients (TDN) and digestible crude protein (DCP) under field condition for group I, II, III and IV were deficit by 1.9 & 22, 7.0 & 30, 1.0 & 34 and 1.0 & 16 % respectively than the standard values. It was concluded that the deficit of DMI, TDN and DCP were observed in feeding practices followed in lactating dairy cattle under field condition in Tamil Nadu.

Keywords: Feeding practices, field condition, lactating dairy cattle.

INTRODUCTION

Dairying is a main source of income for many rural families. Most of the milk obtained is from small, marginal farmers and landless dairy farmers (DAHD & F, 2019). All the small holding rural dairy farmers feed the cattle with agricultural by products with very limited quantity of concentrate feed. The production record of these animals is far below their genetic potential. This is because of deficiency in critical nutrients required for production and reproduction. Low production with higher genetic potential is mainly due to the imbalanced feeding of dairy animals (FAO, 2016). Hence, the present research work was carried out to assess the feeding practices followed in lactating dairy cattle under field conditions in two Villages of Tiruvallur District.

MATERIALS AND METHODS

Data was collected about the feeding practices followed for fifty dairy cattle under field conditions at Puzhal and Kadirvedu Villages of Tiruvallur District. The data regarding the dairy cattle viz., breed of cattle, age, body weight, physiological status and milk yield were recorded. Data regarding the feeding pattern viz., the quantity of dry fodder, green fodder, grains, oil cakes, concentrates and other feed additive given to dairy cattle were also recorded. Feed samples commonly fed to the dairy cattle were collected from the farmers and analysed for their nutrient composition as per the AOAC 2005. Based on the milk yield and feeding practices of the surveyed, dairy cattle were categorized into four groups. The nutrients provided to the lactating dairy cattle under field and the actual requirement was compared as per ICAR (2013). The data was statistically analyzed using SPSS software (version 20.0).

RESULTS AND DISCUSSION

From the present study (Table 1), it was noted that, most of the dairy cattle were fed predominantly on paddy straw with a limited quantity of green grass fodder. De-oiled wheat bran, oil cakes, pearl millet and gram husk were the main concentrate feeds, fed to the lactating dairy cattle. Commercially available concentrate feeds were regularly fed to the dairy cattle with high milk yield. Avijit *et al.*, (2018) reported that, in India and most of the developing countries, livestock were reared on agricultural crop residues with nominal supplementation of concentrate feed sources, leading to improper utilization of feed resources and low in animal production.

The feeding practices followed in lactating dairy cattle were categorized into four groups based on the milk yield and quantity of feed given under field conditions (Table 2). Group I was high yielders fed with roughage and concentrate ratio of 71: 29 with milk production of 18 litre per day and Group II was medium yielders fed with roughage and concentrate ratio of 76: 24 with milk production of 8 litre per day. Group III was average yielders fed with roughage and concentrate ratio of 80: 20 with milk production of 5 liter per day and Group IV was low yielders fed with roughage and concentrate ratio of 88:12 with milk production of < 4 litre per day.

The present study was similar to that of Garg *et al.*, (2013) who studied the feeding strategy of cattle and categorized them based on the milk yield as low(<8 Kg/d), medium (8-12 kg/d) and high (>12 Kg) yielders.

Table 1: Feeding practices and milk yield followed in 50 dairy cattle under field conditions

No. of dairy cattle	Feeding Practices (intake in Kg)							Milk yield per day per animal (L)
	Paddy straw	Green Fodder	Grains	Oil cakes	DORB / Wheat bran	Gram Husk	Others/ Concent rates	
1	Ad lib	12	0.5	0.5	1	0.5	1	20
1	Ad lib	10	0.5	1	1	0.5	1	18
10	4	8	0.5	0.5	0.5	-	-	10
8	4	8	0.5	0.5	0.5	-	-	8
12	3	5	-	0.3	0.5	0.5	-	6
16	3	5	-	0.3	0.5	0.5	-	5
1	2	2	-	-	0.5	0.5	-	3
1	2	2	-	-	0.5	0.5	-	2

Table 2: Categorization of the feeding practices followed in lactating dairy cattle under field conditions

Group	No. of dairy cattle	% of dairy cattle	Ratio of Roughage: Concentrate feeding	Milk yield (litre/day/ Dairy cattle)	Remarks
I	2	4	71 : 29	18-20	High yielders
II	18	36	76 : 24	8-10	Moderate yielders
III	28	56	80 : 20	5-6	Average yielders
IV	2	4	88 : 12	< 4	Low yielders

The nutrients provided to the dairy cattle of all the groups were calculated (Table 3) and it was observed that, they were deficient in DMI, TDN and DCP when compared to actual requirement (ICAR 2013). The DMI, TDN and DCP were deficient by 8.5, 1.9 and 22 % which was very critical for high yielding animals. The DMI, TDN and DCP were deficient by 24, 7 and 30 % in medium yielding dairy cattle.

Table 3: Nutrients provided to dairy cattle under field conditions vis-o-vis with actual calculated requirement

Group	Body weight /Milk yield (L)	Nutrients (Kg)	DMI (Kg/d)	DCP (Kg/d)	TDN (Kg/d)
I (High yielder)	400	Provided	9.15	0.93	9.12
	18	Required	10	1.2	9.3
II (Medium yielder)	350	Provided	6.65	0.49	5.46
	8	Required	8.75	0.7	5.9
III (Average yielder)	300	Provided	5.57	0.33	4.26
	5	Required	7.5	0.5	4.3
IV (Low yielder)	250	Provided	3.7	0.25	3.01
	2.5	Required	6.25	0.3	3.04

The DMI, TDN and DCP were deficient by 25, 1 and 34 % respectively in case of average yielders. The DMI, TDN and DCP were deficient in 40, 1 and 16 % respectively in low yielders. The present study was similar to that of Meenalochini *et al.,* (2015) where they reported, 10,12 and 30 % deficit of DMI, TDN and CP intake in early lactating Jersey cross breed cattle respectively. Singh *et al.* (2008) also reported that, DMI of dairy cows reared in urban and rural areas were deficit by 7 and 5 % respectively. Garg *et al.* (2013) reported that, out of 12,518 cattle 13 % animal received deficit of CP and ME. Hence by providing the minimum concentrate feed along with crop residues will fulfill the nutrient requirement to meet out the maintenance and production requirement under field condition.

CONCLUSION

It was concluded that the deficit of DMI, TDN and DCP were observed in feeding practices followed in lactating dairy cattle under field condition in Tamil Nadu.

REFERENCES

1. AOAC. (2005). Official Methods of Analysis. 19th ed. Association of Official Analytical Chemists, Maryland, USA.

2. Avijit Dey, Singh, R.K,, Dahiya, S.S. and Imaz, J.A. (2018). Ration Balancing for Sustainable Animal Production: Resources and Methodology. *J. Nutr. Biol.,***4**(2): 278-281.

3. DAH, D&F (2019). Department of Animal Husbandry, Dairying & Fisheries, Ministry of Agriculture. Government of India: Krishi Bhawan, New Delhi, India.

4. FAO, (2016). Food and Agriculture Organization of the United Nations. Yield of dairy cattle's statistics.

5. Garg, M.R., Sherasia, P. L., Bhanderi, B. M., Phondba, B. T., Shelke, S. K. and Makkar, H.P.S. (2013). Effect of feeding nutritionally balanced rations on animal productivity, feed conversion efficiency, feed nitrogen use efficiency, rumen microbial protein supply, parasitic load, immunity and enteric methane emissions of milking animals under field conditions. *Anim. Feed Sci. Tech.* **179**: 24–35.

6. ICAR, (2013). Nutrient requirements of cattle and buffaloes. Indian Council of Agriculture Research, New Delhi, India.

7. Meenalochani.V., Chandrasekaran, C., Murugan, M., Purusothaman, M., Vasanthakumar, P. (2015). Research article nutritional status of dairy cows at various physiological stages in Dharmapuri District of Tamil Nadu. *Indian J. Anim. Health.*, **54**(2): 123-132.

8. Singh, V.K., Singh, P., Verma, A.K. and Mehra, M.R. (2008). On farm assessment of lactating cattle and buffaloes in urban, periurban and rural areas of Middle gangetic plains. *Liv Res Rural Dev.*, **20**: 37-42.

Current Status of Researches in Biosciences (2020) : 285-293
Editor : Prof. P.C. Joshi, N. Joshi, Reshman Yasmin, and D. K. Mansotra
Today & Tomorrow's Printers and Publishers, New Delhi - 110 002

EFFECT OF MARIJUANA (*CANNABIS SATIVA*) ON HARDERIAN GLAND OF RABBIT, *ORYCTOLOGUS CUNICULUS*

Surya Prakash Verma and Sunanda Dubey,

Department of Zoology, Udai Pratap College, Varanasi-221002,

Email: sunandadubey@rediffmail.com

ABSTRACT

Marijuana is one of the *Cannabis* preparations popular as Bhang, is produced from the dried and crushed leaves, seeds and stem of the hemp plant *Cannabis sativa*. Marijuana, *Cannabis* or hemp plant is one of the oldest psychoactive plants known to humanity. Harderian gland (HG) is a large bilobed gland. It is located in the orbit of many terrestrial vertebrates. It opens by a duct into the surface of the nictitating membrane. This gland is well developed in some species like rodents, lagomorphs and cetaceans. It may occur as a transit embryological structure as in case of primates, terrestrial carnivores and chiroptera. The Harderian gland has diverse of functions such as lubrication of the eye, thermoregulation, photoreception, photo-protection and as a source of either pheromones or growth factors. It was reported that the Harderian gland possessed receptor for melatonin. No report on Harderian Gland and Marijuana relationship is available till date. The experimental group rabbits were administered daily dose (100 mg/kg body weight) of Marijuana (Bhang). Five rabbits each of experimental and control groups are anaesthetized after 15 days and 45 days administration of Marijuana respectively to take Harderian gland tissue sample for histological studies. The studies showed that Harderian glands become more active in Marijuana fed groups, showing thin follicular layer and large lumen. Apical microvilli like cytoplasmic protrusions were seen in abundance. Nucleus is large, oval and arranged in row on basal lamina. Small cytoplasmic vacuoles were present showing more type II cell as the sections were from white part. Study suggests a relation between Marijuana consumption and Harderian gland function as it accelerates the Harderian gland secretion.

Keywords: Marijuana (*Cannabis sativa*), Harderian gland, rabbit etc.

INTRODUCTION

Marijuana contains more than four hundred sixty known compounds of which more than sixty are cannabinoids. The only cannabinoid that is both highly psychoactive and present in large amount is Tetrahydrocannabinol or THC. Marijuana is one of the Cannabis preparations popular as Bhang. Bhang is produced from the dried and crushed leaves, seeds and stem of the hemp plant *Cannabis sativa*. Marijuana, cannabis or hemp plant is one of the oldest psychoactive plants known to humanity.

The Harderian gland is widespread throughout the animal kingdom, but differs in its structure among various species. Even though it was discovered as early as 1694 by Johann Jacob Harder, a noted physician born on September 17, 1656, in Basel, Switzerland, who first described it in the deer (*Dama vulgaris*), little is known about the function and metabolism of this orbital organ. Since that time, the Harderian gland has been observed in many terrestrial species, absent only in the orbit of primates. However, the Harderian gland is rudimentary in monkeys and may even be found in adults with abnormalities (Kennedy 1970; Olcese and Wesche 1989). The rabbit Harderian gland is irregularly kidney-shaped. It is convex on the outer side and concave on the inner, to comply with the shape of the eyeball. Light Microscopy Examination of numerous sections of the rabbit Harderian gland demonstrates a tubuloalveolar structure, uniform in all areas, with each part divided into lobules separated from each other by a small amount of connective tissue containing fibroblasts, small blood vessels, and nerve fibers (Kuhnel 1971a). Sometimes alveoli from both lobes are mingled and occasionally some alveoli contain pink and white cells side by side (Bjorkman *et al.* 1960; Kuhnel 1971a). The light microscopic picture yields no apparent morphological differences compared with findings of Buschke (1933) and Kuhnel (1971a) with respect to material embedded with paraffin and stained by hematoxylin-eosin, azan, ironhematoxylin or chromalaun-hematoxylin-phloxin following formaline fixation. The pink lobe is characterized by large alveoli with a cuboidal epithelium. The basal nuclei are round with uneven or serrated outlines. In the white part, the alveoli have a smaller diameter and the lumina are narrower. The epithelium is high columnar and the cells are more regular than in the pink lobe. The rounded nuclei are regularly placed in the basal compartment of the cell. The secretion is discharged by a single duct opening on the concave surface of the palpebra tertia, well back and near its inferior border (Davis, 1929; Loewenthal, 1892).

The review of literature reveals that some work has been done on toxicity of Marijuana in humans and few other mammals. However, there is no information available on effect of Marijuana (Bhang) on Harderian gland histology of rabbit so far.

MATERIALS AND METHODS

To study the histopathological alteration inflicted by Marijuana on Harderian gland, healthy male rabbits, *Oryctologus cuniculus* were acclimatized to the laboratory condition after housing them separately in cages. The rabbits were divided in two groups one consisting of control and other experimental group.

The experimental group rabbits were administered daily oral dose (100 mg/kg body weight) of Marijuana (Bhang). Five rabbits each of experimental and control groups were anaesthetized by intravenous administration of Sodium Pentobarbital (150mg/Kg) after 15 days and 45 days administration of Marijuana respectively, to take the tissue sample of Harderian gland. Small pieces of the tissues from anterior end of white lobe were taken and immediately fixed in Bouins fluid. Fixed tissues were embedded in paraffin wax and sectioned at 5 u in thickness with the help of microtome and then stained with Harris Hematoxylene and Eosin stain and mounted in DPX. Sections were visualized and studied using Laczene microscope model HL-23(LacZene Biosciences, India).

RESULTS AND DISCUSSION

Marijuana has long been used as a medicine and recommended to quicken the mind, lower fever, induce sleep, cure dysentery, stimulate appetite, improve digestion, relieve headaches and body pains and cure cough and veneral diseases (O' Shaughnessey, 1838, M' meens, 1860, Hare, 1887). Moderate dose of Indian hemp is useful for patients with 'Senile insomnia' and remains effective for months and even years without an increase in dose (Reynolds, 1980).Harmful effect of cannabis have been reviewed by Hall, (2000), by Farrel and Riston, (2001), and by Aston, (2001). Ingestion of Bhang has led to fatality in a young man with rheumatic heart diseases (Gupta *et al.*, 2001). Farrel, (1999) asserts a clear dependence pattern in a small percentage of those who use cannabis. Use of addictive drugs such as a cocaine, Marijuana and nicotine affects food and liquid intake behaviour taste preference and body weight. During use and /or withdrawal from nicotine, heroine, marijuana and cocaine, major changes, selection and intake occur which result in weight gain or loss (Mohne, 1990). Koshior *et al.,* (2001) have reported two cases of atrial fibrillation following cannabis smoking, suggesting that delta-9- THC may have effects on adrenergic system. Administration of cannabis receptor agonist decreases adrenaline secretion from the adrenal medulla in rabbits, an effect that is probably mediated by pre synaptic CBI receptor (Niederhoffer *et al*; 2001). Khan, (1986) has observed that long term inhalation of marijuana smoke brings about a marked depletion in the granular content of the adrenal medulla cells indicating increase in medullary activity.

Harderian gland (HG) is a large bilobed gland. It is located in the orbit of many terrestrial vertebrates. It opens by a duct into the surface of the nictitating membrane (Sakai, 1981 and Ortiz *et al.,* 2009). This gland is well developed in some species like rodents, lagomorphs and cetaceans. It may occur as a transit embryological structure as in case of primates, terrestrial carnivores and chiroptera (Payne, 1994; Seyama and Uchijima, 2007). The Harderian gland has diverse of functions such as lubrication of the eye (Gans, 1974; Chieffi *et al*., 1996), thermoregulation (Chieffi *et al*., 1996; Thiessen and Kittrell 1980; Harriman and Thiessen, 1983, 1988; Shanas and Terkel, 1996), photoreception (Wetterberg *et al*., 1970), photoprotection (Spike *et al*.,1985, Hugo *et al*., 1987) and as a source of either pheromones or growth factors (Minucci *et al*., 1989; Seyama *et al*., 1996). It was reported that the Harderian gland possessed receptor for melatonin (Lopez-Gonzalez *et al.,* 1991). Harderianectomy in the golden hamster reduced the peak of pineal melatonin occurring during the dark period (Panke *et al*., 1979). Administration of melatonin in the harderian glands of female hamsters reduce porphyrin content and lead to the appearance of type II cells typical of the male gland (Rodriguez-Colunga *et al*., 1992).The Harderian gland in rabbit was identified to be localized ventromedial to bulbus oculi. The horseshoe shaped gland consisted of two lobules of pink or white colour. The pink lobule had a pale eosinophilic cytoplasm that recessed and separated by a slight stroma while the white lobule showed a dark eosinophilic cytoplasm. The nucleus of type II cells was situated basally and had an oval shape Type II cells had apical microvilli-like cytoplasmic protrusions, longer than those of type I cells.

Our present studies showed that Harderian glands of Marijuana administered rabbits become more active than the control one showing thin follicular layer and large lumen. Nucleus was large oval and arranged in row on basal lamina while in control groups they were small. Cytoplasm contains small vacuoles. Harderian gland function is characterized by well demarcated alveoli lined by increased numbers of large cells. This study suggests a relation between Marijuana consumption and Harderian gland function as it accelerates the Harderian gland secretary function.

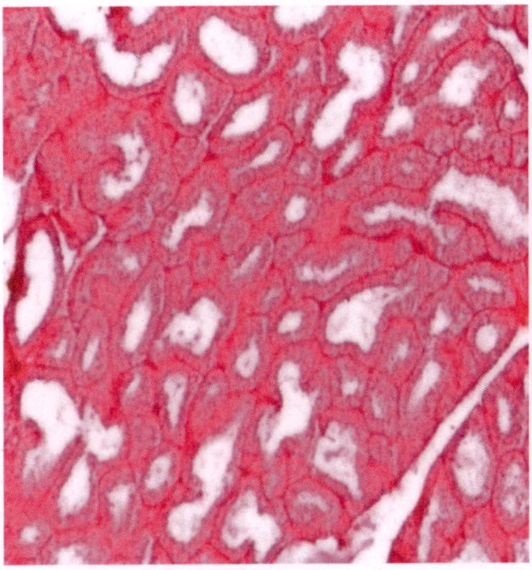

Fig. 1

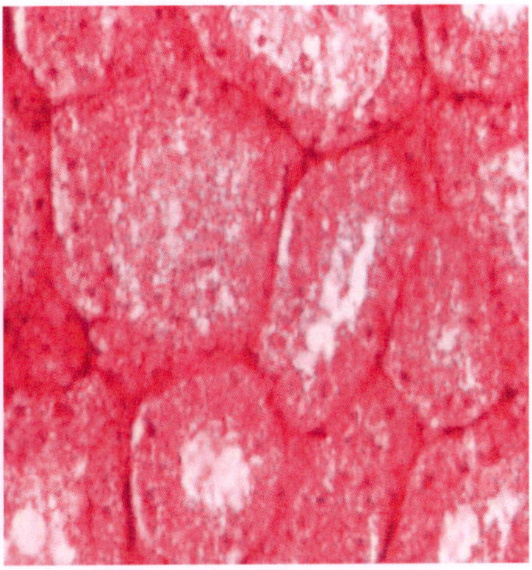

Fig. 2

Fig. 1 Histomicrograph of Control HG for 15 days showing rounded nucleus with acini and small vacuole in cytoplasm. (X10) and Fig. 2 Histomicrograph of Control HG for 15 days showing acini epithelial cells, rounded nucleus arranged at basal lamina and small vacuoles. (X40).

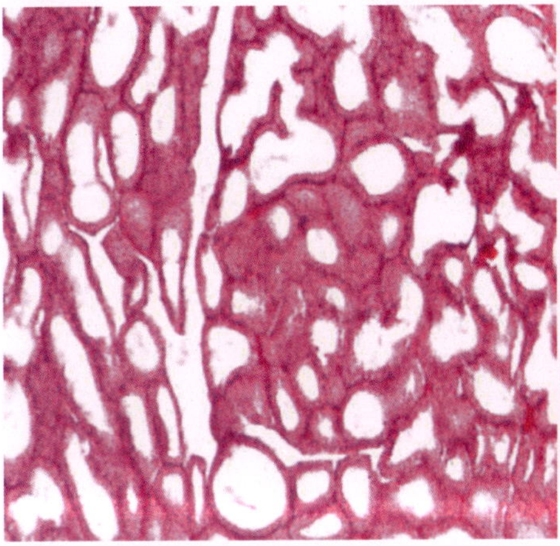

Fig. 3

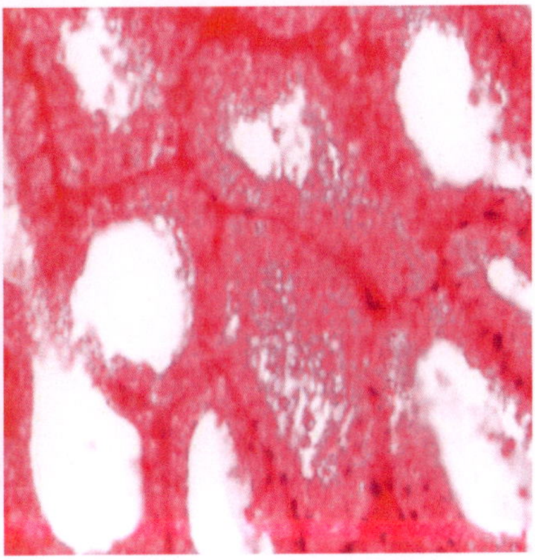

Fig. 4

Fig. 3 Histomicrograph of Marijuana treated HG for15 days showing acini with thin follicular layer with large lumen. (X10) and Fig. 4 Histomicrograph of Marijuana treated HG for 15 days showing acini with thin epithelial cells, rounded nucleus arranged at basal lamina and small vacuoles. (X40).

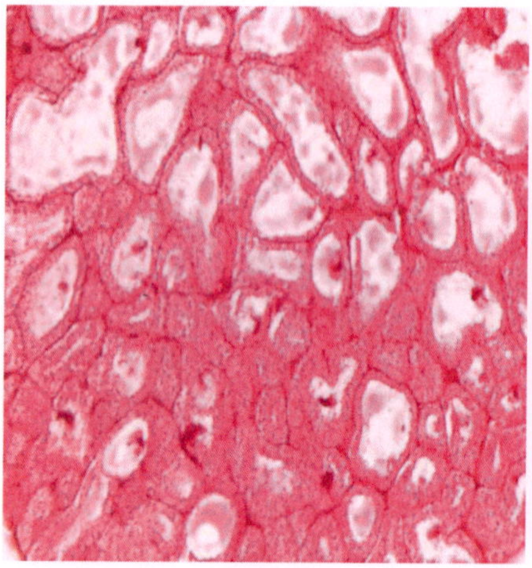

Fig. 5

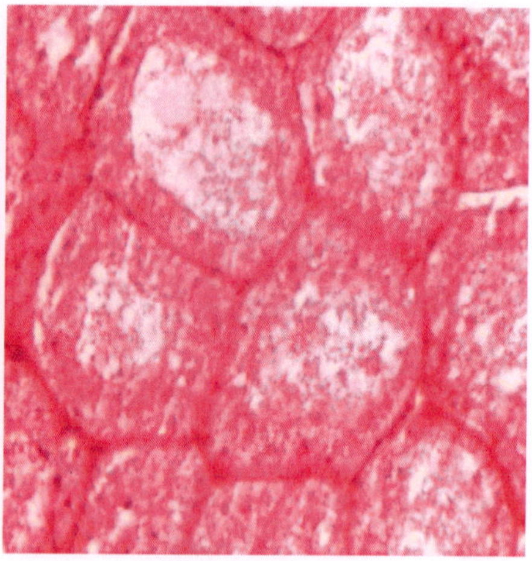

Fig.6

Fig. 5 Histomicrograph of Control HG for 45 days showing rounded nucleus in acini and small vacuoles in cytoplasm. (X10) and Fig. 6 Histomicrograph of Control HG for 45 days showing acini with rounded nucleus and small vacuoles in epithelial cells. (X40).

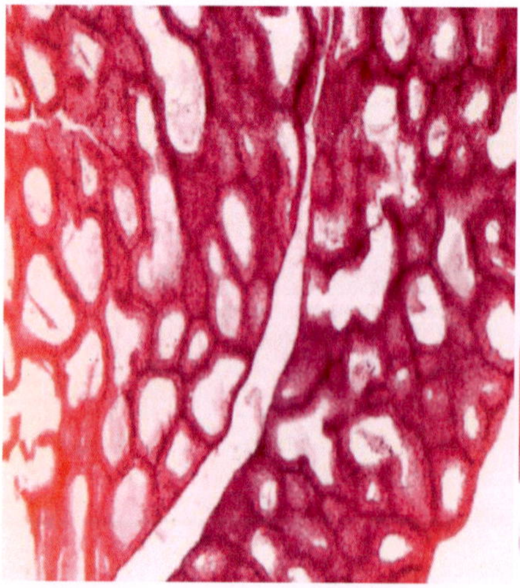

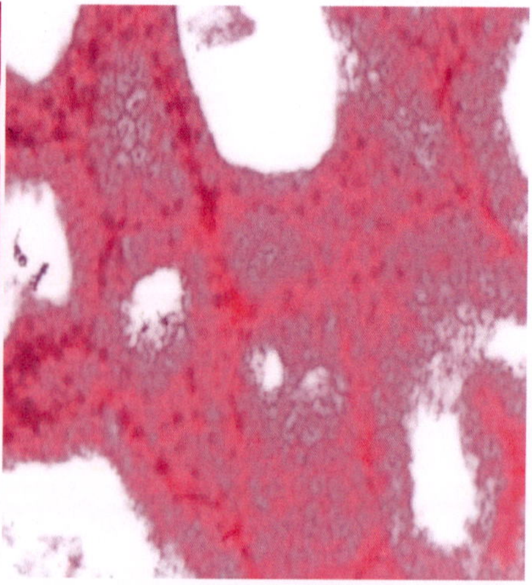

Fig.7 Histomicrograph of Marijuana treated HG 45 days showing acini in various size and shape with prominent nucleus. Enlarged lumen and inter lobular space. (X 10) and Fig.8 Histomicrograph of Marijuana treated HG for 45 days showing thin layer of columnar epithelial cells with prominent nuclei in acini. Cytoplasmic vacuoles are large. (X 40)

ACKNOWLEDGEMENTS

Authors gratefully acknowledge the assistance received from Department of Zoology, Udai Pratap College Varanasi, for providing necessary infrastructural facilities. The authors are also thankful to Dr. S. R. Barai and Mr. B. Yadav for providing Harderian gland from their preserved experimental tissues.

REFERENCES

1. Aston, C.H. (2001). Pharmacology and effects of cannabis: a brief review. *Br. J. Psychiatry,* 178: 101-106.

2. Björkman N., Nicander L., Schantz B. (1960). On the histology and ultrastructure of the Harderian gland in rabbits. *Z Zellforsch Mikrosk Anat* 52:93–104

3. Buschke W. (1933). Die Hautdrüsenorgane (Hardersche Drüsen, Inguinaldrüsen, Analdrüsen, Kieferdrüsen) der Laboratoriumsnagetiere und die Frage ihrer Abhängigkeit von den Geschlechtsdrüsen. *Z Zellforsch Mikrosk Anat* 18:217–243

4. Chieffi G., Baccari G.C., Di Matteo L., d'Istria M., Minucci S., Varriale B. (1996). Cell biology of the harderian gland. *Int Rev Cytol* 1996; 168: 1-80.

5. Davis F.A. (1929). The anatomy and histology of the eye and orbit of the rabbit. *Trans Am Ophthalmol Soc* 27:401–441

6. Farrel, M. (1999). Cannabis dependence and withdrawal addiction, 94: 1277-1278.

7. Farrel, M. and Riston, B. (2001): Cannabis and Health. *Br. J. Psychiatry,* 178:98.

8. Gans, C. (1974). Biomechanics: An Approach to Vertebrate Biology, J. B. Lippincott, Philadelphia, USA, pp261, 6. Di Matteo L, Minucci S, Chieffi Baccari G, Pellicciari C, d'Istria M, Chieffi G. The harderian gland of the frog, *Rana esculenta*, during the annual cycle: histology, histochemistry and ultrastructure. *Basic Appl Histochem* 1989; 33(2): 93-112.

9. Gupta, B.D., Jani, C.B., and Shah, P.H. (2001). Fatal bhang poisoning, *Med Sci.* Law, 41: 349-352.

10. Hall, H.A. (2000). Cannabis use and public health: assessing the burden. *Addiction.,* 95: 485-490.

11. Harder J.J. (1694). Glandula nova lachrymalis una cum ductu excretorio in cervis et damis. *Acta Erudit Publ Lipsiae*, pp 49–53

12. Hare,H.A. (1887). Clinical and physiological notes on the action of *Cannabis indica. Therapeutic Gazette,* 11: 225-226.

13. Harriman A.E., Thiessen D. (1983). Removal of Harderian exudates by sandbathing contributes to osmotic balance in Mongolian gerbils. *Physiol Behav.*; 31(3): 317-23.

14. Hugo J., Krijt J., Vokurka M., Janousek V. (1987). Secretory response to light in rat Harderian gland: possible photoprotective role of Harderian porphyrin. *Gen Physiol Biophys;* 6(4): 401-4.

15. Kennedy G.Y. (1970). Harderoporphyrin: a new porphyrin from the Harderian gland of the rat. *Comp Biochem Physiol* 36:21–36

16. Khan, N.A. (1986). Impact of Marijuana smoke inhalation on normal and stressed rats. *Indian J. Pharmcol.*, 18: 215-222.

17. Koshior, D.A. Filipiak, K.J., Stilarz, P. Opolski, G. (2001). Proxysmalatrial Fibrillation following Marijuana intoxication: a two case report of possible association. *Int. J. Cadiol.* 78: 183-184.

18. Kühnel W. (1971a). Struktur and Cytochemie der Harderschen Drüse von Kaninchen. Z Zellforsch Mikrosk Anat 119:384–404

19. Loewenthal, N. (1892). Beitrag zur Kenntnis der Harderschen Drüse bei den Säugetieren. *Anat Anz* 7:546–556

20. Lopez-Gonzalez M.A., Calvo J.R., Rubio A., Goberna R., Guerrero J.M. (1991). Characterization of melatonin binding sites in the harderian gland and median eminence of the rat. *Life Sci*; 48(12): 1165-71.

21. M' Meens, R.R. (1860). Report of the Committee on *Cannabis indica* Transactions. Fifteenth Annual Meeting of the Ohio State Medical Society, Colombus.

22. Minucci S., Baccari G.C., Di Matteo L., Chieffi G. (1989). A sexual dimorphism of the harderian gland of the toad, Bufo viridis. *Basic Appl Histochem*; 33(4): 299-310.

23. Mohne, M.E., Waston, R.R. and Leonard Green, T. (1990). Nutritional effects of marijuana, heroin, cocaine and nicotine. *J. Am. Diet. Assoc.* 90: 1261-1267.

24. Niederhoffer, N., Hansen, H.H., Farnandez, Ruiz, J.J., Szobo, B. (2001). Effects of cannabinoids on adrenaline release from adrenal medullary cells. *Br. J. Pharmcol*, 134: 1319-1327.

25. O' Shaughnessey, W. B., (1838). On the preparation of the Indian Hemp, or Gunjah (Cannabis indica): The effects on the animal system in health, and their utility in the treatment of tetanus and other convulsive Diseases. Transactions of the Medicinal and Physiological Society of Bengal (1838-1840): 421-461.

26. Olcese J., Wesche A. (1989). The Harderian gland (mini review). *Comp Biochem Physiol* 93A:655–665

27. Ortiz, G.G., Feria-Velasco, A., Pacheco-Moisés, F.P., Rodríguez-Reinoso, S., Cruz-Ramos, J.A., Rosales-Corral, S.A., Reiter, R.J. (2009). Scanning electron microscopy of the orbital Harderian gland in the male Atlantic bottlenose dolphin (Tursiops truncatus). *Anat Histol Embryol*; 38(4): 279-81.

28. Panke E.S., Reiter R.J., Rollag M.D. (1979). Effect of removal of the Harderian glands on pineal melatonin concentrations in the Syrian hamster. *Experientia*; 35(10): 1405-6.

29. Payne, A.P. (1994). The harderian gland: a tercentennial review. *J Anat.*; 185 (Pt 1): 1-49.

30. Reynolds, J.R. (1980). Therapeutic Uses and toxic effect of *Cannabis indica*. *Lancet*, 1: 637.

31. Rodriguez-Colunga M.J., Fernandez C., Rodriguez C., Tolivia D., Menendez-Pelaez A. (1992). Female Syrian hamster Harderian gland: development and effects of high environmental temperature and melatonin injections on histology and porphyrin deposits. *Anat Rec*; 232(2): 293-300.

32. Sakai, T. (1981). The mammalian Harderian gland: morphology, biochemistry, function and phylogeny. *Arch Histol Jpn*; 44(4): 299-333.

33. Seyama Y., Hida A., Hayashi S., Buzzell G.R. (1996). Androgenic control of 1-alkyl-2, 3-diacylglycerol in the harderian gland of the golden hamster, *Mesocricetus auratus*. *J Biochem*; 119(4): 799-804.

34. Seyama Y., Uchijima Y. (2007). Novel function of lipids as a pheromone from the Harderian gland of golden hamster. *Proc Jap Acad*; 83 (3): 77-96.

35. Shanas U., Terkel J. (1996). Grooming secretions and seasonal adaptations in the blind mole rat (*Spalax ehrenbergi*). *Physiol Behav*; 60(2): 653-6.

36. Spike R.C., Johnston H.S., McGadey J., Moore M.R., Thompson G.G., Payne A.P., (1985). Quantitative studies on the effects of hormones on structure and porphyrin biosynthesis in the Harderian gland of the female golden hamster: I. The effects of ovariectomy and nitrogen administration. *J Anat*; 142: 59-72.

37. Thiessen D.D. (1988). Body temperature and grooming in the Mongolian gerbil. *Ann N Y Acad Sci;* 525: 27-39.

38. Thiessen D.D., Kittrell E.M. (1980). The Harderian gland and thermoregulation in the gerbil (*Meriones unguiculatus*). *Physiol Behav;* 24(3): 417-24.

39. Wetterberg L., Geller E., Yuwiler A. (1970). Harderian gland: an extraretinal photoreceptor influencing the pineal gland in neonatal rats? *Science*; 167(3919): 884-5.

Current Status of Researches in Biosciences (2020) : 295-299
Editor : Prof. P.C. Joshi, N. Joshi, Reshman Yasmin, and D. K. Mansotra
Today & Tomorrow's Printers and Publishers, New Delhi - 110 002

CHICKEN CESTODE (*RAILLIETINA TETRAGONA)* AS A NOVEL SOURCE FOR THE SYNTHESIS OF NANO FIBRE AND NANOPARTICLES

Padmavathi Sriram and Sunila Kumari

Department of Zoology, Osmania University College for Women, Hyderabad, Telengana, 500095

Email: anjanipadmavathi@gmail.com

ABSTRACT

In nanotechnology, nanoparticles, nano membranes, nanofibers, nanorods, are frequently used in almost every walk of life, ranging from, industrial, biological, pharmacological and biomedical fields. Metals, and heavy metals, are commonly used for the synthesis of nanoparticles, which are frequently used in many biological and medical applications. Though they are extensively used, they have deleterious effects upon human body. In this regard, a number of plant based nanoparticles and other nano products, are being synthesized. Apart from plant materials, there are insatnces, where in microorganism, such as bacteria, fungi yeast are also used to synthesize nano products. These are very much eco friendly and also less expensive to metal nanoparticles. It is quite evident that nanoparticles, synthesized from a natural source are more appealing as they are both eco friendly and inexpensive. Based upon these conclusions, we hypothesize that nanoparticles and nano fibre can also be produced from natural and some sort of a novel source *i.e.,* an intestinal parasite, which is often considered as a waste, or as an intruder or infestation of human/chicken intestines. This discarded piece of parasitic life may be an appropriate and cheaper alternative source for the production of nano products. In the present study we have for the first time, attempted to synthesize nano particles and nano fiber using an intestinal parasite *Raillietina tetragona* as a potential source. These particles may be referred to as para/bio nano particles, as they originate from a parasite, or from a living animal source, *Raillietina tetragona* is a common, cestode, found in chicken intestines. The parasite measuring approximately 15cm in length, was carefully removed from the intestine of chicken was suitably stored at room temperature, and later processed for the synthesis of nano particle and nano fiber. Nanoparticles were synthesized using the method of acid hydrolysis and nanofiber was prepared using electro spinning technique. The nanofibre which was synthesized, was hydrophilic in nature, and measured about 6cm x 3cm in size. This study, strongly suggests that wealth can be reaped out of waste. These processed para nano materials are cost effective, eco friendly and hence may be suggested to be used frequently in biological and medical applications.

KEYWORDS: Nanoparticle , nanofiber, *Raillietina tetragona*, electro spinning, chicken cestode

INTRODUCTION

Nanoscience and nanotechnology has certainly revolutionized biological and medical sciences in a big way. Nanoparticle, nanomembranes, nanofibers, nanorods, find applications in almost every walk of life, ranging from, industrial, biological , pharmacological and biomedical fields[(1)]. The unique versatility, of nanoparticles, is attributed to their, size and dimensions. (100nm) and materials, exhibit certain unique physical properties, when they are downsized to nano proportions, in contrast to heavier materials, which show constant physical properties [(2)]. The main driving force, for the development of nanotechnology, stems, from the understanding of biological processes as they are often happening at a nanoscale [(3)] and living cells are typically 10 µm in size and their components are much smaller in size. Metals and heavy metals are commonly used for the synthesis of nanoparticles, which are frequently used in many biological and medical applications. Though they are extensively used, they have their own drawbacks as there are evidences where in nanoparticles, made from metals, have deleterious effects upon human body[(4)]. To avoid the risk of using metal nanoparticles, scientists are constantly exploring, different materials, which can be both eco friendly and also cost effective. In this regard, a number of plant based nanoparticles, have been synthesized and this method of developing plant based nanoparticles, is considered to be green chemistry approach [(5)]. Apart from plant materials, being used as nanoparticles, there are evidences, where in microorganism, such as bacteria, fungi yeast are also used to synthesize nano products[(6)]. It is quite evident from these studies, that nanoparticles, synthesized from natural sources are more appealing as they are eco friendly, biocompatible and also least expensive. Based upon these evidences, we hypothesize that nanoparticles and nano fibre can also be produced from natural or some sort of a unusual source *i.e.,* an intestinal parasite. The present study is an attempt to develop nanoparticles and nanofibre from much cheaper source or to be precise, from waste material *i.e* from an intestinal parasite (chicken cestode).

MATERIALS AND METHOD

For the purpose the experimentation, chicken intestines, were collected from the source. The intestines, infested with parasitic infection, were thoroughly washed, and then dissected for parasitic infection, chicken cestode, identified as *Raillietina tetragona*, was indentified, which was white in color and was carefully collected in to the petridish. The worm measured approximately 15 cm in length was collected and dried overnight in an incubator at a temperature of 37 degree Celsius. The parasite was thoroughly dried and pounded into a fine powder using a motor and pestle, and was further processed for nanoparticles and nanofibre.

Preparation of nanoparticles: Cestode nanoparticles were produced from the purified cestode by repeated acid hydrolysis. Cestode powder was soaked in 3M HCl for acid hydrolysis and incubated for 90 min at 90°C in a thermostat water bath. The sample was centrifuged at 6000 rpm for 10 min and the pellets were collected. The acid hydrolysis step was repeated thrice and the final pellet was suspended in distilled water to dilute the acid concentration. The suspension was dialyzed against distilled water until it reaches pH 6.

Preparation of nanofibre: Cestode powder of 2% was dissolved in 80 %(w/v) of acetic acid. To the mixture 1 ml of 10% PVA (Poly vinyl alcohol) polymer was added as additive to obtain thick fibre consistency. The mixture was electro spinned in 18kv potential using electro spinning apparatus, and tip to plate distance of 15 cm at a flow rate of 0.5 ml/hour.

RESULTS

Nanoparticles and nanofibre synthesized from Cestode:

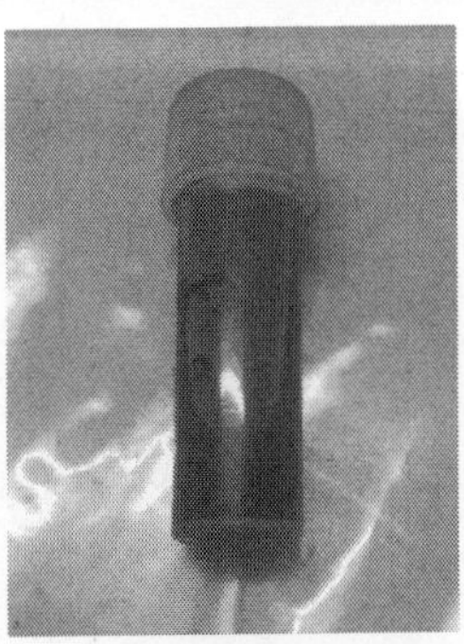

Fig 1: Nanoparticle

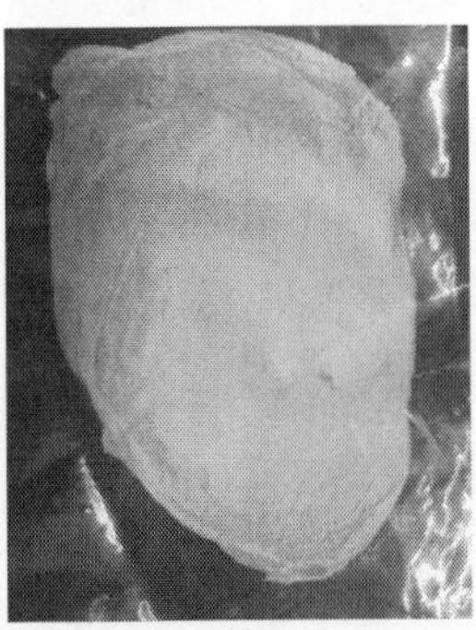

Fig:2 Nanofibre

DISCUSSION

In the present study, we have successfully synthesized nanoparticles and a hydrophilic nanofiber, using a rather unusual source, chicken intestinal cestode (*Raillietina tetragona*).As hypothesized, this seems to be a cost effective, biocompatible, nano material. In recent times, plethoras of different materials are being constantly tested, for effective drug delivery system, in cancer therapy and many other pathological applications. As mentioned earlier, metals and other non metal ions, though can be effectively molded in to various nanostructures, they don't seem to be best suited for medical and biological applications, attributing to their deleterious effects upon human body. Nanoparticle based drug delivery systems, are the most sort after treatment options, for many pathological conditions, in particularly in cancer treatment, because of their ability to directly deliver to the desired cells [(7)] and also to target precisely the tumor cells amongst the healthy ones [(8)]. Nanofibers are more frequently used in tissue engineering, bone remodeling and skin regeneration. Apart from these, they are also being efficiently used in drug delivery systems, [(9, 10)]. In conclusion it may be said, that, it is often advisable and advantageous to use nano structures, which are derived from natural components, for biological and medical applications, as they are more eco friendly, cost effective and more importantly biocompatible. Therefore in the present study we have successfully derived or molded nano particles and nano fibre from a more natural or unusual source, a cestode.

CONCLUSIONS

This experiment, distinctly illustrates, the dogma, that there can never be any material described as waste in nature, and also strongly re asserts the law of thermodynamics, as energy cannot be created or destroyed, but can only be converted from one form to another, similarly mass cannot be destroyed , but can only be converted from one form to another.

REFERENCES

1. Edina C. Wang and Andrew Z. Wang (2014). Nanoparticles And Their Applications In Cell And Molecular Biology *Integr Biol (Camb).* (1): 9–26.

2. Ananya Mandal (2018). Properties of nanoparticles Medical life sciences

3. Kipp JE (2004). The role of solid nanoparticle technology in the parenteral delivery of poorly water-soluble drugs. *Int. J. Pharm.*, 284: 109-122.

4. Megha Agarwal, M.S. Murugan, Anupama Sharma, Rinita Rai, Aman Kamboj Heena Sharma and Sanjeev K. Roy (2013). Nanoparticles and its Toxic Effects: A

Review *Int. J. Curr. Microbiol. App.Sci* (2013) 2(10): 76-82

5. Khadeeja Parveen, Viktoria Banse, and Lalita Ledwani (2016). Green synthesis of nanoparticles: Their advantages and disadvantages: *AIP Conference Proceedings* 1724, 020048.

6. Xiangqian Li, Huizhong Xu,Zhe-Sheng Chen, and Guofang Chen (2011). Biosynthesis of Nanoparticles by Microorganisms and Their Applications: Journal of Nanomaterials Volume 2011, Article ID 270974, 16 pages

7. Hammel E, Tang X, Trampert M, *et al.* (2004). Carbon nanofibers for composite applications. *Carbon*. 42:1153–8.

8. Hartgerink J.D., Beniash E., Stupp S.I. (2001). Self-assembly and mineralization of peptide-amphiphile nanofibers. *Science*. 294:1684–8.

9. Kenawy el-R, Bowlin G.L., Mansfield K., Layman J., Simpson D.G., Sanders E.H., Wnek GE J Control Release. 2002 May 17; 81(1-2):57-64.

10. Verreck G, Chun I, Rosenblatt J, Peeters J, Dijck AV, Mensch J, Noppe M, Brewster ME J Control Release. 2003 Oct 30; 92(3):349-60.

Current Status of Researches in Biosciences (2020) : 301-308
Editor : Prof. P.C. Joshi, N. Joshi, Reshman Yasmin, and D. K. Mansotra
Today & Tomorrow's Printers and Publishers, New Delhi - 110 002

IN VITRO THROMBOLYTIC ACTIVITY OF EXTRACT OF POSTERIOR REGION OF THE EARTHWORM, *EUDRILUS EUGENAE*

Yogita Y. Falak, Ashwini R. Patil, Namrata G. Mahajan and Manojkumar Z. Chopda

Department of Zoology, Animal House Division, Moolji Jaitha (Autonomous) College, Kavayitri Bahinabai Chaudhari North Maharashtra University, Jalgaon, Maharashatra, India 425002

Corresponding author email: mzczoo@gmail.com

ABSTRACT

Atherothrombosis is the main cause of cardiovascular diseases. Cardiovascular diseases (CVDs) take the lives of 17.7 million people every year, 31% of all global deaths due to the heart attacks and strokes caused by a vessel blockage by thrombus. Thrombolytic drugs are available in market, however, they cause some side effects and expensive. Therefore, present research work carried out on, *In vitro* thrombolytic activity of extract of posterior part of earthworm: *Eudrilus eugeniae* (PPEE) using plasma clot method. Clot lysis is check at various concentrations of extract against the artificially prepared plasma clot. Streptokinase used as standard and sterilized distilled water used as control. The percentage of clot lysis was analysed by computing the difference between weights of clot. The earthworm: *Eudrilus eugeniae* extract (PPEE) at various concentrations exhibit significant activity. Group III at 300mg/ml, of concentration exhibit moderate activity as compare to that of standard, which is 6.74 ± 1.26 %, 9.49 ± 0.26 % and 8.31 ± 0.74% respectively. All three concentrations showed significant activity as compared to control. Hence, it may conclude that, the earthworm extract have the potential to dissolve clot, therefore, it is urgent need to isolate the active molecule which is therapeutically beneficial in future use.

KEYWORDS: Thrombosis, *Eudrilus eugeniae*, Earthworm extract, Thrombolytic activity.

INTRODUCTION

Earthworm plays a major role in the proper functioning of the soil ecosystem. It acts as scavenger and helps in recycling of dead and decayed plant material by feeding on them. Earthworm increases the soil fertility and is often referred to as a farmer's friend. Earthworms have been used in medicine for various remedies since 1340 AD (John *et al.*, 1972). Earthworm ashes are then applied to the head with oil of roses to promote the hair growth. Earthworms have been used to treat smallpox in Burma and Laos. The worms are then roasted, powdered, mixed with the course of the disease and reduces mortality from 100 percent to 25 percent (Cooper *et al.*, 2004). Earthworm surface excreta were found to have potent antimicrobial activity. It is also having anti-cogulatory or fibrinolytic activity which results in the facilitation of blood circulation. The earthworm has been suspected to contain proteases which selectively interfere with the intrinsic pathway of blood coagulation. Recently, there has been increased interest in the anti-tumor or anti-coagulation activity of EFE, isolated from earthworm extract. The earthworm fibrinolytic enzyme (EFE) is a complex protein enzyme that is widely distributed in the earthworm digestive cavity. EFE not only has a direct effect on fibrin, but also can activate plasminogen. Fibrinolytic enzymes were first isolated from earthworms in 1980's and since that time, the medical value of earthworms has been given much more consideration. It was founded that earthworm extracts could significantly diminish the coagulation of platelets and promote the dissolution of thrombi in the blood.

Astrup *et al.*, (1952) described an improved fibrin plate method for the estimation of plasmin and other Fibrinolytic enzymes and activity.

Sawyer *et al.*, (1959) reported thrombolytic activity of human plasma. The use of unaltered plasma and performed clots mimics *in vivo* condition and permits the obtaining of physiologically relevant measurements. The results indicated that plasma obtained from non-stressed normal adults exhibits thrombolytic activity due primarily to presence of a plasminogen activator and that this activity varies as a result of stress and disease.

Mihara *et al.*, (1991) worked on a Novel Fibrinolytic Enzyme extracted from the earthworm, *Lumbricus rubellus* a strong fibrinolytic enzyme was readily obtained in saline extract of the earthworm, *Lumbricus rubellus*. It hydrolyzed not only plasminogen-rich fibrin plates, but also plasminogen-free fibrin plates.

Nakajima *et al.*, (1993) investigated characterization of potent

Fibrinolytic Enzymes in Earthworm, *Lumbricus rubellus*, fibrinolytic enzymes lyse fibrin, the main protein, component of blood clots. They demonstrated that the enzyme also orally administrated could be absorbed from the intestine into blood and activate the endogenous fibrinolysis system.

Qiao *et al.,* (2001), studied of some features of intestinal absorption of fibrinolytic enzyme III-1 extracted from the *Lumbricus rubellus.* They stated that the isolated enzyme has capacity to transporting in to the blood and keeping biological function as it is. After rising of antibody against isolated enzyme they showed that the 10 - 15% intact enzyme absorbed by intestinal epithelium which is detected by immune-histochemistry.

Purnima *et al.,* (2015), applied two methods for preclinical assessment of fibrinolytic agents namely Streptokinase and Plasmin. The methods used that is plasma clot lysis and fibrin clot lysis; which is analyzed *in vitro*. In which they used the fibrinolytic agent as positive control and sterile distilled water as negative control. They conclude that these methods are the novel models *i.e.* Plasma clot lysis and fibrin clot lysis models are the better method for analyzing new fibrinolytic drug as well as well-known drugs *in vitro* experimentation.

Kotb *et al.,* (2012) worked on Fibrinolytic Bacterial Enzymes with Thrombolytic Activity. This study described the fibrinolytic enzymes of microbial origin that are able to dissolve endogenous thrombin *in vivo*. The methods used for assessment of clot lysis *in vitro* are fibrin plate methods and streptokinase lysis methods.

Sugimoto *et al.,* (2001) experimentally deduced Molecular Cloning, Sequencing, and Expression of CDNA Encoding Serine Protease with Fibrinolytic Activity from Earthworm. Fibrinolytic enzymes dissolve Fibrin, the main component of blood clots. The molecular clonning and Sequencing of two earthworm protease genes to clarify the protein structure, as well as expression of the Fibrinolytic protease secreted in the culture medium of *Pichia pastoris* for application to the therapeutics or the protection of the environment.

Sumi *et al.,* (1990) reported the existence of a potent Fibrinolytic Enzyme (nttokinase, NK) in the traditional Fermented Food Called "natto". It was confirmed that oral administration of NK (or natto) produced a mild and frequent enhancement of the Fibrinolytic activity in the plasma, as indicated by the Fibrinolytic parameters. The results obtained suggest that NK represents a possible drug for the use not only in the treatment of embolism but also in the prevention of the disease, since NK has a proven safely and can be mass produced.

Popoviae *et al.,* (2001) carried out an Effect of Earthworm (G-90) Extract on formation and lysis of clots originated from venous blood of dogs with Cardiopathies and with Malignant Tumour. The stability of homeostasis is important to keep a balance between coagulation and fibrinolysis. A disorder of homeostasis leads to different physiological changes and causes different diseases such as Cardiopathies and malignant tumours. In this study the venous blood as well as the blood and the fibrin clots, derived from healthy dogs, the dogs with Cardiopathies and with malignant tumour, were examined for the time of coagulation and fibrinolysis by adding different substance. The result suggested that a possibility that G-90, along with the Fibrinolytic enzymes and other biologically active factors, also contains a factor that deceleration the formation of clots in a specific medium, such as the blood from the dogs malignant tumour.

Cooper *et al.,* (2004) reported alternative sources of fibrinolytic, anticoagulative, antimicrobial and anticancer molecules. The medicinal properties of earthworm in various remedies date bake to 1340 A.D and have been extended to other countries and cultures. Assays of tissue homogenates of earthworm (*Eisenia foetida*) have revealed a glycolipoprotein mixture referred to as G-90 that is composed of macromolecules with medical and pharmaceutical application. The bacteriostatic effect is significantly greater for non-pathogenic species. Finally G-90 also participates in tissue regeneration and wound healing. Taken together, component of earthworm could be tested in certain clinical trials.

Patrick *et al.,* (1982) reported Inhibitions of human blood clotting by extract of *Ascaris suum*. Whole worm supernatant solutions prolonged the whole blood clotting time and the kaolin- activated, partial throboplastin time but it did not alter the prothrombin time. The data indicate impairment of the intrinsic pathway of blood Coagulation. Activity was also noted in the cuticle and secretory / excretory products. Perhaps inhibition of blood clotting by helminths may facilitate their passage through the blood stream.

Very recently, Trisina *et al.,* (2011) worked on a protein (DLB51033) extract from *Lumbricus rubellus*, possesses Antimicrobitic and Thrombolytic Activities as a promising therapeutic agent for thrombosis associated disease. This present data indicate that DCBS1033, A unique agent generated from *Lumbricus rubellus* earthworm posse's antithrombotic and thrombolytic activites.

Hence, a hypothesis has been made to evaluate the activity of native earthworm *Eudrilus eugenae* as thrombolytic agent.

MATERIALS AND METHODS

Collection of Earthworm: Earthworm brought from vermicompost unit which was established in the laboratory of Department of Zoology, Moolji Jaitha (Autonomous) College, Jalgaon.

Earthworm saline

Sodium Chloride (NaCl): 6.00 g/L

Potassium Chloride (KCl): 0.12 g/L

Calcium Chloride ($CaCl_2$): 0.20 g/L

Sodium Bicarbonate ($NaHCO_3$) : 0.10 g/L

Preparation of earthworm sample: Earthworms were washed with tap water. Kept in annelid saline for 2- 3 hrs allow to removing the cast from alimentary canal of earthworm. Separate the posterior region and weigh this region. Then proceed for further homogenate procedure.

Earthworm homogenate: Homogenate the Posterior region of cleaned earthworms in 20mM phosphate buffer (pH 7.4) at 2000 RPM for 15 min. This homogenate was used for further study at various concentrations *i.e.*, I-100 mg, II-200 mg III-300mg, while 0.2 M EDTA used as standard.

Preparation of Plasma clot: The method for plasma clot was adapted from Purnima *et al.,* (2015). The blood was collected from the volunteers and mix immediately with 3.8% tri- sodium citrate in volume ratio of 9: 1. The mixture was centrifuged at 2500 rpm for 15 min. to obtained citrated plasma. 400 μl supernatant was added to pre-weighed micro centrifuge tube followed by addition of 100 μl 0.2 M $CaCl_2$. The micro centrifuge tube incubates at 37^0C for 30 min. in water bath, allowed to stand until firm clot was obtained.

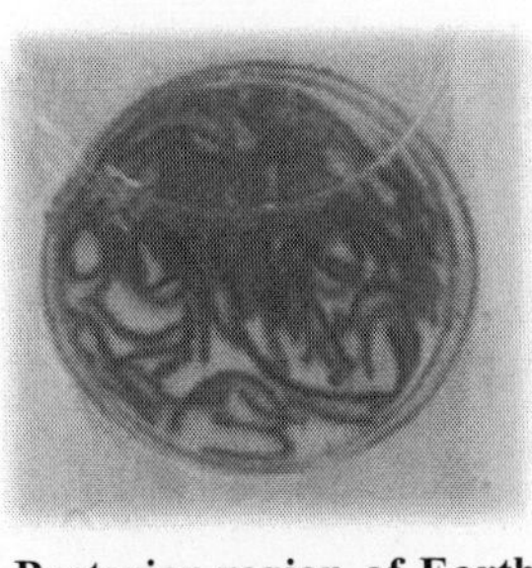

Fig. 1 Posterior region of Earthworm

Determination of Thrombolytic activity: Thrombolytic activity determine as per method mentioned by Prasad *et al.,* (2006). The clot bearing micro centrifuge tube then labeled as, Control- sterilized distilled water, standard- 0.2 M EDTA, Test : I- 100mg, II- 200 mg III- 300 mg. All these tubes were incubated for 90 min, at 37^{0}C in water bath. After incubation released fluid was discarded from tube and tube was again weighed to observe the weight of clot released. Compute the clot lysis percentage by applying following formula;

% of clot lysis = (weight of released clot / weight of clot) × 100.

Where, Weight of released clot = deduction of weight of eppendorf tube after clot lysis from weight of Eppendorf tube before clot lysis.

Weight of clot = subtracts the weight of empty eppendorf tube from the weight of clot containing eppendorf tube.

Statistical analysis: Statistical analysis was carried out by one way analysis of variance (ANOVA) followed by Boneferroni compare test at a $p<0.05$ significance level using GraphPad Prism. The level of significance compared with control. The data was expressed as mean ± standard error.

RESULTS AND DISCUSSION

The Posterior region of earthworm showed moderate thrombolytic activity. There are three concentration of homogenate of Posterior region namely; I, II, III. The group I at 100 mg/ml, group II at 200 mg/ml and group III at 300 mg/ml, of concentration exhibit moderate activity as compare to that of standard, which is 6.74 %, 9.49 % and 8.31 % respectively. All three concentrations showed significant activity as compared to control.

Table 1 Percentage clot lysis of Posterior region of *E. eugenae*

Group	Clot lysis %
Control	0.00± 0.88
Standard	16.47 ± 0.35
Test I	6.74 ± 1.26
Test II	9.49 ± 0.26
Test III	8.31 ± 0.74

Values expressed as Mean ± Standard error compared with control, **P< 0.001

Mihara *et al.,* (1991), studied on fibrinolytic activity enzyme extracted from the earthworm, *Lumbricus rubellus*, they evaluated the fibrinolytic activity of body pieces of earthworm. They observed that the anterior piece of earthworm body has good fibrinolytic activity. They concluded that the fibrinolytic enzyme secreted from the anterior region of the earthworm body; the fibrinolytic enzyme appeared to be secreted from the pharyngeal region, crop, gizzard and anterior portion of the intestine. The fibrinolytic activity measured by the plasimnogen-rich fibrin plate method. The activity of EPE rose sharply during the first 10 days and continued to rise gradually until and after 75 days. In present study we use homogenate of middle region of earthworm and results are in accordance with that of Mihara *et al.*, (1991). Therefore, we may conclude that the fibrinolytic enzyme secreted from the anterior region of the earthworm body; the fibrinolytic enzyme appeared to be secreted from the pharyngeal region, crop, gizzard and anterior portion of the intestine as stated by various researchers and these results are also supports the results of Mihara *et al.*, (1991).

CONCLUSION

The potential results reveal that further, study is needed for isolating the particular active principle and characterizes the specific molecule, which is responsible for clot lysis from earthworm, *Eudrilus eugeniae.* This may also supports pharmaceutical industries and farmers for their sustainable livelihood.

REFERENCES

1. Reynolds, J. W., & Reynolds, W. M. (1972). Earthworms in medicine. *AJN The American Journal of Nursing, 72*(7), 1273.

2. Mihara, H., Sumi, H., Yoneta, T., MizuMOTO, H., Ikeda, R., Seiki, M., & Maruyama, M. (1991). A novel fibrinolytic enzyme extracted from the earthworm, *Lumbricus rubellus. The Japanese journal of physiology, 41*(3), 461-472.

3. Sugimoto, M., & Nakajima, N. (2001). Molecular cloning, sequencing, and expression of cDNA encoding serine protease with fibrinolytic activity from earthworm. *Bioscience, biotechnology, and biochemistry, 65*(7), 1575-1580.

4. Sumi, H., Hamada, H., Nakanishi, K., & Hiratani, H. (1990). Enhancement of the fibrinolytic activity in plasma by oral administration of nattokinases. *Acta haematologica, 84*(3), 139-143.

5. Popoviæ, M., Enjak, T. M. H., Babiæ, T., Kos, J., & Mira, G. A. (2001). Effect of earthworm (G-90) extract on formation and lysis of clots originated from venous blood of dogs with cardiopathies and with malignant tumors. *Pathology Oncology Research, 7*(3), 197-202.

6. Cooper, E. L., Hrzenjak, T. M., & Grdiša, M. (2004). Alternative sources of fibrinolytic, anticoagulative, antimicrobial and anticancer molecules. *International Journal of Immunopathology and Pharmacology, 17*(3), 237-244.

7. Crawford, G. P. M., Howse, D. J., & Grove, D. I. (1982). Inhibition of human blood clotting by extracts of Ascaris suum. *The Journal of parasitology*, 1044-1047.

8. Sawyer, W. D., Fletcher, A. P., Alkjaersig, N., & Sherry, S. (1960). Studies on the thrombolytic activity of human plasma. *The Journal of clinical investigation*, *39*(2), 426-434.

9. Astrup, T., & Müllertz, S. (1952). The fibrin plate method for estimating fibrinolytic activity. *Archives of biochemistry and biophysics*, *40*(2), 346-351.

10. Nakajima, N., Mihara, H., & Sumi, H. (1993). Characterization of potent fibrinolytic enzymes in earthworm, *Lumbricus rubellus*. *Bioscience, biotechnology, and biochemistry*, *57*(10), 1726-1730.

11. Trisina, J., Sunardi, F., Suhartono, M. T., & Tjandrawinata, R. R. (2011). DLBS1033, a protein extract from *Lumbricus rubellus*, possesses antithrombotic and thrombolytic activities. *BioMed Research International*, *2011*.

12. Fan, Q., Wu, C., Li, L., Fan, R., Wu, C., Hou, Q., & He, R. (2001). Some features of intestinal absorption of intact fibrinolytic enzyme III-1 from *Lumbricus rubellus*. *Biochimica et Biophysica Acta (BBA)-General Subjects*, *1526*(3), 286-292.

13. Sarkar, P. D., Rajput, M. S., & Nirmal, N. P. (2015). Tailored methods for preclinical assessment of fibrinolytic agents in vitro. *Journal of Chemical and Pharmaceutical Research*, *7*(5), 1130-1135.

14. Kotb, E. (2012). Fibrinolytic bacterial enzymes with thrombolytic activity. In *Fibrinolytic Bacterial Enzymes with Thrombolytic Activity* (pp. 1-74). Springer, Berlin, Heidelberg.

15. Prasad, S., Kashyap, R. S., Deopujari, J. Y., Purohit, H. J., Taori, G. M., & Daginawala, H. F. (2006). Development of an in vitro model to study clot lysis activity of thrombolytic drugs. *Thrombosis Journal*, *4*(1), 14.

Current Status of Researches in Biosciences (2020) : 309-322
Editor : Prof. P.C. Joshi, N. Joshi, Reshman Yasmin, and D. K. Mansotra
Today & Tomorrow's Printers and Publishers, New Delhi - 110 002

POISSON PROBABILITY MAPPING OF THE ORAL CAVITY CANCER INCIDENCES IN THE WESTERN REGIONS OF TAMILNADU, INDIA

Harathi P.B.[1] and Prashanthi Devi M.[2]

[1]*Department of Zoology, PSGR Krishnammal College for Women, Coimbatore, India*

[2]*Department of Environmental Management, Bharathidasan University, Tiruchirappalli, India*

Email: harathi@psgrkcw.ac.in

ABSTRACT

Cancer being significantly, a lifestyle related disease could very well be prevented. The key mission of WHO Cancer Control Program is to promote national cancer control policies plans and programs, integrated to non-communicable diseases and other related problems. Along with the traditional medical practice, cancer data when coupled with GIS could be an innovative prevention strategy to reduce the impact of the high cancer burden. The present study aims to develop spatial statistical for cancer to study the associated risk factors in the western regions of Tamil Nadu. A total of 14392 cancer cases were collected and segregated in accordance with the age, gender, cancer and district. Poisson Probability Mapping was used to estimate the underlying risk for small regions using regional count. The Poisson Probabilities were calculated for each taluk and the results obtained were integrated into the GIS environment to visualize the burden of oral cancer incidences in the study region.

KEYWORDS: Cancer, GIS, Spatial Analysis.

INTRODUCTION

The typical goal of mapping public health data is to provide insight into geographic variations in disease risk. Risk is simply the probability that an unfortunate event occurs. In public health, the unfortunate event is usually the contraction of or death from a specific disease. When a map of disease counts, proportions or rates is made, it conveys inferences about disease risk. However, a map of raw counts is not the best tool for inference about disease risk, since regions with larger populations is expected to have higher disease counts. Instead, population differences by using rates (disease incidence per person per time) could be accounted as measures of risk. Higher disease rates reflect greater chances for contracting the disease, and thus, rates reflect a person's risk for disease.

However, a map of rates may still obscure the spatial pattern in disease risk, particularly if the rates are based on populations of very different sizes. Since the variability in the estimated local rates depends on population size, some rates may be better estimated than others, and this may obscure spatial patterns in disease risk. Rates based on small populations or on small numbers of disease cases are likely to be elevated artificially, reflecting lack of data rather than true elevated risk. As a simple example, consider a large metropolitan area with 2 million residents and 2 observed cases of a disease.

Suppose that the study area consists of sub-regions each containing 100 persons at risk. For the entire study area the crude incidence rate is 1/ 1,000,000, while two of the small areas contain crude rate estimates of 1/ 100. This is often referred to as the small number problem.

There are several solutions to this problem. First, we could calculate rates over larger areas (e.g., use states instead of counties), although this comes at the expense of giving up some of the geographic information we wish to convey. Second, we could make a comparative map, one that compares each rate to a common measure and, in doing so, adjusts for different population sizes. One such map is a probability map.

Probability mapping: Choynowski (1959) suggested the use of a probability map as an alternative. If the disease counts, $Y_1, Y_2, Y_3 \ldots Y_n$ are considered as independent Poisson random variables, the probability of any observed disease count, "Y_i" is calculated for each region/county using the poisson distribution. Expected count "E_i" for each region/county is calculated. An overall mean risk estimate over the entire domain or an age adjusted estimate is derived.

$$\xi = \sum_{i=1}^{N} \frac{Y_i}{\sum_{i=1}^{N} n_i} \quad \ldots\ldots 1$$

Using the overall mean risk, expected count in each region as $\hat{E}_i = n_i \xi^\wedge$. This is referred as constant risk hypothesis and implies that the risk is the same in all areas.

$$p_i = \begin{cases} \Pr[Y_i \geq y_i | E(Y_i) = \hat{E}_i] \; if y_i > \hat{E}_i \\ \Pr[Y_i \leq y_i | E(Y_i) = \hat{E}_i] \; if y_i < \hat{E}_i \end{cases} \qquad \ldots\ldots\ldots\ldots\ldots\ldots\ldots 2$$

The "p_i" values are based on cumulative probabilities from Poisson distribution and are an index of deviation from the null hypothesis of equal risk. The probabilities are depicted on a choropleth map for visualizations and the regions with probabilities less than 0.05 have rates significantly different from the average.

In the probability maps, the occurrence of an event *x* depends only on the space considered (area, volume, time or inhabitants) and is depicted as

$$P(x) = u^x e^{-x} / x! \qquad \ldots\ldots\ldots\ldots\ldots\ldots\ldots 3$$

Where, *u* is assumed as a constant density and equals: events x space.

To build the maps, the number of expected events *E* in a region *i* are defined as:

$$E_i = n_i \hat{p} \qquad \ldots\ldots\ldots\ldots\ldots\ldots\ldots 4$$

Where, $\hat{p} = \Sigma H_i / \Sigma n_i$; H_i is the number of observed oral cancer cases in a region*i* and n_i is the population of region *i*. For Poisson probability mapping we assume that H_i are independent Poisson random variables with expected values h_i, and that $h_1/n_1 = \ldots\ldots = h_{135}/n_{135} = p$.

An index of deviation from equal h_i/n_i can be defined:

$$\left. \begin{array}{l} ro_i = \Sigma_{x \geq H_i} E_i^x e^{-Ei} / x! \;\; for H_i \geq E_i \\ \qquad\qquad \text{Or} \\ ro_i = \dfrac{\Sigma_{x < H_i} E_i^x e^{-Ei}}{x!} for H_i < E_i \end{array} \right\} \qquad \ldots\ldots\ldots\ldots\ldots\ldots\ldots 5$$

Based on the value of "ro_i", a choropleth map is generated which is called as "probability map". Values of $ro_i < 0.01$ indicate that the region *i*'s oral cancer rates are different from the expected Poisson values, being unusually high (for e") or low (for <). In addition to using the rates as cases per inhabitant (n) as defined previously, other denominators are employed: n defined as population (np), n defined as area (na) and as the product of area and population (nap).

Considering a small area containing a population, *n* and *k* cases of diseases, we need to find out whether presence of those *k* cases in a population of size *n* is unusual i.e. to find out if the actual number of cases that are significantly higher than expected based on the national or regional rate.

Application of Probability Mapping in public health: Choynowski (1959), used the Poisson model to map the probabilities of brain tumor frequencies in sixty counties in Poland. As the tumor occurrences are rare, independent and discrete events they are included under the simple Poisson process. He further assumed that rejection of null hypothesis in a given county (indicating the occurrence of significantly more or less than the expected number of brain tumor) reflected some peculiar characteristic to that county. Cresie and Read (1989) have demonstrated the utility of probability mapping for mapping sudden infant deaths in North Carolina. Choynowski (1959) and Cresie and Read (1989), developed probability maps that were much more informative than maps generated using raw rates. Probability mapping has been used by researchers to show the statistical significance of prevalence rates. Cameron (1998) has discussed in detail the use and application of GIS in animal health data information and further explains that Probability mapping provide a better comparison of incidence of foot mouth disease in Northern Thailand and is a better estimate of risk. For spatial analysis and mapping of Cholera in Kumasi, Ghana, the authors have adopted relative risk mapping and probability mapping along with crude incidence rate mapping as a tool to provide precise and better risk estimate of cholera in the region (Anamzui-Ya, 2012).

Waller and Gotway (2004) discuss the application of probability approach for the analysis of classic Scottish lip cancer data set. Masoomy and Mesgari (2006) applied Probability mapping approach to estimate skin cancer occurrences from a generalised linear model and a comparison probability map that represented change in burden of skin cancer after introducing a 20% increase of fruit and vegetable consumption in the population. Banerjee *et al.*, (2014) discuss data modelling for probability mapping and hierarchical modelling by providing code for a WinBUGS example.

Oral cancer epidemiology: Oral cancer is amongst the most prevalent cancers worldwide and incidence rates are higher in men than women. This chronic disease is a public health problem both in developing as well as developed countries. The incidence of head and neck cancer varies enormously all around the world and especially oral cancer which accounts for 30-40% of all malignant tumors in India, being only about 2-4% in western countries (Soben.P, 2003).

Table 1: Incidence rates for cancer of the lip and oral cavity in men and women in 2012

Male	World	South Central Asia	India	Female	World	South Central Asia	India
Cases	198975	73553	53842	Cases	101398	35098	23161
Crude Rate	5.6	7.9	8.3	Crude Rate	2.9	4	3.8
ASR(W)	5.5	9.9	10.1	ASR(W)	2.5	4.7	4.3
Cumulative Risk	0.64	1.15	1.16	Cumulative Risk	0.28	0.55	0.51

The burden of oral cancer is great because of the associated high cost of treatment, permanent impairment and high mortality (SreeVidya Krishna Rao *et al.,* 2013). Estimated incidence rates of lip and oral cavity cancer, as estimated by GLOBOCAN 2012 for the world, south central Asia and India is summarized in Table.1.

The Indian subcontinent, because of its large population, has been regarded as the global epicentre of oral cancer (Bhawna Gupta *et al.*, 2012). The World Health Organization (WHO) regards oral cancer as a major public health challenge in India (Petersen, 2003). In India, because of cultural, ethnic, geographic factors and popularity of addictive factors, the frequency of oral cancer is high. It ranks number one in terms of incidences among men and third among women. It has one of the lowest 5-year survival rates of all cancers because most lesions are not diagnosed in the early stages. Studies have shown that oral cancers with identified risk factors and effective treatment for early lesions may remain undetected in the early stages because of the dental practitioners' attitude and knowledge (Sadowsky *et al.*, 1988; Schnetler, 1992). However if detected early, the probability of survival from oral cancer is remarkably better than for most of the other cancers (Rajendran and Sivapathasundharam, 2006).

More than 90-95% of all new cases of oral malignancy in most populations, including that in India are squamous cell carcinomas (SCC) arising from inner lining mucosa (Johnson *et al.*, 2011). Various studies have confirmed the roles of tobacco smoking, alcohol drinking and betel quid chewing in the aetiology of oral cancers (Lambert *et al*, 2011). Infection with Human Pappillomavirus (HPV) has also proven to be a major risk factor (Kriemer *et al.*, 2005).

The objective of the present section is to describe the basic epidemiological situations of the oral cavity incidences in the western regions of Tamil Nadu and also to compute the Poisson probability estimates of the incidences at the taluk level.

STUDY AREA

The data for the present cross-sectional study was collected from the National Cancer Registry Program (NCRP) Reports, as well as recognized cancer hospitals from the western region of Tamil Nadu for a tenure of 13 years from 2001 to 2013.

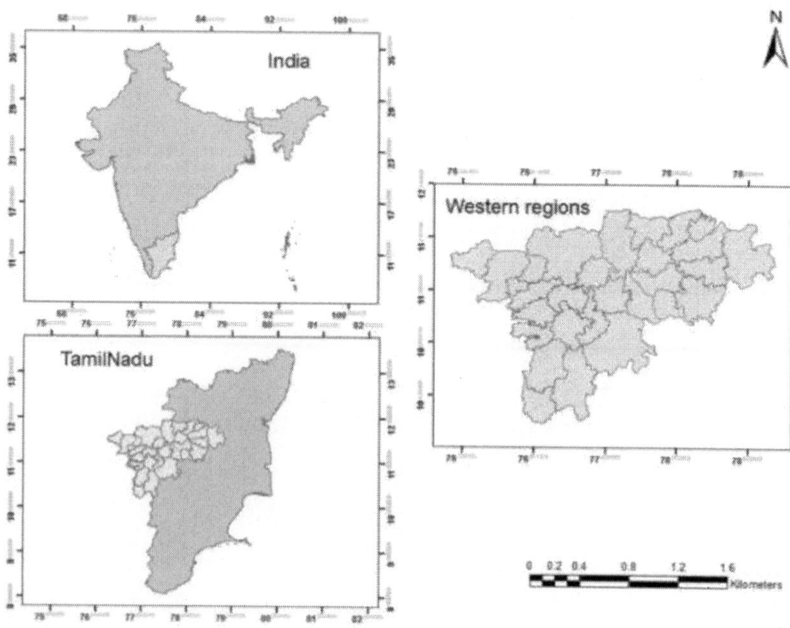

Figure 1. Study Area

The collected data comprised of new registered cancer cases of both male and female patients in the study area. The western regions of Tamil Nadu include districts like Coimbatore, Tiruppur, Erode, Namakkal, Salem, and Nilgiris. The total individual cancer incidences from the 6 districts as mentioned above from the western region of Tamil Nadu were segregated with respect to the each district and were reported to be around total of 14,392 cases. The reported oncology case sheets from the hospital records were added with the necessary attributes and were recorded into a database. The attributes include demographic details, type of cancer, occupation, type of treatment, behavioral factors such as tobacco and alcohol consumption and reproductive factors.

MATERIALS AND METHODS

The oral cancer cases among men were segregated for each district and taluk and were observed for different age-groups and anatomical sites. Patients diagnosed with cancers of the lip, base of tongue , other and unspecified parts of tongue, gum, floor of mouth, other and unspecified parts of mouth, parotid gland, other and unspecified parts of major salivary glands, tonsil, oropharynx, nasopharynx, piriform sinus and hypopharynx (ICD 10:C00-C13) were included in the study. Point density map was prepared to realize the spread of the disease in the study region The age-standardised incident rates (ASIR) for the oral cancer incidences were computed for each district by direct standardisation using the WHO new world standards as the standard population.

The Poisson probability estimates were computed using the R software 3.1 0 and spdep package was used to perform the analysis. The oral cancer cases recorded from the year 2001 to 2013 were segregated at the taluk level and the observed and the expected rates were calculated and entered into the R environment for computing the Poisson probabilities. Thematic maps were prepared for the observed incidences and the resultant Poisson probabilities using ArcGIS 10.3.1 software.

RESULTS AND DISCUSSION

The highest oral cancer incidences were observed in the Coimbatore district contributing 42% of the total oral cancer incidences. High clustering of cancer cases was concentrated in the intersection regions of Coimbatore North and Coimbatore South taluks. Among all oral cancer sites, the prevalence of tongue cancer was found highest among all districts with 442 cases (35%), followed by oropharynx-199(16%) and hypopharynx – 171(14%) with the highest incidences in the Coimbatore district (Figure 2).

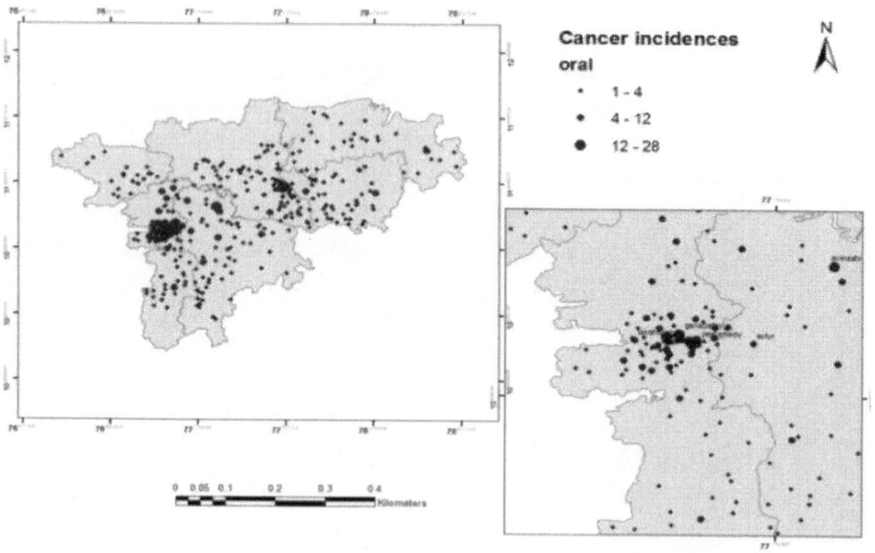

Figure 2: Point density map of the oral cancer incidences

The oral cancer incidences were observed for the five age-groups namely children (0-14), teenagers and young adults aged 15-24, adults aged 25-49, older adults aged 50-74 and elderly aged 75+. The age group 50-74 carry the highest burden of the oral cancer with the highest incidence in Coimbatore district (351 oral cancer cases) and the least in Nilgiris (28 oral cancer cases). The age-standardised incidence rates were (ASIR) were recorded highest in the Coimbatore district (2.5/100000) and the least in Salem (0.3/100000) (Table 2.).

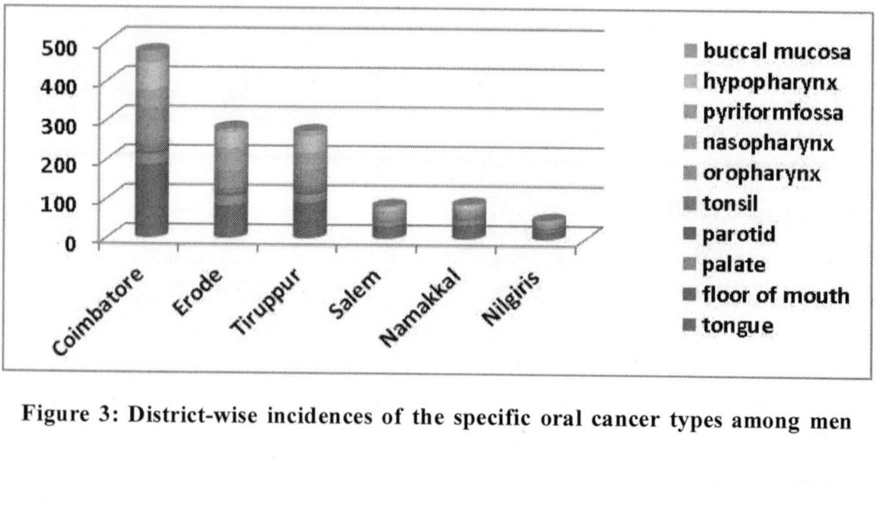

Figure 3: District-wise incidences of the specific oral cancer types among men

Table 2: Age-wise distribution of oral cancer incidences for each district among men

District	N	ASIR	0-14	15-24	25-49	50-74	75+
Coimbatore	508(42%)	2.49	2	4	92	351	58
Erode	266(22%)	1.99	1	0	33	211	20
Tiruppur	250(21%)	1.69	2	2	39	184	24
Salem	74(6%)	0.35	0	0	16	48	6
Namakkal	75(6%)	0.72	0	0	19	51	5
Nilgiris	42(3%)	0.99	0	0	10	28	3

After identifying high risk age groups for oral cancer among the western districts of Tamil Nadu and also specific locations of high cancer burden, analyses for areas with small number and small population differences was carried out. The statistical significance of rates rather than the observed oral cancer rates were used for mapping. Statistical significance is measured by probability values that show the likelihood of a rate occurring given the normal rate of disease in the corresponding national or local population. This rate is referred as the population rate, p. The probability value for an area indicates the likelihood that the rate observed in that area would occur by chance if the underlying risk of disease was equal to p. If the probability values were either close to 0 or 1 they indicate that rates are significantly different from the population rate. The map of the taluk-wise recorded oral cancer incidences is presented in the Figure 4.

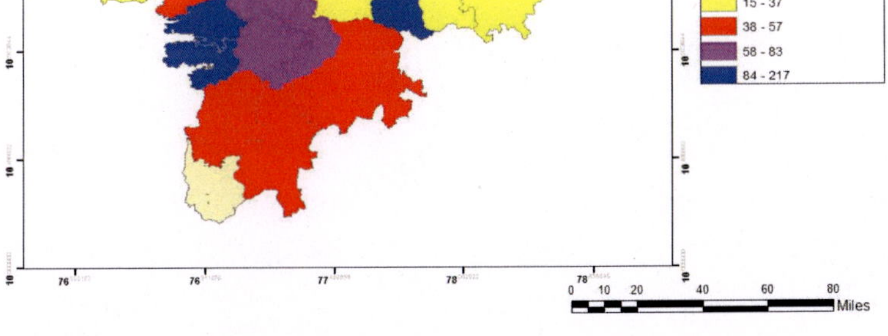

Figure 4. Distribution of observed oral cancer incidences

If the regional incidence rate is p, the expected number of cases in the study area, λ, is the study area population (n), multiplied by the national or regional rate(p):

$$\lambda \; np \qquad \text{.. 6}$$

Where λ is the expected number of cases and x is the observed number of cases

If we know the number of cases, k, occurring in a study region population, we can use the Poisson distribution to determine the probability, P (k), that the observed number of cases would occur in a population of the study region's size. The Poisson distribution states that in a population of size n, the probability of x cases occurring is . From this calculation, we can determine the probability of k or more cases occurring by chance, P(x e" k), if the true rate of disease in the population were p. That probability is calculated as:

$$P(x \geq k) = 1 \quad \sum_{x=0}^{k-1} P(x) \qquad \text{...7}$$

Using the formula, the Poisson probabilities of the oral cancer cases occurring in each taluk was calculated using R- software. The relative Risk and Probability Estimates derived for the 29 taluks are presented in Table 3.

Table 3: Relative risk and Poisson probability estimates of oral cancer in 29 taluks of the western districts of Tamil Nadu

Taluk	Male population	Observed	RR	Poisson Probabilities
Mettur	419887	10	0.345	0.000
Omalur	215850	6	0.403	0.000
Yercaud	21070	2	1.376	0.518
Sankagiri	125250	9	1.041	0.057
Salem	644713	26	0.584	0.000
Attur	217058	14	0.935	0.005
Gudalur	51497	5	1.402	0.387
Ooty	93663	24	3.714	0.999
Kotagiri	52668	12	3.302	0.984
Coonoor	78343	7	1.295	0.252
Sathy	164469	23	0.207	0.000

Gobi	186691	44	3.416	0.999
Bhavani	233314	29	1.801	0.566
Perundurai	134270	20	0.216	0.000
Dharapuram	140576	29	2.990	0.996
Erode	411124	158	5.570	1.000
Paramathivelur	106576	19	2.584	0.953
Nammakal	272644	21	1.116	0.014
Rasipuram	173456	6	0.502	0.000
Avinashi	169340	52	4.450	1.000
Cbe north	320620	198	12.443	1.000
Cbe south	795624	177	3.224	1.000
Mettupalayam	129447	32	3.583	0.999
Palladam	123576	48	5.629	1.000
Pollachi	285835	40	2.028	0.813
Tirupur	466836	57	1.770	0.504
Udumalpet	118141	29	3.558	0.999
Valparai	35204	1	0.412	0.069
Thriuchengode	316604	21	0.961	0.001

The resultant Probabilities were mapped into ArcGIS 10.3.1 environment to obtain a probability map depicting the statistically significant high risk regions. The map produced shows the regions of high oral cancer risk including the taluks of the Coimbatore, Tiruppur, Erode and Nilgiris districts which is presented in Figure 5.

In the present study, while comparing the probability and the thematic maps, of the oral cancer incidences in the western regions of Tamil Nadu, different approaches have been illustrated. The incidence map will just show the observed incidences in the study area with Coimbatore North and Coimbatore South having the highest incidences. On the other hand, in the probability map areas with exceptionally high or low rates are removed which may be due to the small numbers problem. The crude incidences may not be useful while initiating intervention and awareness programs. Hence the probability analysis will be more helpful in delineating the high risk zones for which the community awareness programs for specific groups of people could participate.

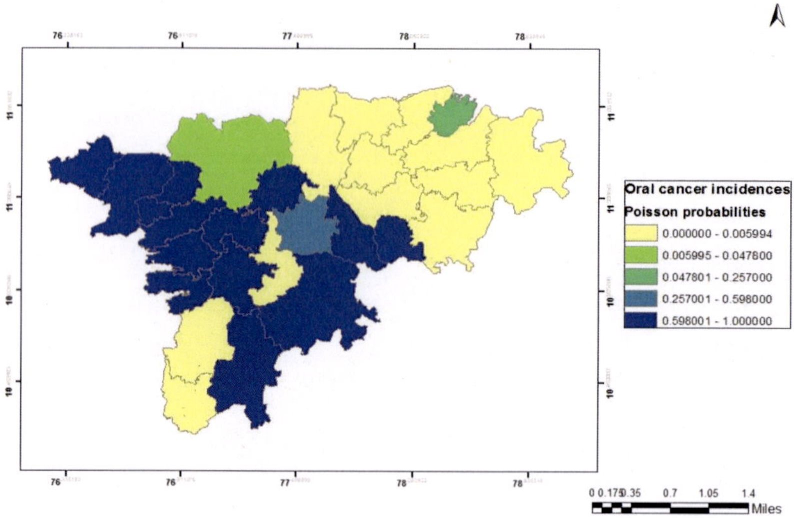

Figure 5. Map depicting the distribution of poisson probabilities

CONCLUSION

Health-related data are often available at the level of areal units, rather than for individuals. Spatial disease mapping or the distribution of health care services is therefore often based on a system of areal units. These units, such as local health wards or taluks or districts are essentially arbitrary in size and shape, such that different sets of results can be obtained by slightly varying their geographical definition. This is the well-known 'modifiable areal unit problem' or MAUP (Openshaw, 1984) which generates particular difficulties for geographical epidemiology. The precise definition of such areal units may become particularly significant, for example, if a potential 'cluster' of cases with a particular disease straddles a unit boundary. Further problems ensue when data are available for sets of areal units which are non-compatible, in the sense that they do not nest neatly into one another, but rather overlap each other in arbitrary ways. By, recognizing the difficulties involved in relating health and population data to other datasets for traditional areal units such as wards or health districts, the Chorley Report on the Handling of Geographic Information (HMSO, 1987) recommends more widespread use of the postcode. However, although many datasets can now be referenced to this level, specific areal boundaries for these units have not as yet been defined. Pyle (1979) discusses some of the earlier uses of such 'geocoded' information for epidemiology and health care applications in the United States, notably DIME (Dual Independent Map Encoding) files developed by the US Bureau of Census.

Poisson probability is one of the methods used for modelling the probability of rare binary (Yes/No) events in large populations. Health conditions like cancers congenital defects, that are rare, occurring in only a small fraction of the population, fit the Poisson distribution. Probability mapping is a useful way of addressing the small number problem when mapping area-level heath data.

Although epidemiological data often exist only for areal units, these data often represent individual level data which have been aggregated to areal units. While interpreting choropleth maps, we are often in danger of committing an ecological fallacy (Langbein and Lichtman, 1980) in other words, of making a causal inference about an individual(s) on the basis of observations on groups. Likewise, data on variables which may be used to explain disease distributions but which are available at the individual level (e.g. smoking habits, alcohol consumption) cannot be incorporated directly into area-based studies.

REFERENCES

1. Anamzui-YA, J.A (2012). Spatial Analysis and mapping of cholera causing factors in Kumasi, Ghana. Thesis submitted to University of Twente.

2. Banerjee, S., Carlin B.P Gelfand A.E., Boca Raton, F.L. and Chapman (2014). *Data sets and associated software programs for Hierarchical Modeling and Analysis for Spatial Data*, 2nd ed. S. Hall/CRC Press.

3. Becker, R. A., Chambers, J. M., and Wilks, A. R. (1988), The New S Language, London, UK: Chapman & Hall.

4. Bhawna Gupta Anura Ariyawardana Newell Johnson The epidemic of oral cancer in India continues unabated: Need for new policy initiatives *Oral Oncology* 48(8):e31-2 · July 2012

5. Cameron, A. R. (1998). Active surveillance and GIS as components of an animal health information system for developing countries: Thailand and Laos as examples. University of Queensland.

6. Choynowski, M. (1959). Maps based on probabilities. *Journal of the American Statistical Association*, 54(286), 385-388.

7. Cressie, N. and Read, T. R. (1989). Spatial data analysis of regional counts. *Biometrical Journal*, 31(6), 699-719

8. Kreimer AR, Clifford GM, Boyle P, Franceschi S (2005). Human papillomavirus types in head and neck squamous cell carcinomas worldwide: a systematic review. *Cancer Epidemiol Biomarkers Prev*, 14, 467-75.

9. Lambert R, Sauvaget C, de Camargo Cancela M, *et al.,* (2011). Epidemiology of cancer from the oral cavity and oropharynx. *Eur J Gastroenterol Hepatol*, 23, 233-41.

10. Langbein, L.I. and Lichtman, A.J. (1980). Ecological Inference, *Sage,* London .

11. Masoomy, Z. and Mesgari, M.S. (2006). Prediction Of Skin Cancer Epidemiology For Decision Making Using Geostatistical Analyses. *ASPRS annual Conference.* URL: http://www.asprs.org/ a/publications/ proceedings/ reno2006/0092.pdf

12. Openshaw, S. (1984). The Modifiable Areal Unit Problem, CATMOG No 38, Geo Books: Norwich.

13. Pebesma, E.J., R.S. Bivand, (2005). Classes and methods for spatial data in R. R News 5 (2), http://cran.r-project.org/doc/Rnews/.

14. Pyle, G.F. (1979). Applied Medical Geography, *Wiley*, New York

15. Rajendran, R, B. Sivapathasundharam Shafer's Textbook of Oral Pathology (sixth ed.) Elsevier, New Delhi, India (2006) pp. 896–897

16. Sadowsky D., Kunzel C., Phelan J. (1988). Dentists' knowledge, case-finding behavior, and confirmed diagnosis of oral cancer. *J Cancer Educ,* 3, 127-34.

17. Schnetler J.F. (1992). Oral cancer diagnosis and delays in referral. *Br J Oral Maxillofac Surg*, 30, 210-3.

18. Soben P, Peter S. Essentials of preventive and community dentistry. 2nd ed. (2003). Sequeira epidemiology, etiology and prevention of oral cancer; p. 471.

19. Sree Vidya Krishna Rao, Gloria Mejia, Kaye Roberts-Thomson , Richard Logan Epidemiology of Oral Cancer in Asia in the Past Decade- An Update (2000-2012) *Asian Pac J Cancer Prev,* 14 (10), 5567-5577

20. Waller, L. A. and Gotway, C. A. (2004) Applied Spatial Statistics for Public Health Data John Wiley and Sons, Inc. New York, NY

Current Status of Researches in Biosciences (2020) : 323-329
Editor : Prof. P.C. Joshi, N. Joshi, Reshman Yasmin, and D. K. Mansotra
Today & Tomorrow's Printers and Publishers, New Delhi - 110 002

POPULATION STUDIES OF PHYTONEMATODES AND THEIR EFFECT ON PLANT GROWTH

DatrikaSwapna., Geeta Rajalingam., and P.Nagaraja Rao

Department of Zoology, Osmania University, Hyderabad

ABSTRACT

The phytonematode diversity changes were studied in Squash and Pumpkin soil samples collected and checked for the population of nematodes. After the identification soil analysis was done to study the plant growth Based on the phylogenetic studies it was identified that species *Tylenchorhynchus brevilineatus,* related to *Belonolaimus longicaudatus, Hoplolaimus indicus*. Thus, our present studies revealed that there is a diversity changes of phytonematode based on type of the plant. This diversity is more at root level as it is effecting the soil physicochemical nature of the root zone. Hence, based on these results, one can develop phytonematode control methods.

KEYWORDS: Diversity, phytonematode, *Hoplolaimus indicus, Tylenchorhynchus brevilineatus*

INTRODUCTION

India is the second largest producer of vegetables after China and a leader in the production of peas and okra (Sophie Mantelin *et al.,* 2017). Besides, India occupies the second position in the production of brinjal, cabbage, cauliflower and onion, and third position in potato and tomato in the world (Bellafiore *et al.*, 2015). The other category of Vegetables that are produced in abundance are especially Pumpkin and Squash. India grows a large variety of vegetables belonging to tropical, subtropical and temperate groupsdue to various agroclimate zones. The most common vegetables grown are potato, eggplant, okra, cauliflower, cabbage, cucurbits, tomato, carrot, onion, chilli and some other legumes (FAO, 2010; Prior , 2015). Vegetable crops usually are among the most susceptible and worst affected by the nematodes (Sharma *et al.,* 2006; Anwar and Mckenry, 2007; Dhaliwal and Koul, 2007; Singh and Khurma, 2007). Phytonematodes are serious and economically most important pest of many cultivated crops around the world (Trifonova *et al.,* 2009). They are particularly damaging vegetables in tropical and subtropical countries (Sikora and Fernandez, 2005; Sophie Mantelin *et al.,* 2017). The plant parasitic nematodes are generally called eelworms, phytonematodes, phytohelminths or simply plant nematodes. They parasitize both animals and plants. Nematodes are widespread in distribution and found in soil, fresh water and salt water wherever organic matter exists, from oceans to mountains, from arctics to tropics and are said to be "Ubiquitous' (Thome, 1961). Phytonematodes are an extremely important limiting factor in vegetable production, therefore crop protection is an integral part of food production and must be considered within the context of modern agriculture and sustainable development. Effective crop protection is essential both to combat the threat of widespread diseases and to provide the effective pest management programmes. The majority of the plant species, which account for the major world's food supply, is susceptible to attack from phytonematodes which are capable of causing sustainable economic losses in the quantity and quality of the crops (Jain *et al.,* 2007; Berry *et al.,* 2008). Once nematodes are established in the field, the possibilities of complete eradication are exceeding remote and impractical on field scale due to soil inhabitant nature of nematodes. However, several measures are adopted to decrease nematode population up to an acceptable level. The main objectives this study was phytonematodes management are usually a matter of reducing the nematode population by one or more methods or integration of one or more methods to a low level so that the damage is negligible or at an economically acceptable level (Rajvasnshi and Sharma, 2007).

METHODOLOGY

The soil is alluvial in nature and affected by salts. Average annual rainfall is 662 mm and temperature ranges from 4°C – 47°C. The average relative humidity ranges from 32 to 82%. The chosen site was naturally infested with phytonematodes viz., stunt nematode (*Tylenchorhynchus brassicae* Siddiqi), reniform nematode (*Rotylenchulus reniformis* Linford and *Oliveira*), filiform nematode (*Tylenchus filiformis* Butschli), lance nematode (*Hoplolaimus indicus* Sher), spiral nematode (*Helicotylenchus indicus* Siddiqi) and root-knot nematode (*Meloidogyne incognita,* (Kofoid and White) Chitwood. The field was divided into two parts, one part received normal ploughing (20 cm deep), whereas the other part received deep ploughing (40 cm deep). Both parts were further divided into 3×3 m 2 beds and a buffer zone of 0.25 meter left between the beds. Different beds, which were replicated five times and this experiment was carried out using the experimental design of Randomized Complete Block Design (RCBD) and made them as the plots for detailed study.

1). Control soil with out plants (CSP)

2). Test soil samples with plants (TSSP)

Five replicates for each treatment were arranged in a random manner according to the table of Khan and Khanum (2004). Three week old seedlings of Squash and Pumpkin were raised in pots filled with autoclaved soil and transplanted after two week of the treatment. Squash were sown directly and necessary caring such as watering, weeding etc. were also done throughout the experiment which was terminated after three months of the treatment. Soil sampling was taken at the depth of 0-15 cm both prior and after termination of the experiment from each beds with the help of a soil sampler to count the initial and final population of phytonematodes. These samples were collected and brought to the laboratory in polythene bags and were mixed thoroughly. A representative soil sub-sample of 250 g was used for isolating the nematodes with the help of Cobb's sieving and decanting method along with modified Baermann funnel technique. The nematodes were counted with the help of counting dish. For each sample aliquots of 100 ml were used for counting and from this, 5 ml suspension was used after thorough stirring. Plant growth (fresh weight of shoot and root in terms of gram) and yield was also measured.

Residual effect of different treatments: For these experiments, different beds were maintained as such and were thoroughly prepared for the next growing season. Three week old seedlings of Squash and Pumpkin were raised in autoclaved soil and transplanted without giving any further treatment to the beds except the ploughing treatment (normal and deep ploughing). The seeds of Squash and Pumpkin were sown directly. The final data including nematode population, plant growth (length of shoot and root in terms of centimeters (cm), fresh weight of shoot and root and fruit weight in terms of grams) were determined after three months at the termination of the crop.

Identification of Nematodes: The classic method of extraction of nematodes from soil is conducted following the method of Jenkins (1964). The soil sample is mixed thoroughly, but gently when tumbling, to homogenize the nematodes within the soil. A measured volume of soil (either 100 cm^3 or 250 cm^3) is rinsed through a 864 μm (20 mesh) sieve into a large pitcher.

The filtrate is mixed with a pressurized water spray to fill the pitcher. After allowing the water and soil in the pitcher to settle for 20 seconds, the suspension is poured over a 38 μm (400 mesh) sieve held at a 45° angle. Material captured on the sieve is rinsed into a 100 mL centrifuge tube and centrifuged for 3minutes at 1,700 rpm. The supernatant is poured off and the pellet is resuspended in a 1.328 M sucrose solution (specific gravity = 1.10) before a repeated centrifugation at 1,700 rpm for 3 minutes. Following centrifugation, the supernatant is poured over a 25 μm (500 mesh) sieve and rinsed with water to remove any traces of sucrose. The resulting material captured on the sieve can be examined under a light microscope for identification and quantification. An alternative method to centrifugation of the soil sample is a modified Baermann tray or funnel. In this case, the required volume of soil is rinsed through a 864 μm sieve and over a 38 μm sieve, just as with the centrifugation method. The captured material is rinsed into a coffee filter placed within a plastic bowl or funnel and supported by a screen. The water level is brought up to at least 1.0 cm above the coffee filter and allowed to incubate for 24 hours. Following incubation, the filter and screen are removed from the bowl and the water left in the bowl or funnel base is poured over a 25 μm sieve. The material contains only live, mobile nematodes and can be observed under a light microscope.

Maintenance of nematode culture: Healthy egg masses of root-knot nematode, M. incognita was collected from the heavily infected roots of eggplant field near G.T Road, Aligarh-Delhi and placed on a small course sieve (1mm pore size) lined with tissue paper and then placed in 10 cm diameter Petri dish containing double distilled water. The second stage juveniles, which hatched out, were collected along with water from the Petri dish. This process was repeated for several days. These second stage juveniles (J2) of root-knot nematode served as the initial inoculum in glass houses and in vitro experiments. The nematode species were asserted by close examination of the perennial pattern of the female from the first egg masses collected. It was confirmed by the similar examination of the randomly collected females from the inoculum raised as above. Stock populations of *Rotylenchulus reniformis* were maintained on cowpea under glass house conditions and mixed vermiform (4th stage) of the nematodes were extracted by processing the soil (250 g) with the Cobb's sieving and decanting method followed by Baermann's funnel technique.

RESULTS AND DISCUSSION

Phytonematode Population studies: The population studies were performed at a distance of 20cm from the root-knot of the selected leafy vegetables. In this studies when compared to control more number of phytonematodes were observed. Among the Squash and Pumpkin, Squash showed less phytonematode population and Pumpkin showed high phytonematode population. *Hoplolaimus indicus* was found to be in low number and *Aorolaimusperscitus* was found to be in high number in vegetables. Less nematodes were observed in Squash and high number of nematodes observed in Pumpkin (Table 1).

Table 1: Mean numbers of nematodes per 473 cc (1 pit) soil-20cm

Soil Type	*Hoplolaimus indicus*	*Belonolaimus longicaudatus*	*Tylenchorhynchus brevilineatus*
Control	06	06	08
Squash	**08**	**09**	**15**
Pumpkin	**15**	**26**	**32**

At a distance of 40cm from the root-knot the phytonematode population was decreased when compared to 20cm distance of root-knot (Table 2).

Table 2: Mean numbers of nematodes per 473 cc (2 pit) soil-40cm

Soil Type	*Hoplolaimus indicus*	*Belonolaimus longicaudatus*	*Tylenchorhynchus brevilineatus*
Control	**04**	**04**	**06**
Squash	**06**	**08**	**12**
Pumpkin	**10**	**17**	**22**

Phytonematode Population studies during 30 days period: The nematode population density was increased from first day to 30th day in all the four leafy vegetable crops. When compared to control (without leafy vegetables), the soil which contain with leafy vegetables have high nematode number (Table 3).

Table 3: Relationship between nematode population density (individual g^{-1} dry soil) at 20cm distance from Root-Knot

Days	Control	Squash	Pumpkin
1	10.8±0.12	11.2 ±0.13	11.4± 0.11
10DAYS	13.0±0.21	14.2± 0.24	14.3±0.26
20DAYS	15.6± 0.16	15.1± 0.15	15.2±0.18
30DAYS	16.7± 0.22	18.3± 0.23	18.5± 0.18

The nematode population density was decreased from first day to 30th day in all the four leafy vegetable crops at 40cm distance from root - knot. When compared to control (without leafy vegetables), the soil which contain with leafy vegetables have less nematode number (Table 4).

Table 4: Relationship between nematode population densities (individual g-1 dry soil) at 40cm distance from Root-Knot

Days	Control	Squash	Pumpkin
1	8.8±0.12	10.2 ±0.13	10.4± 0.11
10DAYS	11.0±0.21	12.2± 0.24	13.3±0.26
20DAYS	12.6± 0.16	13.1± 0.15	14.2±0.18
30DAYS	13.7± 0.22	14.3± 0.23	15.5± 0.18

CONCLUSION

Among the selected vegetables Pumpkin is suitable for the growth of Squash and Pumpkin. The nematode population density was increased from first day to 30th day in vegetable crops along with the growth of the crop plant.When compared to control soil (without vegetables), the soil which contain with vegetables have high nematode number.The nematode population density was decresed from first day to 30th day in all the four leafy vegetable crops at 40cm distance from root -knot. When compared to control (without vegetables), the soil which contain with leafy vegetables have less nematode number.The population studies were performed at a distance of 20cm-40cm from the root-knot of the selected leafy vegetables. In this studies when compared to control more number of phytonematodes were observed. At a distance of 40cm from the root-knot the phytonematode population was decreased when compared to 20cm distance of root-knot. Among the vegetables Squash showed less phytonematode population and Pumpkin showed high phytonematode population.

REFERENCES

1. Tzortzakakis, E.A. and Gowen, S.R. (1994). Evaluation of *Pasteuria penetrans* alone and in combination with oxamyl, plant resistance and solarization for control of *Meloidogyne* spp. on vegetables grown in greenhouses in Crete. *Crop Protection,* 13(6): 455-462.

2. Uma Shankar, Kumar, N., Krishnappa, K., Reddy, B.M.R., Ravichandra, N.G. and Karuna, K. (2005). Intercropping for the management of root-knot nematode, *Meloidogyne incognita* in vegetable-based cropping systems. *Indian Journal of Nematology*, 35(1): 46-49.

3. Husan-Bano, Siddiqui, I.A., Ahmed, W. and Ehteshamul-Haque, S. (1999). Effect

of some medicinal plants on egg hatching of *Meloidogyne javanica* root-knot nematode. *Pakistan Journal of Biological Science*, 2: 1364-1365.

4. Ismail, A.E., Eid, R.A. and El-Nagdi, W.M.A. (2006). Effect of different composts, biofertilizer and olive pomace as soil amendements on *Rotylenchulus reniformis*, growth and chemical analysis of Jasmine in Egypt. *Journal of Applied Sciences Research*, 2(11): 909-919.

5. Zareena, B.M. and Kumar, S.M. (2005). Management of disease complex involving *Heterodera cajani* Koshy, 1967 and *Macrophomina phaseolina* (Tassi.) Goid. ongreengram (*Vignaradiata* L. Wilczek). *Indian Journal of Nematology*, 35(2): 192-194.

6. Widmer, T.L. and Abawi, G.S. (2000). Mechanism of suppression of *Meloidogyne hapla* and its damage by a green manure of sudangrass. *Plant Disease*, 84(5): 652-568.

7. Jada, M.Y., Gungula, D.T. and Jacob, I. (2011). Efficacy of carbofuran in controlling root-knot nematode (*Meloidogyne javanica* Whitehead, 1949) on cultivars of Bambara groundnut (*Vigna subterranea* L. Verdc.) in Yola, Nigeria. *International Journal of Agronomy*, 2011: 1-5.

8. Hussain, S.I., Kumar, R., Kahn, I.A. and Titov, A. (1984). Effect of root-dip treatments of eggplant seedlings with plant extracts, nematicides, oil cake extracts and antihelmintic drugs on plant growth and root knot development. *Pakistan Journal of Nematology*, 2(2): 79-83.

9. Faske, T.R. and Starr, J.L. (2007). Cotton root protection from plant parasitic nematodes by abamectin treated seed. *Journal of Nematology*, 39(1): 27-30.

10. Feldmesser, J., Edwards, D.I., Epps, J.M., Heald, C.M., Jenkins, W.R., Lear, B., Mc Beth, C.W., Nigh, E.L. and Perry, V.G. (1971). Estimated crop losses from plant-parasitic nematodes in the United States. Comm. Crop losses. Special publication No. 1. Soc. Nematologists, Hyattsville, Maryland.

11. Usman, A., Rehman, B. and Siddiqui, M.A. (2012). Effect of some botanical extracts for the management of *Meloidogyne incognita* infecting eggplant. *Biosciences International*, 1(1): 12-15.

12. Segers, R., Butt, T.M., Kerry, B.R. and Peberdy, J.F. (1994). The nematophagous fungus *Verticillium chlamydosporium* produces a chymoelastase-like protease which hydrolyses host nematode proteins in situ. *Microbiology*, 140: 2715-2723.

13. Vadhera, I. and Shukla, B.N. (2002). Management of *Rotylenchulus reniformis* in agriculture crops - a review. *Agriculture Review*, 23 (1): 14-22.

14. Talaveran, M., Itou, K. and Mizukubo, T. (2002). Combined application of *Glomus* sp. and *Pasteuria penetrans* for reducing *Meloidogyne incognita* (Tylenchida: Meloidogynidae) opulations and improving tomato growth. *Applied Entomology and Zoology*, 37: 61-67.

15. Tariq, I. and Siddiqui, M.A. (2005). Evaluation of nematicidal properties of neem for the management of *Meloidogyne incognita* on tomato. *Indian Journal of Nematology*, 35(1): 56-58.

Current Status of Researches in Biosciences (2020) : 331-336
Editor : Prof. P.C. Joshi, N. Joshi, Reshman Yasmin, and D. K. Mansotra
Today & Tomorrow's Printers and Publishers, New Delhi - 110 002

MEIOTIC CHROMOSOMES OF GARDEN SLUG, *LAEVICAULIS ALTE* FROM PRACHI BELT OF RURAL ODISHA, INDIA

Badal Das

Slug Research Laboratory, U.N. Autonomous College of Sc. & Tech., Adaspur, Cuttack, Odisha

E.mail id: badalkumardas@gmail.com

ABSTRACT

Chromosomal studies in *Laevicaulis alte* was conducted in the laboratory on collection of garden slugs from Prachi River bed of rural Odisha. A standardized dose of colchicine was injected to the weighed species prior to dissection of ovotestis to obtain chromatin material. Standardized aceto-orcein squashing technique employed for the purpose recovered several meiotic stages of chromosomes with a diploid set determined as 2n=34. Haploid chromosome complement observed under high resolution microscope revealed the arrangement of 17 bivalents in different micro photographic plates of gonadal meiosis. Spermatogonial metaphase also confirmed the diploid number in the species consisting of metacentric & submetacentric configuration only as reported earlier

KEYWORDS: Chromosomes, Garden slug, Rural area, Prachi river, Gastropoda

INTRODUCTION

The tropical leatherleaf slug, *Laevicaulis alte* (Class: Gastropoda, Order: Systellommatophora, Family: Veronicellidae) is a serious agri-horticultural pest of India and neighboring countries (Raut & Mandal, 1984) where it is invasive (Herbert and Kilburn, 2004). The preliminary morphometric studies during laboratory observations of the eco-biology of the same species were first reported from coastal belt of Odisha (Das &

Parida, 2015). Chromosomal studies are an informative technique to establish phylogenetic relationship amongst species determining their course of evolution in the animal kingdom. The first ever account on the description of the karyotypic analysis of the slug *Laevicaulis alte* was reported from Philippines to confirm its taxonomic position within the family Veronicellidae (Cordova *et al.* 2010). The present investigation also aims at the chromosomal analysis of the exotic species to account for its taxonomic status as well as analyzing the impact of certain molluscicides in controlling the garden pests in a rural setup.

MATERIALS AND METHODS

The experimental animals were collected from the U.N. College garden located on the Prachi river bed of Odisha. Adult species of required length-weight were cultured in specialized rearing tanks constructed for the purpose as well as in laboratory terrarium properly maintained with an artificial environment. Each weighed specimen was injected with 0.02% colchicine (Fig.1) just 24 hours prior to removal of ovotestis meant for chromosomal studies. The dissected gonadal tissue was also undergone hypotonic treatment for 45 minutes using 0.075M KCl and later fixed in freshly prepared aceto-methanol mixture (1:3) with three changes in it. Now following the standardized aceto-orcein squashing technique (La Cour, 1941 and Stern,1975), temporary slides were prepared for examining under 2000x magnification of Microscan B-2 Universal Trinocular research grade microscope using high resolution digital camera with caliper pro software for photographic image analysis of meiotic stages in the species.

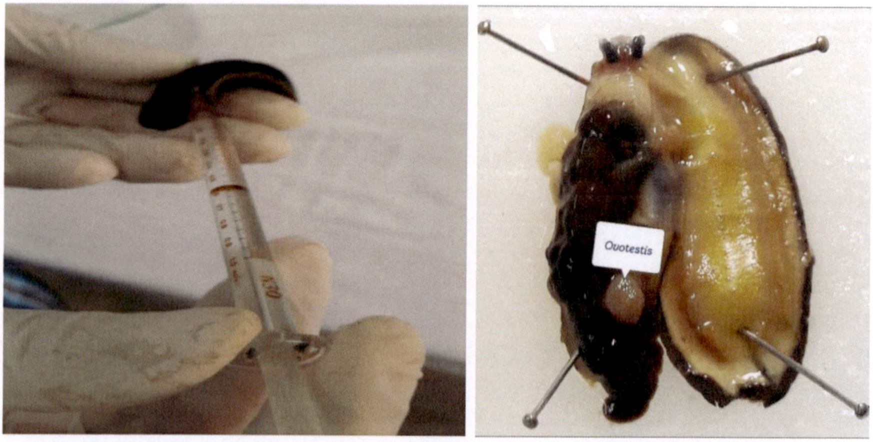

Fig.1. Colchicine pretreatment of *Laevicaulis alte* with ovotestis exposed for chromosome preparation

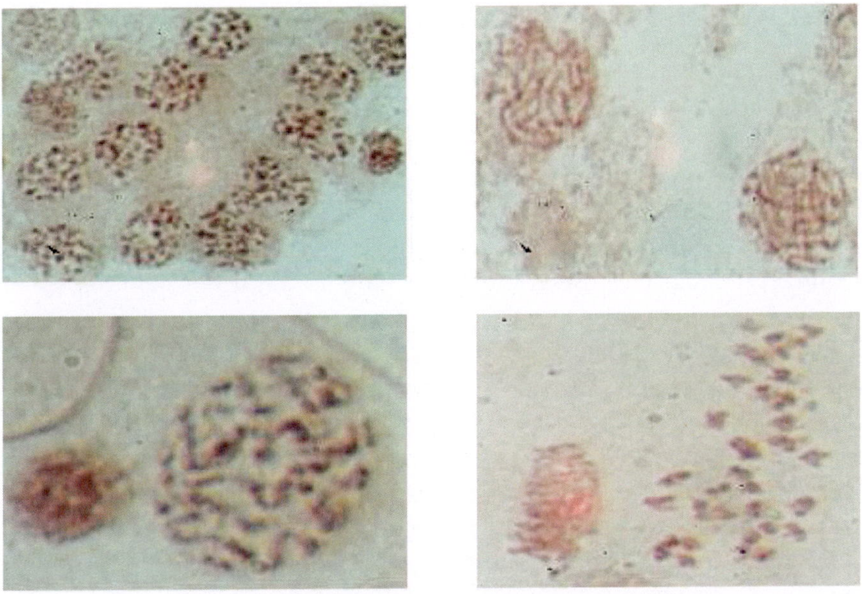

Fig.2. Photographic plates showing chromosomes of *Laevicaulis alte* in different stages of gonadal meiosis

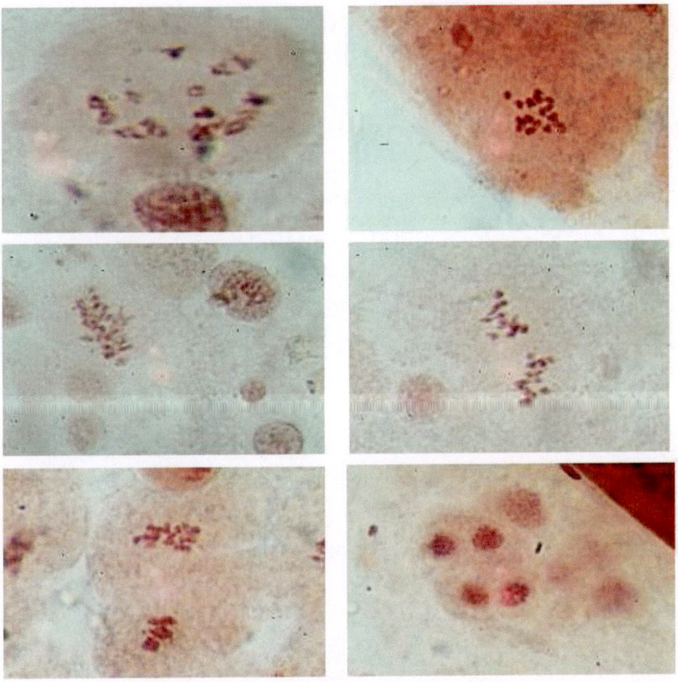

Fig.3. Photographic plates showing chromosomes of *Laevicaulis alte* in different stages of gonadal meiosis

OBSERVATIONS

The technique described above yields several stages of cells showing meiotic activity in gonadal tissue. The most frequent stages of meiotic chromosomes were observed as follows: Leptotene, Pachytene, Diakinesis, Metaphase-I, Anaphase-I, Anaphase-II, and Telophase-II (Fig.2). The chromosomes observed to form 17 bivalents in Diakinesis as well as in Metaphase-I confirming the diploid number to be 2n=34 as a characteristic feature of the species. Such bivalent configuration was supported by the presence of 17 thick and double structures with maximum condensation. The small unstained gaps present in the middle of each chromosome pair also proved their bivalent nature showing prominent chiasmatic regions. Spermatogonial metaphase plates were frequent in occurrence which revealed a diploid number of 34 small chromosomes exhibiting only metacentric and submetacentric configuration. Slug chromosomes at different meiotic stages exhibited a tendency of sticking together causing chromosomal overlapping as a result of which the cytogenetic study of gastropod of interest became more difficult.

DISCUSSION

Chromosome studies in gastropod molluscs have greatly increased since the reviews of Burch (1967), Burch & Natarajan (1967), Patterson (1969), Patterson & Burch (1978), Nagla *et al.* (1994), Nakamura (1986) and Thiriot-Quievreux (2002). As per the latest review the morphological features of gastropod chromosomes may be identical or very different within a family. Even within same species different karyotypes may be observed either due to polymorphism among different geographical populations or because differences in the techniques employed in order to obtain the chromosome preparations. Mostly Gastropod karyotypes include a majority of metacentric and submetacentric chromosomes following the nomenclature of Levan *et al.* (1964) with an exception in Veronicellid, *Sarasinula linguaeformis* (n=17) possessing equal number of different chromosome morphologies i.e.; four metacentric, four submetacentric, five subtelocentric and four telocentric (Majo A.D, 1996 and Majo A.D.et.al.1996). The haploid chromosome number in Veronicellid slug *Laevicaulis alte* (n=17) has already been reported previously by Natarajan (1960) and Majo A.D. (1996) but without any karyotypic detail. The first ever karyotypic analysis of Veronicellids like *Sarasinula plaebia* and *Laevicaulis alte* were reported with a haploid set to be n=17. The detailed karyotype of *Sarasinula plaebia* resembled to that of *Sarasinula linguaeformis* but with a minor difference in chromosome morphology as four metacentric, five submetacentric, four acrocentric and four telocentric (Cordova *et al.* 2010). Another observation on meiotic chromosomes of Bean Slug, *Sarasinula plebeia* from Jammu & Kashmir also reported the haploid complement to be seventeen as specific in veronicellids but karyotypic differences from earlier consisting of three metacentric, six submetacentric, five subtelocentric and three telocentric configuration (Poonam *et al.* 2019). The present investigation on *Laevicaulis alte* also supported presence of 17 bivalents in metaphasic arrangements of gonadal meiosis. The first ever karyotypic formula in *Laevicaulis alte* has already been described as consisting of only four metacentric and thirteen submetacentric chromosomes in meiotic metaphase-II (Cordova *et al.* 2010). The present investigation emphasizing the impact of molluscicides on the karyotype of rural garden pest *Laevicaulis alte* is with a view to support the previous observation as similar considering the inclusion of the species under same family Veronicellidae with the haploid chromosome number falling within the established range as n=16-18 described in earlier chromosomal reviews.

ACKNOWLEDGEMENTS

The author is extremely thankful to the President GB and Principal of U.N. Autonomous College for provision of necessary infrastructural facilities required for research work with the financial support extended by UGC, New Delhi under the scheme of Minor Research Project. The author is highly indebted to the Director, Zoological Survey of India, Kolkata for taxonomic identification of the specimen.

REFERENCES

1. Alderete-D.M, A.M., Mercado, L. A.C. & Usandivaras, E. M. (1996). The chromosomes of *Sarasinula linguaeformis* (Semper, 1885) (Gastropoda, Veronicellidae). *Iberus*, 14: 155–160.

2. Alderete-De Majo, A.M. (1996). The Veronicellidae (Mollusca, Gastropoda) in the light of new techniques for the karyotype and gametogenesis analysis. *Iberus*, 14: 147–154.

3. Burch, J.B. and Natarajan, R. (1967). Chromosomes of some Opisthobranchiate mollusks from Eniwetok Atoll, Western Pacific. *Pacific Science*, Vol.XXI (2), 252-259.

4. Burch, J.B. (1967). Cytological relationships of some pacific gastropods. *Venus*: 25,118-135

5. Cordova K. A. T., Batomalaque G. A. and Fontanilla I. K. C. (2010). Karyotypic Analysis of Two Veronicellid Slugs, *Laevicaulis alte* (Ferussac, 1821) and *Sarasinula plebeia* (Fischer, 1868) Tropical Natural History, Supplement 3, 17th International Congress of Unitas Malacologica, World Congress of Malacology

6. Das,B. and Parida, L. (2015). Morphometric studies of the tropical leatherleaf slug *Laevicaulis alte* from prachi belt of Odisha. *Journal of Entomology and Zoology Studies* 3: 132-134

7. Herbert D and Kilburn D. (2004). Field guide to the land snails and slugs of eastern South Africa. Pietermaritzburg: *Natal Museum*; Pp 336.

8. La Cour. L, (1941). Aceto-orcein: a new stain-fixative for chromosomes. *Stain Tech.* Vol.16.169 -174

9. Levan A., Fredga K. and Sandberg A.A., (1964). Nomenclature for centromeric position on chromosomes. Hereditas, 52: 201-220.

10. Nagla, Z. El-Alfy, K.A.Al-Ali and A.H. Abdel-Rehim, (1994). Karyotype, meiosis and sperm formation in the land snail, *Macrochlamys indica*. *Qatar Univ. Sci. J.* 14(1),122-128

11. Nakamura, H.K. (1986). Chromosomes of Archaeogastropoda (Mollusca Prosobranchia) with some remarks on their cytotaxonomy and phylogeny. Publications of the Seto Marine Biological Laboratory, 31: 191–268.

12. Natarajan R., (1960). Further cytological studies in Pulmonata (Mollusca-Gastropoda). *J. Zool. Soc. India.*, 12: 69-79.

13. Patterson, C.M. and Burch, J.B. (1978). Chromosomes of pulmonate molluscs. In: Pulmonates: Systematics, evolution and ecology, 2A. (V. Fretter & J. Peake, eds), 171–217. Academic Press, New York.

14. Patterson, C.M. (1969). Chromosomes of molluscs. In: Proceedings of the 2nd Symposium of Mollusca, Ernakulam, Cochin, India, 2: 635–689. Marine Biological Association of India.

15. Poonam, Tripathi, N.K. and Kour, P. (2019). First report of meiotic chromosome number and Karyotype analysis of Bean slug, *Sarasinula plebeia* from Jammu & Kashmir. Proceedings of 106th Indian Science Congress, Jalandhar, p-369

16. Raut, S.K. and Mandal, R.N. (1984). Natural history of the garden slug *Laevicaulis alte*. *Journal of the Bengal Natural History Society*; 3: 104-105.

17. Stern, E.M. (1975). A technique for the preparation of gastropod chromosomes. *Veliger*. 17.296-298

18. Thiriot-Quiévreux, C. (2003). Advances in chromosomal studies of gastropod molluscs. *Journal of Molluscan studies*, 69: 187-201

Current Status of Researches in Biosciences (2020) : 337-344
Editor : Prof. P.C. Joshi, N. Joshi, Reshman Yasmin, and D. K. Mansotra
Today & Tomorrow's Printers and Publishers, New Delhi - 110 002

EFFECT OF *DAUCUS CAROTA* SEED EXTRACT ON SPERM CHARACTERISTICS & TESTIS HISTOLOGY IN MALE ALBINO RATS

Shamshun Nehar* and Hena Kausar

Department Of Zoology, Ranchi University, Ranchi

Email: shamshun1805@gmail.com

ABSTRACT

In the present study seeds of *Daucus carota* commonly known as wild carrot were evaluated for its effect on Sperm characteristics and testis histology in male albino rats. The *Daucus carota* seed extract (DCSE) was orally administered at dose of 200 and 400mg/kg b.w for 90 days after which body weight and reproductive organs weight, spermatozoa characteristic, histological changes in the testis were observed. The result revealed significant decrease (P<0.001) in weight of primary and secondary reproductive organ. Remarkable changes like vacuolization and degeneration of somniferous tubules were observed in the testis of the rats treated with high dose. The treatment also caused significant (P<0.001) decrease in sperm motility, sperm count. Large numbers of sperms were found to be morphologically divested. The results from the present study suggest that seed extract of *Daucus carota* possess the potential to adverse effect on testicular function in rats.

KEYWORDS - Antifertility, *Daucus carota*,

INTRODUCTION

Population growth is one of the major factors behind environmental problems. As of July 2018 the population of India reached to daunting 1,355,145,202 this number is only second to china's population. With an effective, largely acceptable contraceptive population growth can be controlled. Herbal medicine is highly acceptable in general masses because they are cheap, having less or no side effects and safe for human consumption and environment. Various plants and plant products have been used for contraceptives purposes from ancient times. *Daucus carota* belonging to family *Apiaceae* used as/in traditional medicines for the treatment of many diseases such as kidney dysfunction, asthma and inflammations (Kritikar and Basu, 1991). The plant has undergone extensive phytochemical studies and a large number of active ingredients have been isolated. These include volatile oils, steroids, triterpenes, carbohydrates, glycerides, tannins, flavonoids, amino acid, carotene and hydrocarotene (Mahran *et al*., 1991; Gilani *et al*., 2000; Mazzoni *et al*., 1999). Seeds of *Daucus* carota have been used for abortion and infertility in female (Majumder *et al.*, 1997; Keenan *et al*., 1997). nevertheless its antisteroidogenesis in male is not studied. In the present study we are going to evaluate the effect of two different doses of *Daucus carota* seed extract (DCSE) on sperm characteristics and testis histology in male albino rats.

MATERIALS AND METHODS

Preparation of seed extract: The seeds of *Daucus carota* were obtained from Birsa Agricultural University, Ranchi. Seeds were cleaned and milled into powder by a mechanical grinder. The powdered seed were extracted with 95% ethanol using a soxhlet extraction unit. The ethanolic extract was filtered and concentrated by distillation process. A brown green colored residue was obtained which was kept in desiccators. This ethanolic extract of *Daucus carota* seed was used for further study.

Animals: 30 male albino rats (130-180) were divided into three groups (n=10). Each group was kept in a separate cage. Group I served as control and treated with vehicle only. Group II and Group III were treated with carrot seed extract at the rate of 200mg/kg b.w. and 400 mg/kg b.w. respectively for 90 days.

Determination of body and reproductive organ weight: Initial and final (after 24 hours of last treatment) boy weight of the animals was recorded. The animals were dissected after 90 days of treatment and the various organs of the reproductive tract excised were cleared of adhering fat and connective tissue and weighed to the nearest mg in a top pan balance.

Spermatozoa characteristics: Scheduled sacrifices were used for spermatozoa analysis by chipping off cauda epididymis in 1 ml of normal saline. Sperm count, motility, viability and abnormality were recorded according to WHO method manual (1999). Ejaculated spermatozoa collected from vaginal smear following successful mating were washed thoroughly and examined by Papanicolaou staining for morphological observation

(WHO, 1999).

Histological Examination: A portion of testis, was fixed in Bouin's fluid, dehydrated in ethanol, cleaned in xylene and embedded in paraffin wax. Five micron thick sections were stained with haematoxylin and eosin for light microscopic observations.

Statistical analysis: The data were analyzed using ANOVA and Student t test and p< 0.05 was considered statically significant.

RESULTS

Body and reproductive organ weight: After 90 days of treatment the changes in body weight (Table 1) and weight of the accessory reproductive organs (Table 2) were recorded. The low dose did not bring any significant change while at high dose the weight of testis as well as all the accessory sex organs were significantly reduced(P<0.005) Group III animals also showed significant variation in body weight (P<0.05).

Table1. The Effect of seed extract from *Daucas carota* on the body weights of male rats.

Dose	Initial	Final
Control	170.25±6.21	225.14±13.22
DCSE 200mg/kg b.w	171.98±6.56	223.65±11.55
DCSE 400mg/kg b.w	170.47±6.15	221.57±11.48***

The weights of the individual organs (g) were computed per 100g body weight of the animal. Each value represents mean ± SE of five animals. All comparison were made with the corresponding control groups (Student's t- test *P<0.05, **P<0.01, ***P<0.001)

Spermatozoa characteristics: Following 90 day of administration of *D. carota* seeds extracts sperm motility results showed that rats treated with higher dose 0f DCSE had significantly low motility compared to the other group. Sperm concentration significantly (p<0.01) decreased in the same group. Large numbers of sperms were found morphologically devastated. Visible deformities were noted as coiled tail (fig 1a), deformed head, twisted neck (1b), cytoplasmic droplets (fig 1c), and completely separated head (fig 1d) mid-piece abnormality (Fig 1). Group III animals showed higher deformities comparing to group II.

Table 2. The influence of seed extracts from *Daucas carota* on the weights of the testes and accessory sex organs of male rats

Dose	**Testis**		**Epididymis**		**Prostate**	**Seminal vesicle**
	Left	Right	Left	Right		
Control	465.31±22.56	466.31±22.97	125.88±5.79	126.39±5.81	100.25±4.19	290.66±13.48
DCSE 200mg/kg b.w	461.55±21.55	462.45±22.51	122.11±5.74	122.29±5.78	98.47±4.15	285.99±13.47
DCSE 400mg/kg b.w	459.49±21.43***	458.22±22.41***	115.47±5.61***	114.48±5.75***	95.12±4.12***	281.47±13.37***

Table 3: Cauda epididymal sperm characteristics of animals treated with *D. carota* seed extract during treatment at various doses. (Group I: n=30; Group II-III: n=10) (Value±SD)

Treatment Groups	Sperm motility (%)	Sperm count (million/mL)	Sperm viability (%)	Sperm abnormality (%)
Group I	72.34±2.49	61.37±07.48	71.39±1.96	21.05±1.54
Group II	51.39±4.96	50.48±03.69	35.12±4.46	44.45±1.79
Group III	45.91±2.45**	44.14±07.58**	32.42±2.36**	55.35±1.56**

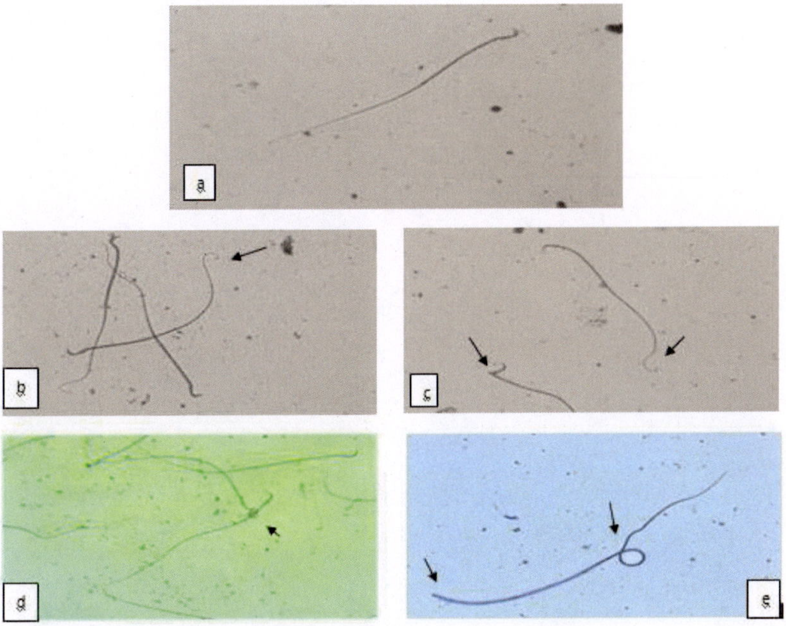

Fig 1. Effect of *Daucus carota* seed extract on sperm morphology of male albino rats after 90 days of treatment.

Histological Examination: Following 90 days of administration of *D. carota* seeds extracts, animals showed extensive damage in the histology of testis. Group II and III animals showed higher histological disorientation as visible in the Fig. 2(d, e & f). Round or oval somniferous tubules with the irregular epithelium containing disoriented Sertoli cells(d) and germ cells were observed. The basal lamina was dilated showing clear disassociation with spermatogonia and Sertoli cell (Fig. 2f). Spermatids appeared to have disassociated from Sertoli cells, amount of sperm present in the lumen of seminiferous tubules lowered extensively (Fig. 2c & 2e). In addition, leydig cells appeared unorganized, showing inconspicuous nuclear and cytoplasmic characteristics (Fig 2d &2f).

Fig 2a & 2b showing histology of testes of control rats (Group I). Fig 2b & 2c showing histology of testis of rats treated with 200mg/kg b.w. of DCSE (Group II) Fig 2e &2f 400mg /kg b.w. (Group III) of the DCSE. Group II & III Showing areas of degeneration(c,d,e& f) and vacuolization(Va) of germinal epithelial cells and less quantity of sperm in the lumen(c & e) of the somniferous tubules.[Lu- lumen, Sg spermatogonia, Sz- Spermatozoa, Lc- Lydig cell, Sr- Sertoli cells, Ge- Germinal epithelium, Bl-Basal lamina, Va- Vacuolization]

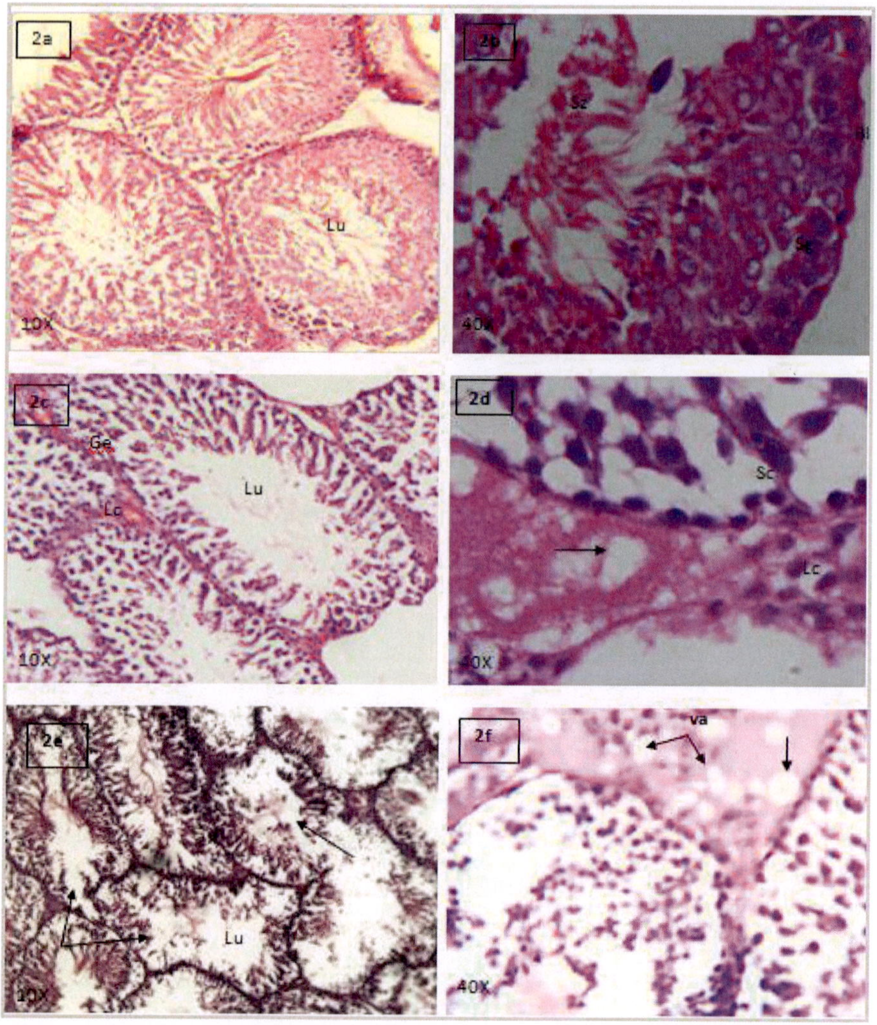

Fig 2: Photomicrograph of the testis histology of the male wistar albino rats.

DISCUSSION

The two different doses i.e 200 and 400 mg per kg body weight of *Daucus carota* seeds were administered to rats. No significant changes were observed in the body weight of treated rats when compared to control. The low dose did not bring any significant change while at high dose the weight of testes, epididymis and seminal vesicle were significantly reduced ($P<0.05$). Our study for the first time recorded that administration of *D. carota* seed extract affects the weight of testis. It gradually lowers weight of testis in treatment group. At higher doses decline in weight of testis was higher than lower doses and found significantly varied from control group. Depression in spermatogenesis is usually accompanied by reduced testicular weight since the bulk of testicular weight is made up of spermatids and spermatozoa (Kenjale *et al.,* 2008; Chauhan *et al.*, 2007). Histological data of our study further confirms a disturbance in spermatogenesis as evident by several somniferous tubules that have few spermatozoa in their lumen. Thus DCSE administration resulted in inhibition of spermatogenesis. Further there was significant decrease in the epididymal sperm count in high dose. The decreased sperm motility observed in the study may be an effect of altered function of the accessory organs (Mann and Ludwak-Mann *et al.*, 1981; Hamilton, 1975). Testosterone plays a major role in the maintenance of spermatogenesis (Mooradian *et al.,* 1987). In the present study, the observed reduction in spermatogenesis may thus, be attributed to testosterone deficiency. Our research showed that oral administration of *Daucus carota* seed extract lead to a dose dependent defects in the testicular spermatogenesis which leads to production of defective sperms. Male rats with such defective sperms quality and quantity are unable to produce the offspring.

CONCLUSION

We conclude from this study that *D. carota* seeds extract have direct influence on male reproductive system. Further studies are essential to determine mechanism of antifertility action of this plant.

Conflict of Interest: There is no conflict of interest

ACKNOWLEDGEMENTS

The authors are thankful to University Grant Commission, New Delhi for awarding Maulana Azad National Fellowship, which provides financial support.

REFERENCES

1. Kritikar, K. R., & Basu, B. D. (1999). Indian medicinal plants, Vol. 1. *International Book Distributors, New Delhi, India*, 632-635.

2. Mahran, G. H., Kadry, H. A., Isaac, Z. G., Thabet, C. K., Al Azizi, M. M., & El Olemy, M. M. (1991). Investigation of diuretic drug plants. 1. Phytochemical screening and pharmacological evaluation of *Anethum graveolens* L., *Apium graveolens* L., *Daucus carota* L. and *Eruca sativa* Mill. *Phytotherapy Research*, *5*(4), 169-172.

3. Gilani, A. H., Shaheen, F., Saeed, S. A., Bibi, S., Sadiq, M., & Faizi, S. (2000). Hypotensive action of coumarin glycosides from *Daucus carota*. *Phytomedicine*, *7*(5), 423-426.

4. Mazzoni, V., Tomi, F., & Casanova, J. (1999). A daucane type sesquiterpene from Daucus carota seed oil. *Flavour and fragrance journal, 14*(5), 268-272.

5. Majumder, P. K., Dasgupta, S., Mukhopadhaya, R. K., Mazumdar, U. K., & Gupta, M. (1997). Anti-steroidogenic activity of the petroleum ether extract and fraction 5 (fatty acids) of carrot (*Daucus carota* L.) seeds in mouse ovary. *Journal of ethnopharmacology*, *57*(3), 209-212.

6. Keenan, D. L., Dharmarajan, A. M., & Zacur, H. A. (1997). Dietary carrot results in diminished ovarian progesterone secretion, whereas a metabolite, retinoic acid, stimulates progesterone secretion in the in vitro perfused rabbit ovary. *Fertility and sterility*, *68*(2), 358-363.

7. World Health Organisation. (1999). *WHO laboratory manual for the examination of human semen and sperm-cervical mucus interaction*. Cambridge university press.

8. Kenjale, R., Shah, R., & Sathaye, S. (2008). Effects of *Chlorophytum borivilianum* on sexual behaviour and sperm count in male rats. *Phytotherapy Research: An International Journal Devoted to Pharmacological and Toxicological Evaluation of Natural Product Derivatives*, *22*(6), 796-801.

9. Chauhan, A., Agarwal, M., Kushwaha, S., & Mutreja, A. (2007). Suppression of fertility in male albino rats following the administration of 50% ethanolic extract of Aegle marmelos. *Contraception*, *76*(6), 474-481.

10. Mann, T., & Lutwak-Mann, C. (1981). Male reproductive function and the composition of semen: general considerations. In *Male Reproductive Function and Semen* (pp. 1-37). Springer, London.

11. Hamilton, D. W. (1975). Structure and function of the epithelium lining efferentes, ductus epididymidis and ductus deferens in the rat. *Handbook of physiology, Sect 7*, 259-301.

12. Mooradian, A. D., Morley, J. E., & Korenman, S. G. (1987). Biological actions of androgens. *Endocrine reviews*, *8*(1), 1-28.

Current Status of Researches in Biosciences (2020) : 345-353
Editor : Prof. P.C. Joshi, N. Joshi, Reshman Yasmin, and D. K. Mansotra
Today & Tomorrow's Printers and Publishers, New Delhi - 110 002

IRON FORTIFICATION IN KULFI

Kuldeep Mishra*[1]; Panchaal Bhattacharjee[2] C N Dharaiya[3] and Suneeta V. Pinto[4]

[1,3,4]Dairy Technology Department, SMC College of Dairy Science,

[2] Department of Horticulture, BA College of agriculture, Anand Agricultural University, Anand.

**Corresponding author email: mishraypikuldeep@gmail.com*

ABSTRACT

Kulfi was prepared from milk standardized to 2% fat and 9% MSNF. Iron fortified kulfi formulated to contain 30 ppm iron from external added iron salts (based on the approximate iron content and bioavailability of different sources of iron) was prepared using six iron salts viz., Ferrous gluconate (FG), Iron reduced (IR), Ferric chloride (FC), Ferric pyrophosphate (FP), Ammonium ferric citrate (FA) and Ferrous sulphate (FS). The experiment samples were compared with control (C). Fortification of kulfi with FA and FC did not significantly ($P>0.05$) affect the acidity, viscosity and meltdown time of the samples. From amongst the six types of iron sources only FA was found to be acceptable. The effect of level of addition of FA at different levels viz. 15, 30, 40 and 50 was studied. It was found that addition of iron at the highest level i.e. 50 ppm resulted in significant reduction in overall acceptability of the resultant product. In shelf life studies, it was found that fortification of iron at 40 ppm resulted in significant reduction in flavour and overall acceptability scores after 30 days of storage whereas iron fortified samples at 30 ppm remained acceptable up to 90 days of storage. Hence, it can be concluded that acceptable quality iron fortified kulfi can be prepared by addition of ferric ammonium citrate (30 ppm available iron) just before freezing of kulfi mix is acceptable up to 90 days at -18±2°C.

KEYWORDS: Kulfi, Iron fortification.

INTRODUCTION

It is well known that iron deficiency (anemia) is still the most prevalent nutritional problem in the world. Calcium and iron are two nutrients that are essential for growth. Both are also needed in higher amounts by women, especially because of increased requirements during pregnancy and because of their higher risk for osteoporosis and iron-deficiency anemia. Dairy products contribute many nutrients to the diet (calcium, riboflavin, protein, magnesium, and vitamins A, B_6, and B_{12}) but provide very little iron (0.6 mg of iron/kg of milk). Fortification of dairy products with iron has thus been frequently considered as a potential means for preventing iron deficiency. Compatible and non-reactive iron compounds are needed for fortification of foods as they have less of an 'iron taste' compared to soluble iron. There is a paucity of information on iron fortified frozen dairy products.

Kulfi is very popular in our country even amongst lower income groups. Halwais and small scale manufacturers usually prepare kulfi. Several research workers have described procedures for manufacture of kulfi (Salooja and Balachandran, 1982, Ghosdekar and Rao, 1982, Ashokraju *et al.,* 1988, Rajor and Vani 1991, Ramachandran *et al.,* 2005). Kulfi is characterized by its slightly whitish to brownish appearance, compact and smooth to sometimes coarse texture and nutty, caramel flavour. In the traditional process, cow, buffalo or mixed milk is concentrated in a large pan over the fire. The concentrated milk is then cooled, sugar is added and the blend is thoroughly stirred. Sometimes, a small quantity of khoa is added during boiling. After adding nuts and saffron the mixture is filled in aluminum or plastic cones leaving about 5% headspace. The lids are put on and sealed with wheat flour dough. After sealing, the cones are immersed in ice-salt mixture (4:1 w/w) in an earthenware pot (*MATKA*) for freezing (Aneja *et al.,* 2002). Problems associated with iron fortification such as development of oxidative rancidity will be lessened or minimum due to the low temperature of storage of the product. Hence, Kulfi was selected as a product for iron fortification. The study was conducted with to select a suitable source of iron and to select the level of iron fortification compounds which will be acceptable based on sensory analysis and give the most acceptable shelf-life and to study the shelf life of the fortified product.

MATERIALS AND METHODS

Kulfi was prepared from milk standardized to 2% fat and 9% MSNF. Sucrose was added at 12% by weight of milk. The blend was concentrated tohalf its volume. Sodium alginatewas added at 0.2% by weight of milk at the final stages of concentration. The mix was cooled to 4 ± 1° C and aged overnight. Freshly ground cardamom (2 g/kg mix) was added before freezing. Iron fortified kulfi using different sources of iron were also prepared using the same method as described above. The iron salts were added after concentration in the hot mix (at 85-90 °C) to ensure that it was properly blended. The mixes were formulated to contain 30 ppm available iron. To study the effect of stage of addition of iron, ferric ammonium citrate (at 30 ppm available iron content) was added at different stages of manufacture i.e. before heating and before freezing. The effect of level of addition of ferric ammonium citrate was assessed by fortifying kulfi with different levels of ferric ammonium citrate (to give 0, 15, 30, 40 and 50 ppm available iron) before freezing. Shelf life studies of iron fortified kulfi were conducted in the last part of the study. Iron fortified Kulfi was prepared using ferric ammonium citrate at 0, 15, 30 and 40 ppm. Ferric ammonium citrate was added after heating to the aged kulfi mix. The samples were evaluated at regular intervals of 0, 30, 60, 90 and 120 days to study the effect of fortification of iron on shelf life of kulfi.

Analysis: The fat, titratable acidity and total solids content of kulfi mixes were determined bythe standard method as suggested in ISI Handbook (ISI Handbook of food analysis, 1989). The protein content of mixes were determined by Semi – Micro Kjeldahl method (ISO: 8968-1 (2014), using Kjel-plus Digestion System (Model-KPS 006L, M/s.Pelican Instruments, Chennai) and Kjel-plus Semi-Automatic Distillation System (Model-Distil M, M/s.Pelican Instruments, Chennai). As per this method the total nitrogen was determined and the value so obtained was multiplied by a standard factor 6.24 to get the protein content. Ash content of processed bottle gourd forms were determined by procedure described in Indian Standards (SP: 18 Part XI – Dairy Products). Total carbohydrate content was calculated on difference basis from total solid content

Sensory Analysis: All the products (-18$\pm$2 °C) were evaluated by a panel of 10 judges for its sensory characteristics using a 9-point hedonic scorecard (Stone and Sidel, 2004). The product was subjected to the sensory evaluation by semi expert trained panel of ten judges for colour and appearance, body and texture, melting quality, flavour and overall acceptability criteria. Kulfi samples were tempered to -12$\pm$2 °C for 1-2 h before judging. The fresh and stored samples were tempered to -12±2 p C. The samples were labelled with random 3 digit codes. The order of presentation of samples was randomized across judges.

Statistical analysis: All experiments were in triplicate. .Mean scores obtained for the experimental samples were compared with those of the control samples and the results were analyzed statistically for test of significance according statistical method of one way-ANOVA and mean ± SE as per the procedure given by Steele and Torrie (1980). Critical difference was determined at the 5% significance level.

RESULTS AND DISCUSSION

Iron fortified kulfi was prepared using six iron salts viz., Ferrous gluconate (FG), Iron reduced (IR), Ferric chloride (FC), Ferric pyrophosphate (FP), Ammonium ferric citrate (FA) and Ferrous sulphate (FS). The mixes were formulated to contain 15 ppm iron from external added iron salts based on the approximate iron content and bioavailability of different sources of iron All the experimental samples were compared with control (C) containing no external iron source. The average chemical composition of kulfi mix was Total solids: 42.5%, fat: 3.5%, protein: 5.6%, sucrose: 21.5% and ash:1.31%. Addition of iron in any form did not influence the body and texture and melting resistance of the products. However, adverse effect on flavour was noticed when FG, FS, FP and IR were used as external sources for iron fortification. From amongst the six types of iron sources screened only FA and FC were found to be acceptable and used in the next part of study.

Iron fortified kulfi was prepared using FA and FC. The experiment samples were compared with control (C). Fortification of kulfi with FA and FC did not significantly (P>0.05) affect the acidity, viscosity and meltdown time of the samples (Table 1).

Table 1: Effect of iron fortification on selected physico-chemical properties of kulfi

Parameter	Type of Kulfi			CD
	C	FA	FC	
Acidity (% lactic acid)	0.30	0.31	0.29	NS
Viscosity (cP at 4±2 °C)	258.3	263.2	268.4	NS
Meltdown time, min	35.5	36.5	36.3	NS

Each observation is mean of 6 replications

The average sensory scores of C, FA and FC are presented in Table 2. Addition of iron salts (FA and FC) had no significant (P>0.05) effect on colour and appearance, body and texture, melting quality scores of samples. However, the overall acceptability and flavour score of control and FA were found to be at par (P>0.05) with each other. Whereas, the overall acceptability score of FC was significantly (P<0.05) lower than control and FA. Samples containing FC were found to contain a slight metallic flavour and some very sensitive judges criticized it for chemical flavour and hence scored lower in flavour compared to control and FA.

Table 2: Effect of iron fortification on average sensory scores of kulfi

Parameter	Type of kulfi			CD
	C	FA	FC	
Colour and appearance	8.00	7.93	7.82	NS
Flavour	7.98^{a}	7.83^{a}	7.37^{b}	0.17
Body and texture	7.99	7.88	7.63	NS
Melting quality	7.90	7.96	7.93	NS
Overall acceptability	7.90^{a}	7.85^{a}	7.48^{b}	0.11

Each observation is mean of 6 replications; a–b Superscript letters following numbers in the same row denote significant difference (P <0.05)

From this part of the study FA was found to be the most suitable source for iron fortification in kulfi. The next part of the study was conducted to study the effect of stage of addition of iron on acceptability of kulfi. Kulfi was prepared by addition of FA to contain 15 ppm available iron at different stages of manufacture viz. before concentration and after concentration, before freezing. The effect of stage of addition of iron source on sensory score of kulfi is presented in Table 3.

Table 3: Effect of stage of addition of iron salts on sensory scores (1-9) of iron fortified kulfi

Stage of Addition	Flavour	Body and texture	Colour and appearance	Melting quality	Overall acceptability
Control	8.27^{a}	8.17	8.23^{a}	8.08	8.16^{a}
Before Freezing	8.03^{a}	7.92	7.78^{a}	7.96	8.02^{a}
Before Heating	4.12^{b}	7.88	6.42^{b}	7.91	4.71^{b}
CD (0.05)	2.526	NS	1.004	NS	2.481

Each observation is mean of 6 replications; a–b Superscript letters following numbers in the same row denote significant difference (P <0.05)

It can be seen in Table 3 that there all the sensory attributes of iron fortified samples viz. flavour, body and texture, colour and appearance and overall acceptability were at par (P>0.05) with control when the iron source was added before freezing. However, a significant reduction (P<0.05) in flavor, colour and appearance as well as overall acceptability score was observed when iron was added before heating. The samples developed strong metallic flavor and browning in colour was observed in such samples. Hence, in the next part of study iron was added before freezing.

In next part of the study, effect of level of addition of FA at different levels was studied and the result pertaining to sensory scores of such samples is presented in Table 4. It can be seen from Table 4 that addition of iron at the highest level i.e. 50 ppm resulted in significant reduction in overall acceptability of the resultant product.

Table 4: Effect of level of addition of iron source on sensory score of kulfi

Level of Addition	Flavour	B&T	C&A	MR	OA
Control	8.27 [a]	8.17	8.23 [a]	8.06	8.16 [a]
15 ppm	8.03 [a]	8.04	7.78 [a]	7.96	8.02 [a]
30 ppm	7.58 [a]	7.98	7.44 [a]	7.88	7.48 [a]
40 ppm	6.92 [a]	7.84	7.21 [a]	7.86	6.96 [a]
50 ppm	4.22 [b]	7.68	6.92 [b]	7.83	4.71 [b]
CD (0.05)	2.314	NS	1.119	NS	2.228

Each observation is mean of 6 replications; a–b Superscript letters following numbers in the same row denote significant difference ($P < 0.05$)

The shelf life of iron fortified kulfi was also determined by evaluating the samples at 0, 30, 60, 90 and 120 days of storage. The effect of concentration of iron on sensory attributes of kulfi during storage at -18+2 C is presented in Table 5. It can be seen from the Table that fortification of iron at 40 ppm resulted in significant reduction in flavour and overall acceptability scores after 30 days of storage. Iron fortified samples at 30 ppm remained acceptable up to 90 days of storage.

Table 5: Effect of different levels of FA on sensory attributes of kulfi during storage at -18+2 °C

Concentration of iron (ppm)	Storage period (d)				
	0	30	60	90	120
	Flavour				
0	8.17	7.83	7.58	7.25	7.08
15	8.08	7.67	7.33	7.08	6.91
30	7.83	7.58	7.25	6.75	5.25
40	7.33	7.00	5.67	-	-
C = 0.68; S = 0.59; CxS = 0.46					

	Body and texture				
0	8.17	8.17	8.17	8.17	8.00
15	8.00	7.83	7.85	7.50	7.43
30	8.00	8.00	7.92	7.90	7.83
40	8.00	8.00	7.75	-	-
C = NS; S = NS; CxS = NS					
	Colour and appearance				
0	8.00	8.00	8.00	8.00	8.00
15	7.92	7.83	7.83	7.83	7.83
30	8.08	7.92	8.00	7.92	7.83
40	8.02	8.17	7.92	-	-
C = 0.68; S = 0.59; CxS = 0.46					
	Overall acceptability				
0	8.17	7.83	7.58	7.25	7.08
15	7.91	7.50	7.25	7.00	6.67
30	7.83	7.50	7.25	6.67	5.25
40	7.33	7.01	5.67	-	-
C = 0.69; S = 0.60; CxS = 0.47					

Each observation is mean±SD of 6 replications, C= level of iron (ppm), S= storage period (days)

The data obtained for changes in flavour score of kulfi during storage at -18±2 °C are presented in Table 5. Incorporation of iron showed significant (P <0.05) effect on flavour scores of kulfi. The results indicated that the flavour score of fresh samples of kulfi was in the order of 0>15 ppm>30 ppm> 40 ppm. The flavour score of control was higher than samples containing iron. The flavour score of C declined from beginning of the storage. Amongst the experimental samples the highest rate of decrease in flavour score was observed in samples containing 40 ppm iron and the lowest rate were observed in 15 ppm iron. The flavour score of samples containing 40 ppm iron went below acceptable level (6.0) after 30 days of the storage. It became unacceptable due to pronounced oxidized metallic flavor and bitter taste. The effect of storage on flavour scores and the interaction effect between concentration of iron and storage period was found to be significant (P<0.05).

Incorporation of iron had no significant (P >0.05) effect on body and texture and colour and appearance scores of kulfi. The effect of storage on body and texture scores and colour and appearance and the interaction effect between level of iron and storage period was found to be non-significant (P >0.05).The overall acceptability score of fresh samples of kulfi was in the order of 0>15 ppm>30 ppm> 40 ppm. Incorporation of iron showed significant (P <0.05) effect on over all acceptability scores. The effect of storage on overall acceptability scores and the interaction effect between iron concentration and storage period was found to be significant (P <0.05). The overall acceptability score of kulfi containing 30 ppm iron went below acceptable level after 90 days of the storage.

The effect of different levels of FA on acidity and melting resistance of kulfi during storage at -18+2 °C is presented in Table 6. Incorporation of iron had no significant (P >0.05) effect on acidity and melting resistance of kulfi. The effect of storage on acidity and melting resistance and the interaction effect between level of iron and storage period was also found to be non- significant (P >0.05).

Table 6: Effect of different levels of FA on selected physico-chemical attributes of kulfi during storage at -18+2°C

Concentration of iron (ppm)	Storage period (d)				
	0	30	60	90	120
	Acidity (% lactic acid)				
0	0.293	0.300	0.297	0.297	0.297
15	0.297	0.297	0.297	0.298	0.30
30	0.297	0.297	0.297	0.293	0.297
40	0.303	0.300	0.293	-	-
C = NS; S = NS; CxS = NS					
	Melting resistance				
0	36.00	37.00	36.67	37.00	37.00
15	36.67	36.67	36.33	37.00	36.33
30	36.67	36.00	37.00	36.00	36.67
40	37.00	36.00	37.00	-	-
C = NS; S = NS; CxS = NS					

Each observation is mean±SD of 6 replications, C= level of iron (ppm), S= storage period (days)

CONCLUSION

It can be concluded from the study that ferric ammonium citrate is a suitable iron salt for manufacture of iron fortified kulfi when added just before freezing. Fortification of kulfi with ferric ammonium citrate in kulfi at the rate of 30 ppm iron was found to be acceptable sensorily. Levels greater than 40 ppm adversely affected the acceptability of product and shelf life of product fortified with 40 ppm iron was only 30 days. Among the levels studied, addition of 30 ppm iron was found to be most suitable and provides shelf life of Kulfi of 90 days at -18±2°C storage.

REFERENCES

1. Aneja, R.P., Mathur, B.N., Chandan, R.C., Banerjee, A.K. (2002). Technology of Indian milk products, Dairy India Yearbook, A Dairy India Publication, New Delhi, pp: 74- 96, 99–101.

2. Ashok raju A, Pasha A.M., Reddy, K., Reddy C.R., Rao M. (1989). Studies on preparation and quality of kulfi. *Indian Journal of Dairy Science;* 42:127-129.

3. Dodake S.R., Talekar S. S., Thombre B.M. (2015). Sensory Properties of kulfi prepared from ashgourd pulp. *International Journal of Tropical Agriculture;* 33(2): 313-316.

4. FSSAI (2011). Food Safety and Standard Authority of India, 2006 (www.fssai.gov.in).

5. Ghosdekar D.R., Rao K.S. (1982). Microbiology of kulfi. *Indian Dairyman.* 34: 257-262.

6. ISI Handbook of food analysis (1989). SP: 18 (Part XI – Dairy Products). Bureau of Indian Standards, Manak Bhavan, Bahadur Shah Zafar Marg, New Delhi, India.

7. IS: 1547 (1985) Specifications for Infant Milk Foods. Bureau of Indian Standards, Manak Bhavan, New Delhi.

8. ISO: 8968-1 (2014) Determination of nitrogen content - Part 1: Kjeldahl principle and crude protein calculation.

9. Rajor R.B., Vani B., (1991) A new approach for manufacture of kulfi. *Indian Dairyman*, 43: 256-259.

10. Ramachandran L, Singh S., Rathour A. (2005) Preparation of kulfi from admixtures of partially de-oiled groundnut meal and milk/milk powders. *Natural Product Radiance* 4(2): 90-96.

11. Salooja M. K., Balachandran R. (1982) Studies on the production of kulfi Part I. The acceptable level of total milk solids. *Journal of Food Science and Technology;* 19: 116-118.

12. Steel R. G. D., Torrie J. H. (1980) A biometrical approach. In: Principal and Procedures of statistics. 2[nd]Edn. McGraw Hill Kogakusha Ltd., Japan, p.137-150.

13. Stone, H., Sidel, J.L. (2004). In Sensory Evaluation Practices, 3rd edn. Elsevier Acad. Press, NY, p 88.

Current Status of Researches in Biosciences (2020) : 355-360
Editor : Prof. P.C. Joshi, N. Joshi, Reshman Yasmin, and D. K. Mansotra
Today & Tomorrow's Printers and Publishers, New Delhi - 110 002

AMELIORATIVE EFFECTS OF POLYHERBAL FORMULATION AGAINST GENTAMICIN INDUCED NEPHROTOXICITY IN RAT

B. D. Talele, N. G. Mahajan, R. T. Mahajan and M. Z. Chopda

Department of Zoology, Moolji Jaitha (Autonomous) College, Kavayitri Bahinabai Chaudhari North Maharashtra University, Jalgaon

Email:bdt72zoo@gmail.com

ABSTRACT

Herbal plants known for its various medicinal properties are also a natural antioxidant and a free radical scavenger with no documented evidence as a nephroprotective agent. To evaluate the nephroprotective activity of polyherbal formulation in a rodent model of gentamicin induced nephrotoxicity. Materials and Methods: Wistar albino rats of either sex, weighing 150–200 g was divided into 6 groups; normal saline, CMC as placebo, gentamicin 80mg/kg, intraperitoneally for 14 days, herbal mixture at 250, 500, and 1000 g/kg, per oral for 14 days, the test extract administered 3 days prior. Blood urea, serum creatinine, and blood urea nitrogen analyses and microscopic examination of kidney were performed. Gentamicin treatment caused nephrotoxicity as evidenced by marked elevation in, Serum protein, serum creatinine and blood urea nitrogen (7.51±0.14 mg/dl, 1.56±0.08 mg/dl,76.13±3.86 mg/dl) respectively when compared to the saline treated groups. Co-administration of herbal mixture with gentamicin decreased the rise in these parameters in a dose dependent manner. Histopathological analysis revealed epithelial loss with intense granular degeneration in gentamicin treated rats, whereas herbal mixture mitigated the severity of gentamicin-induced renal damage. Conclusion: Our data suggest that polyherbal formulation exhibits renoprotective effect in gentamicin induced renal damage and further studies on its mechanism of action are warranted.

KEYWORDS: Gentamicin, Nephroprotective, polyherbal, Nephrotoxicity.

INTRODUCTION

Nephrotoxicity is one of the most common kidney problems and occurs when body is exposed to a drug or toxin. Aminoglycoside antibiotics, especially gentamicin. Despite rigorous monitoring, nephrotoxicity appears in 10-25% of therapeutic courses. Gentamicin mainly causes tubular toxicity, it is an aminoglycoside antibiotics used for the treatment of Gram negative bacterial infections. Overdose of gentamicin causes renal damage. It may give serious side effects while continuous consuming at higher concentrations. (Suji Arivazhagan 2014) When kidney damage occurs, body unable to rid of excess urine and wastes from the body and blood electrolytes (such as potassium and magnesium) will all become elevated (Ramya, 2011). The use of herbs as nephroprotective is a major avenue in Indian perspectives particularly for treating kidneys damage, which require to be explored more successfully as there are many literatures available on these aspects. Ayurveda, an indigenous system of medicine, offers a vast scope of renal treatment for renal failure. Plants and other natural substances have been used as the rich source of medicine. Also, herbal drugs are more easily accessible. Medicinal plants have curative properties and therapeutic values due to the presence of various complex phytochemical compounds. This traditional medicines are assuming greater important because of very effective, safer, locally available, and no side effects. Wide variety of plants and its active principles, with minimal side effects, provides an alternate therapy for kidney diseases. In present study we will try to evaluate the neproprotective activity of polyherbal formulation. (Talele 2012)

MATERIALS AND METHODS

Collection of plants: Following listed plants were collected from North Maharashtra region during June 2009 to September 2009. The plants were identified by the Dr. Tanveer Khan taxonomists of Department of Botany, Moolji Jaitha College, Jalgaon, Maharashtra (India).

Table 1 List of selected plants for nephroprotective activity

Sr. No.	Botanical Name	Family	Vernacular name	Part used
1.	*Bauhinia recemosa* (Lam)	Caesalpiniaceae	Aapata	Stem bark
2.	*Dolichos biflorus* (Linn.)	Fabaceae	Kulitha	Seed
3.	*Sphaeranthus indicus* (Linn)	Asteraceae	Gorakhmundi	Flower
4.	*Tectona grandis* (Linn)	Verbenaceae	Sag	Seed
5.	*Tephrosia purpurea* (Linn)	Fabaceae	Sharpunkha	Leaves
6.	*Tribulus terrestris* (Linn)	Zygophyllaceae	Gokhru	Fruit

Poly herbal preparation (Herbal Mixture – HM): Shade dried plant material pulverized using an electrically operated grinder. Equal quantities of the respective herbal powders were blended and then mixed with 1% Carboxy Methyl Cellulose for preparing the doses.The doses were prepared fresh daily, be for administration, the rats were dosed at approximately the same time each day.

Animals(Wistar rat): Animals for the study were obtained from the National Biosciences, Pune. The animals were acclimatized to laboratory conditions (temperature: 22^{0}C - 24^{0}C, humidity: 65 - 70% and 12 h light and 12 h dark rhythm) prior to the start of the experiment in the CPCSEA registered (1062/cpcsea/2007) Animal House of the Department of Zoology, Moolji Jaitha College, Jalgaon.

METHODOLOGY

Study plan: Animal experimentation

Eight to twelve week old, either sex (weighing between 200-300 gm.) Wistar rats were divided into following 6 groups.

Sr. No	Group	Types of group
1	Control	Treated with Saline
2	CMC	Treated with 1 % Carboxy methyl cellulose
3	GM induced nephrotoxicity(GAG)	Introduced Nephrotoxicity,with gentamicin at 80 mg/kg b.w.
4	Herbal mixture I	Treated with Hm250 mg/kg b.w.
5	Herbal mixture II	Treated with Hm 500 mg/kg b.w.
6	Herbal mixture III	Treated with Hm1000 mg/kg b.w.

On 14th day,blood samples from each group of animals were collected for serum biochemistry.

Blood Collection: The animals were mildly anesthetized using anesthetic ether before blood collection. Blood was collected from the retro-orbital sinus using heparinised capillaries (Remi, Mumbai). A portion of blood was collected in heparinsed vials for biochemical analysis of blood urea nitrogen (BUN), creatinine (CRE), protein (PRO), using kits of Span Dignostics (Mumbai, Maharashatra, India) / Nirmal Laboratories (Chopda, Maharashatra, India) on a Chariot Prince Biochemistry Analyser

Body weight and feed consumption: Body weight and feed consumption of animals were recorded daily throughout the experimental period. The animals were also observed for clinical signs daily throughout the experimental period.

RESULTS AND DISCUSSION

Clinical signs: No clinical signs of toxicity and mortality were observed in animals belonging to any group.

Changes in biochemical values of rat after 14th day of drug treatment: Decreased inproteinin GAG group of animals 5.82 ± 0.18 g/dlas compared to control group of animals. 6.51±0.19g/dlreached to normal level in herbal mixture treated group at the dose 1000 mg/kg b.w 6.53±0.13g/dl. An increase BUN, and Creatinine 152.1±2.50, 3.56±0.11mg/dl, in gentamicin alone group as compared to the control group of animals 16.91±0.83 mg/dl and 0.70±0.02 mg/dl confirming the nephrotoxicity induced by gentamicin. BUN 89.00 ± 5.31mg/dl, CRE 0.93±0.01mg/dl were reach to normal level in herbal mixture treated group at the dose 1000 mg/kg b.w.

Table 2 Changes in Biochemical values in rats after 14th day of drug treatment.

Group	BUN $_{(mg/dl)}$	CRE $_{(mg/dl)}$	PRO $_{(g/dl)}$
Control	16.91±0.83	0.70±0.02	**6.51±0.19**
CMC	15.88±1.36	0.73±0.01	**6.55±0.03**
GAG	152.1±2.50	3.56±0.11	**5.82±0.18**
Hm – I	143.6±2.42	2.88±0.20*	**6.05±0.13**
Hm – II	136.1±2.37**	1.98±0.17***	**5.49±0.13***
Hm – III	89.00±5.31***	0.93±0.01***	6.53±0.13**

Hm – I = 250 mg/kg, Hm – II = 500 mg/kg, Hm – III = 1000 mg/kg

Values are expressed as Mean ± S.E. (N = 6), *p<0.05, **p<0.01***P< 0.001

The results revealed that signs of nephrotoxicity manifested by significant increases in BUN 94% and CRE 78% in rats administrated with gentamicin alone treated group as compared to control group of animals, however, significantly approach to the normal level BUN 75% and CRE 68% on co-administration with the herbal mixture at the dose 1000 mg/kg b.wAcute toxicity studies are the platform for risk detection phase of safety evaluation. As per the OECD guideline 425. (OECD 2004)

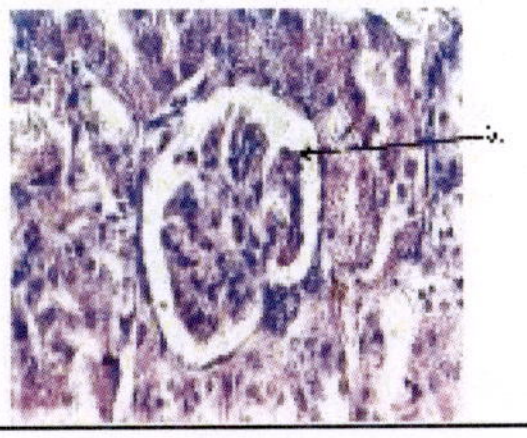

Fig 1: T.S. of kidney of Control group(15th Day).GL = Glomerulas (100X; H-E Stain)

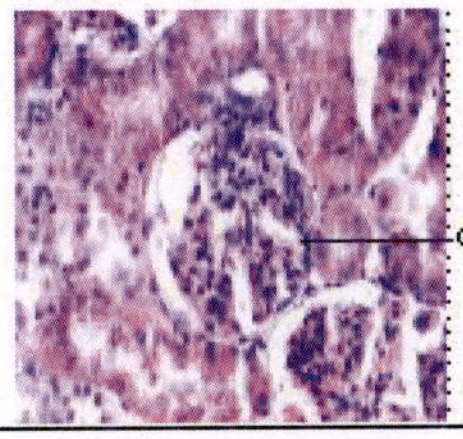

Fig 2: T.S. of kidney of CMCgroup (15th Day). Normal Glomerulus (GL) (100X; H-E Stain)

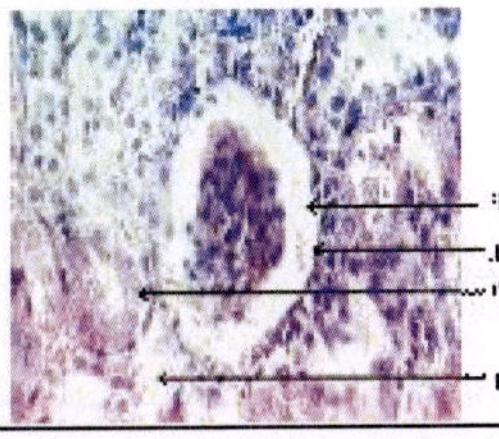

Fig 3: T.S. of kidney of gentamicin 80 mg/kg b.w. group (15th day). Tubular Necrosis (TN), Increased Urinary Space (IUS), Loss of Epitheilal Lining (LEL) and Inconspicuous Nucleoli (IN). (100X; H-EStain)

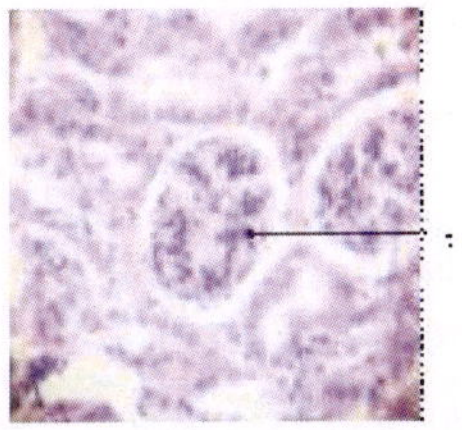

Fig 4: T.S. of kidney of Hm 250 mg/ kgb.w. group (15th Day). Inflammatory cell (IC), (100X;H-EStain)

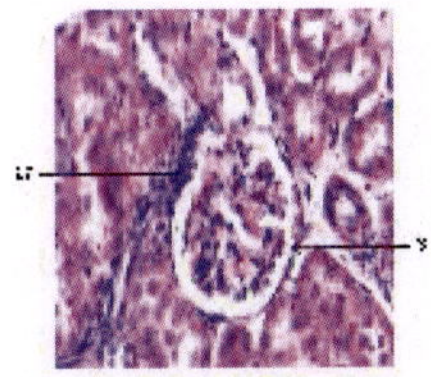

Fig 5: T.S. of kidney of Hm 500 mg/kgb.w. group (15th Day). Restored Epitheilal Lining (REL) and Nuclear Debris (ND). (100X;H-E Stain)

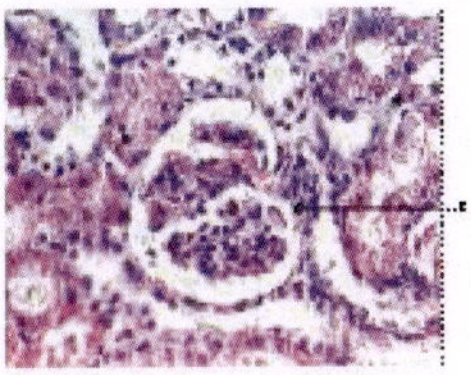

Fig. 6: T.S. of kidney of Hm 1000 mg/ kg b.w. group (15th Day).Normal Glomerulus (GL) (100XH-E Stain)

Histological examination of kidney reveled normal structure GM induced histomorphologal changes in kidney of rat. After 14 days of treatment histopathological examination of kidney section of rat revealed normal architecture in control group of rats Fig.1 Whereas, severe and generalized tubular changes such as tubular Necrosis, (Vasita and Bhargava 2014) Increased urinary space, Loss of epithelial lining, and Inconspicuous nucleoli seen in gentamicin treated rats. Fig.2 [3] while, Co-administration of herbal mixture at a dose 250mg /kg shows mild tubular necrosis and nuclear debris Fig.3. Whereas, Hm dose increased at 500 mg /kg in male restored epithelial lining and nuclear debris.Whereas, the Hm at a dose of 1000mg/kg showing normal architecture of glomerulus of kidney seen in rats. Fig.4 and 5

CONCLUSION

The present study showed that the polyherbal prepration has nephroprotective activity at the dose level of 500mg/kg orally similar to the standard compound. Also itshows good antioxidant activity. Further investigationof individual compounds and characterization of bioactive compounds responsible for the observed significant efficacy is needed.

REFERENCES

1. Arivazhagan J.J. and Vimalastalin R. (2014). Nephroprotective activity of *Aristolochia indica* Leaf extract against gentamicin induced renal dysfunction. *International Journal of Research in Biochemistry and Biophysics.* 4(2): 13-18.

2. Jain PK, Agrawal RK. (2008). Antioxidant and free radical scavenging properties of developed mono- andpolyherbal Formulations. *Asian J. Exp. Sci.,* 22(3):213-220

3. Matthew JE, Mantri A,Vachala SD, Srinivasan KK, Movaliya V. (2005). Effect of *Sphaeranthus indicus* ethanol extract on tissue antioxidant activity in gentamicin induced nephrotoxic rats. *Herbaprolonica,* 5(4):87-95.

4. Padi R., Rajalakshmi I., Indumathy, Kavimani S. (2011). Nephroprotective medicinal Plants. A Review *International Journal of Universal Pharmacy and Life Science.* 1(1):266-274.

5. Talele BD, Mahajan RT, Chopda MZ, Nemade NV. (2012). Nephroprotective plants: A review. *Int J Pharm Pharm Sci* 4:8-16

6. Vasita A. and Bhargava S. (2014). Effect of different leaf extracts of *Bombax ceiba* on gentamicin induced nephrotoxicity in albino rats. *IJARPB* 4(1):1-7.

7. World Health Organization (2004). WHO Guidelines on Safety Monitoring of Herbal Medicines in Pharmacovigilance Systems.

Current Status of Researches in Biosciences (2020) : 361-367
Editor : Prof. P.C. Joshi, N. Joshi, Reshman Yasmin, and D. K. Mansotra
Today & Tomorrow's Printers and Publishers, New Delhi - 110 002

CONTROLLED BREEDING AND REPRODUCTIVE MANAGEMENT IN BUFFALOES - USING EAZI BREED CIDR

S. S. Hiremath

Karnataka Milk Federation, Training Cetnre, Rayapur, Dharwad-580009 Karnataka, India.

E-mail: hshivayogi@yahoo.com

ABSTRACT

The profitability of buffalo farming mainly depends upon their regular and efficient breeding. Buffalo reproduction is considerably affected on account of late maturity, poor estrus, symptoms and long post partum intervals. The present study was undertaken with the objective of buffaloes to evaluate the efficiency of EAZI Breed CIDR – an intra vaginal progesterone release device at a controlled rate into the blood stream of buffaloes in relation to estrus and fertility. For the study purpose 500 buffalo cows of age group 4-6 years in 10 villages of Dharwad in Karnataka state in India were randomly selected. True anestus buffaloes were inserted the intra vaginal device EAZI Breed CIDR and remained in the vaginal cavity for 9 days. On 10th day CIDR was removed and the 2 ml of CIDIROL (1 mg estradiol benzoate, InterAg) intramuscular injection was given to all animals. Ninety percent of the animals showed the pronounced estrus symptoms and they were inseminated at mid heat period. Those buffaloes which failed to show estrus were again inserted the device and inseminated after the expression of estrus. After 45-60 days animals are inseminated by AI and NS, later on buffaloes were examined per rectally for pregnancy and observed that 80-85 percent were pregnant and again after 90 days buffaloes were re-examined for reconfirmation of pregnancy. The present study revealed that the use of EAZI Breed CIDR device in buffaloes could be one of the possible ways to improve anoestrus reproductive performance, thus helping in profitable dairy farming with buffaloes.

KEYWORDS: Buffalo cows, Anoestrus, Progesterone, EAZI Breed CIDR. Intra vaginal device, Cidirol injection.

INTRODUCTION

In India, buffaloes contribute to food security through 60 million tones of milk and more than 1 million tones of meat, besides work energy for agricultural purposes (Indian livestock history) They are found in widely differing geographical conditions, which suggests that this species is adaptable to wide range of environmental conditions. Buffalo reproduction is considerably affected on account of late maturity, poor oestrus, symptoms, and long post partum intervals. Besides genetic makeup, several factors such as nutrition, management, environment, physiology, pathology and psychology affects the conception rate in buffaloes both in farm and field conditions and pose serious threat to profitable dairy farming. Ideally, buffalo cow should produce a calf once in 14-16 months. The profitability of buffalo farming mainly depends upon their regular and efficient breeding.

Anoestrus due to ovarian inactivity is considered to be most important cause for lowered fertility in buffaloes and is responsible for tremendous economic losses to the farmers by decreasing milk yield besides the number calves produced in her lifetime. In true anoestrus condition both ovaries are small, smooth, and inactive with the absence of corpus luteum and characterized by cessation of sexual cycle and psychic manifestation of oestrus, this condition is more in buffalo as compared to cows (Tanwar *et al.,* 2003). Chede (1990) reported the pattern of oestrus, oestrus behavior and oestrus synchronization in buffaloes with varying success. Agrawal, *et al.,* (1999) reported progesterone profile in anoestrus cattle using a systemic progesterone regimen-Crestar. Kathiresan *et al.* (1995) observed the influence of ovarian status and lactational stress on induction of oestrus through norgestomet treatment in buffaloes. Totewad *et al.,* (2009) reported induction of oestrus using Cloprostenol by Intra vulvo-sub mucosal route in sub-oestrus buffaloes. Patel *et al.,* (2003) reported induction of oestrus in buffaloes using norgestromet ear implant. Hormonal treatment like, estrogen, progesterone 75 and GnRH alone or in combination has been tried with variable success (Rao and Rao, 1984; Rao, *et al.,* 1985; Hafez, 2000; Singh *et al.*, 2004). Under this background, the present study was undertaken on true anoestrus buffaloes to see the efficiency of EAZI-BREED CIDR intra-vaginal progesterone release device at a controlled rate into the blood stream of animals in relation to oestrus and fertility.

MATERIALS AND METHODS

The present study was conducted at Dharwad milk shed area. Dharwad district is situated in the Western sector of the northern half of Karnataka State. The District encompasses an area of 4263 sq. kms lying between the latitudinal parallels of 15002' and 15084 51' North and longitudes of 73043' and 75085 35' East. A baseline survey was conducted and nutritional and reproductive conditions of each buffalo cow was examined and recorded. The buffaloes which did not show oestrus symptoms for 8-10 months after calving were examined per rectally and inactive ovaries with no cyclic activity were considered as true anoestrus. 500 anoestrus buffaloes aged between 4 to 6 years belongings to 10 different villages with 45-55 buffaloes from each village were selected as experimental animals for the study. These animals were inserted EAZI-BREED CIDR (1.9gms of natural hormone progesterone) device were easily administered using the specially designed applicator. The wings of the device fold inward and were held in the folded position by the applicator. Squeezing of the plunger handle then expelled the device when in the correct position in the anterior of the vagina. When in place, the tail of the device protrudes from the valve lip.This device remained in the vaginal cavity for 9 days. In some animals second insertion was required due to incorrect positioning in the vagina or other animals in the group / cow boys pulling on the tail of the device and thereby removing them. The intra-vaginal progesterone release devices deliver 98 progesterone at a controlled rate into the blood stream of the animals. This progesterone is released by diffusion from a silicon rubber easterner molded over a nylon spine which is shaped to retain the device in the vaginal cavity. Removal at the end of the insertion period was easily achieved by gently but firmly pulling on the tail and the oestrus symptoms were observed. Twenty-four hrs after the removal of device, all the animals were injected with 2 ml CIDIROL (1mg estradiol benzoate, InterAg, Hamilton, Newzealand) and the pronounced external oestrus symptoms were more pronounced. oestrus symptoms were detected both by external symptoms and per rectal examination. Buffaloes showing oestrus symptoms were inseminated during mid heat using fresh semen brought from KMF Nandini sperm station at Hessaraghatta, Bangalore. oestrus was detected by observing behavioural symptoms and confirmed by rectal examination of gentile. oestrus intensity was recorded as intense (standing discharge and mucus flow), intermediate and weak. All the inseminated buffaloes were examined per rectally on 12th and 15th 110 day post artificial insemination, for the presence of corpus luteum. Pregnancy was confirmed by per-rectal examination on 60th day after artificial Insemination while in doubtful cases reexamined on 90th 112 day. Those buffaloes which failed to show oestrus symptoms were repeated again with the same treatment.

RESULTS AND DISCUSSION

On the removal of implants, different external oestrus symptoms were observed (Table 1). Majority of the animals showed hyperemia of vagina and frequent urination. Per cent of animals which showed different oestrus symptoms are depicted in Table 1. Normally buffaloes does not exhibit pronounced external oestrus symptoms, however, after the injection of cidirol (2ml), the external symptoms of bellowing, free flowing mucus, frequent 120 urination and hyperemia of vagina got pronounced (Figure 1).

After the removal of the device and 2 ml of injection of cidirol, all the animals exhibited oestrus with duration of 36 to 48 hrs in intense oestrus 67.8 %, intermediate oestrus 25.8 % and weak 6.4 % as shown in table 2, with highest conception rate 85.16 %, 60.46 % and 40.75 % (table 3) respectively in intense, intermediate & weak oestrus.

The better oestrus symptoms and fertility could be attributed to the EAZI-BREED CIDR cattle device and cidirol injection which stimulated the follicular development and ovulation. The present study findings are in agreement with the earlier reports of Chede (1990) and Nayak *et al.,* (2009) who observed 38.70 % and 32.25 % buffaloes in intense and intermediate oestrus.

The present study showed that buffaloes which are having problem of true anoestrus could be easily treated by the use of EAZI-BREED CIDR device which will improve the reproductive efficiency in buffaloes and economical conditions of the rural dairy farmers. Controlled breeding under scientific management conditions could be one of the possible way to improve reproductive performance of buffaloes.

Table 1. Classification of buffaloes based 14 on the oestrus symptoms in different villages

Sl.No	Name of the village	No of animals	Animals showed oestrus symptoms		
			Intense	Intermediate	weak
1	Byhatti	52	37	13	02
2	Hebbasur	46	28	16	02
3	Kusgal	48	30	15	03
4	Lokur	45	32	12	01

5	Mugali	53	34	15	04
6	Madhanbhavi	56	36	16	04
7	Padesur	54	38	12	04
8	Shirol	54	38	10	06
9	Yadawad	45	32	11	02
10	Yamanoor	47	32	09	06
	Total	500	337	129	32
		Percentage	**67.80**	**25.80**	**6.40**

Table 2. Percent of animals showing 1 the particular oestrus symptom after the injection of cidriol

SL.No	Estrus symptoms	Percentage
1	Hyperemia of vagina	86
2	Frequent urination	83
3	Free flowing mucus	76
4	Bellowing	68
5	Raised tail	63
6	Excitement/Restlessness	62
7	Reduced feed & water intake	50
8	Milk yield low	40
9	Swollen valve	38
10	Allowing mounting	20
11	Licking other animals	12
12	Mounting on other animals	10

Table 3. Conception percentage in buffaloes showing intense, intermediate and weak oestrus signs

	Conception percentage	
Intense	**Intermediate**	**Weak**
28	07	01
23	08	01
25	08	01

27	07	-
30	08	20
30	12	01
32	10	02
34	06	03
28	07	01
30	05	03
287	78	14
85.16%	**60.46%**	**40.55%**

ACKNOWLEDGMENTS

The authors are thankful to Dr. Peter Jellinek, International Livestock Breeding Consultant, Australia, Dr. A. S. Premanath, Managing Director KMF Bangalore, Dr. Dayanand, Director (AH), KMF, Bangalore and Dr. Bernad Earnest (Additional Director) KMF, Bangalore, Dr. D.N.Hegde, (Additional Director) KMF, Bangalore, Sri. Suresh Kulkarni (Additional Director) KMF, Bangalore, for all the help extended for the study.

REFERENCES

1. Agrawal, S. K., Shankar, U., Kumar, U., & Mohan, G. (1999). Studies on induction of ovarian cyclicity and progesterone profile in anoestrus cattle using a systemic progesterone regimen-Crestar. In *XVth 146 ISSAR Convention Compendium* (pp. 21-22).

2. Thesis, Punjabrao Krishi Vidyapeeth, Akola, India. Hafez, E.S.E. (2000). Reproduction in Farm Animals, 7th 149 ed., Lippincott Williams and Wilking Co., Philadelphia.

3. Kathiresan, D., Napolean, D. E., Antoine, J., Dawson, L., & Pattabiraman, S. R. (1995). Influence of ovarian status and lactational stress on the effect of norgestomet, treatment of buffaloes. *The Blue Cross Book*, *4*, 25-27.

4. Patel, D. M., Sarvaiya, N. P., Patel, A. V., Parmar, A. P., & Dugwekar, Y. G. (2003). Induction of estrus and hormonal profile in buffalo treated with norgestomet ear implant. *Indian J. Anim. Reprod*, *24*(1), 67-68.

5. Rao, A. R., & Rao, K. S. (1984). Improved conception rate in buffaloes after administration of receptal. *Indian Veterinary Journal.*

6. Rao, A. N., Sreemannarayana, O., & Rao, K. P. (1985). Oestrous response and fertility in post-partum anoestrous buffaloes treated with a progestagen, pregnant mares' serum gonadotrophin and prostaglandin during the low breeding

season. *Animal Reproduction Science*, *8*(1-2), 129-135.

7. Singh, A., M.S. Saxena and S. Prasad. (2004). Efficacy of Crestar and its combination with folligon on postpartum anestrus buffaloes. *Indian Journal of Animal Research.*, 25(1): 43-49.

8. Tanwar, P. S., Rakha, N. K., & Phogat, J. B. (2003). Challenges in buffalo infertility. *Intas Polivet*, *4*(11), 121-127.

9. Totewad, G. D., Dhoble, R. L., Sawale, A. G., Naik, P. M., & Ambore, M. N. (2009). Induction of Oestrus using Cloprostenol by Intra vulvo-sub mucosal route in Sub-Oestrus Buffalo. *Veterinary World*, *2*(10).

Current Status of Researches in Biosciences (2020) : 369-375
Editor : Prof. P.C. Joshi, N. Joshi, Reshman Yasmin, and D. K. Mansotra
Today & Tomorrow's Printers and Publishers, New Delhi - 110 002

A COMPARATIVE STUDY ON OXIDATIVE STATUS AND TOTAL ANTIOXIDANT CAPACITY OF MILK IN DIFFERENT CATTLE AND BUFFALO BREEDS

Sushma A, Venkataram K, Pooja Ch, Jayasri K, Padmaja K, Eswara prasad P and Raman D*

Department of Vety. Biochemistry, College of Vety. Science, SVVU, Tirupati, AP, India

**NKBC, Chintaladeevi, A.P*

ABSTRACT

The antioxidant activity of milk plays an important role in preventing lipid peroxidation and maintaining the milk quality. Hence, this study was conducted to evaluate oxidative status and Total Antioxidant Capacity (TAC) of milk in samples collected during early lactation period from different breeds of buffaloes and cows located at National Kamadhenu Breeding Centre, Chintaladeevi, A.P. Malondialdehyde (MDA) levels were highest in Jafarabadi and lowest in Banni , whereas TAC (DPPH and DCIP assay) was highest in Banni buffaloe milk and lowest in Jafarabadi. In cattle, MDA levels were highest in Ongole, lowest in Gir whereas TAC was lowest in Ongole and highest in Gir. High TAC may be responsible for lower oxidative stress.

KEYWORDS: Oxidative status, TAC, Breed, Milk.

INTRODUCTION

Milk is a complex biological fluid produced by the mammary glands of all animals. Milk plays a tremendous role in building a healthy society and can be used as a vehicle for rural development, employment and slowing down the migration of the rural population (Sarwar *et al.*, 2002). Milk exhibits various activities such as antioxidant, antimicrobial and immune modulatory effects (Tidona *et al.*, 2011). The ingestion of oxidated or increased lipid peroxidized milk may be a significant problem for newborn baby to maintain oxidative balance, because antioxidant defence mechanisms are still poorly developed. Increased production of ROS in premature infants are involved in causing serious diseases such as necrotizing enterocolitis, chronic lung disease, retinopathy of prematurity and intraventricular haemorrhage. When metabolic reactions are overloaded and/or the antioxidant defence mechanism decreases, the free radicals get accumulated (Juranek and Bezek, 2005). Milk contains plenty of enzymatic and non-enzymatic antioxidant components that probably account for the vital antioxidant protection of the infants at early stages of life against the development of complications induced by oxygen free radicals.

Oxidative stress is the imbalance between the concentration of Reactive Oxygen Species (ROS) and the antioxidant defence mechanisms (Lykkesfeldt and Svendsen, 2007). Lipid peroxidation in milk causes the off-flavour formation that gives the typical rancid odour to milk. Malondialdehyde is considered to be the end product of lipid oxidation and it is most widely used as a marker for peroxidation (Andrei *et al.*, 2009). The bioactive compounds in milk are known to possess putative antioxidant capacity with excellent health promoting properties. These bioactive compounds are mainly represented as vitamin C, vitamin E, beta-carotene, superoxide dismutase, catalase and glutathione peroxidase. These antioxidants help to protect the body against over production of free radicals or ROS a phenomenon commonly associated with oxidative tissue damage (Jakson *et al.,* 2002).Total antioxidant capacity can be used as a novel marker for assessment of antioxidative status.

According to Chrzczanowicz *et al.,* (2008) TAC is modulated either by ROS overload or by antioxidants concentration and it is measured recently using DPPH and DCIP assay. The DPPH assay is more commonly used for determination of TAC in milk (Smet *et al.*, 2008). DPPH assay is an appropriate method and it is used to determine the free radical scavenging activity based on the acceptance of hydrogen atom or electron donated by antioxidants (Zarban *et al.*, 2009) . DCIP method is based on the use of 2,6- Dichlorophenol Indopenol which looks dark blue in oxidized form and colourless in reduced form and it is used to estimate overall redox status by measuring antioxidants such as glutathione. Mann *et al.*, (2016) studied the alterations in free radical scavenging activity in milk samples during different stages of lactation in different cattle breeds and buffaloes and found higher concentration of antioxidants during early lactation period compared to other stages of lactation. Hence the present study was aimed to evaluate the oxidative status and TAC during early lactation among different breeds of cattle and buffaloes.

MATERIALS AND METHODS

1. Collection of milk samples: Milk samples were collected during early lactation stage from cattle (Ongole, Tharparkar, Kangayam, Deoni, Kankrej and Gir) and buffalo (Murrah, Mehasana, Pandharpuri, Banni and Jafarabadi) breeds. Six samples were collected from each breed maintained by National Kamadhenu Breeding Centre, Chintaladeevi, Andhra Pradesh.

2. The milk samples were analysed for TBARS by method of Roy and Deepak (2014). The TAC was estimated by DPPH (Diphenyl Picryl Hydrazyl) method according to Mann *et al.*, (2016) and DCIP (Di Chlorophenol Indophenol) method according to Ciuti and Liguri, (2017)

RESULTS AND DISCUSSION

Oxidative stress is the result of imbalance between the concentration of ROS and the antioxidant defense mechanisms (Lykkesfeldt and Svendsen, 2007). Lipid peroxidation in milk causes the off-flavour formation that gives the typical rancid odour to milk. The unsaturated fatty acids in raw milk may favour oxidative reactions and forms hydroperoxides (Kondyli *et al.*, 2005). These hydroperoxides inturn rapidly decomposes to form a complex mixture of secondary lipid oxidation products such as pentanal, malondialdehyde (MDA) etc., (Manglano *et al.,*2005). Malondialdehyde is considered to be the end product of lipid oxidation and it is most widely used as a marker for peroxidation (Andrei *et al.*, 2009). According to (Fenaille *et al.,*2001) fatty acids with 3 or 4 double bonds produce MDA, a major product of lipid autoxidation. Hence, TBARS (MDA levels) is commonly used to monitor milk products oxidation.

Table 1 and 2 shows different MDA and TAC levels in cattle and buffaloe breeds respectively. The MDA levels were found to be highest in Ongole cow milk (20.4 ± 0.04) followed by Kankrej, Kangayam, Tharparkar, Deoni and lowest in Gir (10.33 ± 0.14). Jafarabadi buffalo milk has highest MDA (31.16 ± 0.47) levels followed by Mehasana, Murrah , Pandharpuri and lowest was observed in Banni milk (11.28 ±0.07). Al-Helaly *et al.* (2013) studied the oxidative status in human, buffalo, cow, sheep and goat mature milk samples and reported higher levels of MDA in buffalo and human milk, whereas lower levels in sheep and goat milk compared to cow milk.Similar results were observed in the present study where the oxidative status was highest in buffaloes compared to cattle. Kristensen *et al.* (2004) studied that the degree of unsaturation in milk fat has its significant influence on oxidative stability of milk as the higher proportion of unsaturated fatty acid results in the inferior oxidative stability of raw milk.

Table1 : Oxidative status and Total Antioxidant capacity of milk in cattle breeds

Breed	TAC (DCIP) (mM of GSH equivalents/lit)	TAC (DPPH) (%)	MDA levels (µmoles/ml)
Deoni	0.12 ± 0.002^{B}	64.50 ± 1.25^{D}	16.78 ± 0.02^{C}
Tharparkar	0.12 ± 0.000^{B}	47.08 ± 0.97^{C}	16.88 ± 0.01^{C}
Kankrej	0.05 ± 0.013^{C}	46.36 ± 1.07^{C}	16.45 ± 0.07^{B}
Gir	0.22 ± 0.004^{D}	74.33 ± 1.40^{E}	10.33 ± 0.14^{A}
Ongole	0.04 ± 0.002^{A}	32.50 ± 0.76^{A}	20.42 ± 0.04^{D}
Kangayam	0.14 ± 0.000^{C}	42.80 ± 1.03^{B}	16.39 ± 0.008^{B}

Means with different superscripts (A, B, C, D, E) with in a column differ significantly ($p < 0.01$).

Table 2: Oxidative status and Total Antioxidant capacity of milk in buffaloe breeds

Breed	TAC (DCIP) (mM of GSH equivalents/lit)	TAC (DPPH) (%)	MDA levels (µmoles/ml)
Murrah	0.14 ± 0.000^{C}	53.80 ± 1.41^{C}	17.53 ± 0.10^{C}
Banni	0.15 ± 0.001^{D}	66.80 ± 1.12^{D}	11.28 ± 0.07^{A}
Mehasana	0.14 ± 0.001^{C}	53.33 ± 1.33^{C}	19.19 ± 0.04^{D}
Pandharpuri	0.03 ± 0.003^{B}	46.00 ± 0.60^{B}	12.31 ± 0.10^{B}
Jafarabadi	0.01 ± 0.002^{A}	28.51 ± 0.40^{A}	31.16 ± 0.47^{E}

Means with different superscripts (A, B, C, D, E) with in a column differ significantly ($p < 0.01$).

Cervellati *et al.*, (2014) reported that TAC assessment is an established methodology to measure different elements of antioxidant defence system. The presence of antioxidants helps to neutralize the ROS and prevent the formation of free radicals and inhibit the lipid peroxidation, thus reducing the severity of oxidative stress (Mann *et al.*, 2016). Higher antioxidant potential can protect against undesirable effects of ROS and prevents disease incidence whereas, the lower TAC may be due to the depletion of the antioxidant barrier as an effect of long term oxidative stress (Chrzczanowicz *et al.*, 2012). Antioxidant systems of milk can effectively inhibit superoxide radical, hydroxyl radicals and peroxide radicals as reported by Zarban *et al.,* (2009). According to Usta and Yilmaz-Ersan, (2013) casein, whey, sulphur containing amino acids, selenium, zinc, catalase, glutathione peroxidase, superoxide dismutase, vitamin E, C and beta-carotene has antioxidant activity in milk. The anti-oxidative status of biological samples can be analysed by the determination of single components of the system or by measurement of TAC. Due to the fact that antioxidative systems can act synergistically, a single analysis may not exactly reflect the total potential.Hence in the present study for measuring the TAC different methods (DPPH and DCIP) were adopted to compare and support the results obtained.

The TAC was found to be highest in Gir (0.022 ± 0.004, 74.33 ± 1.40) and it is lowest in Ongole cow milk (0.04 ± 0.002, 32.50 ± 0.76) by DCIP and DPPH assays respectively. Among buffaloes, Banni milk has highest TAC (0.15±0.001; 66.80±1.12) followed by Mehasana, Murrah and Jafarabadi has lowest TAC (0.01±0.002; 28.51±0.40) by DCIP and DPPH asssays respectively where similar trend was observed by both the methods. Our results were in accordance with Ryan and Petit (2010), who found that bovine whole milk with higher fat content showed high antioxidant potential.

It was concluded that higher TAC of milk may be responsible for low oxidative status during early lactation stage in different breeds of cattle and buffaloes.

REFERENCES

1. Andrei, S., Pintea, A., Bunea, A., Groza, I., Bogdan, L., Ciupe, S., Matei, S., and Crainic, D. (2009). Non-enzymatic antioxidants concentration and lipids peroxidation level in milk from cows with subclinical mastitis. *Veterinary Medicine*, 66(1):196-201.

2. Al-Helaly , L. U., Rashed, S.H., and Bdaiwi, L. F. (2013). A comparative study of oxidant and antioxidant levels between human milk with othr types of ruminant nimals. *Iraqi National Journal of Chemistry*, 49:86-99

3. Cervellati, C., Romani, A., Seripa, D., Crmonini, E., Bosi, C., Magon, S., Bergamini, C.M., Valacchi, G., Pilotto, A., and Zuliani, G. (2014). Systemic oxidative stress and conversion to dementia of elderly patients with mild cognitive impairment. *Bio Med Research International*, 2014 :1-7.

4. Chrzczanowicz, J., Gawron, A., de Graft- Johnson, J., Krajewski, W., Krol, M., Markowski, J., Kostka, T., and Nowak, D. (2008). Simple method for detrmining human serum 2,2-diphenyl-1-picryl-hydrazyl (DPPH)radical scavenging activity – possible application in clinical studies on dietary antioxidants. *Clinical Chemical Laboratory Medicine* ,46(3):343-349..

5. Chrzczanowicz, J., Gawron-Skarbek, A., Kostka, J., Nowak, D., Drygas, W., Jegier,

A., and Kostka, T. (2012). Physical activity and total antioxidant capacity across an adult lifespan of men. *Medicine and Science in Sports and Exercise*,44(4):575-582

6. Ciuti, R. And Liguri, G. (2017). A Novel Assay for Measuring Total Antioxidant Capacity in Whole Blood and Other Biological Samples. *Journal of Biomedical Science and Engineering,*10,60-76.

7. Fenaille, F., Mottier, P., Turesky, R.J., Ali, S., and Guy, P. A.(2001). Comparision of analytical techniques to quantify malondialdehyde in milk powders. *Journal of Chromotography* A, 921(2):237-245.

8. Jackson, M. J., Papa, S., Bolanos, J., Bruckdorfer, R., Carlsen, H., Elliott, R. M., Flier, J., Griffiths, H. R., Heales, S., Holst, B., Lorusso, M., Lund, E., Moskaug, J. O., Moser, U., Di-Paola, M., Polidori, M. C., Signorile, A., Stahl, W., Vina Ribes, J., and Astley, S. B. (2002). Antioxidants, reactive oxygen and nitrogen species, gene induction and mitochondrial function. *Molecular Aspects of Medicine*, 23(1-3):209-285.

9. Juranek, I., and Bezek, S. (2005). Controversy of free radical hypothesis: reactive oxygen species-cause or consequence of tissue injury. *General physiology and biophysics*, 24:263-278.

10. Kondyli, E., Andersen, M. L., and Skibisted, L. H. (2005). Evaluation of oxidative

11. stability of goat milk by Electron Spin Resonance Spectroscopy (ESR). *Milchwissenschaft*, 60(3):271-274.

12. Kristensen, D., Hedegaard, R.V., Nielsen, J. H and Skibisted , L.H. (2004). Oxidative stability of buttermilk as influenced by the fatty acid composition of cow s milk manipulated by diet. *Journal of Dairy Research*, 71(1):46-50

13. Lykkesfeldt, J., and Svendsen, O. (2007). Oxidants and antioxidants in disease: Oxidative stress in farm animals. *The Veterinary Journal*, 173(3):502-511.

14. Mann, S., Shandilya, U. K., Sodhi, M., Kumar, P., Bharti, V. K., Verma, P., Sharma, A., Mohanty, A., and Mukesh, M. (2016). Determination of antioxidant capacity and free radical scavenging activity of milk from native cows (*Bos indicus*), exotic cows (*Bos taurus*), and riverine buffaloes (*Bubalus bubalis*) across different lactation stages. *International Journal of Dairy Processing and Research*, 3(4):66-70

15. Manglano, P., Lagarda, M. J., Silvestre, M. D., Vidal, C., Clemente, G., and Farre, R. (2005). Stability of the lipid fraction of milk-based infant formulas during storage. *European Journal of Lipid Science and Technology*, 107(11):815823

16. Roy, T., and Deepak ,D. (2014). Antioxidant properties of milk oligosaccharides from various ruminants. *International Journal of Pharm Bio-Sciences*, 5 (2):400-408.

17. Ryan, L., Petit, S. (2010). Addition of whole ,semi-skimmed ,and skimmed bovine milk reduces the total antioxidant capacity of black tea. *Nutrition Research*, 30(1):14-20

18. Sarwar, M., Khan, M. A., and Mahr-Un-Nisa, Z. I. (2002). Dairy industry in Pakistan: A Scenario. *International Journal of Agricultural Biology*, 4:420-428.

19. Smet, K., Raes, K., De-Block, J., Herman, L., Dewettinck, K., and Coudijzer, K. (2008). A change in antioxidative capacity as a measure of onset to oxidation in pasteurized milk. *International Dairy Journal*, 18(5):520-530

20. Tidona, F., Sekse, C., Criscione, A., Jacobsen, M., Bordonaro, S., Marletta, D., and Vegarud, G. E. (2011). Antimicrobial effect of donkey's milk digested invitro with human gastrointestinal enzymes. *International Dairy Journal*, 21(3):158-165.

21. Usta, B., and Yilmaz –Ersan, L. (2013).Antioxidant enzymes of milk and their biological effects. *Journal of Agricultural Faculty*,27(2):123-130.

22. Zarban, A., Taheri, F., Chahkandi, T., Sharifzadeh, G., and Khorashadizadeh, M. (2009). Antioxidant and radical scavenging activity of human colostrum, transitional and mature milk. *Journal of Clinical Biochemistry and Nutrition*, 45(2):150-154.

Current Status of Researches in Biosciences (2020) : 377-390
Editor : Prof. P.C. Joshi, N. Joshi, Reshman Yasmin, and D. K. Mansotra
Today & Tomorrow's Printers and Publishers, New Delhi - 110 002

CONVENTIONAL ELECTRON MICROSCOPY - A WEAPON FOR RESEARCH AND DISEASE DIAGNOSIS

Mekala Lakshman

Ruska Labs Department of Veterinary Pathology, College of Veterinary Science,

PVNarsimha Rao telangana Veterinary University (PVNRTVU) Rajendanagar, Hyderabad- 500 030. T.S. India.

Email: mekala_bry@yahoo.com

ABSTRACT

The conventional Electron Microscopy {(EM -Transmission Electron Microscope (TEM) and Scanning Electron Microscope (SEM)} is one of the gold standard techniques to study the patho-morphological changes in cells of different diseases in the field of life science. Examination of tissues / cells under EM is otherwise called as "cell pathology" which deals with all membranous structural deviations, localization of disease etiological agents {Bacteria, virus, virus like particles (VLP's), fungi, rickettesia like organisms (ROL's), mycoplasma like organisms (MLO's) and parasitic infections}. Visualization of localized and /or attached pathogens and therapeutic nanoparticles in different structures of the cells of different experimental animals and out breaks of different diseases in animals is possible only with EM studies. Despite of accessibility to various laboratory techniques for accurate diagnosis and specific pathogen identification but, the EM technique is having its significance to support all molecular techniques due to the meticulous observation of all sub-cellular structural changes in response to the disease causing agents which target different structures in a cell or group of cells. The conventional EM is a multi-stepped, time consuming and expertise technique which needs to have continuous practice and up-date. In general, the diseases in animal diseases are wide range beginning from nutritional, bacterial, viral, neoplastic, parasitic, RLO's MLO's and many other conditions are being recorded all over the Globe. Most of the diseases are being diagnosed on the basis of pathognomonic lesions and conventional light microscopy (LM) techniques like histopathology (HP), immunohistochemistry (ICH) or immunocytochemistry (ICC) and flow cytometry (FCM). Advanced diagnostic tools like qRT-PCR techniques, Genomics and Proteomics were also being used. But demonstration of minute pathogens like virus and sub cellular structural alterations due to disease process is possible only with EM.

KEYWORDS: TEM, SEM, LM, IHC, HP, FCM, qRT-PCR, RLO.

INTRODUCTION

In general, animal diseases are wide range beginning from nutritional, bacterial, viral, parasitic, mycotoxic, neoplastic, and management related abnormalities are being recorded all over the Globe. Most of the diseases are being diagnosed on the basis of path gnomonic gross lesions and conventional techniques like histopathology & IHC by using light microscope (LM) and also through application of advanced tools like QRT-PCR techniques, microarray, genomics and proteomics etc. Demonstration of minute pathogens like virus, and sub cellular structural alterations due to disease process are not possible with conventional methods and advanced tools, but it is possible only with Electron Microscopy (EM), which is rapid (direct sample analysis for virus), specific, and accurate with high resolution[(6,7,11, 12,13,14,21,22)]. However, it is necessary to value every tool used in the field of diseases diagnosis which are complementary each other and ancillary to the core techniques. Different viral pathogens like avian influenza, avian infectious bronchitis, avian pox, Newcastle disease, infectious bursal disease, chicken anemia, mycotic granuloma11 etc., were demonstrated under EM by different authors[(2,4,6,7,11,15,16,20)] in avian species. The pathogenic bacteria, mycoplasma (MLO's), probiotics attachment with host cells (epithelial cells of respiratory and gastrointestinal tracts), localization of therapeutic nanoparticles in aquatic animals[14] and sub cellular alterations are being demonstrated in experimental specimens and in tissues of natural out breaks at Ruska Labs [(13,14)]. Different types of viruses were also demonstrated under EM in different specimens like plant tissue, experimental cell lines, tissue homogenates, experimental embryonic fluids, stool, tracheal swabs; intestinal fragments. Apart from this the complexity of multiple pathogen interaction with host cell or experimental cell lines is possible only through EM. Hence, it is more relevant to brief about EM and different methods used in EM specimen preparation like resin embedding method and direct electron microscopy (rapid preparation/negative staining technique) [(17)] may benefit the heterogeneous Scientific Professionals assembled under one shade to boost Indian GSDP through increased production performance of livestock by curtailing the spread of pathogens and adopting preventive measures by using available diagnostic techniques.

1. Introduction to Electron Microscopy (EM): Electron Microscopy (EM) is a dynamic specialized tool to study the sub cellular structures and surface morphology of biological and non biological specimens by using a beam of electrons [1, 10, 11, 12, 13, 14]. In Light Microscopy (LM) visible light and optical (glass) lenses are used as a source of illumination to magnify the specimens (10 to 1,000). The EM operates in vacuum and electron beam (e.g., λ = 0.005 nm) will act as a source of light to magnify the processed specimen through electromagnetic lenses when compared to the wavelengths of visible light (e.g., λ = 400 nm to 700 nm) [5]. In EM, resolving power is inversely proportional to the wavelength. In other words, increasing the velocity of electrons results in a shorter wavelength and increased resolution [5]. Research in development of electron microscopes began in the year 1920s. Under the guidance of Max Knoll, Ernst Ruska began work on the development of electron lenses (Germany, 1928). The first functional TEM was developed in the early 1930s by Ruska for which he was honored with Nobel Prize in Physics in 1986 [1, 5, 9, 18]. Early EM studies primarily focused on optical behavior of electron beams under various conditions. Thus, no biological applications were initially envisioned. However, due to superior magnifying power of an EM it soon became clear that they could be applied to the study of various biological specimens [9, 17, 18, 19]. Presently EM remains an important and significant tool in diagnostic ultrastructural pathology besides physical and material science [9, 18, 21].

There are two basic types of electron microscopes (TEM and SEM) were invented within the same decade (SEM was invented by Manfred von Ardenne in 1938), but they differ fundamentally in their usage [9]. In brief, the TEM projects electrons through an ultrathin section of the specimen and produces a two dimensional image (up to 1000kx) while SEM generates three dimensional image (up to 100kx) image with the help of secondary electrons. Extreme high magnifications above 200,000 are rarely used by biologists [3]. EM allows investigators to detect the specimens in much greater detail than those examined under LM. Conventional electron microscopy is used today in many research laboratories, which are attached with computers to make a digital photography with CCD (charge coupled devise) which makes analysis easy [9, 10]. There is ample evidence in the published literature that EM has significantly contributed to our understanding of the ultrastructure of a variety of specimens, including those of pathogenic and nonpathogenic agents [19].

2. Sample preparation protocols and procedures at Ruska Labs: Standard protocols practiced at Ruska Laboratory are familiar with many users and gained expertization in the field of EM studies and helped more than 4300 PG and PhD thesis (Veterinary, Agriculture, Horticulture, and other Life Science courses) apart from routine diagnostic and research work of different samples belongs to ICRISAT, ICAR, CISR, ICMR, IIT's, VIT, CLRI, IISc, Biotech Institutes, Different Vaccine (Human and animal) production MNC's, Corporate Hospitals (Diagnostic and R&D) from different parts of India. Thousands of samples were analyzed and hundreds of publications were supported by the Ruska Labs with the following sample preparation practices.

Resin embedding method protocols (multi-stepped protocol) for Transmission Electron Microscopy (TEM): Specimen preparation for TEM includes eight major steps (multi-stepped): slicing & cleaning, primary fixation, washing, secondary fixation, dehydration, infiltration with a transitional solvent & resin and embedding, polymerization, sectioning and staining.

a) **Slicing and cleaning the surface of the specimen:** Proper cleaning of the surface to remove variety of unwanted deposits if any (not removed may get permanently fixed) otherwise impossible to remove later [1]. Bozzola and Russell suggested that the specimen should be quickly washed or rinsed in a suitable buffered solution (physiological pH) by the way of gentile swiping or washing for three times for 10 minutes at room temperature. The size of the specimen should be in mm^3.

b) **Primary fixation or stabilization of the specimen:** There are various stabilizers / fixatives (Aldehydes, osmium tetraoxide, tannic acid, or thiocarbohydrazide) are available to achieve this step [1,19]. For all biological samples including poultry simple chemical fixation is advocated (0.1M PBS/cacodylic buffer based 2.5% glutaraldehyde solution 10 times to the volume of tissue) and can be stored at room temperature or 4 °C overnight / few hours / few days / few weeks / few months [1,19].

c) **Washing /Rinsing of the specimen:** In order to remove excess fixative from the samples, the specimen should be thoroughly washed twice or thrice with 0.1 M PBS/cacodylic acid buffer (pH 7.3) for 20 minutes [1].

d) **Secondary fixation / postfixation / staining of the specimen:** Secondary fixation is crucial to stain the specimen (block color) and to protect the specimen during other steps employed such as embedding, sectioning. This step also helps in avoiding artifacts and helps in conductivity of electrons to generate quality image [(1)]. All biological samples can be successfully stabilized for TEM investigation by post fixation with 1% aqueous osmium tetraoxide or 0.1M cacodylic acid buffer (pH 7.3) for 1 to 2 hrs at room temperature (rehydration). It is important to realize that no fixation procedure is ideal; any type of fixation is likely to cause some alterations in the specimen [(1)].

e) **Dehydration of the specimen:** The rehydrated sample should be dehydrated in a graded series of ethanol. More specifically, the following protocol is useful: A series of 50, 70, 80, 90, and 99.9 percent ethanol, each for 45 minutes at room temperature. If time does not permitting these can be stopped at 70% level and stored at 4^0C. This process allows the water in the samples to be slowly exchanged through liquids with lower surface tensions [(1, 10)].

f) **Infiltration of the specimen with a transitional solvent:** The ethanol is not miscible with the plastic embedding medium hence it should be replaced with another intermediary solvent like propylene oxide is necessary[(1)]. Immersion in propylene oxide twice for 20 minutes at room temperature. Alternatively 1:1, 1:2 and 1:3 ratio of ethanol or acetone and resin can be used each for 30 minutes finally replaced with pure resin for overnight and keep it in 4^0C. This process facilitates better sectioning and EM details will be more clear [(1,10)].

g) **Infiltration with resin and embedding of the specimen:** All biological specimens can be embedded in a variety of different media depending on the purpose (conventional TEM or immuno TEM). For conventional TEM, the epoxy resin EMbed, araldite or LR white or Durcupan ACM are quite suitable. Next day of post transitional step, the specimens should be immersed in a freshly prepared pure resin and left for 2 hrs at room temperature followed by embedding (free of air bubbles), block making and keep it for polymerization at 500C to 600C for 48 to 72 hrs. Store the samples in desiccater for 3-5 days or place it for 1-2 weeks at room temperature which will improve the subsequent trimming and sectioning quality as the resin blocks continue to harden during this time[(1)].

h) Sectioning and staining of the specimen: The procedure of cutting of the specimens into semi thin and ultrathin slices (sections) is known as microtomy and ultramicrotomy, respectively [(1)]. Semi thin sections (about 600 to 800 nm) should typically be stained with toluidine blue for 1 min on a hot plate (700 C to 900 C), examined under LM, and used for identifying the specimen within the resin block before proceeding with ultramicrotomy. Ultrathin sections (about 50 nm to 60 nm) should typically be stained with saturated uranyl acetate (20 minutes) followed by lead citrate (for 5 minutes)[(1)]. As mentioned before, TEM sample preparation is multi-stepped; every step can virtually affect the quality of the final electron micrograph. It is therefore important that the expert should plan and execute meticulously. I believe that these procedures involve a significant time commitment and require knowledge, patience and skills that come only through practice. It is important to note that most of the chemicals used in EM are dangerous. All steps should be under fume hood on orbital shaker. Bozzola and Russell wrote an laudable chapter on safety in the EM laboratory. They emphasized on the importance of training in the proper usage of all equipments and reagents in the EM laboratory. They also mentioned that the investigator must be aware of potential hazards such as fire, chemical, electrical, and physical harm associated with these items. EM facilities usually offer training and orientation programs all over the world. In my opinion it is exceedingly recommend not only for the apprentice in EM (first learning) but also for the experienced investigator (continuous learning).

4. Scanning electron microscopy: Scanning Electron Microscopy (SEM) is also a powerful tool for investigation of surface structures and multiple pathogen interactive studies of samples. This technique is advantageous to observe a large area of field at different focal points [(5)]. SEM also facilitates relatively wide range of magnification allowing the investigator to easily focus in/on an area of interest on a specimen that was initially scanned at a lower magnification. Furthermore, the 3-D images may be more appealing to the human eye than the 2-D images (obtained with TEM). Therefore, an investigator may find it easier to interpret SEM images. Finally, the number of steps involved for preparing specimens for SEM investigation is lower and less time consuming than the TEM sample preparation. However, SEM specimen preparation harbors various risk factors that can easily distort the integrity and ultrastructure of the sample. The basic steps involved in SEM sample preparation include thin slicing, surface cleaning, stabilizing the sample with a fixative, washing /rinsing, dehydrating, drying, mounting of the specimen on a metal holder over double sided carbon conductivity tape, and coating the sample with a layer of a material(e.g., gold, gold-palladium or platinum) that is electrically conductive[1]. Because each of these steps are crucial and will affect the outcome of the study, they are described in more detail below. The first four steps are essentially the same as those described for TEM specimen preparation. These steps are therefore only briefly mentioned below.

a) **Slicing and cleaning of the surface of the specimen:** As discussed earlier, the best way to clean the surface of biological samples from contaminants is to carefully washed or rinsed. The sample size should in cm^3.

b) **Fixation / Stabilizing of the specimen:** All biological samples can be chemically prefixed by immersing the specimens in a 2.5% glutaraldehyde solution prepared in 0.1 M PBS /cacodylic acid buffer (pH 7.3) and stored as that of TEM samples.

c) **Washing / rinsing of the specimen:** In order to remove excess glutaraldehyde from the samples, should be subjected to a thorough washing or rinsing procedure in 0.1 M PBS /cacodylic acid buffer (pH 7.3) as explained for TEM sample preparation.

d) **Dehydration of the specimen:** The dehydration process of a biological sample needs to be done very carefully. It is typically performed with either a graded series of acetone or ethanol. The protocol that proved most suitable for dehydrating the biological specimens for SEM includes the immersion of the specimens in 50% , 70% , 80% , 90% and 100% acetone (dried with $CaCl_2$) of each for 45 minute at room temperature under fume hood on orbital shaker. This process allows the water in the samples to be slowly exchanged through liquids with lower surface tensions [1,5,10,11,12,13,14].

e) **Drying of the specimen:** The scanning electron microscope also operates in a vacuum, for which specimens must be dry otherwise the sample will be destroyed in SEM chamber. Many electron microscopists consider a procedure called the Critical Point Drying (CPD) as the gold standard for SEM specimen drying by using liquid carbon dioxide in which specimen is dried without structural damage [1,10]. It is very important to follow exact instructions of the manufacturer of the CPD apparatus, to avoid significant structural alterations. In my experience, I tried a specimen drying process called Simple Desiccation (SD) and also Vacuum Desiccation (VD), which are giving superior results as excellent as that of CPD. This technique is essentially a simple air-drying procedure after fixation, rinsing, and dehydration of the specimens. SD/VD is risky as the specimen may collapse, flatten, or shrink or be rolled and become uncontrollable under these conditions(1,10). Although SD / VD is faster and cheaper, this method is like *"walking on a tight rope."* For an EM beginner, I would suggest that the safer method is CPD. I would like to recommend the same procedure (But I have performed it only few times by using liquid carbon dioxide as the transitional fluid).

f) **Mounting of the specimen:** After complete drying of the samples, they must be mounted on metallic (Aluminum) stubs using a double sided sticky carbon conductivity tape. It is important that the investigator first decides about the best orientation of the specimen to be mounted before placing on carbon conductivity tape. A re-orientation proves difficult and can result significant damage to the sample. If it is not properly stuck to the tape, a drop of pure silver paste can be used for its adherence.

g) **Sputtering (coating) of the specimen:** The idea of coating the specimen is to increase its conductivity in the SEM and to prevent the build-up of high voltage charges on the specimen by conducting the charge to the ground [1]. Typically, specimens are coated with a thin layer of (approximately 20 nm to 30 nm) a conductive metal (e.g., gold, gold-palladium, or platinum) for 180 seconds. In the Ruska Lab's, (College of Veterinary Science, PVNRTVU, Hyderabad, T.S), using gold, and found it most suitable. To guarantee best results (i.e., to achieve an even layer of metal coating over the sample), one should carefully follow the instructions of the sputter coater manufacturer. It is important to remember that each step has to be performed to perfection in order to achieve quality images that can be interpreted without the influence of artifacts caused by specimen handling.

5. Suspensions for EM [8,17,18,19]

a) Centrifuge the suspension at a speed that will yield a solid pellet of the material under study.

b) Add the fixative slowly down the wall of the tube taking care not to dislodge the pellet. 3.Allow to fix for 10 min at room temperature and then release the pellet using a wooden

c) cocktail stick and leave for a further 20 min. The material can now be treated as tissue blocks

d) If the pellet resuspends, the pellet can be recentrifuged after each part of the process. Or:

e) Resuspend in 1% low-gelling temperature agarose (37°C) in buffer, centrifuge to pellet, cool and cut into blocks and then proceed like embedding method for TEM and mounting on carbon conductivity tape stubs for SEM.

6. Direct Electron microscopy - negative/ positive staining: Different materials like intestinal contents, stool, embryonic fluids etc., were demonstrated under direct EM by various methods described by different authors [8,15,16,19,20,21,22] by using negative sating and positive stains.

a) A drop of about 10 μL of the virus suspension to be studied is applied to the hydrophobic surface of a parafilm square in a Petri dish.

b) A formvar-coated grid is floated onto this drop for one minute, with the formvar side of the grid in contact with the liquid.

c) The excess liquid is removed from the grid by touching its border with a cut piece of filter paper.

d) The grid is immediately floated in a drop of 1.5% phosphotungstic acid, 2% ammonium molybdate or 1% aqueous uranyl acetate, depending on the specimen.

e) For a better assessment of the samples, two or three grids should be prepared, each stained with one of these stains.

f) After staining for one minute, the excess stain is removed with filter paper and the grid left to dry for a few minutes, before insertion into the microscope column.

7. General precautions in EM for interactive studies

Many factors will influence the EM studies of multiple interactive specimens:

a) Laboratory procedures need to be carefully reviewed as some drugs can affect the function and structure of host cells.

b) The procedures for the establishment of cell cultures and explants cultures can influence the EM data (e.g., improper collection of in-vivo/biopsy material by using surgical procedures)

c) Most host cells have surface exposed structures (e.g., cilia and microvilli) that can be easily damaged during handling; this damage may lead to misinterpretation of EM data.

d) Infection of host cells with specific pathogens may results in accumulation of exudates and cell debris which requires a balanced washing protocol.

Pathogen and host cell interactive studies may provide a voluminous knowledge about pathogenesis of any disease or apoptotic/necrotic changes due to ageing of the cell. The noteworthy feature in multiple pathogen interactive studies is that there is a direct relation between membranous structures of the host cells and pathogens which is very sensitive. Hence, proper care should be taken during processing of samples for EM studies.

CONCLUSIONS AND FUTURE OUTLOOK

A descriptive approach of interactions between pathogens and host cells besides an exhaustive ultrastructural examination of different types of biological samples would not be possible without the electron microscope (EM). Both TEM and SEM have proven over the years to be valuable paraphernalia in this regard. Although TEM generates a different set of electron images than SEM, and thus provides different scientific data, the combination of these two methods in a single investigation can be extremely powerful besides other advanced techniques. Each technique will create its own milestones in the field of science hence; no technique should undermine the other. As per my experience, EM is not any easy technique unless one should train and gained handful experience to become expert which is possible with continuous practice only. But any biologist essentially a pathologist should have complete idea about EM techniques to narrate healthy and diseased cells. EM techniques are gaining pivotal importance in cancer biology, tissue culture biology, toxicological studies and many other emerging and challenging fields. There are many possibilities that an investigator may not get usable electron micrographs for interpretation. The reason behind this is multi-step specimen preparation for EM; a single mistake in one of these steps will affect all remaining steps, and outcome of entire work will go futile exercise. Despite the risk factors involved, TEM and SEM techniques provide fascinating images of biological specimens, in particular of the smallest free- living and self-replicating life forms on planet

Earth. Finally, I wish that diversified specialists should enter the field of electron microscopy as this would allow exchanging of exigent ideas and thoughts. There are many new recent advances in electron imaging technology providing numerous new tools for viewing and characterizing pathogens (e.g., TEM with 3D tomography, Cryo EM and STEM). This should be enough reason to attract young scientists who develop an interest in *"playing"* with these powerful machines and applying the generated data to the fields of life science and pathology to address the challenges in the field of disease diagnosis (Both animal and Human as a part of One Medicine concept).

REFERENCES

1. Bozzola, J. J. and Russell, L. D. (1992). Electron Microscopy. 2nd ed. Jones and Bartlett Publishers Inc., Boston.

2. Catroxo, M. H. B. Martins, A. M. C. P. F. Petrella, S. Milanelo, L. Aschar, M. Souza, F. Nastari, B.D.B.Souza, R.B. (2012). Avian Paramyxoviruses. Detection by Transmission Electron Microscopy Techniques. *Int. J. Morphol.* 30(2). 723-730.

3. Dykstra M. J. and. Reuss, L. E. (2003). Biological Electron Microscopy, Kluwer Academic/Plenum Publishers, NewYork.

4. Enver Beytuti, Mehmet Haligur. (2007). Pathological, Immunohistochemical, and Electron Microscopic Findings in the Respiratory Tract and Skin of Chickens NaturallyInfected with Avipox virus. *Turk. J. Vet. Anim. Sci.*31-5. 311-317.

5. Flegler, S. L. Heckman J. W. Jr. and. Klomparens, K. L. (1993).Scanning andTransmission Electron Microscopy, W.H. Freeman and Company, New York.

6. Faruku Bande, Siti Suri Arshad, Abdul Rahman Omar, Mohd Hair Bejo, Muhammad, Salisu, Abubakar, and Yusuf Abbaa. Pathogenesis and Diagnostic Approaches of Avian Infectious Bronchitis. (2016). Advances in Virology. Article ID 4621659, 1-11.

7. Gough, R.E. Alexander, D.J. Collins, M.S. Lister S.A. and Cox W.J. (1988). Routine virus isolation or detection in the diagnosis of diseases in birds. *Avian Pathology.* 17. 893-907.

8. Goldsmith, C.S. Miller S.E. (2009). Modern uses of electron microscopy for detection of viruses. Clin. *Microbiol Rev* .22. 552-563.

9. Haguenau, F. Hawkes, P. W. Hutchison, J. L. Satiat-Jeunemaître, B. Simon G. T. and Williams, D. B. (2003). *Microscopy and Microanalysis*.9. 96.

10. Heath, J. P. (2005).Dictionary of Microscopy, John Wiley & Sons Ltd., Chichester, England.

11. M.Lakshman, Y.Nagamalleswari and C.Balachandran. (2009). Cutaneous Mycotic Granulo- ma (dermatomycosis) in fowl: A pathomorphological observations. *Indian J. of Vety. Pathology*. 33 (2): 204-206.

12. M.Lakshman, (2015). Diagnostic Electron Microscopy (EM) for Avian Diseases - An Overview: *Intr. J. of. Science and Research.*6 (5): 1478-1483.

13. M.Lakshman, V.Sudha and B.Haripriya, (2017). Ultrastrctural Pathology of Paraquat (PQ) Induced Acute and Subacute Lung Injury in Experimenatal Rats. *Int. J. of Science and Research.* 6 (6): 1454-1458.

14. M.Lakshman, (2019). Application of conventional electron microscopy in aquatic animal disease diagnosis: A review. *Journal of Entomology and Zoology Studies.* 7(1): 470-475.

15. Lickfeld, K. G. Elektronenmikroskopie, (1979). Eugen Ulmer Publisher, Stuttgart, Germany.

16. Marsolais, G. Berthioume, L. DiFranco E. and Marois P. (1971). Rapid Diagnosis by Electron Microscopy of Avian Coronavirus Infection; *Can. J. comp. Med.* 35-Oct. 285-288.

17. McNulty, M.S. Curran, W.L. Todd D. and McFerran, J.B. (1979). Detection of viruses in avian faeces by direct electron microscopy *Avian Pathology.* 8. 239-247.

18. Nasser Hajibagheri, M. A. (2005). Electron Microscopy Methods and Protocols; In: Methods in Molecular Biology. Humana press volume Totowa, NJ.

19. Rosai and Erlandson. (1995). A Realistic Approach to the Use of Electron Microscopy and therAncillary Diagnostic Techniques in Surgical Pathology. *American J. of Surg.Path.* 19: 247- 250.

20. Stadtländer, C. T. K.H. (2003). *Microscopy and Microanalysis* 9, 269.

21. Vale1, F.F. Correia, A.C. Matos, B. Moura Nunes .J.F and Alves de Matos A.P. (2010). Applications of transmission electron microscopy to virus detection and identification. *Microscopy: Science, Technology, Applications and Education*; 128-136.

22. Zhang ying, Hung tao, Song jingdong & He jinsheng. (2013). Electron microscopy: Essentials for viral structure, morphogenesis and rapid diagnosis. *Science China.* 56-5. 421- 430.

Different TEM Images studied at Ruska Labs

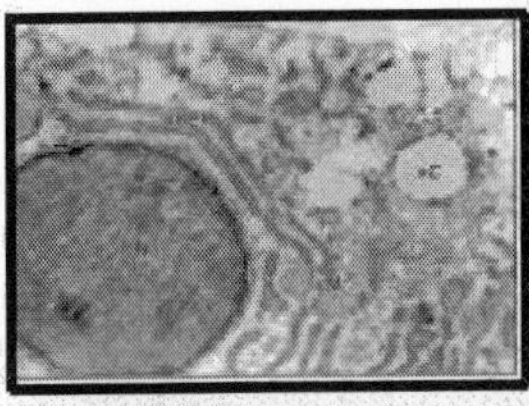

Disrupted RER, cyto-vesicles and condensed mitochondria and nucleus

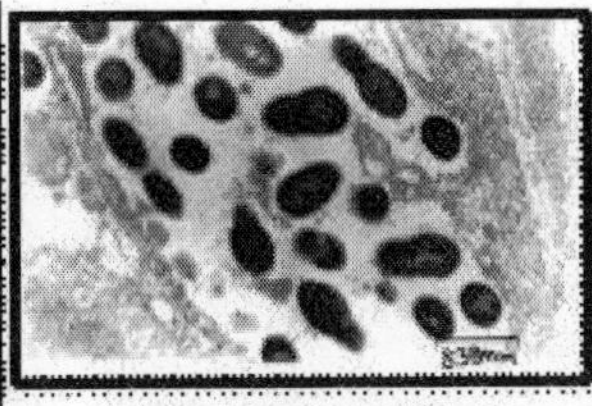

Shrimp gill tissues showed intracytoplasmic pleomorphic RLOs

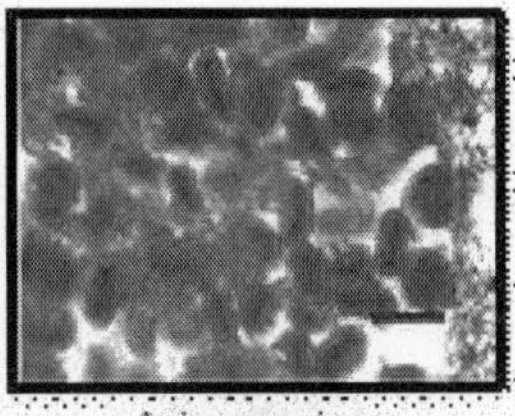

Avian pox virus

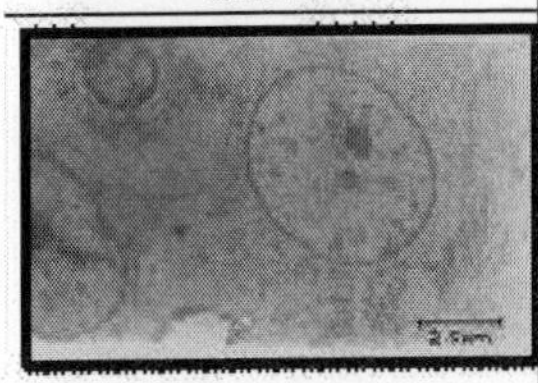

Pyknotic and Swollen nuclei, pertinacious granular cytoplasm

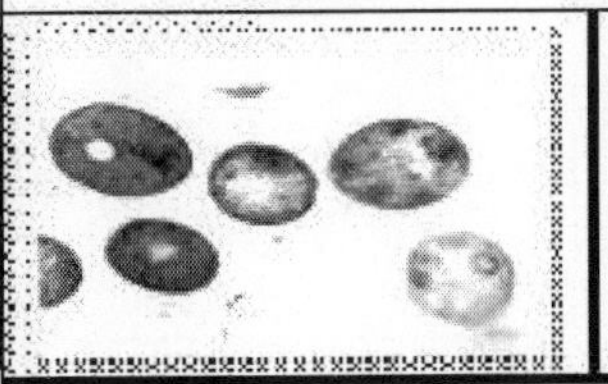

Endospore of Bacteria

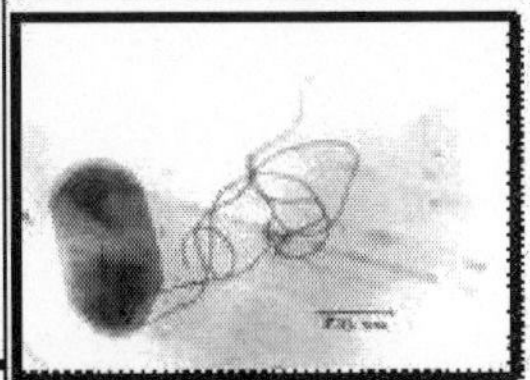

Flagella of Bacteria

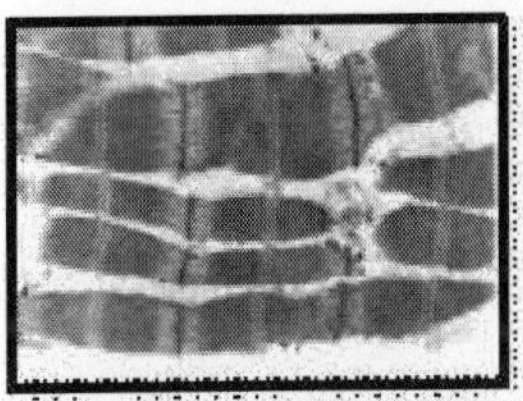

Spent hen muscle fiber x2148

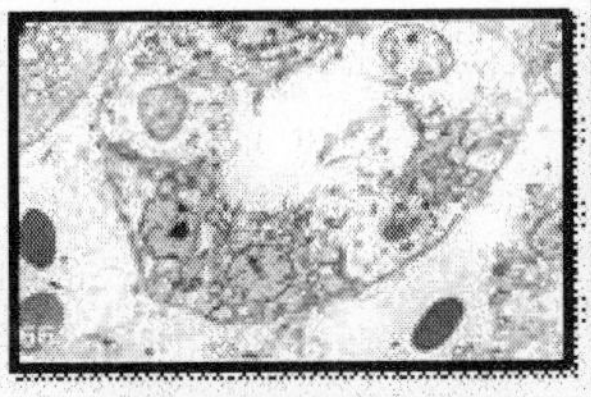

Loss of brush border in PCT and vacuolation of cytoplasm of epithelial cells. x 3580.

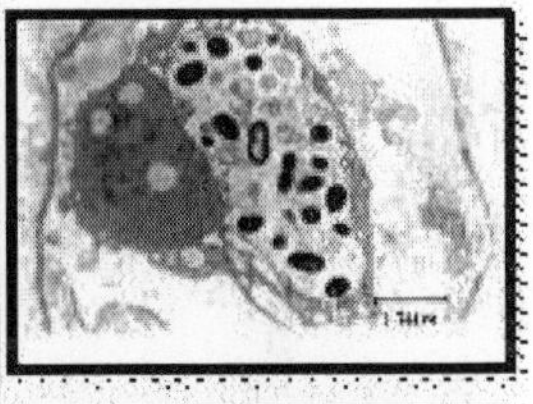

shrimp gill intracytoplasmic pleomorphic RLOs in cytoplasm phagolysosomes and in the.

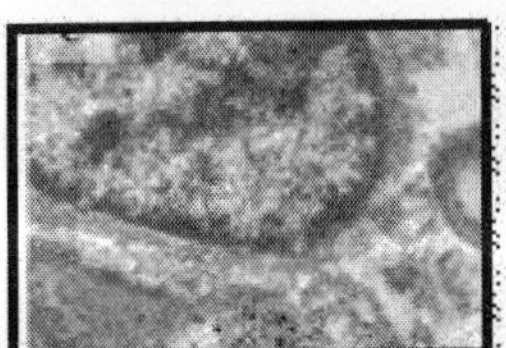

Localization of CuSO4 in cytoplasm of cat Fish testis

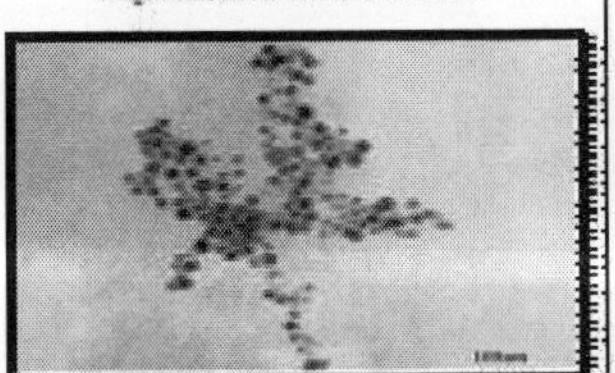

Silver Nano particles

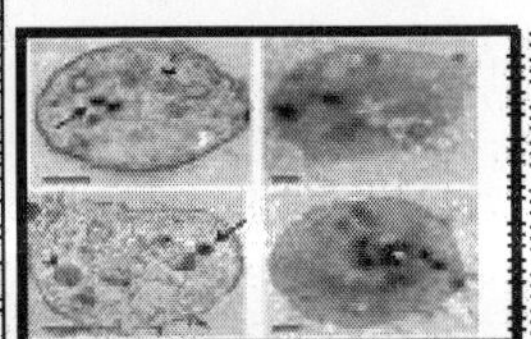

Cross section of Platelets in Rabbits

SEM Images

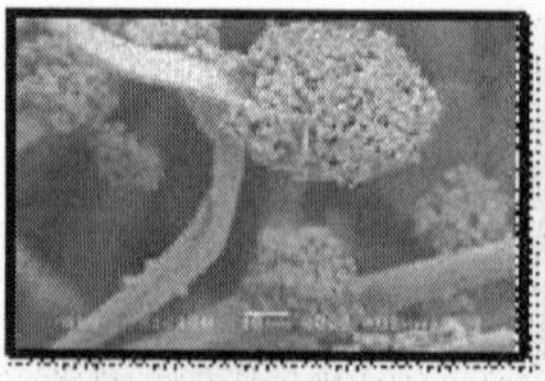

Aspergillums Fumigates

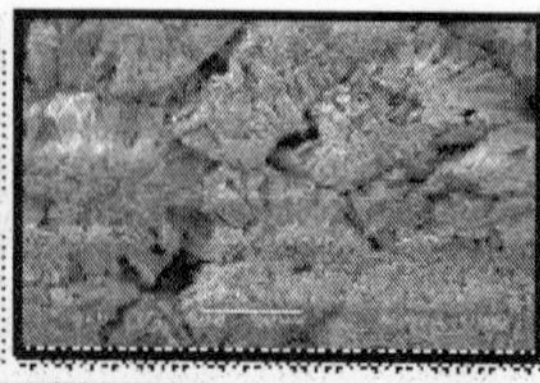

Emu Egg shell

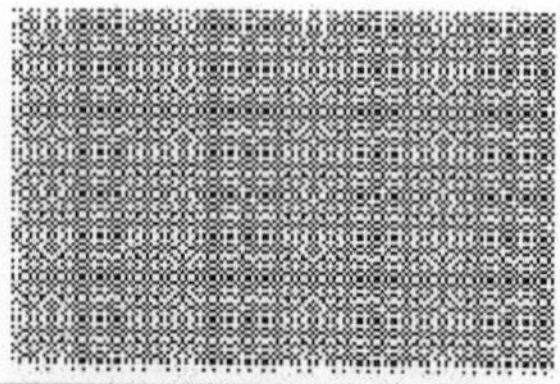

Different Yeast cells

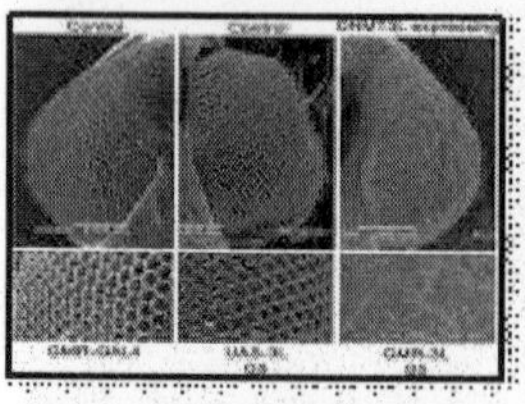

Transagenic Drosophila eyes

Pollen grains

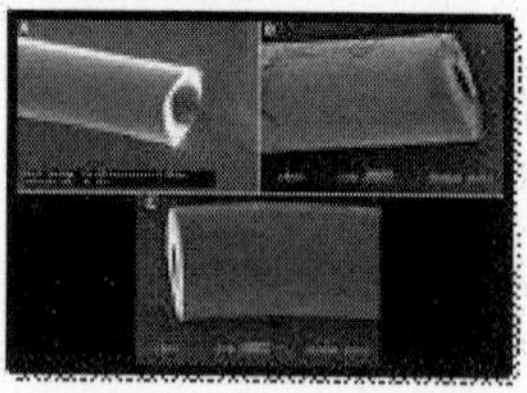

Ocular stents

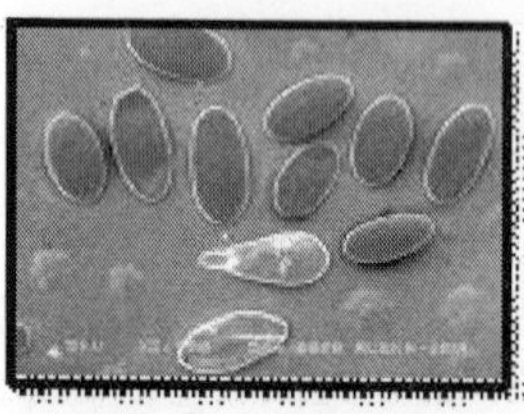

Poultry RBC, Tear drop RBC

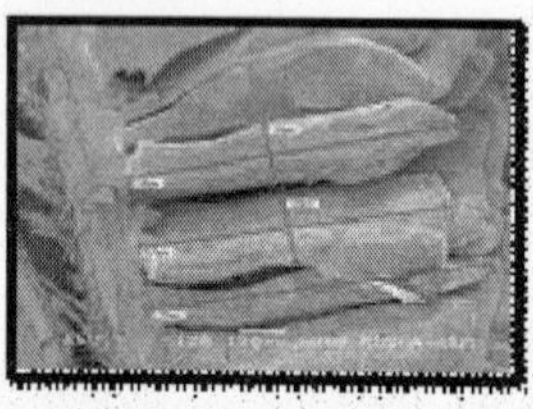

Intestinal villi in poultry

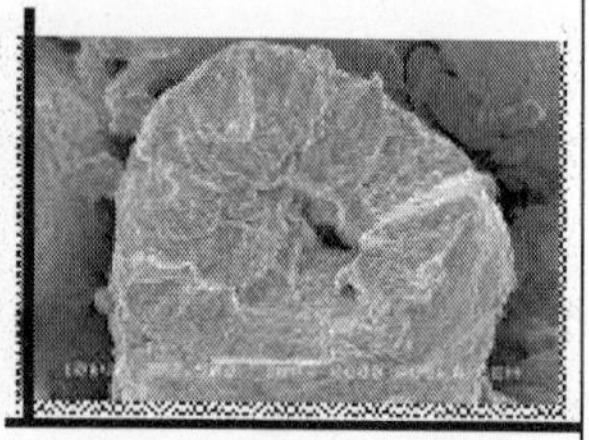

Renal tubule in poultry

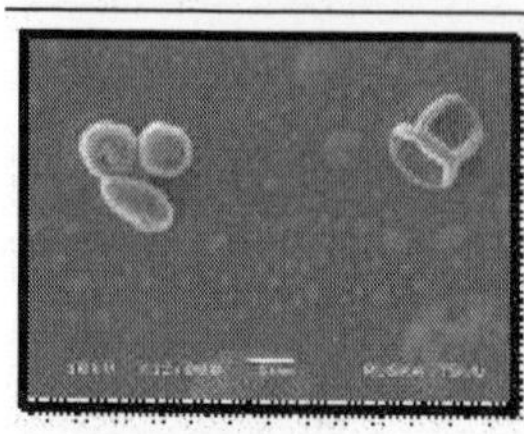

Spores of *Enterocytozoon hepatopenaei* in shrimp

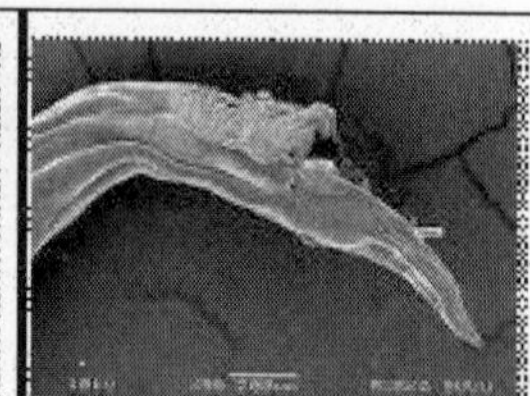

Anterior Surface *Oesoph- agostomum columbi-anum*

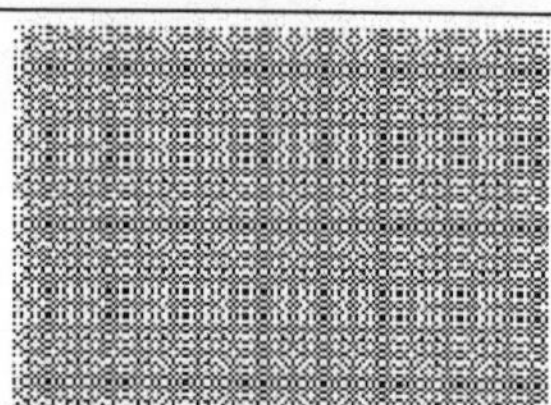

Thick Male and slender female of S.spindale

Current Status of Researches in Biosciences (2020) : 391-395
Editor : Prof. P.C. Joshi, N. Joshi, Reshman Yasmin, and D. K. Mansotra
Today & Tomorrow's Printers and Publishers, New Delhi - 110 002

CYTOMORPHOMETRICAL ANALYSES OF BLOOD OF BLACK RAT *RATTUS RATTUS* (LINNAEUS, 1758)

Aryadhara Das*, Prafulla Kumar Mohanty

Postgraduate Department of Zoology, Utkal University, Vani Vihar, Bhubaneswar -751 004, Odisha, India

Email- aryadharads@gmail.com

ABSTRACT

Cytomorphometry of blood cells, an important aspect of haematology, can reveal the physiological condition of organisms. Quantitative microscopy has strengthened conventional diagnostic scheme through better understanding of microscopic features from clinical perspective. Pathological image analyses have gained immense significance in medical fraternity through visualization and quantitative evaluation of clinical features. Till date, pathological inspection of rodents' blood is solely dependent on subjective assessment which usually leads to significant inter-observer variation in grading and subsequently resulting in late diagnosis of certain diseases. This paper aims at a systematic approach to morphologically characterize five types of white blood cells, and their nuclei from light microscopic image of blood samples. Hence, cellular and nuclei based geometric features are computed and statistically analyzed with t- test to show the discriminating potentiality of the species.

Keywords: *Rattus rattus,* blood cells, differential leucocyte count, morphometry.

INTRODUCTION

Cytomorphometry is a quantitative description of geometrical structures in all dimensions (Baak, 1985; Vandiest *et al*., 1991). Morphometry is the simplest form of image cytometry and refers to the evaluation of cells or tissues by measurement of various cellular features in a two-dimensional view. Abnormal and insufficient white blood cell function is most often reflected in modified cell morphology, and mathematical analysis of morphometrical cell characteristics is very useful for its estimation (Bins, 1985). Furthermore, changes in morphometrical erythrocyte indicators have been detected in humans (Alexandratou *et al.*, 1999; Manjunatha and Singh, 2000) and dog ailments (Berezina *et al.*, 2001). A complete blood count is an ideal indicator of general health, as stress and numerous illnesses can modify haematological parameters, especially with regard to erythrocyte and lymphocyte count (Hinton *et al.*, 1982). A complete blood count is undisputedly the most important diagnostic method available to veterinarians, along with proper anamnesis and physical examination of the animal. The modern computerized geometric, and morphometric methods have been established as efficient tools to quantify differences in the cell shape or morphological structures in particular and can provide a better characterization in describing the complexity of anatomical structures (Grizzi and Chiriva- Internati, 2005; Russ, 2007; Rosioru *et al.*, 2012).

Morphometry is used as the prognosis and diagnosis of diseases of animals. It enables the analysis of changes in the entire cell, changes in cytoplasm, changes in the nucleus, and structures of nucleus (Dalton, 1992; Russack, 1994). Since the haemocytomorphometrical data on rodents are inadequate, the particular study on the blood cells of two species of rats are analyzed, and interpreted.

MATERIALS AND METHODS

The investigation was conducted on black rats *Rattus rattus* (ten from each sex) which were caught by wooden and wire net trapping in evening time from the backyard of the residential complex of the Utkal University campus, Bhubaneswar, Odisha. These were allowed to acclimatize to captive condition prior to experimentation and were carefully handled to minimize the stress. Trapped rats were kept at room temperature and were fed with paddy, cereals, and grains. After a week, the rats were anaesthetized by using chloroform in a jar and two milliliter of blood was drawn from the jugular vein with the help of a disposable syringe. Thin blood smears of peripheral blood were prepared from a small drop of fresh blood directly from the needle and were made for each sample to determine the differential blood count and morphometrical analyses of leucocytes and erythrocytes. The morphometry like cellular and nuclear length and breadth of erythrocytes, and leucocytes (monocytes, lymphocytes, neutrophils, eosinophils, and basophils) was undertaken. Cytomorphometry of cells and nuclei of blood of both the species with sexual dimorphism was carried out with the help of microscope eyepiece digital Camera [CatCam130–1.3 Mega Pixel (MP), Code No. CC130, Catalyst Biotech, Maharashtra, India] attached to Hund Wetzlar Microscope [MICROSCOPE H 600 WILOZYT PLAN, Serial No. 1024980, Helmut Hund GmbH, Wetzlar-Nauborn, Germany] and computer.

RESULTS

In this investigation, the differential leucocyte count and cytomorphometrical analyses are taken into account. The findings of this study reveal the effect of sex on blood parameters. The results of differential leucocyte count are recorded in detail (Table 1). The number of neutrophils is found to be 36.8±1.48 in males and 32±0.59 in females. The number of monocytes is 12.6±0.61 in males and 12.8±0.44 in females. The concentration of lymphocytes is 48.5±1.21 in males and 50.6±0.85 in females. The number of eosinophils is 3±0.33 in males and 3±0.33 in females. The number of basophils is 0.7±0.15 in males and 0.6±0.16 in females.

The morphometrical parameters like length and breadth of both RBC and five categories of WBC such as monocyte, lymphocyte, neutrophil, eosinophil and basophil are measured and expressed in micron meter (μm) (Table 2). The cell length of erythrocyte is 19.8±3.12, and 6.57±0.11 in males and females, respectively. The cell breadth of erythrocytes is 10.80±0.32 in males, and 5.91±0.10 in females. The length of monocytes is 4.47±0.36, and 6.34±0.27 in males and females, respectively. The breadth of monocytes is 3.90±0.34 in males and 5.91±0.24 in females. The cell length of lymphocyte is 8.14±0.80, 5.72±0.29 in males and females, respectively. The cell breadth of lymphocytes is 8.29±0.83, and 5.22±0.30 in males and females, respectively. The nucleus length of lymphocytes is 6.34±0.76 in males and 6.30±0.73 in females. The nucleus breadth of lymphocytes is 5.36±0.68, 5.31±0.63 in males and females, respectively. The cell length of neutrophils is 10.59±0.94, 5.88±0.39 in males and females, respectively. The cell breadth of neutrophils is 9.59±0.91 in males and 5.19±0.39 in females. The cell length of eosinophil is 7.93±0.70, and 4.5±0.32 in males and females, respectively. The breadth of eosinophils is 7.78±0.65 in males and 4.08±0.27 in females. The cellular length of basophils is 7.45±0.86, and 4.39±0.37 in males and females, respectively. The breadth of basophils is 6.59±0.86 in males and 6.14±2.05 in females.

DISCUSSION

The findings of this study reflect the effect of sex on the size of erythrocytes and leucocytes (lymphocytes, monocytes, neutrophils, eosinophils, and basophils) of *R. rattus*. In this investigation, the cytomorphometrical parameters of blood of healthy black rats have been observed. The findings of cytomorphometrical analyses of *R. rattus* can be helpful in clinical investigation and interpretation. In differential leucocyte count, the highest number of lymphocytes is found in males than females and shows significant difference at pÂ0.01 which is due to the variation in species. It is observed that the per cent of monocytes increases in females and males which are due to the variation in sexual dimorphism. The neutrophils are found to be more in females in comparison to males which is possibly due to some pathogenic infections. The number of eosinophils is same for both males and females of *R. rattus*. The basophils are observed to be more in males than females which may due to inflammatory responses such as certain types of skin inflammation, asthma or parasite infections, settings in which basophils may appear in the affected tissues. The parameters like monocytes, neutrophils, eosinophils, and basophils do not show any significant difference.

The length and breadth of leukocytes do not vary much due to their round shape. Among agranulocytes (lymphocytes and monocytes), lymphocytes are larger than monocytes. The cellular length and breadth of erythrocytes are noted to be more in males than females which deviates significantly sex wise at pÂ0.001. The cell length and breadth along with nucleus of lymphocytes have more value in males whereas less in females. The parameters show sex wise significant difference at pÂ0.01. The morphometric parameters of neutrophils are found to be higher value in males and lower in females and deviate significantly sex wise at pÂ0.001. The morphometric parameters of eosinophils are observed to be higher in males than females and deviate with respect to sex which is significant at pÂ0.01, and basophil cell breadth does not show significant difference.

CONCLUSION

The study reveals that as a whole, effect of sex on morphometry of blood cells of black rat, but certain genetic and non-genetic factors such as, onset of maturity, sexual dimorphism, breeding, and environment are believed to affect the shape and size of blood cells. So, it is equally important to consider these factors and detailed investigation as to the stated aspects is suggested to arrive at accurate clinical and physiological interpretations.

REFERENCES

1. Alexandratou E, Yova D, Cokkinos DV. (1999). Morphometric characteristics of red blood cells as diagnostic factors for coronary artery disease. *Clinical Hemorheology and micro circulation,* 21: 383-388.

2. Baak JPA. (1985). The principles and advances of quantitative pathology. *Analytical and Quantitative Cytopathology*, 9: 89-95.

3. Berezina TL, Zaets SB, Kozhura VL, Novoderzhkina IS Kirsanova AK (2001). Morphologic changes of red blood cells during hemorrhagic shock replicate changes of aging hemorrhagic shock replicate changes of aging. *Shock*, 15: 467-470.

4. Bins M (1985). Morphometry of white blood cells. *Pure and Applied Chemistry*, 57: 599-601.

5. Dalton LW (1992). Computer-based image analysis of prostate cancer comments with emphasis on use of commercially available system. *Human Pathology,* 23: 280-286.

6. Grizzi F, Chiriva- Internati M (2005). The complexity of anatomical systems. *Theoretical Biology and Medical Modelling*, 2(26):1-9.

7. Hinton MD, Jones RE, Festing MFW (1982). Haematological findings in healthy and diseased rabbits, a multivariate analysis. *Laboratory Animals*, 16:123-129.

8. Manjunatha M, Singh M (2000). Digital analysis of induced erythrocyte shape changes in hypercholesterolemia under in vitro conditions. *Current Science,* 79: 1588-1591.

9. Rosioru C, Talu S, Talu M, Talu M, Giovanzana S, Craciun C (2012). Morphometric assessments for the healthy rat hepatocytes. *Annals of the Romanian Society for Cell Biology,* 7(1): 74-79.

10. Russack V (1994). Image cytometry: current applications and future trends. *Critical Reviews in Clinical Laboratory Sciences,* 31: 1-34.

11. Vandiest PJ, Baak JPA, Andiest PJ, Baak JPA (1991). Morphometry. In: Comprehensive Cytopathology. WB Saunders Company, Philadelphia, USA. pp 946-964.

Current Status of Researches in Biosciences (2020) : 397-403
Editor : Prof. P.C. Joshi, N. Joshi, Reshman Yasmin, and D. K. Mansotra
Today & Tomorrow's Printers and Publishers, New Delhi - 110 002

UTILITY OF FOLDSCOPE IN CLASSROOM TEACHING OF GENETICS

Bendangtula Walling [1], Kelevikho Neikha, Abuno Thepa, Nukshimenla Jamir, and Sarat Chandra Yenisetti*

Drosophila Neurobiology Lab, Department of Zoology, Nagaland University (Central), Lumami 798627, Nagaland, India.

Email: yschandrays@rediffmail.com

ABSTRACT

Foldscope is an inexpensive optical microscope composed of paper and a lens and it is sufficient for viewing the microscopic world around us. It can be used as an educative tool where the students can individually have access to a lightweight, cheap and durable tool to study/discover objects. The magnification power of a Foldscope is similar to that of a light microscope and it is made up of waterproof origami paper, making them capable for field studies. Moreover, the exposure of students to Foldscope in biology courses could channel student's interest towards further experimental activity and original research. We have optimized this simple paper microscope to observe and understand *Drosophila* mutants and to study developmental stages. I will be presenting and discussing my efforts to teach basic genetics to school and college children and instigate interest and ignite thinking process of these budding scientists.

KEYWORDS: Foldscope, genetics, mutants

INTRODUCTION

Microscopes are pervasive tools in today's scientific world, providing the essential link between the macro and the micro world. As we all know that, organisms are made up of either single cell or billions and billions of cells, while the complex organisms are made up of different set of cells which merge together to form the tissues and organs of the body to carry out the biological processes to maintain life. The invention of microscope has greatly enhanced the mankind to visualize and understand the biological process of each cells [1, 2]. Besides other instruments used in biology, Microscopes has played a vital role in building up the understanding and knowledge of the subject [3]. Microscopes are crucially important public health tools, allowing workers to identify pathogens and correctly diagnose the cause of illnesses. As educational tools, it can help the students identify a lot of different organisms which are not visible to the naked eye. However, this microscopic modality is not always available in all health centers and schools especially in the rural and remote areas [4]. Also in an exigency situation, it is not always possible to use microscope for investigations due to limited resource. Since the microscope is fragile and very expensive, most of the people are either afraid to use or they lack the resources to benefit from this vital tool. In the year 2012, Dr. Manu Prakash and his team came up with the idea of creating a low-cost microscope which people can use without any hesitation and can afford it, yet still have the range of performance necessary to collect useful data. The low cost microscope is called "Foldscope". A Foldscope is an origami based optical microscope that can be assembled from a punched sheet of cardstock, a spherical glass lens and works under any light source. Since cost effective and scalable manufacturing is an integral part of frugal science and engineering, it is designed in such a way that it is portable and cost less than 1$ and small enough to fit in a pocket and requires no power to operate. This new technology is believed to have benefits in education, global health, and medicine [5]. The Foldscope is capable of magnifying over 2,000 times, however because the Foldscope is only in Beta testing, the microscopes sent out to global testers will only have 140 times to 480 times magnification [6].The invention of Foldscope has greatly removed the cost barriers, providing new opportunities for immense use in both science education and field work for science and medicine [7].

In a classroom setting, a Foldscope can be used as an educative tool where the students can individually have access to a lightweight, cheap and durable tool to study/discover objects. The light microscopes are also very useful for understanding organisms in a classroom but are very expensive and in some under-developed countries, the people can't afford it. Also because of their heavyweight we cannot take them for field studies. The magnification power of a Foldscope is similar to that of a light microscope and it is made up of waterproof origami paper, making them capable for field studies. Moreover, the exposure of students to Foldscope in biology courses could channel student's interest towards further experimental activity and original research. Here, we have optimized this simple paper microscope to observe and understand using *Drosophila melanogaster* wild type and mutant fly as a model and to study their developmental stages.

MATERIALS AND METHODS

Teaching method: Foldscope was assembled in-order to carry out the study. *Drosophila melanogaster* fly was used as a sample to study in a classroom environment with every student having the opportunity to touch, experience and hold.

Sample preparation: To start with, *Drosophila melanogaster* mutant fly of white eye and wild type fly was taken on a clean slide, mounted on phosphate buffer saline in order to avoid any tissue attached, transfer them gently on cavity slide and observe under the Foldscope. The entire slides was evaluated for a minimum of 3-5 mins and finally mounted on Samsung Galaxy S8 +. Vestigial wing of the mutant fly was dissected and placed on a clean slide to observe under the Foldscope. Dissected sex comb present on the foreleg of the fly was taken on a clean glass mounted on a drop of NaCl and processed the slides for observation under the Foldscope. Similarly, mouth parts of larvae, egg and antennae was also dissected, observed under the Foldscope which was finally imaged on Samsung Galaxy S8 +.

RESULTS AND DISCUSSION

With a keen observation of the *Drosophila* sample using the foldscope, provided hands on experience to all the students for the study and following images were captured and imaged through the provided smart-phones.

1. Demonstration of *Drosophila melanogaster* eye mutant using Foldscope

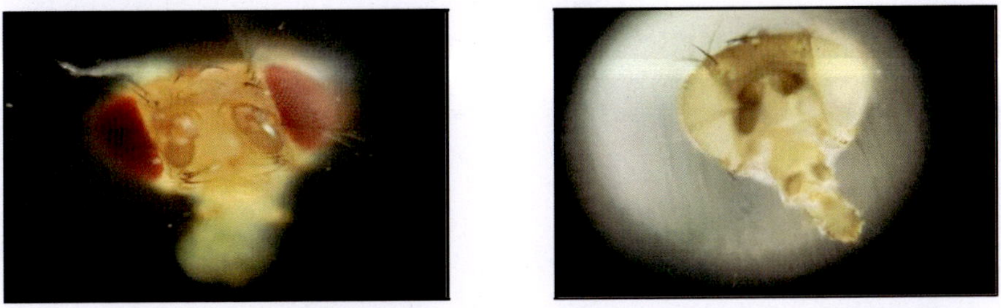

Fig.1(A)Wild type eye **(B) White eye mutant fly**

Many eye color mutants are collected available from *Drosophila* stock centres and they are of great help to demonstrate and teach mutant studies using *Drosophila* model. Here we have prepared and demonstrated wild eye (brick red) and white eye mutant (Fig.1 A and B). White eye mutant was the first sex linked mutation discovered, found in the fruit fly [8]. Protein coded by the white gene functions as an ATP-binding cassette (ABC) transporter. It carries the precursors of the red and brown eye color pigments, guanine and tryptophan, into the developing eyes during pupation [9]. *Drosophila melanogaster* with the white eye mutation typically have reduced life spans than wild-type fly. Further they exhibit many neurological defects such as mobility defects and reduced stress tolerance. They have significantly reduced number of synaptic vesicles of photoreceptors. White eye mutants exhibit lower rate of reproduction due to low sexual arousal during day light [10]. Over expression of White+ induces abnormal male-male courtship in *Drosophila* [11]. Studies also demonstrated that White+ regulates copulation success in *Drosophila* [12].

2. Study of wild type and mutant wing of *Drosophila*

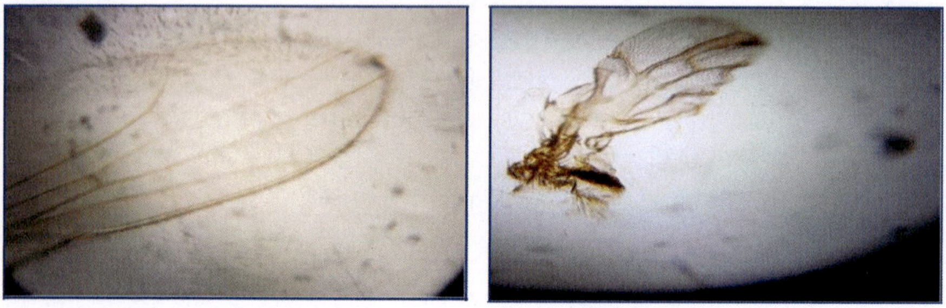

Fig 2. A) Wild type wing B) Vestigial wing of mutant fly

Demonstration of wild type wing and vestigial wing of *Drosophila melanogaster* (Fig. 2 A & B) was performed where the wing represents an appropriate model to study their developmental control of phenotypic variation. It is involved in several functions of ecological and evolutionary importance (e.g., flight and male courtship song). It was observed to have a fine intersection of veins. While, the vestigial wings have crumpled wings which prevent them from flying. This is because of a recessive mutation of the vestigial gene present in the 2nd chromosome. One of the pair of vestigial genes carried by each fly (one from each parent), both have to be altered to produce the abnormal wing shape.

3. Study of Life developmental stages of *Drosophila melanogaster*

Drosophila has a relatively short life cycle about 10 days and it reproduces prolifically. The life cycle consists of four distinct stages: egg, larva, pupa, and adult. Using Foldscope, *Drosophila* egg and larvae was observed and imaged as follows:

Fig. 3. A) Egg of a *Drosophila melanogaster* **B) Mouthparts of *Drosophila* larvae**

Egg of *Drosophila melanogaster* were observed using Foldscope and imaged (Fig:3.A). The egg of the fly measures about 0.5mm in length and is white colour and oval in shape. It has a pair of respiratory filaments and a micropyle. A female may lay as many as 400- 500 eggs in a favourable laying medium. At about 24 hours of laying, the eggs hatch into the 1st star instar larvae.

The larva has a pair of mouth hooks that extends to bring food to the mouth (Fig:3B) *Drosophila* larvae has three instars stages; 1st instar, 2nd instar and 3rd instar. The 1st instar larvae develops in about 24hrs from egg laying. At about 48 hrs the 2nd instar develops. Again after 24 hrs (i.e., 72-96 hrs after egg hatching), this larva molts and matures to 3rd instar larva. During the 3rd instars, the larva loses its spiracles, mouth, and hooks. The larval mouthparts of the wild type are heavily pigmented with what is generally conceded to be melanin. The mouth hooks, mentum and anterior part of the vertical plate are very dark brown, almost black; the posterior, basal prongs are less pigmented than the rest and appear golden brown at the tips.

4. Study of cytotaxonomical characteristics of *Drosophila melanogaster*

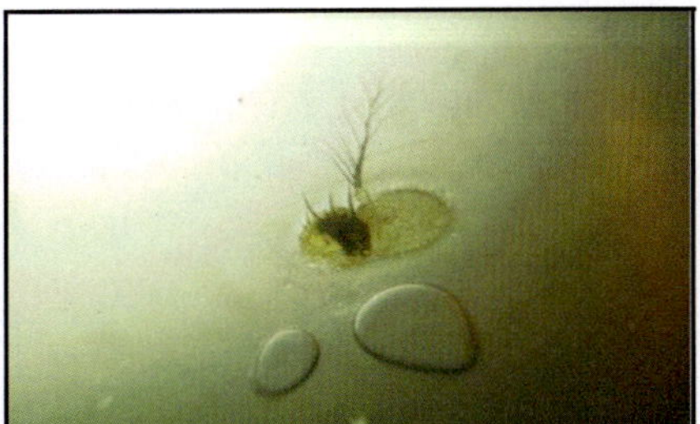

Fig.4. A) Dissected Antenna **B) Dissected Sex comb of *Drosophila***

Drosophila sensory structure- Antenna was recorded using Foldscope (Fig.5 A). *Drosophila* has a pair of antennas on the front of the fly's head, between the eyes which is a sensory element and contains receptor cells (Johnston organ). The Johnston organ allows flies to sense different mechano-sensory stimuli like sound, wind and gravity and neuronal signaling from the antenna to the brain allows for behavioral responses to these stimuli. Each antenna consists of a thin arista and three large proximal segments also known as the scape, pedicel, and funiculus, respectively. The arista is the main receptor of sound and has recently been characterized as analogous to hairs and sensory bristles. The antennae function best within one wavelength of a sound stimulus [13].

Study of sex comb *Drosophila melanogaster* was performed using Foldscope (Fig.4.B). It is a sexual dimorphic character of male *Drosophila* and consists of a row of stout black bristles on proximal tarsal joint of each pair of legs, the sex comb consists of teeth and the number of teeth is species specific. *Drosophila* males use their sex combs to grasp the females' abdomen and genitalia prior to copulation.

CONCLUSION

In view of all the experiment performed using this paper microscope, it gives us an idea that Foldscope not only helps health-care workers around the world but it also places powerful magnification into the hands of students. The mobile phone mounted Foldscope may be a useful diagnostic tool in resource constrained clinical and public health settings if modification is made to improve the sensitivity. Both Dr.Manu Prakash and Jim Cybulski have a vision of each child in the world carrying a Foldscope like a pencil or pen in their pocket. Thus, if more exposure of foldscope is given to the students it will make them more educative and easy access to the environment that surrounds them. It can improvise them practically by obtaining the image compared favourably with that obtained by traditional light microscope. It will also remove the mentality of the students that "Research work is only for scientist in big laboratory". From the above captured images it is evident that foldscope can be used as an educative tool. Simple and inexpensive paper microscope will be of great help to teach basic genetics to school and college children and instigate interest and ignite thinking process of these budding scientists. In simple, this "simple" instrument is contributing significantly in democratizing the science all over the world.

ACKNOWLEDGEMENTS

This work is supported by the DBT (Department of Biotechnology, India) grant No. BT/IN/Indo-US/Foldscope/39/2015, dated 20-3-2018 awarded to corresponding author Dr. Sarat C. Yenisetti, Department of Zoology, Nagaland University, Lumami.

REFERENCES

1. Queller, D. C., & Strassmann, J. E. (2009). Beyond society: the evolution of organismality. *Philosophical Transactions of the Royal Society B: Biological Sciences*, 364(1533), 3143-3155.

2. Pepper, J. W., & Herron, M. D. (2008). Does biology need an organism concept?. *Biological Reviews*, 83(4), 621-627.

3. Bardell, D. (2004). The invention of the microscope. *Bios*, 75(2), 78-84.

4. Waliullah ASM. (2018). Application of Mobile phone based portable microscopy in clinical histopathology:A feasibility study. *International journal of clinical biomedical research*; 4(2):16-20.

5. Cybulski, J. S., Clements, J., & Prakash, M. (2014). Foldscope: origami-based paper microscope. *PloS one,* 9(6), e98781.

6. Prakash Lab.Foldscope:Microscopy for everyone, Stanford University, available online at: http://www.foldscope.com.

7. Judith A, Bazler, Meta VS, Carleigh E.(2016).The Foldscope. *Journal of Modern Education Review*; 6(2): 118-124

8. Morgan, T. H. (1910). Sex limited inheritance in *Drosophila. Science*, 32(812), 120-122.

9. Mackenzie, S. M., Brooker, M. R., Gill, T. R., Cox, G. B., Howells, A. J., & Ewart, G. D. (1999). Mutations in the white gene of *Drosophila melanogaster* affecting ABC transporters that determine eye colouration. *Biochimica et Biophysica Acta (BBA)-Biomembranes*, 1419(2), 173-185.

10. Ferreiro, M. J., Pérez, C., Marchesano, M., Ruiz, S., Caputi, A., Aguilera, P., ... & Cantera, R. (2018). *Drosophila melanogaster* white mutant w1118 undergo retinal degeneration. *Frontiers in neuroscience*, 11, 732.

11. Zhang, S. D., & Odenwald, W. F. (1995). Misexpression of the white (w) gene triggers male-male courtship in *Drosophila. Proceedings of the National Academy of Sciences*, 92(12), 5525-5529.

12. Xiao C, Qui S, Robertson RM (2017).The white gene controls copulation success in *Drosophila melanogaster. Scientific reports* :7(1).

13. Bennet-Clark, H. C. (1971). Acoustics of insect song. *Nature*, 234(5327), 255-259.

Current Status of Researches in Biosciences (2020) : 405-409
Editor : Prof. P.C. Joshi, N. Joshi, Reshman Yasmin, and D. K. Mansotra
Today & Tomorrow's Printers and Publishers, New Delhi - 110 002

BIOMASS YIELD AND NUTRITIVE VALUE OF MAIZE GRAIN SPROUTS

Chethan, K. P*., Shivakumar Gowda, N.K., Prabhu, T.M*., and Giridhar, K.S****

**Veterinary College, Hebbal, Bengaluru, Karnataka.*

***ICAR-NIANP Audogodi, Bengaluru, Karnataka.*

Email: chethanvet@gamil.com

ABSTRACT

The study was aimed to find out the biomass yield and change in nutrient composition of maize grain when sprouted in hydroponic system. Maize grain sprouts (MGS) was harvested on the 10th, 11th, 12th and 13th day. Height of the shoot growth and yield was recorded. Analysis was performed to determine the chemical composition and ME content with *in vitro* gas production technique. It was determined that on 10th day maximum yield of MGS i.e., 4.5 kg per kg of maize grain with DM content 16.5% was obtained. As the number of day was increased there was reduction of biomass yield. Found increase in nutrient content CP (13.0%), EE (4.40%), NDF (32.99%), ADF (15.50%), TA (2.89%) and decreased ME 12.204 MJ/kg DM in MGS compared to the maize grain. As per the results obtained suitable harvesting time to get maximum yield is 10 days and MGS has better nutritive value than maize grain.

KEYWORDS: Biomass Yield, Nutritive Value, Hydroponics, Maize grain sprouts

INTRODUCTION

Livestock rearing is an important income generating and subsidiary activity among agriculture families in rural India. This not only provides constant income throughout the year, also acts as an insurance cover for distress periods like drought, flood and market failure for their product.

Providing good quality, nutritious green fodder to livestock is the key for economic productivity and sustainability. Conventional green fodder production in the field requires large area, lot of hard labour (preparation of land, sowing, weeding, watering etc.) and huge quantity of water. In urban and periurban areas where there is huge demand for livestock products, land is not available for fodder production at all and even if available, cost of the land is so high that it is not feasible to cultivate fodder at economical price. In some areas there is scarcity of fertile cultivable arable land due to unfavorable geographical and topographical conditions. Due to shrinking pasture and grazing lands livestock farmers are facing unpredictable high prices for dry fodder crop residues in summer months. The dairy farmers are in desperate need of dependable and affordable newer alternative feed resource which is water efficient, require less land and immune to weather conditions. Hydroponic technology offers an opportunity to farmers to grow their own fodder in an economical and sustainable way.

The green fodder can be grown by sprouting the grains. This method of growing plants using mineral nutrient solutions, in water, without soil is known as hydroponics technology. Dry grains contains sufficient amount of nutrients to sustain growth for a period of 7-10 days without any additional nutrient supplementation. By extending the sprouting period from 1 or 2 days to 10 days yields a fairly good quantity of green fodder in short time utilizing less water and space.

Nutrients in the maize grain are transformed into more bio available form as sprouting occurs and before the plant is matured. During the sprouting process the beneficial change that occurs are change in enzyme activity, increased total protein content and more bio available amino acid profile, increased fiber level, vitamins and minerals in comparison to the maize grain. Hydroponically grown maize fodder is more palatable, digestible and has improved feeding value than feeding maize grain as such. Thus during inclement weather conditions, in periurban areas and inadequate rainfall years farmers can adopt hydroponic technology to grow maize fodder in short time with improved nutritive value. However biomass yield and nutrient content of sprouted maize grains varies with the local conditions provided *viz.*, temperature, humidity and quality of the maize grain used etc. As green fodder shortage and inclement weather conditions are more common in Karnataka, this experiment was undertaken to assess the biomass yield and nutrient profile of maize grain sprouts grown in hydroponic system.

MATERIALS AND METHODS

Hydroponic system and Grain sprouting: Fresh Hydroponic maize grain sprouts was produced (20 kg/day capacity) in a stainless steel frame of 13 ft length X 6 ft width X 6 ft height which was placed inside a poly house measuring about 25 ft length X 10 ft width X 10 ft height, equipped with automatic sprayer irrigation (at hourly interval) of tap water. Quality grains was soaked overnight (12-14h) in a bucket containing tap water and thereafter transferred to porous gunny bags for sprouting up to the period of 48hours. Sprouted maize grains were uniformly spread on the porous plastic tray placed on the racks of stainless steel frame. Inside the poly house, the germinated grains were allowed to grow into sprouts and harvested on 10th, 11th, 12th and 13th day intervals to record biomass yield. The sample harvested on 10th day was used for nutritive value analysis. The samples were collected in triplicates.

Proximate composition and *In vitro* gas production study: The MGS sampled 10th day was oven dried at 60^0C for 48 h to determine the dry matter content and the sample was ground to 1 mm size for further chemical analysis. The ground MGS sample was chemically analyzed in duplicate for proximate composition as per AOAC (2005), forage fiber fractions, Neutral detergent fiber (NDF) and Acid detergent fiber (ADF) were determined according to Van Soest *et al.* (1991). Samples were incubated *in vitro* with rumen fluid collected from a rumen canulated cow as inoculum in a calibrated glass syringe for gas production following the procedure of Menke and Steingass, (1988) to predict the Metabolisable energy (ME) content.

The equations used to estimate the ME was according to Menke and Steingass (1988).

$$\text{ME (MJ/KgDM)} = 2.2+0.136\text{GP}+0.057\text{CP}+0.002859\ \text{EE}^2$$

Where CP is crude protein in g/kg DM, EE is ether extract in g/kg DM and GP is the gas produced (ml) from 200mg ground MGF sample after 24 h of incubation.

RESULTS AND DISCUSSION

Grain sprouting and Biomass yield: The biomass yield from MGF was recorded at 10th, 11th, 12th and 13th day and it was observed that at on 10th day maximum yield of MGS i.e., 4.5 kg per kg of maize grain with DM content 16.5%. later on as the number of day was increased for further growth there was reduction of biomass yield on dry matter basis and observed that roots of MGS was more slimy to touch showing possible mould contamination. From 11th and 12th day onwards black spots and yellowish green spots were noticed in some of the trays. This may be the reason for reduction in biomass yield and if fed animals may dislike such fodder. In the present study maximum increase in yield was found to be 5 fold from the initial quantity of seed taken, compared to observations of Naik *et al*., (2015) who reported upto 6-8 fold increase in the biomass yield. The height of MGS was found to be 23, 24, 25 and 25 cm at 10th, 11th, 12th and 13th day respectively, showing there was less advantage in terms of shoot growth by keeping for more than 10 days of sprouting. The fodder was fresher and green in colour on 10th day without any sliminess and discoloration. Feeding of MGS to animals would be better on 10th day of harvesting rather extending for two more days though yield may be less at this stage. Water consumed for growing one kg of fodder in the present study was 4.2 litre which includes for soaking, sprinkling at three hourly intervals for ten days on the rack and washing of trays etc.

Table: Chemical composition (%DM Basis) and *in vitro* gas production at 24 h of maize grain and MGS

Attributes	Maize Grain	Maize grain sprouts
Dry matter (on fresh Basis)	92.50	16.50
Organic Matter	98.58	97.11
Crude Protein	8.50	13.00
Ether Extract	2.44	4.40
Neutral Detergent Fiber	13.26	32.99
Acid Detergent Fiber	3.42	15.50
Total Ash	1.42	2.89
Acid Insoluble Ash	0.17	0.461
Gas production (GP-24, ml/200 mg DM)	76.71	63.23
Metabolisable Energy MJ/Kg DM	13.26	12.20

Chemical composition and *In vitro* gas production: The results of nutrient composition and ***In vitro* gas production** study are presented in Table 1. The MGS contained higher CP (13.0%), EE (4.40%), NDF (32.99%), ADF (15.50%), TA (2.89%) and lesser ME of 12.204 MJ/kg DM compared to the maize grain.

CONCLUSION

From the present study it can be concluded that Hydroponic maize grain sprouts can be grown in 10 days period with less water and space requirement and it has a better nutritive value in terms of protein and fibre.

This technique has a potential to grow green feed during scarcity and contingency situations.

REFERENCES

1. A.O.A.C. (2005). *Official Methods of Analysis*, Association of Official Analytical Chemists, 15th Ed. Arlington, VA.

2. Menke, K. H. and Steingass, H. (1988). Estimation of the energetic feed value from chemical analysis and *in vitro* gas production using rumen fluid. *Anim. Res. Dev.*, **28**: 47-55.

3. Naik, P.K., Swain, B. K. and Singh, N.P. (2015). Production and utilization of hydroponics fodder. *Indian J. Anim. Nutr.* **32** (1): 1-9.

4. Van soest, P. J., Robertson, J. B. and Lewis, B. A. (1991). Methods for dietary, neutral detergent fiber, and non starch polysaccharides in relation to animal nutrition. *J. Dairy Sci.*, **74**:3583-359.

Current Status of Researches in Biosciences (2020) : 411-417
Editor : Prof. P.C. Joshi, N. Joshi, Reshman Yasmin, and D. K. Mansotra
Today & Tomorrow's Printers and Publishers, New Delhi - 110 002

MONTH – WISE PREVALENCE OF GASTROINTESTINAL NEMATODE PARASITES OF GOATS IN LUCKNOW, UTTAR PRADESH, INDIA

Kamal Jaiswal*, Suman Mishra and Savita

Department of Zoology, Babasaheb Bhimrao Ambedkar University (A Central University) Vidya Vihar, Raebareli Road, Lucknow-226025

Email: drkamaljaiswal@gmail.com, parasitologylabbbau11@gmail.com

ABSTRACT

Goat is an important livestock species all over the world. They are reared under a wide variety of production systems ranging from traditional smallholder and village production system to large-scale intensive commercial farms. Parasites are major public health problem of the world. Recurring losses in productivity is due to the overwhelming infection of parasite. Nematode is the most widespread among all the gastrointestinal helminthes infecting in population of goat. Intestinal parasitic infections remain a major public health problem in many parts of Lucknow, particularly in rural areas. A cross sectional survey was carried out from May 2017 to April 2018 which included a total sample of 260. Overall 65.76% samples were found infected by the nematode parasites. Female goats were found to be more infected than the male goats.

KEYWORDS: Goat, gastrointestinal tract, nematodes parasites, prevalence.

INTRODUCTION

In India, livestock is one of the most potential sub-sectors of agriculture which plays an essential role in promoting human health and national economy of the country (Alemayehu, *et. al.,* 1995). However, livestock is an essential part of a farming system which is enhancing the economy of the country. Goats are the most important livestock species all over the world and an incorporated in the agricultural production system for the developing countries (Mir, *et. al.,* 2013). In India 27% livestock and 5 million people are engaged in goat farming (Honmade, 1998 and Anon, 2012). Mostly smallholder and village farmer are depend on the goat rearing milk and meat purpose (Hossain, *et. al.,* 2004). Demand of goats for consumption is increasing speedily as they provide the astonishing source of animal proteins. Goats harbour a variety of helminthes protozoan's and bacterial infection but helminthes parasites are most harmful for the production of the meat and milk. Helminthosis having three classes nematode, cestode and trematode. Nematode is most prevalent parasites following to the trematode and cestode in this area; these parasites are recognized the most prevalent infections in subtropical and tropical areas due to complimentary conditions obtain in these regions. Nematodes remain the severe problem to goats industries. They are responsible for losses in productivities resulting in economically heavy losses in goat industries. Some major nematode parasitic infection is found to be prevalent in gastrointestinal tract of goats (*Haemonchus* sp., *Trichuris* sp., *Oesophagostomum* sp., and *Bunostomum sp.*). Various climatic conditions are responsible for the parasitic infection for the development, survival and spreading of nematode parasites, they are confirmed in many studies (Odoi *et. al.,* 2007). However, gastrointestinal nematode parasite infections, diverse or solitary species, are major problems mainly depend on the optical conditions to the affecting of goat (Rahman, 1994). Some major gasterointestinal parasitic disease are Gastritis, general weakness, anorexia, enteritis, diarrhoea, anaemia, affecting on production of growth (Urquhart *et al.*, 1987). The present study was this taken up to investigate major the gastrointestinal nematode parasites to assess the major risk factors and associated infection in goats.

MATERIALS AND METHODS

Study Area and period: The gastrointestinal tracts were collected from different slaughter houses of Lucknow and brought to the parasitology laboratory of Department of Zoology, Babasaheb Bhimrao Ambedkar University, Lucknow for detection of parasites. The study was conducted during May 2017 to April 2018.

Sample collection and preservation: A total number of 260 local breed goats were selected for carrying out the present research, conventionally, hosts included were demanded as kids, young and adults. Collected nematode parasites were preserved within the help of glycerol and alcohol (3:1) for further studies. These parasites were mounted in glycerol and images were taken under the stereomicroscope (10x and 40x). Parasites were identified with the help of the technique derived by Soulsby key (1982).

Statistical Analysis: Data exploration was done using within the help of SPSS software version -20. The prevalence of parasites was calculated using the formulae as given by Margolis *et. al.,* (1982).

$$\text{Prevalence (\%)} = \frac{\text{Total number of individuals infected with a particular parasite species}}{\text{Total number of hosts examined}} \times 100$$

RESULTS AND DISCUSSION

A total number of 260 gastrointestinal tracts were dissected and collected parasites during the study period (May 2017 to April 2018) in which 171 gastrointestinal tracts were observed to be infected. The overall prevalence was found to be 65.76%. The prevalence was highest in the month of August (96.00%) followed by September (95.00 %), while the lowest infection was recorded during month of January (28.57%) and February (35.00%) (Table-1). Statistical analysis using Chi square test revealed non-significant difference between the prevalence rates across all months. Similar findings have been reported by Raza *et. al.*, (2007) and Hawlader *et. al.,* (2002). The different species of nematodes recovered during this investigation were already reported by various researchers in different parts of the World (Ahmed and Ansari, 1987, Asanji and Williams, 1987, Guiomaraes and Walter, 1987 and Njau, 1987) the other important factor was found to be reported rainfall in the study area (Raza MA *et. al.,* 2007). Many major causative factor are also responsible for high rate incidence of parasitic infestation concludes that malnutrition, overstocking and in turn, weak immunity. Higher prevalence of gastrointestinal parasite infection in August and September (wet months) compared to winter months December to February(winter month) has been reported by various researchers (Nakanishi *et. al.,* 2011, O'Connor *et. al.,* 2007, Shirale *et. al.,* 2007 and Skykes, 1994).

Table 1: Month-wise prevalence of gastrointestinal nematode parasites in goat

S.no.	Name of months	Total intestines	Infected intestines	% of infected intestine
1.	May	20	14	70.00%
2.	June	23	15	69.56%
3.	July	20	18	90.00 %
4.	August	25	19	96.00 %
5.	September	20	18	95.00 %
6.	October	21	16	76.19%
7.	November	22	11	50.00%
8.	December	23	9	39.13%
9.	January	21	6	28.57%
10.	February	20	7	35.00%
11.	March	20	8	40.00%
12.	April	24	10	41.66%
	Total	260	171	**65.76%**

p value = 0.252 (>.05 non-significant).

Table 2: Gender -wise prevalence of nematode parasites of goat

S. no.	Gender	No. Of intestine	No. of Infected Intestine	Percentage
1.	Male	132	78	45.61%
2.	Female	128	92	54.38%

p value = 0.157 (>.05 non- significant).

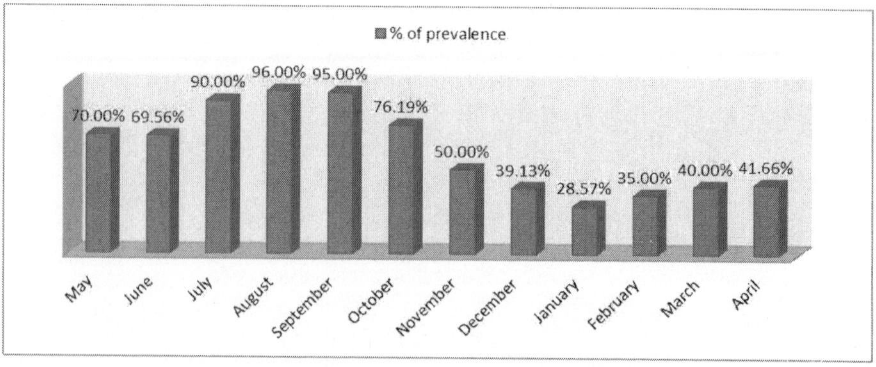

Fig1: Month-wise prevalence of nematodes parasites of goat

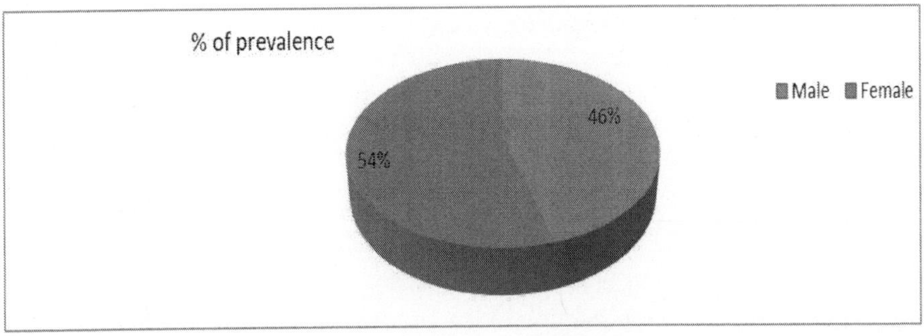

Fig 2: Gender wise prevalence of nematodes parasites

The sex-wise incidence of nematode infection in goats is depicted in Table-2. The higher prevalence of nematode infection was found in females (54.38%) as compared to males (45.61%). The presence of sex difference in infection is also conflicted with other reports (Ghanem *et. al.,* 2009 and Hassan *et. al*., 2013) the studies reported that attributable to the stall feeding of female animals during pregnancy, revelation to pasture contamination is generally reduced and normally females are assumed to be more heavily infested due to hormonal differences and stress during pregnancy (Dagnachew 2011, Ayaz *et. al.*, 2013, Shirale *et. al.,* 2007 and Miller, 1984).

CONCLUSION

Gastrointestinal nematodes parasites in small ruminants are highly prevalent and which is found to be the crucial cause of economic losses in the goat production. Practice of separate grazing of an animal with low stocking rate and proper deforming at regular intervals may be adopted to overcome the losses and reduce the infestation rate. Furthermore, during the wet season (July, August and September) climatic factors like temperature and humidity are found to be most good for the development and survival of parasitic stages of nematodes, for that reason, it is suggested that anthelmintic treatment on quarterly basis (regular deworming) may be implemented to reduce the risk of infection. Necessary steps should be taken to combat parasitic infection in timely manner to improve the productivity from these animals.

ACKNOWLEDGEMENTS

This work has been supported by University Grant Commissions, India under RGNF (SRF) Fellowship program within the frame work of the doctoral study. We are also thankful to our Honorable Vice- chancellor, Prof. Sanjay Singh and Other authorities of the University for providing necessary infrastructure facilities to carry out the research work.

REFERENCES

1. Ahmed M. and Ansari J A, (1987). Prevalence of gastrointestinal nematodes of sheep and goats in Aligarh (India). *Ind. Vet. Med. J.* 11:165-170.

2. Alemayehu, Z. and I. Fletcher, (1995). Small ruminant productivity in the central Ethiopian mixed farming system. Institute of Agricultural Proceeding, 4: 1941-1947.

3. Anon. 19th Livestock census, (2012). Department of animal husbandry, dairying and fisheries. Ministry of Agriculture, Government of India.

4. Asanji M.F. and Williams M.O .(1987). Variables affecting the population dynamics of gastrointestinal helminthes parasites of small farm ruminants in Siera Leone. Bull. Anim. Hlt. Prod. Afr. 35:308-313.

5. Ayaz M.M., Raza M.A., Murtaza S. and Akhtar S., (2013). Epidemiological survey of helminths of goats in southern Punjab, *Pakistan. Trop. Biomed*, 30, 62-71.

6. Dagnachew S., Amamute A. and Temegen W., (2011). Epidemiology of gastrointestinal helminthiasis of small ruminants in selected sites of North Gondar zone, Northwest Ethiopia, *Ethiopian Veterinary Journal*, 15, 57-68.

7. Ghanem Y.M., Naser M.H., Abdelkader A.H. and Heybe A. (2009). An Epidemio coprological study of protozoan and nematode parasites of ruminants in tropical semi-arid district of Somalilan D (North of Somalia). *Veterinary Medical Journal.* 3rd Science Congress. 10-12 May, pp. 768-787.

8. Honmade, (1998). Facter like breeding, feeding and management and disease control affect the meat potential.

9. Haq S. and Shaikh H., (1968). A survey of helminth parasiting the gastro-intestinal tracts of goats and sheep in East Pakistan. *East Pakistan Journal Veterinary Science*, 2, 54.

10. Hassan D.I., Mbap S.T. and Naibi S.A., (2013). Prevalence of Worm Infection in Yankasa sheep and West African dwarf goats in Lafia Town and Environs, Nigeria. IOSR *Journal of Agriculture and Veterinary Science* (IOSR-JAVS) ISSN: 2319-2372, Volume 4, Issue 4, 4 84-90.

11. Hawlader M.M.R., Mahbub-E-Elahi, A.T.M., Habib S. and Bhuyan M.J.U., (2002). "Gastro-intestinal nematode infestations in the black Bengal goats of Sirajgong district of sBangladesh," *J. Biol. Sci.*, 2, 556-557.

12. Margolis L., Esch G.W., Holmes J.C., Kuris A.M. and Schad G.A. (1982). The use of ecological terms in parasitology (report of an ad hoc committee of the American Society of Parasitologists). *J. Parasitol.*, 68: 131-133.

13. Miller H.M., Scheurman E.A. and Holmes J.H.G., (1984). The effect of heat and humidity on pregnant feral goats. *Animal Production*, 15, 41-544.

14. Mir M.R., Chishti M.Z., Dar S.A. and Rashid M., (2013). Bionomics of Helminth Parasites in Goats of Subtropical Jammu Area of J&K, *India. Agric. Biol. J. N. Am.,*

4(3): 316-319.

15. Njau, B.C., (1987). Gastrointestinal nematodes of small ruminants of king “roi” in Northern Tanzania. *Bull. Anim. Hlth. Prod.* Afr. 35:298-303

16. O’Connor L.J., Kahn L.P., Walkden-Brown S.W. (2007). Moisture requirements for the free-living, 142 (1-2):1-15.

17. Odoi A., Gathuma J. M., Gachuiri C. K. and Omore A. (2007). Risk factors of gastrointestinal nematode parasite infections in small ruminants kept in smallholder mixed farms in Kenya. *BMC Veterinary Research*.3:6:10.1186/1746-6148-3-6.

18. Rahman M.H., Soliman K.N., and Shaikh H., (1975). Proceedings of Pakistan Scientific Conference, University of Peahwar. 17.

19. Raza M., Younas and Schlecht E. (2014). Prevalence of gastrointestinal helminths in pastoral sheep andgoat flocks in the cholistan desert of Pakistan. *The Journal of Animal Plant Sciences*. 24(1):127-134.

20. Shirale S.Y. and Made D.K. (2007) Bionomics of helminth parasites in goats, *Indian Vet. J.*, 84, 1237-1239.

21. Sutar A.U., Kengar S.B., Patil S.S., Khan M.R. (2010). Prevalence of Gastrointestinal Parasites in Goats of Ahmed Nagar district of Maharashtra. *Veterinary World*, 3 (10): 456-457. Skykes AR 1994, Parasitism and production in farm animals. Animal Parasitology, 59, 155-172.

22. Soulsby E.J.L. (1982). Helminths, Arthropods and Protozoa of Domesticated Animals.

23. Urquhart, G. M., Armour, J., Duncan, J. L., Dunn, A.M. and Jennings, F.W. (1996).Veterinary Parasitology. Second Edition, Blackwell Science Ltd, Oxford, U. K. Blackwell Publishing Company.

Current Status of Researches in Biosciences (2020) : 419-425
Editor : Prof. P.C. Joshi, N. Joshi, Reshman Yasmin, and D. K. Mansotra
Today & Tomorrow's Printers and Publishers, New Delhi - 110 002

IN VITRO ANTI – INFLAMMATORY ACTIVITY OF *SYZYGIUM CUMINI*

Manojkumar Z. Chopda, Pinjari N.A., Patil A.S., Khare V.V., B. D. Talele and Namrata G. Mahajan

Department of Zoology, Animal House Division, Moolji Jaitha (Autonomous) College, Kavayitri Bahinabai Chaudhari North Maharashtra University, Jalgaon, Maharashatra, India 425002

E mail: mzczoo@gmail.com

ABSTRACT

Syzygium cumini has number of traditional uses; however anti-inflammatory action is not explored. In the present study, anti-inflammatory activity of hydroalcoholic extract of *Syzygium cumini* was evaluated. The prepared Jamun Juice Residue (JJR), Jamun Pulp Filtrate (JPF) and Jamun Pulp Residue (JPR) were tested for *in vitro* anti-inflammatory activity. Two different concentrations of hydroalcoholic extract JJR, JPF and JPR of *Syzygium cumini* have been evaluated for their *In vitro* anti-inflammatory activity. All test samples exhibited significant membrane stabilization activity, when compared with control. Therefore, it is noteworthy that, purification of active principle from extract will probably enhance the membrane stabilization activity than standard.

KEYWORDS: *Syzygium cumini*, Anti-inflammatory

INTRODUCTION

Inflammation is a complex process, which is frequently associated with pain and involves occurrences such as: increase of vascular permeability, increase of protein denaturation and membrane alteration. When cells in the body are damaged by microbes, physical agents or chemical agents, the injury is in the form of stress. Inflammation of tissue is due to response to stress. It is a defensive response that is characterized by redness, pain, heat, and swelling and loss of function in the injured area. Loss of function occurs depends on the site and extent of injury. Since inflammation is one of the body's nonspecific internal systems of defense, the response of a tissue to an accidental cut is similar to the response that results from other types of tissue damage, caused by burns due to heat, radiation, bacterial or viral invasion (Tortora and Reynolds, 1993). When tissue cells become injured they release kinins, prostroglandins and histamine. These work collectively to cause increased vasodilation (widening of blood capillaries) and permeability of the capillaries. This leads to increased blood flow to the injured site.

Inflammation is the reaction of living tissues to injury, infection or irritation. Lysosomal enzymes released during inflammation produce a variety of disorders which leads to the tissue injury by damaging the macromolecules and lipid peroxidation of membranes which are assumed to be responsible for certain pathological conditions as heart attacks, septic shocks and rheumatoid arthritis etc. The extra cellular activity of these enzymes is said to be related to acute or chronic inflammation. Stabilization of lysosomal membrane is important in limiting the inflammatory response by inhibiting the release of lysosomal constituents of activated neutrophil such as bactericidal enzymes and proteases, which cause further tissue inflammation and damage upon extra cellular release or by stabilizing the lysosomal membrane. HRBC or erythrocyte membrane is analogous to the lysosomal membrane and its stabilization implies that the extract may as well stabilize lysosomal membranes (Chippada *et al*., 2011). Stabilization of human red blood cell membrane (HRBC) by hypo tonicity induced membrane lysis can be taken as an *in vitro* measure of anti inflammatory activity of the drugs or plant extracts.

Plants have the ability to synthesize a wide variety of chemical compounds that are used to perform important biological functions, and to defend against attack from predators such as insects, fungi herbivorous mammals. At least 12,000 phyto compounds have been isolated so far; a number estimated to be less than 10% of the total. The chemical compounds in plants mediate their effect on the human body through processes identical to those already well understood for the chemical compounds in conventional drugs. Herbal medicines to have beneficial pharmacology activities, but also gives them the same potential as conventional pharmaceutical drugs to cause harmful side effects (Tapsell *et al*., 2006 and Lai and Roy, 2004). *Syzygium cumini* has number of traditional uses; however anti-inflammatory action is not explored. In the present study. Hence, an attempt has been made to investigate anti-inflammatory activity of hydroalcoholic extract of *Syzygium cumini*.

Collection of plant: Fruits of *Syzigium cumini* (Jamun) were collected from Campus of Moolji Jaitha (Autonomous) College, Jalgaon, (MS). The plant was identified by Dr. Maruti Deshattiwar, Department of Botany, Moolji Jaitha (Autonomous) College, Jalgaon.

Preparation of extract: Fruits washed thoroughly with tap water and then distilled water to remove the debris. Then seeds were separated from fruits. Obtained pulp was grind and from that juice was separated by filtration. The juice and pulp were shade dried.

After complete drying, juice was crushed and grinded to form coarse powder. Dried powder of juice was exhaustively extracted in Soxhlet apparatus with hydroalcohol (50:50). The solvent extracts so obtained were then filtered to remove any suspended impurities. Extract was concentrated under reduced pressure and controlled temperature (55^0C to 60^0C). At the time of experimentation extract was dissolved in aqueous solvent. The dissolved extract was filtered through whatmann's paper no. 46. The obtained filtrate labeled as Jamun Juice Filtrate. Collected residue later dissolved in hydroalcoholic solvent and labeled as Jamun Juice Residue (JJR).

After complete drying, pulp was crushed and grinded to form coarse powder. Dried powder of pulp was exhaustively extracted in Soxhlet apparatus with hydroalcohol (50:50). The solvent extracts so obtained were then filtered to remove any suspended impurities. Extract was concentrated under reduced pressure and controlled temperature (55^0C to 60^0C). At the time of experimentation extract was dissolved in aqueous solvent. The dissolved extract was filtered through whatmann's paper no. 46. The obtained filtrate labeled as Jamun Pulp Filtrate (JPF). Collected residue later dissolved in hydroalcoholic solvent and labeled as Jamun Pulp Residue (JPR). The extracts were preserved in dry, cool condition in desiccator. In present study Jamun Juice Residue (JJR), Jamun Pulp Filtrate (JPP) and Jamun Pulp Residue (JPR) extracts were screened for their *in-vitro* Anti-inflammatory activity.

Protocol for Anti-inflammatory activity

RBC membrane stabilization method: The anti-inflammatory activity was evaluated using RBC membrane stabilization method (Phanse *et al.*, 2012). Blood sample (2 ml) was collected from volunteer in a heparinized tube and mixed with equal volume of sterilized Alsever solution (2% dextrose, 0.8% sodium citrate, 0.5% citric acid and 0.42% sodium chloride). The blood was centrifuged at 3000 rpm and cell were washed with isosaline

(0.85%, pH 7.2) and a 10% (v/v) suspension was made with isosaline. The assay mixture contained the drug, 1ml of PBS (0.15M, pH 7.4), 2ml of hyposaline (0.36%) and 0.5ml of HRBC suspension. Instead of hyposaline 2ml of DW was used in the control. All the assay mixture were incubated at 37ºC for 30 min and then centrifuged. The haemoglobin content in the supernatant solution was estimated using spectrophotometer at 560 nm. The percent haemolysis was calculated by assuming the haemolysis produced in presence of DW as 100%.

Statistical analysis: All data were expressed as mean ± SE and the ANOVA was applied to determine the significance of the difference between the control group and experimental groups.

Table 1 *In Vitro* Anti-inflammatory activity of the JJR, JPP and JPR extracts.

Group (μg/ml)	**Percentage protection**
Control	00.00 ± 00.00
Standard	98.66 ± 0.08
JJR100	8.31 ± 0.17***
JJR200	7.59 ± 0.16***
JPR100	7.38 ± 0.13***
JPR200	11.65 ± 0.15***
JPF100	20.92 ± 0.22***
JPF200	11.69 ± 0.19***

Std. 100 μg/ml, each value expressed as mean ± SE, ***P<0.001 Vs Control

RESULTS AND DISCUSSION

Two different concentrations of hydroalcoholic extract JJR, JPF and JPR of *Syzygium cumini* exhibited excellent results (Table1). JJR 200 μg/ml and JPR 100 μg/ml exhibited near about equal membrane stabilization activity, 7.59 ± 0.16 and 7.38 ± 0.13 respectively. This is also true for, high concentration of JPR and JPF which showed equivalent membrane stabilization activity, 11.65 ± 0.15 and 11.69 ± 0.19 respectively. Among all test extracts, JPF100 μg/ml was most potent *i.e.* 20.92 ± 0.22% protection.

High concentration (200 μg/ml) of JJR and JPF were less effective (7.59 ± 0.16 and 11.69 ± 0.19) as compared to low concentration of extract. The JJR and JPF extracts exhibited membrane stabilization activity is not dose dependent. This may indicates that the activity is enzymatic hence; substrate concentration is responsible at the 100 and 200μg/ml.

The JPR extract exhibited membrane stabilization activity as dose dependent. At the dose 200 μg/ml of JPR extract revealed a near about double % of inhibition activity (11.65 ± 0.15%) as that of dose 100 μg/ml (7.389 ± 0.13%); are significantly different than standard (98.66 ± 0.08%). However, it is interesting to note that, the standard is a pure compound, whereas, the all extracts contains complex of active ingredients. Therefore, it is noteworthy that, purification of active principle from extracts will probably enhance the membrane stabilization activity than standard.

Jain *et al.,* (2010) evaluated carrageenan induced paw edema in Wistar rats at the dose level of 200 and 400 mg/kg administrated ethyl acetate and methanol extracts of *S. cumini* leaves orally. Both the extracts exhibited significant anti-inflammatory activity, which supports the traditional medicinal utilization of the plant. Very recently, Ajiboye *et al.,* (2018) investigated anti-hyperglycemic and anti-inflammatory activities of polyphenolic-rich extract of *Syzygium cumini* leaves in alloxan induced diabetic rats. All concentrations were significantly ($P<0.05$) reduced glucose level and inflammation with the administration of polyphenolic-rich extract of *Syzygium cumini* leaves. On the other hand significant ($P<0.05$) increase in glycogen and insulin concentrations, pancreatic β-cell scores, antioxidant enzymes and hexokinase activities, as well as glucose transporter levels in diabetic animals administered with polyphenolic-rich extract of *S. cumini* leaves.

The methanol extract of leaves from *S. cumini* (MESC) was evaluated for anti-inflammatory activity in experimental acute (carrageenan, histamine and serotonin induced rat paw oedema) and chronic models (cotton pellet induced rat granuloma) by Roy *et al.,* (2011). In all the models, the MESC (100 and 200 mg/kg body wt. p.o.) exhibited significant anti-inflammatory activity ($P<0.001$) in a dose dependent manner. These findings revealed that the *S. cumini* leaf had remarkable acute and chronic anti-inflammatory actions in the tested rodent models.

Our results are in concord with the results of Jain *et al*., (2010); Roy *et al.,* (2011) and Ajiboye *et al.,* (2018), only the difference is that, they use leaves and in present study we use fruit pulp.

The essential oils from *S. cumini* and *P. guajava* were confirmed as useful to inhibit inflammatory diseases, by mechanisms that include the inhibition of eosinophil and, to a lesser extent, neutrophil and mononuclear cell migration (Siani *et al.,* 2013). In correlation with this we may conclude that, pulp of fruit contains partial presence of oily moiety, which exhibits same mechanism as described by Siani *et al.,* (2013).

Chanudom and Tangpong, in 2015; studied the effects of *Syzygium cumini* (L.) Skeels aqueous extract (SCC) on anti-free radical, anti-inflammation, and antiulcer on indomethacin induced acute gastric ulceration. SCC showed restoration of the imbalance of oxidative damage leading to amelioration of cyclooxygenase enzyme (COX). In conclusion, SCC acts as an antioxidant, anti-inflammation, and antiulcer against indomethacin. Our results are in agreement with that of Chanudom and Tangpong (2015).

Hence, *Syzygium cumini* plant supports its use in various formulations and in control of various diseases.

CONCLUSION

In present study *Syzygium cumini* showed significant antiinflammatory activity *in vitro* which could partly contribute to ethnomedicinal uses of plants. However, further investigation is required for isolation and its mode of action is suggested for the development of a new drug in the treatment of control of inflammation in general. It is well understood that, the rich diversity of forest and field biota provides a fruitful source for the discovery of life saving drugs.

REFERENCES

1. Tortora G. J., Reynolds S. (1993). eds. Principles of Anatomy and Physiology. Harper Collins College Publishers, 7th edition: pp 695

2. Chippada S. C., Volluri S. S., Bammidi S. R. and Vangalapati M. (2011). *In vitro* anti inflammatory activity of methanolic extract of *Centella asiatica* by HRBC membrane stabilisation *Rasayan J Chem* 4(2), 457-460.

3. Tapsell L.C., Hemphill I., Cobiac L. *et al.,* (2006). Health benefits of herbs and spices the past, the present, the future, *Med. J. Aust,* (4 Suppl).;185: S4-24.

4. Lai P.K., Roy J. (2004). Antimicrobial and chemopreventive properties of herbs and spices, *Curr Med Chem*. June.; 11(11):1451-60.

5. Ajiboye B. O., Ojo O. A., Akuboh O. S., Abiola O. M., Idowu O., and Amuzat A. O. (2018). Anti-Hyperglycemic and Anti-Inflammatory Activities of Polyphenolic-Rich Extract of *Syzygium cumini* Linn Leaves in Alloxan-Induced Diabetic Rats. *J Evidence-Based Integrative Medicine.* 23: 1-8.

6. Chanudom L. and Tangpong J. (2015). Anti-Inflammation Property of *Syzygium cumini* (L.) Skeels on Indomethacin-Induced Acute Gastric Ulceration. *Gastroenterology Research and Practice* 12.

7. Phanse M. A., Patil M. J., Abbulu K., Chaudhari P. D. and Patel B. (2012). *In-vivo* and *in-vitro* screening of medicinal plants for their anti-inflammatory activity: an overview. *Journal of Applied Pharmaceutical Science* 02 (06):19-33.

8. Jain A, Sharma S, Goyal M, Dubey S, Jain S, Sahu J, Sharma A, Kaushik A. (2010).

Anti-inflammatory activity of *Syzygium Cumini* leaves. *Inter J Phytomedicine,* 2:124-126.

9. Roy A., Bhattacharya S., Pandey J. N., Biswas M. (2011). Anti-inflammatory activity of *Syzygium cumini* leaf against experimentally induced acute and chronic inflammations in rodents. *Alternative Medicine Studies;* (1) e6: 23-25.

10. Siani A.C., Souza M. C., Henriques M. G. M. O., and Ramos M. F. S. (2013). Anti-inflammatory activity of essential oils from *Syzygium cumini* and *Psidium guajava. Pharm Biol*, 51(7): 881–887.

Current Status of Researches in Biosciences (2020) : 427-440
Editor : Prof. P.C. Joshi, N. Joshi, Reshman Yasmin, and D. K. Mansotra
Today & Tomorrow's Printers and Publishers, New Delhi - 110 002

MATERNAL NUTRITIONAL SUPPLEMENTATION ATTENUATES THE PROTEIN MALNUTRITION INDUCED REACTIVE GLIOSIS AND COGNITIVE DYSFUNCTION IN F1 RATS

Shrstha Sinha

School of Studies in Neuroscience and Zoology, Jiwaji University, Gwalior

Email: shrsthasinha@gmail.com

ABSTRACT

Poor nutritional status is the leading cause to impaired fetal growth in terms of low body weight and body length. We designed a rat model to mimic the conditions of low resource population with pregestational, gestational and lactational protein malnutrition and supplement the pregnant females throughout gestational and lactational period to assess glial cells turnover, oxidative stress and cognitive functioning in F1 rats. *Spirulina* supplementation reduces the reactive gliosis, MDA level, accelerates the acquisition of neurological reflex and improves cognitive abilities in protein deprived rats. Thus, commencing nutritional supplementation may provide therapeutic strategy to counteract malnutrition induced deleterious effects.

Keywords: Maternal malnutrition; glial cells; behavior and cognition; *Spirulina*

INTRODUCTION

Malnutrition is a prime non genetic factor influencing the developing nervous system. It influences cellular differentiation, migration and myelination. Maternal and infant malnutrition is found to be associated with behavioural and cognitive impairment throughout the childhood and adulthood that makes them more vulnerable to neuropsychiatric disorders. Glial cells are known to be the major mediators of neuroinflammation. Pathophysiology of many neurodegenerative diseases have been associated with chronic neuroinflammation and activation of glial cells. Inhibition of microglia and astrocyte activation may be an effective therapeutic approach against neurodegeneration.

Both *in utero* and *ex utero* protein malnutrition affects CNS development and results in progression of neuropsychiatric disorders including depression, Schizophrenia, aggression and antisocial behavior (Duran *et al.*, 2005; Galler *et al.*, 2005). According to Global Hunger Index 2017, 13% of world's population is undernourished with 27.8% stunted and 9.5% wasted children. Prevalence of both under nutrition and malnutrition is very high in under developed and developing countries particularly in Africa and South Asia. Various environmental and genetic factors play an important role in intrauterine growth retardation (IUGR). Amongst many variables (genetic and/or environmental factors), nutrition is a key regulator governing proper fetal growth and development. A strong association exists between early life nutritional insult and later life health and metabolic disorders, beginning before gestation, continuing in utero and extends throughout childhood and adulthood (Bale *et al.*, 2010; Brown and Susser, 2008). Retrospective studies in human populations support a link between early life nutritional insult and an increased risk for schizophrenia development. Developmental origins of health and disease (DOHAD) hypothesis states that developmental patterns of early life programs individual health at older ages (Barker *et al.*, 2002). Moreover, food quality during pregnancy and lactation directly affects brain development, children's behavior and mental performance, which may leave long lasting permanent effects. Accumulating evidence suggests that excessive generation of reactive oxygen species and reactive nitrogen species stimulates pathway involved in neurodegeneration and neuroinflammation (Gemma *et al.*, 2002).

We need to develop economically viable nutraceuticals to combat the nutritional deficiency. For this, we used *Spirulina* as a source of protein. *Spirulina* is progressively used as a dynamic nutraceutical against aging and neurodegeneration. *Spirulina* has enormous therapeutic potential and is known to affect glial activation and oxidative stress, which is a central element underlying pathophysiology of many neurodegenerative disorders (Sinha *et al.*, 2018). This is the first study of its kind emulating the clinical conditions of under privileged low socioeconomic group presenting a larger restriction protocol with pregestational, gestational and lactational protein malnutrition and assessing how nutritional rehabilitation with Spirulina during gestation and lactation period would positively influence the development in F1 malnourished rats. In the present study, we analyzed the effects of dietary supplementation with *Spirulina* during pregnancy and lactation to protein malnourished rats. We have investigated the effects of nutritional supplementation using *Spirulina* on glial cells, dendritic arborization, neurobehavioral and cognitive development of F1 malnourished rats by using immunocytochemical procedures, Golgi staining and neuro-cognitive tools at 3, 6 and 12 month of age.

Diet Composition: The constituent of normal control (NC) and Low protein diet (LP) procured from National Institute of Nutrition (NIN) Hyderabad, India, were as follows:

Table 1: Diet composition

S. No.	Constituents of Diet	8% Protein (LP)	20% protein (NC)
1	Wheat	56.20%	22.50%
2	Bengal gram	0	60.00%
3	Groundnut oil	30.00%	4.00%
4	Soyabean oil	3.00%	0
5	Casein	0	4.00%
6	Skimmed milk powder	6.00%	5.00%
7	Vitamin mixture	0.50%	0.50%
8	Mineral mixture	0.50%	0.40%
9	Starch	23.90%	0
10	NaCl	0.30%	0
11	Di calcium phosphate	0.40%	0
12	Calcium carbonate	0.90%	0
13	Choline chloride	0.20%	0
14	L-Cysteine	0.10%	0

Experimental Model: Three-month old healthy Sprague Dawley rats of approximately 200-220 gm body weight were used in this study. The rats were housed under standard laboratory conditions of 12: 12 hours of light: dark cycle, 23 ± 2°C room temperature and 50 – 65% humidity with ad libitum access to food and water. The animals were retained in translucent propylene cages with dust free husk bedding. All the protocols followed

CPCSEA guidelines and were pre-approved by Institutional Animal Ethics Committee, Jiwaji University, Gwalior (M.P.).

Female rats were switched to low protein **(8% protein; n=16)** or normal protein **(20% protein; n=16)** diet, 15 days before conception. The male and female rats were put for mating in the ratio 1:2 and left overnight. Vaginal smear test was used to confirm time pregnancy and marked as Gestational Day 0 (GD0) only when estrous stage with sperm was observed under light microscopic examination. Half of the low protein and normal protein diet fed pregnant females were grouped as low protein maintained rats **(LP group; n=8)** and normal control rats **(NC group; n=8)** while other half set of females were supplemented with *Spirulina* in both LP and NC group and marked as *Spirulina* supplemented malnourished rats **(LP SPI; n=8)** and *Spirulina* supplemented normal rats **(NC SPI; n=8)** respectively. Rats were maintained on the same diet throughout gestational and lactational periods. *Spirulina* was orally administered to LP SPI and NC SPI pregnant females from gestational day 0 to the entire lactational period till they were weaned from their mother (400 mg/kg/ b.wt.). Pups born from these groups (F1 generation) were used in the study.

Vaginal smear test for time pregnancy: For this, vagina of mated females was washed using small amount of (approx. 0.2 ml) PBS with the help of pasture pipette or by a disposable dropper. The tip of the pipette was introduced gently into the entrance of the vagina and drawn back into the pipette 2 or 3 times by gently squeezing and releasing the bulb of the pipette. A small amount of this vaginal washing was expelled onto a labeled glass slide for the preparation of smear. Then these smears were examined under the microscope at earliest. Presence of estrous stage with sperms in the vaginal smear confirmed the fertilization process and the day was considered as gestational day 0 (GD0).

Whole body Perfusion: Transcardial perfusion is a process commonly used for tissue fixation for Histopathological studies. The aim of the perfusion is to preserve the tissue in life like condition with the help of fixative. Fixative protects the ultra-structure of the tissue by forming protein- protein cross linking and minimizes the degradation of the tissue.

Perfusion setup: Perfusion setup consists of two cylindrical containers A and B. Perfusion pressure was obtained by placing the containers at 120 cm above the rodent to match the physiological pressure. A butterfly valve was attached at the distal end to which tubes from both containers were attached.

Container A was filled with chilled 2% paraformaldehyde prepared in phosphate buffer (pH= 7.4). Container B was filled with phosphate Buffer Saline (pH= 7.4).

Method of perfusion: Sprague Dawley rats of different time points were deeply anesthetized by diethyl ether vapours in a desiccator. The rats were placed on to a dissecting tray and an incision was made through abdomen to reach the thoracic segment in order to expose the heart and PBS was flushed through a needle injected to left ventricle of heart. Concurrently, a cut was made in right atrium to flush out all the blood by pumping mechanism of heart. Later when all blood was flushed out and organ turned pale in colour the PBS flow was stopped and fixative (2% paraformaldehyde) was allowed to flow through the needle to fix the tissues. As soon as the paraformaldehyde is flushed a clear rigor mortis was seen in the limbs of rat.

Dissecting the Brain Tissue: After the fixation of tissues, the rats were placed facing their dorsal side on top. The cervical bone was cut with bone cutter to isolate the brain from the body. After that through slow incision the brain was isolated by removing the meningeal layers slowly and firmly. Then the spinal cord was dissected out from rats of different time points. The dissected out spinal cord was post fixed in 2% paraformaldehyde at 4°C, overnight.

Tissue processing for Cryosectioning: After overnight fixation at 4°C, three washes of Phosphate Buffer (0.001M) for 30 minutes each was given to remove excess amount of fixative (2% paraformaldehyde). The tissues were then shifted to sucrose gradient series of 10%, 20% and 30% (prepared in 0.01M PBS) till the tissues sank in each of the gradients.

Cryosectioning: Cryosectioning was done with the help of Leica CM 1900 Cryotome. The machine was handled as per the manual.

Cryoprotected tissues were taken and fixed on the tissue holder with the help of tissue freezing medium (cryomount). Sections of spinal cord were cut at a thickness of 15μm. These sections were collected on the gelatin coated slides and were stored at -20°C for further studies.

Time points of the study: According to various time points, brain of rats was dissected out after perfusion procedure. The selected time points are postnatal day 6, 30, 90, 180 and 365.

Immunohistochemical localization of Astrocytes and Microglia: Pups from each group were perfused transcardially with 0.01M phosphate buffer saline (PBS) followed by 2% paraformaldehyde in 0.01M phosphate buffer (PB). The brain was dissected out carefully and weighed. The region of interest (cerebral cortex) was post fixed overnight at 4°C in the same fixative. Then, it was washed with PB, cryoprotected with sucrose gradients (10%, 20% and 30% sucrose) and finally coronal sections of 14 μm thickness were cut serially using cryostat (Microm HM 525, Thermo Scientific), collected on Chromalum gelatin coated slides and stored at -20°C until used for immunohistochemical analysis.

Localization of astrocytes and microglia (Iba-1, OX-6, OX-42): Randomly selected cryocut sections from each parameter were air-dried for 1 hour. Three washing were given to the slides in PBS (Phosphate Buffer Saline) to remove excess of cryomount from the section. Sections were kept in 1% Triton X 100 for 20 minutes to permeabilize membrane. Subsequently, three washes were given in PBS to remove excess Triton X 100. Then the slides were placed in 1% H_2O_2 to block endogenous peroxidase activity for 20 minutes followed by 3 consecutive PBS washings of 5 minute each to remove excess hydrogen peroxide. For non-specific protein blocking, sections were incubated with 1% normal serum in PBS (Vector Rabbit IgG kit, PK6101) for 2 hours in a humid chamber at room temperature. Then these sections were incubated with anti-GFAP and anti-S100β antibody for astrocytes or anti Iba-1, OX 6 and OX 42 antibody for microglia and left overnight at 4°C. Next day, the slides were brought to room temperature and three washes for 5 minutes each were given in PBS to remove the excess unbound primary antibody. Subsequently, the sections were incubated with biotin labelled secondary antibody (1:100 in 1% BSA in PBS, Vector Rabbit IgG kit, PK 6101) for 2 hours at room temperature. Then, three washes of PBS were given to remove the unbound secondary antibody. This was followed by incubation with tertiary reagent streptavidin-biotin HRP complex (1:200 in 1% BSA in PBS, Vector Rabbit IgG kit, PK 6101) for 2 hours at room temperature. The sections after washing with PBS were finally visualized with solution containing DAB (3'3-diaminobenzidine tetra hydrochloride; Sigma Aldrich) and hydrogen peroxide in PBS for 20 minutes. The sections were thoroughly washed with running tap water, air dried, dehydrated with ethanol, cleared in Xylene and mounted in DPX.

Quantification and image analysis: All immunostained slides were observed under Leica Laborlux microscope equipped with digital camera (DFC420 C, Leica, Germany). Images were captured from CA1, CA3 and DG region of Hippocampus. The regions for quantification were includes three frames per area per section in the three sections per animal in all the groups. The area of each frame in all groups was fixed by 21670.9 μm^2.

Cell counting was performed using Leica Qwin software (Version 3.1) and data were presented as total number of cells/mm^2. Both GFAP and Iba-1 immunostained sections were also viewed with Leica DM6000 microscope for analyzing the response of astrocytes and microglia in maternally protein deprived and *Spirulina* supplemented rats.

Estimation of Lipid peroxidation: Malondialdehyde (MDA), a major end product of lipid peroxidation, was estimated by the method of Jacobson *et al.*, A 10% of tissue homogenate was prepared in Tris HCl buffer. Prior to homogenization, 10 μl of 0.5 M butylated hydroxytoluene (BHT) in acetonitrile was added per 1 ml of tissue homogenate. After homogenization, the homogenates were centrifuged at 3000 g at 4°C for 10 min and clear supernatant was used for the assay. Briefly 200 μl of supernatant was transferred to 650 μl of 10.3 mM 1-methyl-2-phenylindole in acetonitrile and vortex mixed. To assay MDA alone, 150 μl of 37% HCl was added, vortexed and incubated at 45°C for 60 min. After incubation samples were kept on ice, centrifuged at 9500 g for 5 min and absorbance was measured at 586 nm. The level of MDA is expressed as nmol g^{-1} tissue using extinction coefficient 1.1×10^5 M^{-1} cm^{-1}.

Estimation of superoxide dismutase (SOD) and catalase (CAT)

Superoxide dismutase (SOD) activity was assayed by estimation of inhibition of auto oxidation of epinephrine. A 10% of tissue homogenate was prepared in 0.9% NaCl, centrifuged at 15,000 g for 15 min and the corresponding supernatant was used for enzyme assay. Reaction mixture containing 0.5 ml sodium carbonate buffer (0.3 M, pH 10.2), 0.5 ml EDTA (0.6 mM), 0.5 ml homogenate and 1.0 ml distilled water was incubated at room temperature for 10 min. Reaction was initiated by addition of 0.5 ml epinephrine (1.8 mM) and absorbance change per min was recorded for 5 min at 480 nm. Specific activity is expressed as % inhibition of auto oxidation of epinephrine by the enzyme min^{-1} mg^{-1} protein.

Catalase (CAT) activity was estimated by the method of Aebi *et. al.*, 1984. A 10% of tissue homogenate was prepared in 1.15% KCl, centrifuged at 5,000 g for 10 min and the supernatant thus obtained was used for enzyme assay. Reaction mixture containing 0.8 ml phosphate buffer (K_2HPO_4/ NaH_2PO_4, 50 mM, pH 7.0), 0.1 ml homogenate and 0.1 ml triton X-100 (0.02%) was incubated at room temperature for 10 min. Reaction was initiated by addition of 2.0 ml H_2O_2 (0.03 M prepared in potassium phosphate buffer, pH 7.0) and absorbance change per min was recorded for 5 min at 240 nm. Specific activity is expressed as micromole H_2O_2 decomposed min^{-1} mg^{-1} protein.

Estimation of Lipid peroxidation: Malondialdehyde (MDA), a major end product of lipid peroxidation, was estimated by the method of Jacobsson *et al.* A 10% of tissue homogenate was prepared in Tris HCl buffer. Prior to homogenization, 10 μl of 0.5 M butylated hydroxytoluene (BHT) in acetonitrile was added per 1 ml of tissue homogenate. After homogenization, the homogenates were centrifuged at 3000 g at 4°C for 10 min and clear supernatant was used for the assay. Briefly 200 μl of supernatant was transferred to 650 μl of 10.3 mM 1-methyl-2-phenylindole in acetonitrile and vortex mixed. To assay MDA alone, 150 μl of 37% HCl was added, vortexed and incubated at 45°C for 60 min. After incubation samples were kept on ice, centrifuged at 9500 g for 5 min and absorbance was measured at 586 nm. The level of MDA is expressed as nmol g^{-1} tissue using extinction coefficient 1.1×10^{5} M^{-1} cm^{-1}.

Early biological changes assessment: Daily based body weight gain in pregnant females from all the four groups were observed between 9:00 to 10:00 AM in the morning. The total number of pups born (litter size) and number of dead pups (mortality) were also recorded. Body weight of F1 generation rats was examined daily up to PND 30 and twice a week after 1 month to 6 month of age.

Behavioral and Cognitive Studies: A sequence of behavioral and cognitive tests was performed at 3 and 6 month of age to study spontaneous activity, neuromuscular strength, anxiety and spatial learning and memory.

Behavioral and cognitive test batteries

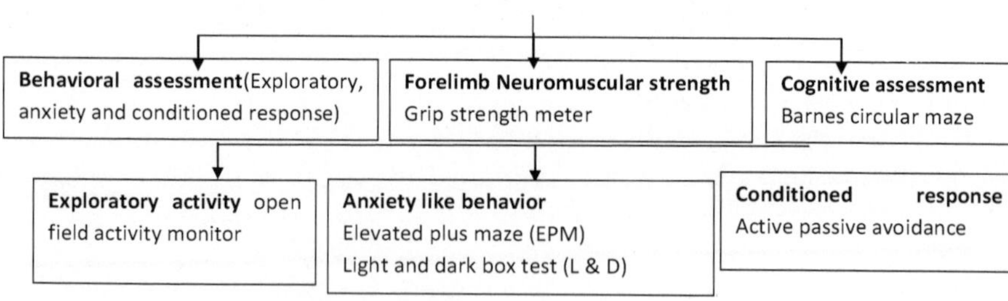

RESULTS

Spirulina supplementation to malnourished dams significantly ameliorated both viability and lactation indices as compared to low protein fed **(LP)** dams thus depicting beneficial effects of *Spirulina* supplementation against protein malnutrition **(PMN)** induced poor reproductive performance and low survival rate.

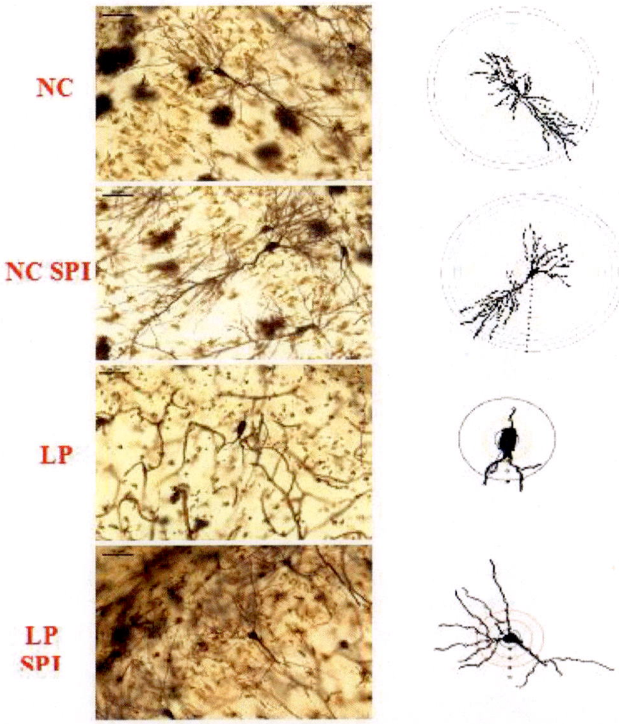

Fig 1: Light microscopic images of Golgi stained sections showing increased dendritic arborization in *Spirulina* supplemented protein deprived rats (LP SPI) with elongated apical dendrites as compared to age matched low protein diet fed rats (LP).

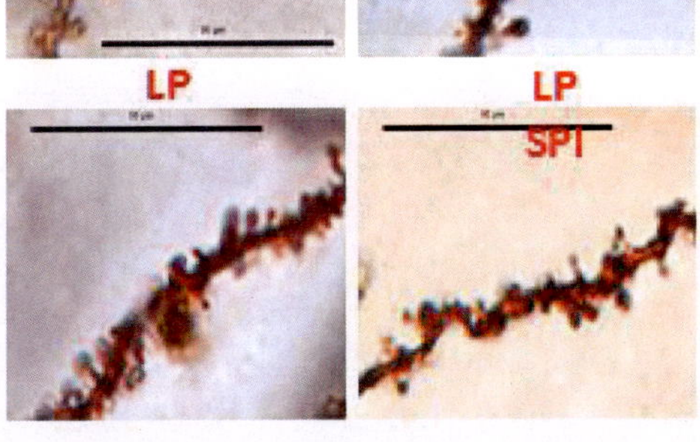

Fig 2: Microscopic images showing increased mushroom spine (healthy spine) density in *Spirulina* supplemented rats (LP SPI), whereas LP rats showed increased density of thin and stubby spines as revealed by Golgi staining

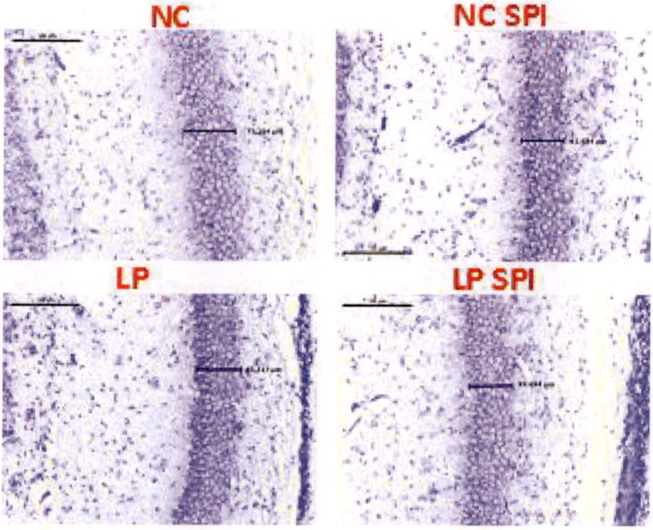

Fig 3: Photographs showing results of cresyl violet staining where, LP SPI group showed high density of functional neurons whereas LP rats showed crowded population of non-functional neurons

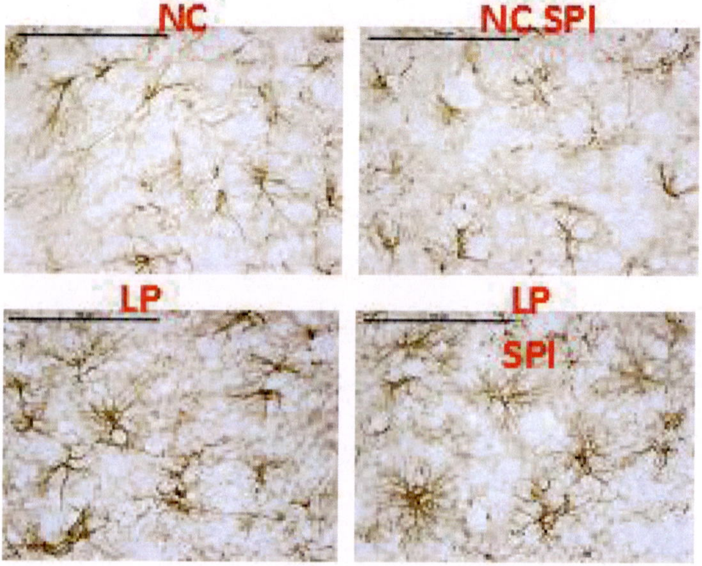

Fig 4: GFAP immunolabeling showed reactive astrogliosis in LP rats wheres LP SPI group showed comparatively reduced population of active astrocytes

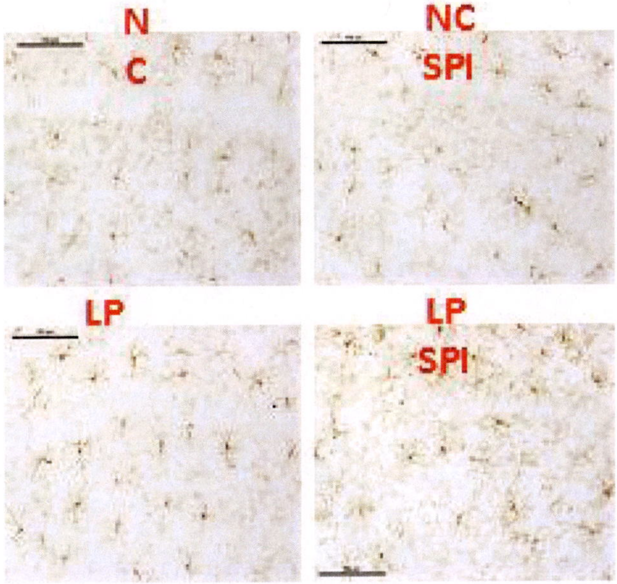

Fig 5: Iba-1 immunolabeling showed reduced microglial activation in LP SPI rats whereas, LP rats showed significantly increased population of reactive microglial cells

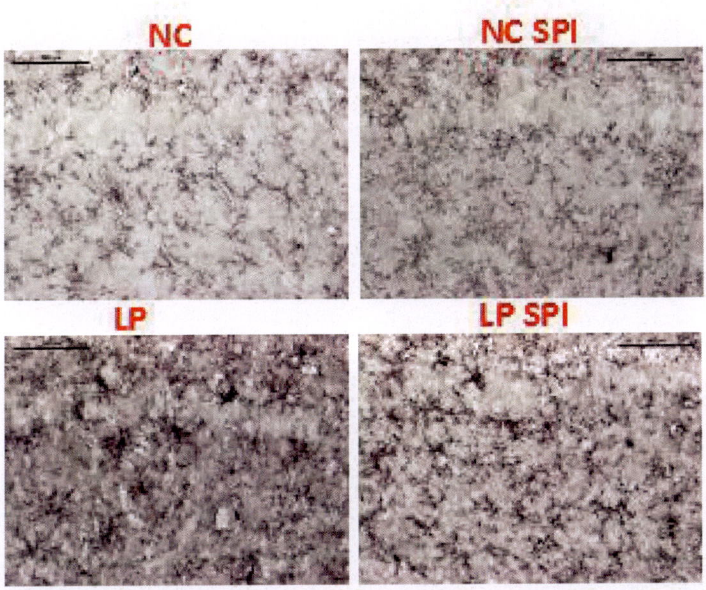

Fig 6: OX42 immunolabeling showed significantly reduced population of OX-42 positive cells in LP SPI rats, whereas LP rats showed drastic increase in OX-42 immunopositive cells, further validating the existence of reactive microgliosis in LP rats

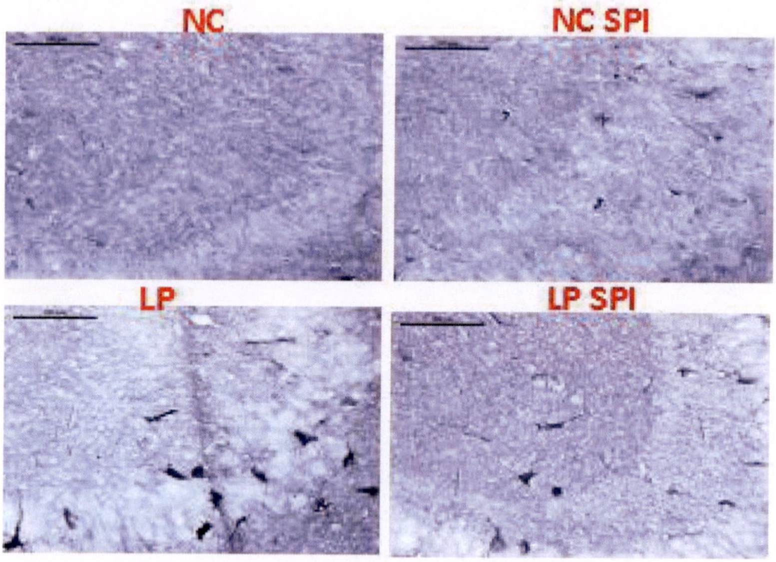

Fig 7: Photographs showing results of OX 6 immunolabeling where, LP SPI rats showed significantly reduced population of OX 6 positive cells as compared to age matched LP rats

The main findings of the present study are-

- Maternal *Spirulina* supplementation to LP group significantly reduced percent mortality from 43% to 10%.
- LP F1 rats presented severe fur loss with stunted body growth prominently up to 3 months of age whereas maternal *Spirulina* supplemented F1 malnourished rats **(LP SPI)** showed better physical appearance in terms of both hair and body growth compared to age matched LP F1 group.
- Body weight, brain weight and brain to body weight ratio was affected by maternal protein level, supplementation and in interaction of these two factors.
- Better neurological reflex development (cliff avoidance and negative geotaxis) in F1 progeny of *Spirulina* supplemented malnourished rats
- LP SPI rats showed comparatively reduced level of MDA thus, suggesting reduced oxidative stress in *Spirulina* supplemented malnourished rats.
- LP SPI rats also displayed significantly increased dendritic

arborization in hippocampal region as compared to LP rats suggesting possible therapeutic nuroprotective property of Spirulina supplementation in protein deprived rats.

- Comparatively reduced reactive gliosis in LP SPI rats as revealed by significantly reduced expression of GFAP, Iba-1, OX 6 and OX 42 immunolabeling might contribute in the amelioration of observed behavioral and cognitive benefits.

DISCUSSION

Spirulina has enormous therapeutic potential and neuroprotective properties and is known to affect glial activation and oxidative stress, which is a central element underlying pathophysiology of many neurodegenerative disorders. Glial cells including astrocytes and microglia are known to be the major mediators of neuroinflammation. Pathophysiology of many neurodegenerative diseases including Alzheimer's disease (AD), Parkisons's disease (PD), Huntington's disease (HD), amyotrophic lateral sclerosis (ALS) and age related muscular degeneration (ARMD) have all been associated with chronic neuroinflammation and activation of both astrocytes and microglia. Inhibition of microglia and astrocyte activation may be an effective therapeutic approach against neurodegeneration.

The present study demonstrates that maternal supplementation with 400mg/kg b.wt. of *Spirulina* reduces (upto some extent) the effects arising from protein malnutrition. The data showed that *Spirulina* supplementation reduces the astrocytic and microglial activation with improved behavioral and cognitive abilities suggesting a possible link between glial activation and cognitive impairment. LP group showed intense GFAP and Iba-1 labeling with astrocytic and microglial activation in terms of their phenotypic changes with hypertrophic and dystrophic cell body whereas LP SPI group presented normal resting morphology. In a similar way, LP SPI group showd reduced MDA level, indicator of lipid peroxidation. The pro-oxidant/anti-oxidant balance is necessary for proper body functioning. *Spirulina* has marked antioxidant properties and its consumption attenuates PMN induced oxidative stress as indicated by decreased MDA content in our study. High protein content and low cost of *Spirulina* makes it suitable adjunct to combat malnutrition. Our study opens up new possible window using *Spirulina* to treat neurological disorders arising from protein malnutrition.

CONCLUSION AND FUTURE DIRECTIONS

Dietary supplementation with *Spirulina* exerts positive effects on F1 malnourished rats in terms of reduced activation of astrocytes and microglia, increased dendritic arborization, improved spatial learning and memory, partly normalized anxiolytic and hyperactive behavioral profile, enhanced neuromuscular strength and passive coping strategy against anxiogenic stimuli as compared to low protein fed rats. The study revealed the potent role of maternal supplementation in combating neurodegenerative and neuropsychiatric disorders resulting from malnutrition in progeny's later life and suggests possible role of *Spirulina* in modulating the interaction between nutrition, glia and behavior in aging rats.

REFERENCES

1. Aebi H (1984). Catalase in vitro- In Methods in enzymology. *Academic Press, 105*(1), 121-126).

2. Bale, T.L., Baram, T.Z., Brown, A.S., Goldstein, J.M., Insel, T.R., McCarthy, M.M., *et al.* (2010). Early life programming and neurodevelopmental disorders. *Biol. Psychiat.* 68, 314-319. doi: 10.1016/j.biopsych.2010.05.028.

3. Barker, D.J., Eriksson, J.G., Forsén, T., and Osmond, C. (2002). Fetal origins of adult disease: strength of effects and biological basis. *Int. J. Epidemiol.* 31, 1235-1239. doi: 10.1093/ije/31.6.1235.

4. Brown, A.S., and Susser, E.S. (2008). Prenatal nutritional deficiency and risk of adult schizophrenia. *Schizophr. Bull.* 34, 1054-1063. doi: 10.1093/schbul/sbn096.

5. Duran, P., Cintra, L., Galler, J.R., and Tonkiss, J. (2005). Prenatal protein malnutrition induces a phase shift advance of the spontaneous locomotor rhythm and alters the rest/activity ratio in adult rats. *Nutr. Neurosci.* 8, 167-172. doi: 10.1080/10284150400026117.

6. Galler, J.R., Waber, D., Harrison, R., and Ramsey, F. (2005). Behavioral effects of childhood malnutrition. *Am. J. Psychiatry*. 162, 760-b. doi: 10.1176/appi.ajp.162.9.1760-b.

7. Gemma, C., Mesches, M.H., Sepesi, B., Choo, K., Holmes, D.B., and Bickford, P.C. *et al.* (2002). Diets enriched in foods with high antioxidant activity reverse age-induced decreases in cerebellar β-adrenergic function and increases in proinflammatory cytokines. *J. Neurosci.* 22, 6114-6120. doi: 20026622.

8. Jacobsson SO, Cassel GE and Persson SA (1999). Increased levels of nitrogen oxides and lipid peroxidation in the rat brain after soman-induced seizures. *Archives of toxicology, 73*(5), 269-73.

9. Sinha, S., Patro, N. and Patro, I.K., (2018). Maternal Protein Malnutrition: Current and Future Perspectives of *Spirulina* Supplementation in Neuroprotection. Front. Neurosci. 12. doi: 10.3389/fnins.0096.

Current Status of Researches in Biosciences (2020) : 441-446
Editor : Prof. P.C. Joshi, N. Joshi, Reshman Yasmin, and D. K. Mansotra
Today & Tomorrow's Printers and Publishers, New Delhi - 110 002

MORPHOLOGICAL STUDIES OF AVIAN MYCOPLASMA

Amit Kumar Ghoshal and S. P. Ghoshal

Charaktala, Duilya, Dist. Howrah (W. B.) Pin – 711302

ABSTRACT

Avian pleuro pneumonia-like organism is one of the etiological agent of chronic respiratory diseases of poultry. In the present study 40 isolates from 909 live, dead birds and dead in cell embryos. Size of avian mycoplasma colony varies from 0.05 to 0.2mm out of 44 isolates only 10 isolates were a nipple, while the rest of the isolates were having a nipple in the center of the colonies stained by dines stain.

KEYWORDS: Mycoplasma, Pneumonia, Etiological agent, Birds.

INTRODUCTION

Avian pleuro Pneumonia like organism is one of the etiological agents of respiratory diseases of poultry. It is not only a problem in India but also throughout the world causing great economic losses. In the present study, an attempt has been made to study the avian mycoplasma with special reference to their biological character.

MATERIAL AND METHODS

1) **Isolation Studies:** In the present study Isolates were Isolated from 963 live and dead birds and also in dead in shall embryos from different poultry farms.

2) **Preparation of culture media:** Heart infusion broth (PPLO Broth) and PPLO Agar. The medium were similar to that described by Agato *et al*. (1967).

Isolation of pleuro Pneumonia like organism in liquid media propagation. The Inoculum was directly put into the sterile PPLO Broth. The swab was shaken vigorously into the PPLO Broth to mix the material thoroughly and then swab removed. The inoculated PPLO broth was incubated at 37 ^{0}C for 48 hours. About 0.5 ml of inoculum was transferred to another tube containing sterile PPLO broth. In this 3 or 4 blind passages was given. The culture which showed turbidity inoculated on Agar plates.

Solid media propagation: The tubes showing signs of broth was subcultured and kept burning to decrease Oxygen tension for 3 to 4 days and examine daily. The plates did not show any growth as discarded. The plates showing broth were examined under Stereoscopic microscope or under low power of ordinary optical microscope (X50—100 magnification) for the detection of typical PPLO Colonies.

Preparation of Stock Culture: The Stock Culture was prepared by putting the agar blocks containing the colonies in the PPLO broth at 37 ^{0}C for 3 — 4 days.

Maintenance of PPLO Strains: The PPLO strains were sub-cultured for every 8 days interval when they were stored at 4 ^{0}C but when they were kept at 20^{0}C the subculturing was done every 4 - 4 months.

Characterization of the Isolates: To characterize the isolated PPLO Strains morphological and pathogenicity studies were carried out. Morphological studies — The liquid culture containing the PPLO organisms was centrifuged at 2000 rpm for one hour and smear was stained with gimsa's stain, Wright's stain or with gram's stain to study the merphological exanation of the organism and also see the contaminating bacteria if present.

Study of Colony Characterisation: To study the characterization of the PPLO Colonies, the same of the colonies were stained by Diame's method. The procedure of staining of the colonies of PPLO was described by Diane's and by Mortem *et al.* 1951.

Table-1: Morphological studies of PPLO colonies

S. No.	No. of Isolates	Size in mm.	Colony character	Staining character (Diene's stain)
1.	2.	3.	4.	5.
1.	B – 15	.05 - .07	WN	Faint blue throughout the colony
2.	B — 25	.07 - .1	N	Dense blue in the center and light at the periphery
3.	P – 12	.07 - .15	N	-do-
4.	Pa - 18	.05 – .1	WN	Faint blue throughout the colony
5.	P – 31	.08 - .1	WN	-do-

6.	B – 3	.06 - .1	N	Dense blue in the center and light at the periphery
7.	B – 9	.08 - .1	N	-do-
8.	B – 7	.05 - .07	WN	Faint blue throughout the colony
9.	B - 5	.05 - .15	N	Dense blue in the center and light at the periphery
10.	B – 19	.08 - .12	N	-do-
11.	Pa- 21 21	.05 – .1	WN	Faint blue throughout the colony
12.	Pa – 9	.06 - .2	N	Dense blue in the center and light at the periphery
13.	B - *3	.08 - .1	WN	Faint blue throughout the colony
14.	B - *11	.05 – .1	WN	-do-
15.	I – 27	.05 - .07	WN	-do-
16.	I – 21	.05 - .07	N	Dense blue in the center and light at the periphery
17.	I – 55	.05 – .1	N	-do-
18.	I – 29	.06 - .1	N	-do-
19.	I – 25	.08 - .1	N	-do-
20.	I – 39	.08 - .1	N	-do-
21.	I – 40	.05 - .07	WN	Faint blue throughout the colony
22.	I –41	.07- .15	N	Dense blue in the center and light at the periphery
23.	I – 72	05 - .1	N	-do-
24.	I – 105	.06 - .1	N	-do-
25.	I – 110	.06 - .1	N	-do-
26.	P – 3'	05 - .07	WN	Faint blue throughout the colony
27.	P – 9'	05 - .07	WN	-do-
28.	P – 3"	05 - .07	N	Dense blue in the center and light at the periphery
29.	B – 9"	.06 - .1	N	-do-
30.	S - 5	.06 - .1	N	-do-
31.	S – 3	.08 - .1	WN	Faint blue throughout the colony
32.	S - 15	.05 - .07	WN	-do-

33.	S – 25	.05 - .07	WN	-do-
34.	S – 23	.07- .1	WN	-do-
35.	S – 45	.1 - .15	N	Dense blue in the center and light at the periphery
36.	S – 71	.08 - .12	N	-do-
37.	S – 109	.07 - .1	WN	Faint blue throughout the colony
38.	P – 18	05 - .1	WN	-do-
39.	S – 149	.05 - .1	WN	-do-
40.	S – 177	.08 -.12	N	Dense blue in the center and light at the periphery
41.	**V - 76	.1 - .15	N	-do-
42.	**V – 77	.07 - .12	N	-do-
43.	**V - 79	.1 - .14	N	-do-
44.	**S - 6	.2 - .8	N	-do-

a) S = From the dead in shell embryos.

b) I = Indore Poultry Farm.

c) P = Poultry farm of Veterinary College, Mhow.

d) B = Poultry fam Of Institute Of Biological Products, Mhow.

e) Pa = Pathology Department, College of Veterinary Science and Animal Husbandry, Mhow (for dead birds) .

f) ** = The isolates obtained from the Veterinary College, Jabalpur.

Cultural Studies: The mycoplasma Isolates grew in PPLO broth at 27 ^{0}C to 37 ^{0}C but grew well at 37 ^{0}C when enriched with 1 percent PPLO serum fraction or 20 percent of sterile horse serum. All of these isolates of PPLO produced slight turbidity granularity in the liquid medium.

RESULTS

Isolation of Avian Mycoplasms: In this study 909 birds and dead in shell embryos were examined for isolation purpose :

i) Poultry farm No. 1 - 189

ii) Poultry farm No. 2 - 216 (trached swabe)

iii) Trached Swabe from live birds — 215 (Poultry farm No. 3)

iv) Tracheal Swabe from the live birds — 19 (Poultry Farm No. 4)

v) From dead in shell embryos (from yolk sae) - 100

The percentage of PPLO isolation from poultry farm No. 1 was 3.17 (6 isolates), poultry farm No. 11 - 4.63 (10 isolates) poultry farm No. 111 — 5.16 (10 isolates) whereas out of hundred dead birds only 3 percent was positive.

Morphological Studies: In the present study, 40 Isolates were isolated from 909 living, dead birds and dead in shell embryos.

Two types of colonies : -

a) Smooth, flat entire - without nipple.

b) Smooth entire having nipple in the center of the colony.

Then these PPLO Colonies were stained with Dieoe's stain. The colonies without nipple were uniformly stained while the colonies having nipple revealed dark blue center and light blue periphery.

Attempt to determine their relative Pathogenicity. Only 10 (ten) of the isolates caused embryos death within 7 (seven) days post-inoculation stunting of the embryos Stunting of embryos . Stunting of embryos sub—culaneous Lacmorr—lages, and odem were frequently noted.

The dead embryos were opened just after death while the live embryos were opened on the seven-day post-inoculation. The lesions of both inoculated and control embryos were studied. The yolk sae of both inoculated and control were studied. The yolk sae of both inoculated and control were studied. The yolk sae was further cultured for further isolation of the organism was not possible from yolk culture of control eggs.

SUMMARY

A total of 909 live and dead birds were used for the isolation purpose. Deed in shell embryos 191 were also used for isolation purpose from different poultry farms. The percentage of isolation of Avian PPLO from live birds was 4.35 while from dead birds was 3 percentage. It was also been demonstrated that the percentage of Avian PPLO from dead in shell was 5.23. The size of the avian mycoplasma colony varied from 0.05 — 0.2 mm.

Out of 44 isolates 20 isolates were without nipple, while rest of the isolates were having nipple in the centre of the colony. When these PPLO Colonies were stained with Diene's Stain, colonies with nipple revealed dark blue centre and light periphery while the colonies without nipple took faint blue through the colonies.

Different workers have suggested different media for the isolation of Avian PPLO, but in the study enriched heart infuation broth was used and give satisfactory results. For the study of Patnogenioity chick and chicken embryos inoculation were performed. Out of 44 isolates only 4 isolates produced moderate and 22 isolates showed mild lesins in the chick.

REFERENCES

1. Morton, H.E & Smith, P , William, N.B. and EiCkenbérd, C. F. (1951) Cited by Pathak R. D. and Singh C.M. (1961).

2. Ogata, M., Ota, T., Obara, A., & Pan, I. Z. (1967). Investigation on growth media for Mycoplasma: evaluation of infusions, peptones, sera, yeast extracts and other supplements. *Nihon juigaku zasshi. The Japanese journal of veterinary science*, *29*(5), 259.

Current Status of Researches in Biosciences (2020) : 447-453
Editor : Prof. P.C. Joshi, N. Joshi, Reshman Yasmin, and D. K. Mansotra
Today & Tomorrow's Printers and Publishers, New Delhi - 110 002

PERFORMANCE OF HF CROSSBRED CATTLE UNDER FIELD PROGENY TESTING PROGRAM IN PUNJAB, INDIA

A K Das, Ravinder Kumar and N V Patil

ICAR-Central Institute for Research on Cattle, Grass Farm Road, Meerut Cantt. Meerut -250001, Uttar Pradesh

E mail: achintya137@yahoo.com[1]

ABSTRACT

ICAR-CIRC executed Field Progeny Testing Program at GADVASU, Ludhiana and the 1st set of crossbred bulls was inducted in April 1995 and subsequent 13 batches were executed in January 1997 to June 2017. A total of 309 bulls in fourteen different batches have so far been inducted. A total 148283 Artificial Inseminations have been done and total 19862 female calves has been borned out of which 6153 female progenies have reached age at first calving. The 1st lactation milk yield of daughters in 1st set was recorded as 2697.8 kgs and in 12th set it was recorded as 3836.5 kgs.

KEYWORDS: Field Progeny Testing, Age at First Calving, First Lactation Milk Yield, Frieswal

INTRODUCTION

A synthetic strain of cattle named Frieswal is developed at ICAR-Central Institute for Research on Cattle in collaboration with Military farms. This Institute executed a Field Progeny Testing Program (FPT) under AICRP on cattle in collaboration with Agricultural Universities and Non-Government Organization at different agro-climatic regions of the country to evaluate crossbred cattle under field condition. At present this programme is running in association with Kerala Veterinary and Animal Sciences University (KVASU), Thrissur, Guru Angad Dev Veterinary and Animal

Sciences University (GADVASU), Ludhiana, BAIF Development Research Foundation, Pune, and G. B. Pant University of Agriculture & Technology (GBPUA&T), Pantnagar for the evaluation and genetic improvement of HF crossbred cattle. This HF Crossbred bulls have 50-75% exotic inheritance (Das *et al*., 2017). Holstein-Friesian crosses have gained most popularity among all crossbreds due to their highest milk production (Singh *et al.* 2000, Upadhyay *et al.* 2002, Akhter *et al., 2003).* Keeping in view the availability of large HF crossbred population in the country Field Progeny Testing (FPT) program was initiated in the year 1992. In Guru Angad Dev Veterinary and Animal Sciences University (GADVASU), Ludhiana the 1st set of crossbred bulls was inducted in April 1995 and subsequent 13 batches were executed in January 1997 to June 2017. The insemination work was undertaken through 31 Artificial Insemination Centers. The total 4451 daughters have completed their 1st lactation 305 days milk yield. A large number of cows are being inseminated in the program so that records of sufficient daughters are achieved for sire evaluation despite significant data loss (Kumar *et al.* 2015). Evaluation of bulls through progeny testing and their extensive use has been a major source of genetic improvement in dairy animals as reported by Ravinder *et. al.,* (2017). Selection of sires on the basis of daughter's first lactation milk production trait would be a better choice in crossbred cattle Singh *et al.* (2017). The present investigation was undertaken to study the performance of HF crossbred progeny born at GADVASU, Ludhiana under field condition.

MATERIALS AND METHODS

The present study was undertaken to determine the performance of HF crossbred cattle born under FPT program at Guru Angad Dev Veterinary and Animal Sciences University (GADVASU), Ludhiana, field unit from young bulls having 50-75% exotic inheritance in different sets. The insemination work was undertaken through 31 Artificial Insemination Centers. The Field Progeny Testing program envisages testing of 30 Frieswal (HF X Sahiwal) bulls in each batch. Fourteen different batches of crossbred bulls have so far been inducted since inception. A total of 6563 daughters from 14 batches of bull have completed their 1st lactation 305 day's milk yield records. The data was recorded from daughters of 309 bulls covering a period of 22 years from 1995 to 2017 under the field progeny-testing program. Each year of calving was divided into 4 seasons i.e. winter (December to February), summer (March to May) rainy (June to August) and post monsoon (September to November) based on climatological conditions. Data was also divided into 4 groups on the basis of year of calving. Bulls with less than 14 daughter's record were excluded from the study. The bulls were evaluated based on first lactation 305 days milk yield of their daughters in field conditions at farmers herd. The milk yield of cows was estimated using test day records. Bulls' breeding values were estimated through best linear unbiased prediction (BLUP) procedure using Model 8 in least squares and maximum likelihood computer package (Harvey, 1990).

RESULTS AND DISCUSSION

The Field Progeny Testing programme is covering adult population of about 13345 cows and 5550 breedable heifers. The insemination work of the Field Progeny Testing project is undertaken through 31 A.I. centres in Ludhiana district. Four of these centres are operated by Punjab State Department of Animal Husbandry and others by trained inseminators. The frozen semen of 14 different set of bulls, belonging to FPT of Frieswal Project, executed by ICAR-CIRC, Meerut. A total of 309 bulls have so far been inducted in the program. A total 148283 Artificial Inseminations have so far been done and a total 19862 female calves has been borned out of which 6563 female progenies have reached age at first calving through the use of semen of bulls under test (Table-1).

A total of 184, 154, 397, 361, 361,278, 351, 326, 647, 576, 608 and 208 progenies of 1st, 2nd, 3rd 4th 5th, 6th, 7th, 8th,9th, 10th, 11th and 12th set of test bulls completed their first lactation 305 days milk production, their average first lactation 305 days milk production was 2697.8±40.1, 2827.09±48.7, 2878.7±25.5, 2896.1±26.0, 2855.9±25.9, 3051.8±24.5, 3305.4±28.9, 3556.6±31.2, 3714.7±22.9, 3751.8±22.9, 3771.7±22.7 and 3836.5±34.5 kg. respectively (Annual Report 2018-19, ICAR-CIRC). The corresponding average age at first calving of the daughters was 1192.1±23.6, 1145.6±20.3, 1191.9±13.2, 1168.4±14.9, 1174.9±14.6, 1115.9±14.2, 1082.2±12.2, 1068.8±15.9, 1044.2±8.5, 1032.7±8.0, 1012.8±8.2 and 885.3±8.4 days respectively (Table 2).

Table1. Information regarding different sets of Bulls

Set No.	Date of start	Total bulls used	Total inseminations	Total A.I.' s followed	Pregnancies confirmed	Conception rate %	Followed for calving	Females calves	
								Born	Reached AFC
1	1.04.95	18	7595	7355	3065	41.7	3000	855	227
2	1.01.97	10	5150	4865	2132	43.8	2000	789	210
3	1.01.99	23	18006	17159	8258	48.1	8000	1844	562
4	16.12.01	30	12548	11504	5720	49.7	5720	1368	490
5	1.04.03	22	10409	10154	4362	43.0	4362	1497	478
6	1.02.05	25	8265	8105	3476	42.9	3476	1181	359
7	1.08.06	22	9710	9710	3999	41.1	3999	1120	448
8	1.01.08	16	9611	9611	3898	40.6	3898	1186	461
9	1.07.09	24	14581	14581	5679	38.9	5679	1671	885
10	1.03.11	20	12971	12971	5604	43.2	5604	2072	688
11	1.08.12	28	15662	15662	7008	44.7	7008	2500	752
12	1.02.14	15	6662	6662	3039	45.6	6662	1321	542
13	1.08.15	30	8299	8299	3851	46.4	8299	1571	352
14	1.06.17	26	8814	6551	3059	45.6	4352	887	109
Total	309	148283	143189	63150	43.9	72059	19862	6563	

Table 2. Set wise progeny performance at GADVASU, Ludhiana

Set No.	Progeny completed 1st lactation	Average 1st lactation 305 days milk yield (kg)	Average age at first calving (days)	Average fat percentage
1	184	2697.8±40.1	1192.1±23.6	3.7±0.02
2	154	2827.1±48.7	1145.6±20.3	3.8±0.02
3	397	2878.7±25.5	1191.9±13.2	3.8±0.02
4	361	2896.1±26.0	1168.4±14.9	3.8±0.01
5	361	2855.9±25.9	1174.9±14.6	3.7±0.01
6	278	3051.8±24.5	1115.9±14.2	3.6±0.10
7	351	3305.4±28.9	1082.2±12.2	3.6±0.03
8	326	3556.6±31.2	1068.8±15.9	3.7±0.00
9	647	3714.7±22.9	1044.2±8.5	3.7±0.00
10	576	3751.8±22.9	1032.7±8.0	3.6±0.00
11	608	3771.7±22.7	1012.8±8.2	3.7±0.00
12	208	3836.5±34.5	885.3±8.4	3.4±0.02

The average fat percentage of milk of the daughters of ninth, tenth, eleventh and twelfth set of bulls was 3.7±0.00, 3.6±0.00, 3.7±0.00 and 3.4±0.02 respectively. A total 10,337 farmers have so far been registered and benefited through this project in Ludhiana. In 1st set of bulls (1995), the average 305-days yield was 2697.8 kg and in 12th set of bulls it was 3836.5 kg indicating a sharp increase of 1138.7 kg milk (42.2 %) in the daughters. The Age at First Calving was 1192 days in progenies of first set of bulls (1995) which reduced to 885 days in current set showing a sharp decrease of 307 days (25.7 %). The overall least square means for first lactation 305 days milk yield and peak yield of crossbred progenies born under field condition were 2854.33±27.52 and 12.58±0.12 kg, respectively. This was almost similar to that reported by Singh *et al.*, (1990) and Rao *et al.*, (2000) and higher than that reported by Iype *et al.*, (1993), Garcha and Dev (1994), Kiran *et al.*, (1995) and Rajeev *et al.*, (2017)

The cows calved during July to October (2482 kg) produced the highest yield followed by those calved from March to June (2375 kg) in 1st lactation. The same trend was also observed for the peak yield. Similar findings also reported by Singh *et al.*, (2018). This might be due to the better availability of the green fodders during rainy season. Progeny born during November to February have shorter age at first calving followed by the progeny born during March to June and highest age at first calving was observed among the progeny born during July to October. Present finding was disagreement with the finding of Kiran *et al.,* (1995) who observed that under field condition season had no effect on age at first calving in the crossbred cattle. The performance under the present study is quite satisfactory and reflected that the program has moved in desired direction and positive trends may be expected in near future when more number of sets will be completed in the program.

ACKNOWLEDGEMENTS

The authors are thankful to Director, ICAR- Central Institute for Research on Cattle, Meerut, for providing necessary facilities for conducting the study. We also acknowledge Head, Department of Animal Genetics and Breeding, GADVASU, Ludhiana, for providing the information and data.

REFERENCES

1. A. K. Das, Ravinder Kumar, S .K. Rathee, B. Prakash, K. Anilkumar, P. P. Dubey, R. L. Bhagat, C. B. Singh, Umesh Singh, Arun Kumar, S. Tyagi and Vineet Bhasin. (2017). Genetic Improvement of Cattle through Field Progeny Testing Programme: An Evaluation of Achievement. *Indian Journal of Animal Science, 87 (12): 1445-1451*

2. Akhter J., Singh H., Kumar D, Sharma K., Akhter j and Singh H. (2003). Factors effecting economic traits in crossbred cattle. *Indian Journal of Animal Sciences* 73: 464-65.

3. Annual Report. (2018-19). ICAR-Central Institute for Research on Cattle, Meerut Cantt.

4. Garcha, D.S. and Dev, D.S. (1994). Effect of genetic and non-genetic factors on 1st lactation milk yield records in HF crossbreds. *J. Dairying, Food and Home Sci.*, 13: 3-4.

5. Gaur G K, Kumar A, Beniwal B K and Praveen. (2008). Genetic evaluation of Frieswal bulls under farm and field conditions. *Indian Journal of Animal Sciences* 78 (2): 218–21.

6. Harvey W R. (1990). Guide for LSMLMW, PC–1 Version, mixed model least squares and aximum likelihood computer programme, January 1990. Mimeograph Ohio State Univ., USA.

7. Iype, Sosamma; Radhakrishanan, J. and Raghawan, K.C. (1993). Influence of socio-economic factors on milk production of crossbred cows in Kerala. *Indian J. Anim. Sci.* 63 (12):1299-1301.

8. Kiran, Ravi; Rao G. N.; Jayaramakrishan, V. and Styanarayana, A. (1995).

Performance of Ongole and crossbred cows under village conditions of Andhra Pradesh. *Indian J. Anim. Sci.* 65 (7):780-784.

9. Kumar S, Alex R, Singh U, Kumar A and Das A K. (2015). Comparative performance evaluation of Frieswal bulls in organized farms and farmers herd. *Indian Journal of Animal Sciences* 85 (3): 316–19.

10. Rajeev, Kumar R, Singh R, Raja T V, Singh Y P and Ali N. (2017). Certain factors affecting first lactation individual monthly milk yields in Frieswal cattle. *Indian Journal of Animal Sciences* 87 (2): 245–48.

11. Rao, J., Rao, B.V.R. and Rao, G.N. (2000). Performance of crossbred cows and Buffaloes under village conditions of Visakhapatnam District of Andhra Pradesh. *Indian J. Dairy Sci.* 53(3): 222-226.

12. Ravinder Kumar, A K Das, T V Raja, S K Rathee, P P Dubey And B Prakash. (2017). Performance of crossbred cattle (HF × Sahiwal) under tropical farming conditions of Punjab. *Indian Journal of Animal Sciences*, 87 (11): 1402–1405.

13. Simran Singh, Dibyendu Chakraborty and A K Das. (2018). Factors Affecting First Lactation Production Efficiency Traits of Frieswal Traits. *International Journal of Advanced Biological Research.* 8 (2) 2018: 231-233.

14. Simran Singh, Dibyendu Chakraborty, A. K. Das, R. K. Taggar. (2017). Sire Evaluation Based on First Lactation Production Efficiency Traits in Crossbred Cattle. *Environment & Ecology* 35 (3B) : 2110-2113.

15. Singh K, Khanna A S, Kanaujia A S and Singh K. (2000). Factors affecting lactation performance and persistency in crossbred cattle. *Indian Journal of Dairy Science* 53: 354-60.

16. Singh, R.N., Singhal, R.A., Verma, B.S. and Mudgal, V.D. (1990). Frieswal cattle to enhance milk production in India. XXIII International Dairy Congress, October 8-12, 1990, Montreal, Canada.

17. Upadhyay P K, Tiwari R P, Dwivedi HB, Dwivedi P and Prasad M., (2002). Factors influencing first lactation and lifetime production traits in Holstein-FriesianxZebu cattle. *Progressive Agriculture* 2: 145-147.

Current Status of Researches in Biosciences (2020) : 455-473
Editor : Prof. P.C. Joshi, N. Joshi, Reshman Yasmin, and D. K. Mansotra
Today & Tomorrow's Printers and Publishers, New Delhi - 110 002

EFFECT OF BRYOPHYTE EXTRACT AS ANTHELMINTIC AND ITS IMPACT UPON GENE EXPRESSION PATTERNS OF *ASCARIDIA GALLI*, HELMINTH PARASITE OF CHICKEN

Yennamaneni Sunila Kumari * and Padmavathi Sriram

Department of Zoology, University College for Women, (O.U.), Koti, Hyderabad

Email: sunilapamara@yahoo.com

ABSTRACT

In view of public concern over perceivable drug residues in animal products, the increasing prevalence of anthelmintic resistant strains of nematodes, there is an urgent need for development of sustainable alternatives to conventional anthelmintics. The aqueous extract of Bryophyte, *Pottia lanceolata* was investigated for anthelmintic activity, using gastrointestinal helminth of most economic significance in poultry, *Ascaridia galli* were collected from naturally infected chicken. Significant anthelmintic activity was observed at the concentration of 20 mg/ml. The present study confirms that an effective natural anthelmintic could be developed using bryophyte extract. To know the effectiveness of herbal drug to target the parasite genome, studying the gene expression patterns before and after the drug administration by mRNA sequencing using Illumina NGS services will be of great value to ascertain potential therapeutic targets and dose rate of anthelmintic and to check drug resistance. The transcriptome analysis, has shown that approximately 2092 genes have been expressed in the chicken nematode samples. The gene expression pattern, of the control and bryophyte treated nematodes of chicken, show a significant up regulation, down regulation of certain specific genes, which play a major role in the apoptosis, cuticle damage, neuronal destruction, suppression of mRNA synthesis and also inhibition of major signaling pathways.

KEYWORDS: Bryophyte, Herbal Anthelmintic, Nematode, Gene expression.

INTRODUCTION

Helminthosis is a disease of animals caused by gastro-intestinal helminths and has been recognized as constituting a major constraint to a profitable production of livestock throughout the world particularly to poor livestock owners of developing countries (Cheijina, 2001). In their prevalence studies, Permin *et al.,* and Hane (1999) revealed a higher risk of helminth parasite infections in free range systems of poultry. Helminth parasites of poultry and sheep are usually controlled by repeated anthelmintic treatments of the flock. Regular and preventive use of such chemically synthesized drugs is not compatible with organic regulations and therefore effective preventive management strategies need to be developed in poultry, and in ruminants (Maurer *et al.*, 2007). Alternate treatments for gastrointestinal helminthes in poultry and sheep have been investigated by several authors, like Lal *et al.*, 1976; Javed *et al,*, 1994; Singh and Nagaich, 2002. However, significantly validated data on the efficacy of herbal treatments against adult and larvae of *Ascaridia galli, Heamonchus contortus, Railleitina tetragona, Multiceps multiceps, Moneizia expansa and Echinostoma sps.* remain scarce and often limited. Antimicrobial activity and preliminary phytochemical screening of epiphytic moss *Stereophyllum ligulatum* was studied by Kumar *et al.,* (2011). These studies suggest that bryophytes can be effective against certain pathogens and so we speculate that bryophytes can also be effective anthelminthic agents against certain helminthes of poultry. Of lately it is interesting to learn about the adaptive and resistant strategies employed by these parasites. This is because, it is found that in spite of use of many known chemical anthelminthics, there is no complete control or elimination of these nematode parasites, particularly *Ascaridia galli* . This nematode parasite is known to infect the poultry in a big way, after seemingly to be eliminated after a period of 10 years .This clearly indicates that the parasite seems to have adapted some new survival strategies, by developing new physiological pathways, or may be by developing certain new anatomical structures. The control of helminth parasites is central to the improved health and productivity of the livestock. Based upon these studies and observations, it is quite reasonable to assume that if a new plant based anthelminthic drug can be developed, which can effectively target the parasite genome, thereby reducing its ability to transform the genome and inhibit the parasite from developing new structures and pathways, will be of great value. Therefore we hypothesize that the bryophyte extract can be effective in inhibiting the nematode activity by impacting the genome and gradually leading to the extermination of the parasite. The present study is an attempt to analyze the anthelminthic activity of the bryophyte extract, by analyzing the transcriptome before and after treatment.

MATERIALS AND METHODS

Plant material collection and extraction: The bryophyte, *Pottia lanceolata* was collected from Osmania University college for women, Koti, Hyderabad and Golconda fort Hyderabad, Telangana, India. The plant material was taxonomically identified by the taxonomists of Botanical Survey of India, Hyderabad. A voucher specimen has been preserved in our laboratory for future reference. The plant material was dried in shade, pulverized, passed through sieve number 40 and stored in air tight container and used for further extraction.

Preparations of Aqueous extract (Maceration method): Bryophyte, *Pottia lanceolata* samples were cleaned removing soil debris, by thorough washings with water. The active ingredient was extracted, using cold maceration method dried in shade, powdered using mortar and pestle. The crude aqueous extract was centrifuged, at 5,000 rpm for 5 minutes, using Remi centrifuge. The clear supernatant was used for the treatment and the precipitate was discarded.

Collection of parasites (Nematodes of Poultry): For the present study, the desired parasites are *Ascaridia galli* which parasitizes chicken. The intestines of freshly slaughtered fowls, were collected from local slaughterhouse and washed with normal saline solution. These intestines were then dissected and the nematodes were extracted. The average size of the nematode was found to be 5 cm. These parasitic worms were used to test the anthelmintic activity with bryophyte extracts *in vitro*. The identification of nematodes was done in the department of Zoology, OUCW, Koti, Hyderabad by using Phase contrast microscope.

Piperazine citrate (glaxo smithkline) was used as the standard anthelmintic drug during the experimentation.

Based upon the results, it was decided to study the gene expression studies. For this purpose, the control and treated (bryophyte extract *Pottia lanceolata*) parasites were preserved in RNA later solution, and were subjected to RNA extraction and isolation using RNA mini easy kit. The Extraction was performed as per the protocol, provided by the manufacturer. Extracted RNA was subjected to quality and quantity check, prior to library construction.

The quality of total RNA is determined using Tape station. Tape station fluorescent marker approached to perform capillary electrophoresis to analyze RNA. They are then detected by their fluorescence and translated into gel-like images (bands) and electropherograms (peaks). Depending on the peaks the integrity of RNA is determined by RNA integrity number (RIN). RIN values are measured from 1 – 10, where RIN Value 1 - 5 indicate complete degradation and 5-7 indicate partially degraded RNA and RIN value above 7 indicate good quality RNA. Quantity check was performed using, Qubit Florometer. Qubit® Fluorometer is designed with highly specific and sensitive Qubit® RNA quantitation assays. The reagent is a very sensitive detection dye for the quantitation of RNA in solution and the detection range is 1-100ng per ul of RNA sample. Library preparation was performed following NEB Next Ultra RNA Library Prep Kit (#E7490L). In brief, approximately 110-350 ng of Total RNA was used to isolate mRNA using Poly dT beads. mRNA was fragmented and converted to cDNA according to the protocol. cDNA was end repaired and purified using Ampure XP beads. Cleaned DNA was adapter ligated and purified using Ampure XP beads. These adapter ligated fragments were subjected to 12cycles of PCR using primers provided in the kit. The PCR products were purified using Ampure XP beads. Quantification and size distribution of the prepared library was determined using Qubit flourometer and Agilent Tape station D1000 Kit (Agilent Technologies) according to the manufacturer's instructions. The steps followed to generate the FastQ reads from Total RNA libraries are as follows-

1. The pooled/multiplexed libraries were put into the Illumina Hiseq Sequencer for sequencing which generated raw data in the bcl format.
2. Bcl files were De-multiplexed using sample wise Barcode information to get the sample specific sequences.
3. De-multiplexing resulted into raw reads in FastQ format, which was used for downstream analysis.

RESULTS

Anthelmintic activity assay was carried out as per the method of Ajaiyeoba *et al.*, (2001). The assay was performed *in vitro,* using nematodes of poultry. Test samples of the extract were prepared at the concentrations, 10, 20 and 50 mg/ml of *Pottia lanceolata* bryophyte extract dissolved in distilled water. Six worms, each of chicken nematodes *Ascaridia galli* were placed in petridish measuring 9 cm in diameter. 25 ml of bryophyte extract was added to the petridishes. Piperazine citrate of concentration 10mg/ml was prepared and used as reference standard and double distilled water was used as control. Extract solution and standard solution were freshly prepared, prior tò experimentation. Observations were made for the time taken for paralysis was noted when no movement of any sort could be observed. Time of death of worms were recorded after ascertaining that worms neither moved when shaken vigorously nor when dipped in warm water. The plant extract exhibited more potent activity at lowest concentration 10mg/ml against *Ascaridia galli.* Anthelmintic activity of the extract was compared with standard drug piperazine citrate. These results suggest that bryophyte extract of *Pottia lanceolata* exhibited significant anthelmintic activity at 20 mg/ml. As the extract had an effect upon, parasitic nematode, it was intended to study the gene expression pattern of chicken nematode parasite, *Ascaridia galli.* For these studies, the mRNA was extracted, isolated and further analyzed as per the protocol, described earlier.

Anthelmintic activity of Bryophyte extract-

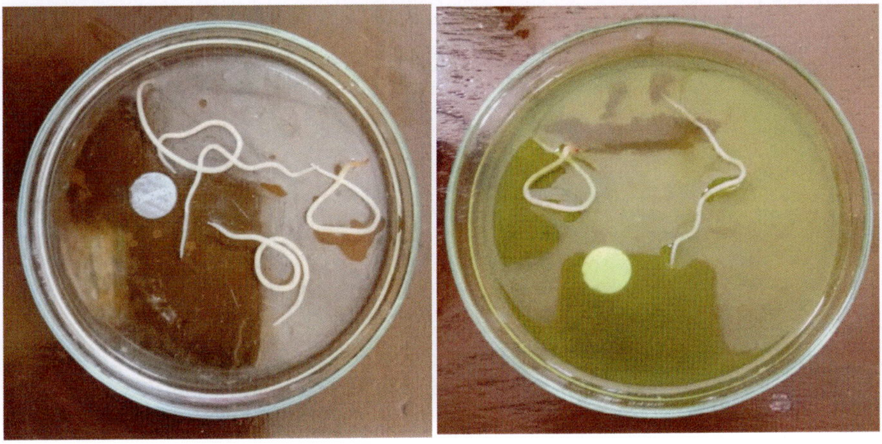

Nematode normal **Nematode treated**

Anthelmintic activity of Bryophyte *Pottia lanceolata* aqueous extract

S.No	Groups	Concentration Mg/ml	*Ascaridia galli* Time taken for paralysis in minutes (mean and SEM)	Time taken for death in minutes (mean and SEM)
1.	Control- Distilled water			
2.	Bryophyte extract	102050	22±1.20 14±1.82 08±1.36	55±1.2 245±1.40 32±0.82
3.	Standard -Piparazine citrate	10	15±1.13	40±1.20

Library Qubit Quantification Check

The samples were quantified by using Qubit flourometer.

Table 1. Test Result

S.No	Sample ID	Qubit Conc (ng/ul)	Volume of Library (ul)	Total Lib. Yield (ng)	Index	Nature of Library
1	Chick Nematode (Control)	24.4	20	488	19	Transcriptome
2	Chick Nematode (Treated)	6.86	20	137.2	20	

Transcriptome Data Needed Statistics

Sample_ID	Chicken Control	Chicken Treated	Sheep Control	Sheep Treated
Total Transcripts Length	87594424 (87.6 MB)	78783915 (78.8 MB)	35073583 (35.1 MB)	61131800 (61.1 MB)
		ATGC Composition (Raw reads level)		
%_base_A	27.2	26.2	27.3	27.3
%_base_T	26.8	25.7	26.7	26.8
%_base_G	23.2	24.3	23.2	23.2
%_base_C	22.8	23.8	22.8	22.8
GC_content_%	46.1	48.1	46.0	46.0
AT_content_%	53.9	51.9	54.0	54.0

ATGC Composition (Assembled transcripts level)				
%_base_A	**27.8**	**25.7**	**27.8**	**28.2**
%_base_T	**27.8**	**25.6**	**27.8**	**28.0**
%_base_G	**22.3**	**24.6**	**22.2**	**22.0**
%_base_C	**22.0**	**24.2**	**22.2**	**21.9**
GC_content_%	**44.4**	**48.8**	**44.4**	**43.8**
AT_content_%	**55.6**	**51.2**	**55.6**	**56.2**

Analysis Report for Transcriptome Data

NGS Experiment Overview

Organism Chicken nematodes

Sequence length	:	150 nt
Adapter	:	NEB Adapters
Protocol	:	Illumina Transcriptome Library preparation
NGS Platform used	:	Illumina NextSeq 500
Library layout	:	Paired End

Raw data QC and statistics

All the 150bp paired end raw reads (IlluminaNextSeq500) were quality checked for low quality bases and adapter sequences (Processing of raw reads). Quality check and processing was done using In-house perl scripts.

Sample	Total raw reads	Total processed reads
Chick Nematode Control	95127396	93744136
Chick Nematode Treated	87118318	86010828

De-novo Assembly

All processed reads were assembled into transcripts without any reference (De-novo) both samples independently using Trinity software (trinityrnaseq-r20140413p1). It represents a novel method for the efficient and robust *de novo* reconstruction of transcriptomes from RNA-seq data.

Sample_ID	Chick Control	Chick Treated
Transcripts Generated :	100750	122729
Maximum Transcript Length :	48529	99643
Minimum Transcript Length :	300	300
Average Transcript Length	869.424	641.934
Total Transcripts Length :	87594424	78783915
Total Number of Non-ATGC Characters :	0	0
Percentage of Non-ATGC Characters :	0	0
Transcripts >= 300 b :	100750	122729
Transcripts > 500 b :	45061	41691
Transcripts > 1 Kb :	23257	15805
Transcripts > 10 Kb :	43	129
n50 value :	1422	708
n90 value :	354	329
Number of reads used :	43848820	34774178
Total number of reads :	46872068	43005414
Percentage of reads used :	93.55	80.86

UNi-Gene generation: Unigenes were generated after clustering the merged sequences from both Control and Treated samples using CD-HIT. It takes a fasta format sequence database as input and gives a cluster file as output. The idea is to reduce the overall size of the database without removing any sequence information by only removing redundant sequences. Unigene is the master control data set which represents common sequences along with unique sequences from both the samples. Chick Unigene contains a total of 210192 sequences

Differential gene expression: The reads for the samples were separately aligned to the corresponding unigene sequences and read count profiles were generated. DESeq "R" package was used for differential gene expression. The package DESeq provides methods to test for differential expression by use of the negative binomial distribution and a shrinkage estimator for the distribution's variance. Transcripts were classified as expressed in both the samples and only present in either one of the samples. The following flow chart illustrates the details.

DGE_Chick_Control_&_Chick_Treated Statistics:				
Category	Total	UP	DOWN	NEUTRAL
Both_Control_&_Treated	150798	32874	15187	102737
*Both_Chick_Control_&_Chick_Treated	48061	32874	15187	0
Only_Chick_Control	20725	NA	NA	NA
Only_Chick_Treated	38598	NA	NA	NA

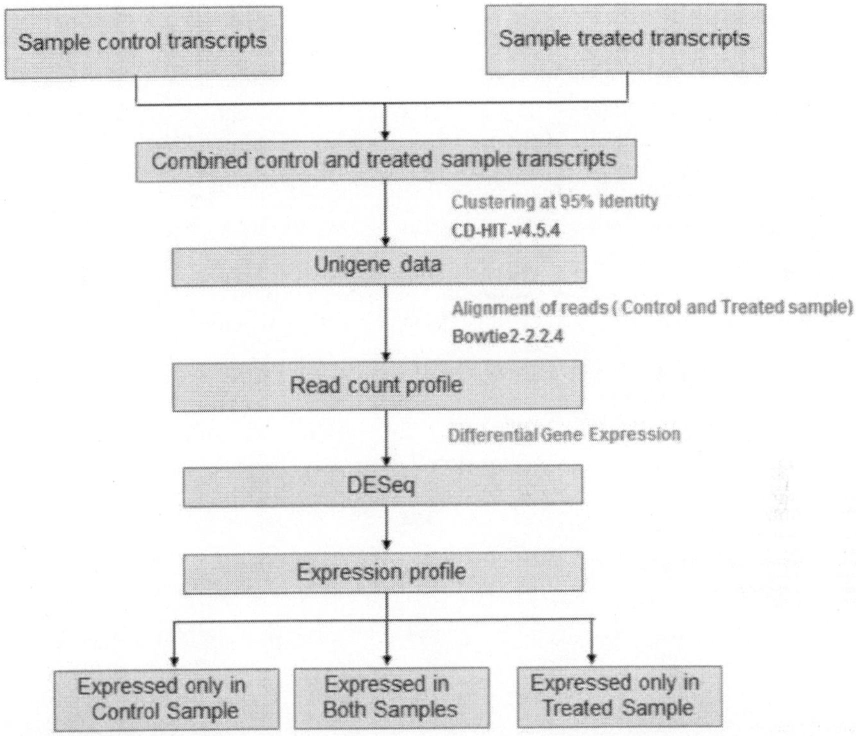

Chicken nematode- Protein Expression in the control and bryophyte treated samples

Protein	Expression	Control	Treated
ATPase, AAA family	NO	Absent	Present
Asparagine synthetase	NO	Absent	Present
Argonaute 1 isoform B	NO	Absent	Present
Bckdha protein	NO	Absent	Present
Bleomycin hydrolase	NO	Absent	Present
Bloom syndrome protein homolog (inferred by orthology to a C. elegans protein)	NO	Absent	Present
CAD protein	NO	Absent	Present
Cuticle collagen dpy-7 (inferred by orthology to a C. elegans protein)	NO	Absent	Present
CRE-CLK-1 protein	NO	Absent	Present
Dynein light chain roadblock	NO	Absent	Present

Dopey protein	**NO**	**Absent**	**Present**
DnaJ central domain protein	**NO**	**Absent**	**Present**
E3 ubiquitin-protein ligase synoviolin B	**NO**	**Absent**	**Present**
F box only protein 11	**NO**	**Absent**	**Present**
FAS-associated factor 2-B	**NO**	**Absent**	**Present**
Filamin C	**NO**	**Absent**	**Present**
Filamin-A (Fragment)	**NO**	**Absent**	**Present**
Glucose-6-phosphate 1-dehydrogenase (inferred by orthology to a C. elegans protein)	**NO**	**Absent**	**Present**
Glutathione reductase	**NO**	**Absent**	**Present**
Heat shock protein 70	**NO**	**Absent**	**Present**
Inosine-5'-monophosphate dehydrogenase	**NO**	**Absent**	**Present**
Isoleucyl tRNA synthetase, cytoplasmic	**UP**	**Present**	**Present**
JNK-interacting protein 1	**DOWN**	**Present**	**Present**
Kinase (EC 2.7.-.-) (Fragment)	**NO**	**Absent**	**Present**
Krueppel-like factor 1	**DOWN**	**Present**	**Present**
Malate synthase (EC 2.3.3.9)	**UP**	**Present**	**Present**
Malcavernin	**NO**	**Present**	**Absent**
Multidrug resistance protein 1b	**NO**	**Present**	**Absent**
Netrin receptor unc-5 (inferred by orthology to a C. elegans protein)	**NO**	**Absent**	**Present**
Neuronal calcium sensor 1	**NO**	**Present**	**Absent**
OmpA family protein	**NO**	**Absent**	**Present**
Oxidoreductase, FAD/FMN-binding protein	**NO**	**Absent**	**Present**
PAB-dependent poly(A)-specific ribonuclease subunit 2	**NO**	**Absent**	**Present**
Papain family cysteine protease	**NO**	**Absent**	**Present**
Quinone oxidoreductase-like protein 2	**DOWN**	**Present**	**Present**

Ribose-phosphate pyrophosphokinase	**NO**	**Absent**	**Present**
RNA binding protein 34	**NO**	**Absent**	**Present**
RuvB-like helicase (EC 3.6.4.12) (Fragment)	**NO**	**Absent**	**Present**
Septin-2 (inferred by orthology to a human protein)	**NO**	**Absent**	**Present**
Serine threonine kinase sad-1	**NO**	**Absent**	**Present**
Temperature-induced lipocalin	**NO**	**Absent**	**Present**
Transcription activator BRG1	**NO**	**Absent**	**Present**
Ubiquinone biosynthesis monooxy-genase COQ6, mitochondrial (EC 1.14.13.-)	**NO**	**Absent**	**Present**
Ubiquitinyl hydrolase 1 (EC 3.4.19.12)	**NO**	**Absent**	**Present**
Very long chain specific acyl coenzyme A dehydrogenase	**NO**	**Absent**	**Present**
von Willebrand factor type D domain protein	**NO**	**Absent**	**Present**
WD repeat-containing protein 37	**NO**	**Absent**	**Present**
Xaa-Pro aminopeptidase 1	**NO**	**Absent**	**Present**
Xylosyltransferase oxt	**NO**	**Absent**	**Present**
ZapA domain containing protein	**NO**	**Absent**	**Present**
Zinc finger SWIM domain-containing protein	**NO**	**Absent**	**Present**

DISCUSSION

The transcriptome analysis, has shown that approximately 2092 genes have been expressed in the chicken nematode samples. The present study showed that the crude bryophyte extract to be quite effective in the treatment of chicken nematodes. The gene expression pattern, of the control and bryophyte treated nematodes of chicken, show a significant up regulation, down regulation of certain specific genes and also expression of unique genes. It was observed that genes related to cuticle development and protein, and certain cadherin related genes, which control the cell adhesion, have been up regulated in the control sample. There was an up regulation of cuticle related genes such as **Cuticle collagen 13** and there was a down regulation of **Cuticle collagen 2 and 40.** The **cadherin** related protein/

genes show a up regulation. This indicates that the cuticle seems to be intact and thus aid in protection of the parasite from damage. The cadherin proteins are important for the parasite, as it is involved in site specific recombination, which is needed during reproduction. As the nematode is not being treated, all the routine functionality of worm proceeds normally and since the worm is in the reproductive stage, these proteins are needed. The untreated worm or the control samples also showed an up regulation of **Multidrug resistant proteins**,which clearly indicates that the untreated worms are fully protected and resistant to anthelminthic drugs, thereby prolonging the life of the parasite. Another set of Proteins and their genes known as the **Fas a apoptotic inhibitory molecule 2** is found to be down regulated giving the parasite, protection against the apoptotic cell death, thus enhancing the survival of the untreated worms. **F box only Protein 11** is another unique protein which causes destruction of the cellular proteins and hence absent in the control samples and thus aiding the parasite to overcome the stress and thus prolonging the lifespan of the worm. In the present study, where a chicken nematode of the species, *Ascaridia galli* was subjected to bryophyte treatment resulted in mortality of the worm. The chicken nematodes, perished after being subjected to moss treatment in dosage dependent manner within two hours of treatment. The mortality of the nematodes, observed during the present investigation, may be attributed, to the expression of certain genes which might have triggered any of the death pathways involving, cuticle, mitochondrial pathways or may be due to certain neuronal related signaling mechanisms. In the bryophyte treated samples, we observed expression of several unique genes, which are as follows. **ATPase, AAA family** is special type of proteins expressed in the treated samples. These proteins are involved in diverse cellular activities like DNA replication, protein degradation, membrane fusion, microtubule severing and also involved in ATP hydrolysis and molecular remodeling events. (Jamie Snider *et al.* 2008). Expression of this gene in the treated samples, indicates that under the influence of the treatment, there may be protein degradation, membrane fusion or there may be microtubule severing, thereby leading to the death of the parasites. There was also expression of **Asparagine Synthetase** in the treated samples. This enzyme plays a role in cellular related stress. It is highly responsive to stress and is the chief enzyme that generates asparagines from asparatate. This enzyme is highly regulated in response to cell stress (Balasubramanian *et al.* 2013). The expression this enzyme in the treated samples, clearly suggests that the bryophyte treated nematode, are under cellular stress and hence the synthesis of this enzyme. **BCKD gene /protein:** This enzyme complex responsible for one step breakdown of 3 protein building blocks

(Leucine, valine and isoleucine). This complex is active in mitochondria and helps in the breakdown of amino acids, to release energy. As these nematodes, are under stress related to treatment, energy production is vital for their survival, and hence the expression of this protein complex. **Bleomycin Hydrolase,** is a cytoplasmic cysteine peptidase, that functions to metabolically inactivate the glycopeptide, bleomycin (drug given to combat cancer). Expression of this protein in the treated samples, is strange. However, it may be suggested that there may be a similar compound which may be synthesized in the treated samples, and to neutralize the effect, there seems to be the expression of this protein. **Bloom syndrome protein homolog,** This protein is known to play an important role in the genomic stability of somatic cells(Shuren, Liao *et al.* 2000). It is also known to participate in DNA replication and repair. As the nematode parasites are under bryophyte treatment, there is a possibility of DNA damage and also instability of the parasite genome and so this protein may be expressed to repair and stabilize the DNA of the treated worms. **C type lectin protein,** which is a key protein, involved in cell-cell adhesion, immune response to pathogens and apoptosis. As the parasite, is exposed to the concentrations of bryophyte extract, these genes /proteins show their expression and may cause cell to cell adhesion, and also initiate apoptosis, which cause the death of the cell. Another important protein/gene which might play a role in the mortality of the worm, is the cuticle related protein known as **cuticle collagen DPY 7 fragment.** This is a gene/protein, which controls the cuticle collagen formation and functionality (John *et al.* 1997). This expression is seen in the treated samples, and may be said that as there may be a damage to the cuticle, because of the bryophyte extract, the expression of this gene is to repair the damaged cuticle. **Cre CLK 1 protein** shows its expression in the treated sample, which is known to play a major role, in mRNA processing of proteins which may be related to apoptosis and cell degradation, and ultimately causing the death of the worms. **E3ubiquitin protein ligase synoviolin B, protein** is expressed in treated samples. This is an enzyme involved in endoplasmic (ER) associated degradation. It removes unfolded protein accumulated during ER stress. It controls ER stress through clearance of misfolded proteins (Sinyi Kong *et al.* 2016) Apart from these genes/ proteins, it was seen that there is expression of **F box only protein 11,** which is a specific protein, which cause the degradation of cellular proteins. There is also presence of **FAS a apoptotic inhibitory molecule 2,**(Mercy pawar, 2017) suggesting that there is degradation of cellular proteins and may lead to cell death, as FAS protein is expressed but not up regulated. **Filamin A and filamin C** are the proteins related to cytoskeleton. There is expression of these in the treated samples, suggesting that the bryophyte extract may have caused cytoskeletal disruption and the expression of these proteins may be useful to restore the disrupted cytoskeleton. **Glucose 6 phosphate dehydrogenase** is an important enzyme in pentose pathway, involved in normal processing of carbohydrates. Its expression in the treated samples, of the parasites, indicates that this metabolism is vital for the survival of the parasite, and hence its expression is appropriate. Glutamate dehydrogenase mitochondrial is yet another enzyme, whose expression is seen in the treated samples of the nematode parasites. This enzyme is important and plays a major role in amino acid metabolism. As the worms, are under the influence of the bryophyte treatment, there is an urgent necessity of the parasite, for survival, therefore, expression and activity of the enzyme play a significant role, in the amino acid metabolism to provide necessary energy. **Glutathione reductase, is an enzyme,** which plays an important role, in protecting haemoglobin and biological cell membranes, against oxidative damage by increasing the levels of reduced glutathione in the process of glycolysis (Chang, *et al.* 1978). Its expression in the treated chick nematodes may be to reduce the oxidative cell damage, which may be caused due to the bryophyte treatment. Apart from these proteins, it was observed that there is expression of stress related proteins, known as **HSP70**. These are exclusive proteins related to stress, and clearly suggests that the bryophyte treatment may have cause stress in the nematode and hence their expression. **Inosine 5 monophosphate dehydrogenase 2,** expression of this protein/gene in the treaded samples of nematodes, may be explained as follows. This is an important enzyme, involved in the conversion of Inosine 5 phosphate to Xanthosine 5 phosphate, the first rate limiting step in the *de novo* synthesis of guanine nucleotides, and plays an important role in the regulation of cell growth. It may play an important role in the development of malignancy and growth progression of some tumors and its activity is tightly linked to cell proliferation as suggested by (Foltie, J. 2016). Its expression in the treated nematodes, may be to synthesize necessary guanine nucleotides, or may be to cause a tumor like growth, and thereby lead to the destruction of the parasite. **Isoleucyl-tRNA synthetase,** this is an enzyme, which along with aminoacyl tRNA synthetases catalyzes the reaction between tRNA with cognate amino acid and are involved in many cellular functions (Elizebeth Kermagard *et al.* 2017). Expression of this enzyme, under the bryophyte treated conditions, clearly illustrates, that it is essential for the synthesis of proteins which might be needed for cellular activities, of the nematodes, needed for their survival. **JNK interacting protein 1-** These are group of MAP Kinases, which are important for many cellular responses including apoptosis, growth, differentiation, embryonic development and immune responses (Emma

A.Willoughby *et al.* 2003). Its expression is seen in the treated and control worms as well. The protein expression in the control samples, is down regulated, however it is merely present in the treated samples. As it seems to play a major role in apoptosis, and embryonic development, its expression in the treated samples, might be to induce the apoptosis and thus lead to the death of the nematodes. **Kinase (EC 27 fragment)** This is one of the key regulatory enzyme in the glycogen metabolism(Khandelwal, and J.H Wang 2012).Expression of this enzyme, in the treated samples, suggests that the nematodes, seem to depend upon glycogen metabolism as they are under treatment stress. **Kruppel like factors,** These are family of transcription factors and are a set of zinc finger DNA binding proteins that regulate gene expression. Presence of these proteins in the treated samples of the nematodes, suggest that as they are under special treatment of bryophyte, there may certainly be necessity of certain unique proteins, which might be essential for the survival of the parasite, and hence the expression of these transcription factors is justified. **Leishmansin and longevity assurance protein** are two unique proteins, whose expression in the treated samples has a pathological significance. These two proteins are known to be involved in the glycolipid synthesis, which might help the parasite, some resistance reaction so as to survive, in spite of the treatment. The bryophyte treated parasites showed the expression of certain genes, which are quite novel and also seems to be expressed under the influence of the treatment, and seem to affect, specific organ systems, such as muscular and nervous systems, and also influence certain pathways, such as glyoxalate pathway. Longevity assurance protein, which is involved in ceramide and spingo lipid synthesis, which plays a crucial role in cell growth, apoptosis and stress response.(Teufel *et.al,* 2009). Expression of this gene, in the treated samples, suggests that the nematode parasites are under stress and the expression of this gene might cause apoptosis. **Malate synthase** an enzyme, which is involved in glyoxalate detoxification, is expressed in the treated worms. This is interesting, because, the presence of this enzyme may help the parasite survive, in the bryophyte soaked medium. This may be identified as new drug target, to eliminate the parasite. (Puckett *et al.* 2017) **Malcaverin** is a protein encoded by CCM2 gene. This gene is involved in CCM pathway (cerebral cervonomous pathway). The CCM2 gene provides instructions for making a protein called malcavernin, which strengthens the interactions between cells that form blood vessels and limits leakage from the vessels. So it may said that the treated samples, under the influence of the extract, there seems to be a destruction of the blood vessels and therefore the expression of this gene, may be to repair the blood vessels and prevent the leakage. Bryophyte treated chick nematodes, show the expression of **Memo like protein, and Metaxin 1 like family protein.** Memo like protein plays a critical role in cell motility, and its expression seems to be essential to overcome the effect of the bryophyte treatment. Metaxin 1 protein is known to be involved in transport of the protein in the mitochondria. Whose expression in the treated suggests its role, in transporting the proteins in to the mitochondria to overcome the stress caused by the treatment. **Multidrug resistant proteins** are expressed in the treated samples, but there is no up regulation of these proteins, in the treated samples, suggesting that the moss treatment t seems to be quite effective against the nematode parasites and causing the destruction of the parasites. **Netrin receptor UnC -5** is a unique protein/gene expressed in the moss treated samples. These are asset of unique proteins, which are known to control the worm movements. Its expression only in the treated samples, suggest that there may be an inhibition of the worm movements, due to the treatment. Expression of this gene/ protein, during this stressful condition may help the nematode parasites to move. Neurofibromin, is a protein / gene which is expressed in the treated samples and the protein is known as to prevent cell overgrowth. This is a tumor suppressor protein. Its expression exclusively in the moss treated nematodes maybe to prevent overgrowth of the cells of the nematodes. **Neuronal calcium sensor 1**, this is a protein/ gene expressed in the treated, and this is known to modulate synaptic transmission and synaptic plasticity. As the nematodes are under treatment, there seems to be a disturbed or altered neuronal transmission, and so the expression of this protein. **OmpA:** There is an expression of this gene/ protein in the treated samples, which plays an important role in cell adhesion, invasion or intracellular survival. This is a heat modifiable surface exposed porin protein. (Confer and Ayalew, 2013) Expression of this protein exclusively in the treated samples suggests that it is expressed to evade any defense mechanisms, as it is under the influence of the bryophyte treatment. These proteins target the immune system, and serve as best target for vaccine production. **Oxidoreductase, FAD/FMN- binding protein:** Its expression is seen in the treated samples, and plays a role in DNA repair and apoptosis. **PAB-dependent complex:** This protein/gene which is exclusively expressed in the bryophyte treated samples, which functions as a catalytic subunit and causes the degradation of cytoplasmic mRNA, may seem to cause destruction of mRNA, which might contribute to the destruction of nematodes. **Pappain family cysteine protease-** This enzyme is involved in the degradation of extracellular matrix, and so its expression in treated may cause damage to the parasite. **Quninone oxidoreductase like protein 2-** This is a protein, involved in oxidoreductase activity, in the respiratory chain and catalyses the transfer of electrons from NADH to Coenzyme Q10.This is a unique protein which is down regulated in the control samples , but merely present in the treated samples., which may or may not play a significant role in the treated condition. **Ribose-phosphate pyrophosphokinase,** This is an important enzyme, which catalyses the synthesis of phosphoribosyl pyrophosphate, which is essential for nucleotide synthesis. **RNA binding protein 34**: Involved in RNA processing and modification, its expression in the treated samples, suggest that there may be synthesis of RNA processing and modification, of certain proteins which may cause the destruction of the parasites. Expression of RUFY2 protein in the bryophyte treated samples, suggests that this expression, might cause neurodegenerative anomalies which might cause the destruction of the parasite. **SEPT 2,** It is a protein involved in the stabilization of cytoskeleton of the cell. It causes actin fibers to disintegrate and cells lose polarity. This protein along with SOCS7 and NCk1 are a part of signaling pathway causing damage response of cytoskeleton. Expression in the treated samples may be to cause the disintegration of the actin fibers and thereby cause damage to the cytoskeleton of the nematode worms. **Serine Threonine Kinase** -This protein is known to regulate several aspects of presynaptic differentiation in *C.elegans*. It regulates cell polarity (Crump *et al.* 2001). There is an expression of this protein/gene in the treated nematodes, suggesting that under the influence of the bryophyte treatment, there might be disturbed cell differentiation and cell polarity and hence the expression of this protein. **Temperature induced lipocalin,** this is a unique gene which is mostly expressed in the plants to tolerate high levels of heat induced stress (Nadai Maria Berterane, 2017).Presence of this gene, in the treated samples, may be explained as follows. Under the influence of the treatment, there might be an increased stress levels due to heat, and so the expression, of this gene may help the parasite to overcome the heat induced stress. **Ubiquitine hydrolase**, and important enzyme involved in protein degradation. Ubiquitination can mediate functions such as protein membrane trafficking, endocytosis and DNA repair. Its expression exclusively in bryophyte treated samples, suggest that there may be endocytosis, which shows the effectiveness of the treatment. (Paul Bishop *et al.* 2016). **Very long chain specific CoA dehydrogenase-** This is an enzyme involved in the mitochondrial Beta oxidation pathway. Its expression in the treated is to aid the nematode in the survival, by providing the necessary energy via beta oxidation. **Von Willebrand factor (VWF),** a protein, which plays a major role in homeostasis and thrombosis in the arterial side of the vascular system (Yan-Feng Zhou *et al.* 2012).Expression in treated may be to maintain homeostasis and cause thrombosis in the vasculature of the nematode parasite. **Xa amino pepetidase 1,** These are a group of enzymes, involved in protein degradation and protein maturation.(Ye Zheng *et al.* 2005) Expression of these enzymes in the treated , may seems to degrade certain proteins, which may cause destruction of the chick nematodes, under the influence of the treatment. **Xylosyltransferase oxt,** A unique enzyme, which catalyses the first step in the biosynthesis of glycosamiglycan, an initial enzyme, in the biosynthesis, of chondroitin sulfate and dermatin sulfate proteoglycan in fibroblasts and condriocytes.(Iain. B.H. Wilson, 2002). Expression in the treated may be for the synthesis of chondriotin sulfate. Condriotin sulfate and dermatin sulfate has been implicated in the neuronal development of the brain (Purushotham *et al.* 2007).Its expression, in the treated samples, may be for the neuronal development, because, under the influence of the treatment of the bryophyte extract, there may be any damage for the nerves, and hence to repair the nerves may be the expression of this gene.

CONCLUSION

In the present study, the bryophyte belonging to the species, *Pottia lanceolata* was tested for the efficacy of being used as the anthelmintic drug against the chicken nematodes, *Ascaridia galli* .The treatment was found to be quite effective, as expression of unique genes was observed , which play a major role in the apoptosis, cuticle damage, neuronal destruction, suppression of mRNA synthesis and also inhibition of major signaling pathways, and hence may be suggested that bryophyte extract may be effectively used as an effective anthelmintic to rid the chicken of helminth infection.

ACKNOWLEDGEMENTS

Thanks to University Grants Commission for funding the project. (UGC F. 30-108/2015, BSR). Special thanks to Sandor Life Sciences Pvt. Ltd, Hyderabad, India for mRNA sequencing using Illumina NGS services.

REFERENCES

1. Andrew Reidenbach, Zachary Kemmerer, Deniz Aydin, Adam Jochem, Molly McDevitt, Paul Hutchins, Emily Wilkerson, Jaime Stark, Jonathan Stefely, Isabel Johnson, Craig Bingman, John Markley, Joshua Coon, Matteo Dal Peraro,(2008). Conserved lipid and small molecule modulation of COQ8 reveals regulation of the ancient UbiB family *Cell Chemical Biology* doi:10.1016/j.chembiol. 2017.11.001;Published online 2008 Feb 1. doi: 10.1621/nrs.06004

2. Confer AW, Ayalew S. (2013). The OmpA family of proteins: roles in bacterial pathogenesis and Immunity *Vet Microbiol.* 2013 May 3; 163(3-4):207-22.

3. Crump JG1, Zhen M, Jin Y, Bargmann CI. EMS Yeast Res. Neuron. (2001). The

SAD-1 kinase regulates presynaptic vesicle clustering and axon termination. 2001 Jan;29(1):115-29.

4. Iain B. H. Wilson (2002). Functional Characterization of Drosophila melanogaster Peptide O-Xylosyltransferase, the Key Enzyme for Proteoglycan Chain Initiation and Member of the Core 2/I N-Acetylglucosaminyltransferase Family; The *Journal of Biological Chemistry* 277, 21207-21212.

5. J S Gilleard, J D Barry, I L Johnstone (2009). cis regulatory requirements for hypodermal cell-specific expression of the *Caenorhabditis elegans* cuticle collagen gene dpy-7.*International Journal of Molecular Medicine.* 2009 Feb;23(2):135-40.

6. Javed I, Akhtar MS, Rahman ZU, Khaliq T, Ahmad M. (1994). Comparative anthelminthic efficacy and safety of Caesalpinia crista seed and piperazine adipate in chickens with artificially induced *Ascaridia galli* infection. Acta Vet Hung. 1994; 42(1):103-9.

7. János András Mótyán, Ferenc Tóth, and József Tőzsér (2013). Research Applications of Proteolytic Enzymes in Molecular Biology; *Biomolecules.* Dec; 3(4): 923–942 Published online 2013 Nov 8. doi: 10.3390/biom3040923

8. Lal J, Chandra S, Raviprakash V, Sabir M. (1976). In vitro anthelmintic action of some indigenous medicinal plants on *Ascardia galli* worms. *Indian J Physiol Pharmacology*; 20: 64- 68

9. Maurer, V., Hoerdegen, P. and Hertzberg, H. (2007). Reducing anthelmintic use for the control of internal parasites in organic livestock systems. In: Cooper, J., Niggli, U. and Leifert, C. (eds.). Handbook of Organic Food Safety and Quality. Woodhead Publishing Limited, Cambridge, England. Pp. 221-240.

10. Mótyán, J. A., Tóth, F., & Tőzsér, J. (2013). Research Applications of Proteolytic Enzymes in Molecular Biology. *Biomolecules*, *3*(4), 923–942. http://doi.org/10.3390/biom3040923

11. Nadine Ruecker,Robert Jansen, Carolina Trujillo, Susan Pucket, Pradeepa Jayachandran, Gerardo G. Piroli, Norma Frizzell, Henrik Molina, Kyu Y. Rhee, and Sabine Ehrt. (2017). Fumarase Deficiency Causes Protein and Metabolite Succination and Intoxicates *Mycobacterium tuberculosis. Cell Chem Biol* 24 (3):306-315

12. P, Kumar *et al* (2011). Anti-microbial activity and preliminary phytochemical screening of epiphytic moss *Stereophyllum ligulatum*, *International Journal of Pharma and Biosciences,* volume 2/issue 4/2011.

13. Permin, A, M. Bisgaard, F. Frandsen, M. Pearman, J. Kold & P. Nansen,(1999). Prevalence of gastrointestinal helminths in different poultry production systems. *British Poultry Science* 40: 439-443·

14. Puckett, S., Trujillo, C., Wang, Z., Eoh, H., Ioerger, T. R., Krieger, I., Ehrt, S. (2017). Glyoxylate detoxification is an essential function of malate synthase required for carbon assimilation in *Mycobacterium tuberculosis. Proceedings of the National Academy of Sciences of the United States of America, 114*(11), E2225–E2232. http://doi.org/10.1073/pnas.1617655114

15. Teufel A Maass T, Galle PR, Malik N. (2009). The longevity assurance homologue of yeast lag1 (Lass) gene family (review). *Int J Mol Med.*; 23(2):135-40.

Current Status of Researches in Biosciences (2020) : 475-484
Editor : Prof. P.C. Joshi, N. Joshi, Reshman Yasmin, and D. K. Mansotra
Today & Tomorrow's Printers and Publishers, New Delhi - 110 002

EVALUATION OF EDIBLE MUSHROOM *PLEUROTUS FLORIDA* (MONT.) SINGER IN PREVENTION AND CONTROL OF DIABETES MELLITUS

Pratima Vishwakarma[1], Pooja Singh[1], Veena B. Kushwaha[2*] and N.N. Tripathi[1]

[1]Bacteriology and Natural Pesticide Laboratory, Department of Botany, DDU Gorakhpur University, Gorakhpur 273009, U.P., India

[2]Department of Zoology, DDU Gorakhpur University, Gorakhpur 273009, U.P., India

Email:veenabatrakushwaha@gmail.com

ABSTRACT

Pleurotus florida is a common edible mushroom that belongs to family Pleurotaceae. Fruiting bodies as well as active mycelia of *Pleurotus* species possess a number of therapeutic properties. It is widely used to treat various ailments. In this study aqueous extract of *Pleurotus florida* was examined for antidiabetic activity (both *in vitro* and *in vivo*). *In vitro* study was performed using dialysis bags while *in vivo* study was done on streptozotocin induced diabetic rats. *In vitro* study revealed that aqueous extract of *Pleurotus florida* (50 g/l) concentration was most effective in controlling the movement of glucose out of dialysis bag (52.65±2.73 mg/dl at 24 h) with respect to control. This effect was observed to be time and dose dependent. Aqueous extract of *P. florida* at 200 mg/kg was able to mimic the drug metformin in reducing the blood glucose in acute and chronic *in vivo* studies. During acute study *P. florida* at 200 mg/kg reduced glucose up to 11.42% while in chronic study it blood glucose level was reduced up to 37.55%.

KEYWORDS: Antidiabetic, *Pleurotus florida,* Streptozotocin, Metformin, Glucose

INTRODUCTION

Diabetes has been known for thousands of years. According to World Health Organization statistics in year 2000 there were about 171 million people worldwide suffering from diabetes and by 2030 this number will be more than double (WHO, 2008). The disease is responsible for 3.2 million deaths every year. Diabetes is characterized with either insulin deficiency or insulin resistance. Treatment of diabetes mellitus involves exercise, diet therapy, insulin therapy and oral antidiabetic agents. Despite of their effectiveness in reducing hyperglycemia, the use of these drugs are associated with undesirable side effects (Adam *et al.,* 2011).

Fungi have been occupying a prominent position in biological world because of their variety, economic and environmental importance (Soosairaj *et al.,* 2012). Mushrooms are the oldest single cell protein (SCP) foods of man, which are grown by bioconversion of post-harvest agricultural waste material like straw, dried leaves, waste from food industries etc. into valuable protein rich foods. These are very popular and valuable food items in modern dietary regimes because of their nutritional value (Goyal *et al,,* 2006). Macrofungi are known to serve as an important source to cure diseases from simple headache to most serious ones like cardiovascular problems. They are reported to show good antiviral, antibacterial, antiparasitic, anti-inflammatory and antidiabetic properties. They contain polysaccharides, glycoproteins and proteoglycans that are reported to modulate immune system and help to inhibit tumor growth (Cui and Chisti, 2003). *Pleurotus florida* (Oyster mushroom) rich in protein and amino acids such as arginine, histidine, lysine, leucin, valine and methionine (Vijaykumar *et al.,* 2007), has many medicinal values. In the present study therefore an attempt is made to evaluate the antidiabetic activity of *Pleurotus florida* , an edible wild mushroom found in North East part of Uttar Pradesh, India.

MATERIALS AND METHODS

Pleurotus florida was collected from Gorakhpur district, in July 2013. The sample was identified by relevant literature (Kong, 2004) and sample was given voucher number DDUNPL249. The sample was dried under shade and ground to fine powder with help of grinder and stored in opaque screw top jar at room temperature. Aqueous extract of macrofungal samples were prepared by the method of infusion (Gallagher *et al.,* 2003). Ability to inhibit diffusion of glucose into external solution from dialysis bags at set interval of time was determined using method described by Edward *et al.,* (1988) with slight modification. Sealed dialysis bags (6 cm X

15 mm) containing 1 ml of aqueous extract of *Pleurotus florida* extract, 1 ml each of NaCl (0.15N) and D glucose (0.22M) were used. Tese dialysis bags were placed in a centrifuge tube containing 45 ml of 0.15 N NaCl and placed on an orbital shaker. The movement of glucose into the external solution was monitored at set interval of times (30 min, 1h, 2h, 3h, 4h, 5h, 6h, 7h and 24h) (Vishwakarma *et al,* 2018). The concentration of glucose within the dialysis bag at the end of experiment was also evaluated. Amount of glucose was measured using anthrone method (Thimmaiah, 1999). These were carried out in triplicate and compared with control that had everything except macrofungal extract. *In vivo* experiment was done on healthy male albino rats weighing approximately 60-160g as described by Kim *et al.,*(2001) with few modifications. The rats were acclimatized to laboratory conditions for 10 days and fed on rat pellets and water *ad libitum*. Each rat was weighed and assigned a number for convenience prior to the onset of experiment. The protocols for these experiments were approved by the Ethical Committee of D.D.U. Gorakhpur University, Gorakhpur, Uttar Pradesh, India, 273009.

Experimental animals were divided into 5 groups of 5 rats in each group.

Grp.1: Normal Control (receiving 0.1M citrate buffer)

Grp.2: Diabetic rat (receiving 0.1M citrate buffer)

Grp.3: Diabetic rat receiving 500 mg/kg aqueous solution of metformin

Grp.4: Diabetic rat receiving 100 mg/kg of *Pleurotus florida* extract

Grp.5: Diabetic rat receiving 200 mg/kg of *Pleurotus florida* extract

Diabetes mellitus (DM) was induced by single i.p. (intraperitoneal) injection of freshly prepared solution of streptozotocin (STZ) (Himedia, CMS 1758) in citrate buffer of 0.1N (3%), pH 4.5. Each test animal was injected with 50 mg/kg volume of freshly prepared STZ for induction of diabetis while normal rats were injected with same volume of citrate buffer. Fasting blood glucose was determined after 3-4 days and animals with blood glucose level higher than 180 mg/dl were used for the experiment. For oral glucose tolerance test (OGTT) the experimental animals were fasted overnight (about 12h) before commencing the experiment. Rat of all groups were loaded with D-glucose (500 mg/ml, 2.5 g/kg) solution after half an hour of metformin and *P. florida* extracts administration. Blood samples were withdrawn from tail vein before and after metformin and *P. florida* administration at different intervals. Glucose level in the blood was measured in two steps, first fasting blood glucose level and second postprandial antihyperglycemia was measured. For fasting blood glucose level experimental animals were deprived from food for about 16h with free access of drinking water before the commencement of experiment and were tested for their blood glucose level at 0, 24, 48 and 72h after treatment of aqueous extract of *P. florida*. For Postprandial antihyperglycemic study two steps were used, acute and chronic. For acute study, rats were fasted for 1h prior to test and a group was administered different concentration of aqueous extract of *P. florida* extracts and while other was administered metformin orally using intragastric gavage. Blood was collected for glucose estimation from the tail vein just before (0 h) and after 2, 4, 6 and 24h administration of extract and metformin. In chronic study glucose level in blood were measured at 0 day and after 5, 10, 15 and 30 days after administering metformin (500 mg/kg) and aqueous extract of *P. florida* (100mg/kg and 200mg/Kg) to experimental rats every day.

RESULTS AND DISCUSSION

Oyster mushroom (*Pleurotus* species) is commercially important in the world of mushroom market and several species are grown commercially on a large and small scale in many countries (Adebayo *et al,* 2012). It possesses both nutritive as well as medicinal value.

Table 1, shows time and dose dependent movement of glucose from dialysis bag to external solution in the presence of aqueous extract of *Pleurotus florida*. Amount of glucose in external solution was found to be 66.44±2.27 mg/dl at 30 min while 125.87 mg/dl at 24 h in the presence of low concentration of (6.25 g/l) aqueous extract of *Pleurotus florida* while at higher concentration (50g/l) aqueous extract of *Pleurotus florida* the amount of glucose in the external solution was found to be 10.49±2.10 mg/dl at 30 min and 52.65±2.73 mg/dl at 24 h. Dialysis bags were used for *in vitro* study as they are semi-permeable membrane and show similarity with small intestine. The aqueous extract of *Pleurotus florida* inhibited diffusion of glucose from the bag to outside. This effect was observed to be concentration and time dependent. Higher the concentration of aqueous extract lower was the diffusion of glucose towards external solution. Most effective concentration in present study was found to be 50g/l, so this concentration was taken for further *in vivo* study. The oral glucose tolerance test (OGTT) measures the body's ability to use glucose that is the body's main source of energy. In present study for OGTT, blood glucose concentration increased rapidly till 30 min. In case of normal control rats after 60 minutes of glucose administration, rapid decrease in the level of glucose was observed in the blood that returned to normal after 120 min. In case of rats treated with metformin blood glucose reduced to194 mg/dl after 120 min from 225 mg/dl while on treatment with aqueous extract of *P. florida* (200 mg/Kg) blood glucose was observed to be 205±1.33 mg/dl after 120 min (Table 2). In the present study on diabetic rats a reduction in the fasting blood glucose level by 5.82% at 24 h, 9.71% at 48 h and 17.96% at 72 h was observed after administration of 200 mg/Kg of aqueous extract of *P. florida* with respect to control (Table 3). In animals administered with 500mg/Kg of metformin reduction of fasting glucose level up to 8.37% at 24 h, 12.78% at 48h and 28.19% at 72h was observed. Administration of metformin (500mg/Kg) reduced glucose up to 20.44% after 24h in acute study while aqueous extract of *P. florida* (200 mg/kg) reduced glucose up to 11.42% (Table 4). In chronic study reduction in glucose level in blood on administration of metformin (500mg/Kg) was up to 46.89% after 30 day while administration of aqueous extract of *P. florida* reduced blood glucose level up to 37.55% (Table 5). Hence, aqueous extract of *P. florida* at 200 mg/kg was able to mimic the drug (metformin) in reducing the blood glucose. Edible mushrooms have been recognized as the ideal food for diabetic prevention of hyperglycemia as they have high content of fibre, protein and a low fat content (Yang *et al,* 2008). Many mushroom species appear to be effective for both the control of blood glucose levels and the modification of the course of diabetic complications. Badole & Bodhankar (2007) have reported that macrofungi significantly reduces blood glucose level in diabetic subjects. According to them after a single administration of aqueous extract of *Pleurotus pulmonarius* (500 mg/kg) significantly reduction in serum glucose level is observed at 2, 4, 6 and 24h. *Pleurotus florida* can act as an adjunct therapy along with first-line therapy in type 2 diabetes mellitus. It helps in lowering glucose concentrations, serum cholesterol, triglycerides and urine glucose and ketones in experimental animals. It also helps in restoring superoxide dismutase, catalase and reduced glutathione (Ganeshpurkar *et al,* 2014). According to Kanagasabapathy *et al* (2014) polysaccharide-rich-glucans from *Pleurotus sajor-caju* prevents the occurrence of hyperglycemia and hyperinsulinemia/insulin resistance in mice fed a high fat diet by upregulating the expression of insulin-response glucose-transporter (GLUT-4) and the antidiabetic, antiatherogenic hormone adiponectin genes and down-regulating the expression of nuclear factor κB (NF-κB) that controls the regulation of genes that encode proteins involved in immune and inflammatory responses, including interleukin-6 (IL-6). In diabetes, some herbal alternatives are proven to provide symptomatic relief and assist in the prevention of secondary complications of the disease. Alternative therapies for diabetes treatment are needed because of inability of current therapies to contribute normoglycemia and prevention of diabetic complications. Moreover herbs are known for their safety, efficacy, cultural acceptability and lesser side effect besides maintaining normoglycemia in diabetes (Pullaiah & Naidu, 2003).

Table-1: *In vitro* effect of aqueous extracts of *Pleurotus florida* on the movement of glucose

Time	Control	Glucose (mg/dl) in external solution (0.15 N NaCl)			
		Concentration of macrofungal species (*Pleurotus florida*)			
		6.25g/l	12.5g/l	25g/l	50g/l
30 min	64.96±1.8	66.44±2.27	34.97±3.47	17.48±2.74	10.49±2.10
1h	70.42±2.09	73.42±3.92	41.96±3.65	38.47±3.65	13.99±3.45
2h	83.92±1.67	83.93±4.31	48.94±1.84	41.96±3.42	17.48±3.51
3h	90.92±1.3	87.43±2.95	55.81±3.61	45.63±2.62	31.46±2.95
4h	97.9±1.58	90.92±2.28	55.98±2.17	49.44±3.54	31.47±2.37
5h	101.39±2.33	94.48±1.64	62.45±4.28	52.95±2.62	34.98±2.42
6h	111.88±2.57	99.41±3.69	87.91±2.35	62.82±3.50	34.77±3.68
7h	118.89±2.8	101.93±1.95	95.15±2.95	69.53±2.98	38.42±2.94
24h	122.38±1.59	125.87±3.45	101.41±3.38	83.92±3.45	52.65±2.73

Values are means ± SEM for groups of 3 observations with their standard errors.

Table-2: Effect of metformin and different doses of aqueous extract of *P. florida* on oral glucose tolerance test (OGTT) activity in diabetic rats

Experimental animal	Blood glucose level (mg/dl) at different time interval after the treatment.				
	0	30 min	60 min	90 min	120 min
Normal (Control)	80±1.02	165±0.99	111±1.28	104±1.23	94±0.89
Diabetic (Control)	276±1.23	325±1.08	315±1.22	290±1.33	278±1.02
Diabetic+Metformin (500 mg/kg)	225±1.28	242±1.33	230±1.80	210±1.58	194±1.30
Diabetic+*Pleurotus florida* extract (100mg/kg)	223±0.98	252±1.03	243±0.89	229±0.98	217±0.91
Diabetic+*Pleurotus florida* extract (200mg/kg)	219±1.33	243±0.97	235±1.32	218±1.27	205±1.33

Values are means ± SEM for groups of 3 observations with their standard errors.

Table-3: Effect of metformin and different doses of aqueous extract of *P. florida* on fasting glucose level in diabetic rats

Experimental animal	Blood glucose level (mg/dl) at different h after the treatment			
	Before treatment	24 h after treatment	48 h after treatment	72 h after treatment
Normal (Control)	98±1.32	95±0.99	95±1.07	95±1.08
Diabetic (Control)	215±1.20	220±1.05	221±1.33	221±1.14
Diabetic+Metformin (500 mg/kg)	227±1.28	208±1.04* 8.37%	198±0.92* 12.78%	163±1.54* 28.19%
Diabetic+ *Pleurotus florida* extract (100 mg/kg)	214±1.37	208±1.68* 2.80%	197±1.08* 7.94%	179±1.02* 16.35%
Diabetic+ *Pleurotus florida* extract (200 mg/kg)	206±1.09	194±1.23* 5.82%	183±1.32* 9.71%	169±0.89* 17.96%

Values are means ± SEM for groups of 3 observations with their standard errors.

*Values in per cent indicate blood glucose reduction relative to 0-hour of the respective treatment group.

Table-4: Effect of metformin and different doses of aqueous extract of *P. florida* on blood glucose (post prandial glucose) levels in streptozotocin induced diabetic rats (mg/dl) (Acute study)

Experimental animal	Blood glucose level (mg/dl) at different h after the treatment				
	0	2 h	4 h	6 h	24 h
Normal (Control)	109±0.98	104±1.09	99±1.23	97±1.22	92±0.98
Diabetic (Control)	265±1.32	271±0.88	277±098	281±1.11	296±1.23
Diabetic+Metformin (500 mg/kg)	269±1.85	255±0.54* 5.20%	248±1.08* 7.81%	230±1.03* 14.50%	214±1.37* 20.44%
Diabetic+ *Pleurotus florida* extract (100 mg/kg)	278±1.23	271±1.25* 2.52%	266±1.25* 4.32%	262±1.58* 5.76%	255±1.09* 8.27%
Diabetic+ *Pleurotus florida* extract (200 mg/kg)	289±1.20	275±1.50* 3.81%	267±0.89* 7.61%	262±1.22* 9.34%	256±1.78* 11.42%

Values are means ± SEM for groups of 3 observations with their standard errors.

*Values in per cent indicate blood glucose reduction relative to 0-hour of the respective treatment group.

Table-5: Effect of metformin and different doses of aqueous extract of *P. florida* on blood glucose (post prandial glucose) levels in streptozotocin induced diabetic rats (mg/dl) (Chronic study)

Experimental animal	Blood glucose level (mg/dl) at different time after the treatment				
	0 day	5 day	10 day	15 day	30 day
Normal (Control)	93±1.09	95±0.99	95±1.05	97±1.32	97±1.91
Diabetic (Control)	266±0.99	281±1.08	294±0.94	301±1.20	304±1.38
Diabetic+Metformin (500 mg/kg)	241±1.23	186±1.23* 22.82%	170±0.83* 29.46%	153±1.65* 36.51%	128±1.02* 46.89%
Diabetic+ *Pleurotus florida* extract (100 mg/kg)	254±1.38	223±1.22* 12.20%	215±1.32* 15.35%	209±1.52* 17.72%	184±1.65* 27.56%
Diabetic+ *Pleurotus florida* extract (200 mg/kg)	269±1.56	234±1.83* 13.01%	222±1.25* 17.47%	201±1.98* 25.28%	168±1.87* 37.55%

Values are means ± SEM for groups of 3 observations with their standard errors. Values in per cent indicate blood glucose reduction relative to 0-hour of the respective treatment group.

CONCLUSION

Macrofungi is very popular food. It has been valued as remedies for various diseases in numerous countries throughout the world. Macrofungi have immense potential and may be developed as effective and safe antidiabetic remedy. For treatment of diabetes mellitus various drugs are available in market, despite of having beneficial effects these drugs are associated with various side effects. In chronic condition insulin therapy is also given but it is very painful. So to combat all these the most effective method to control glucose level in blood is proper diet and regular exercise. Mushrooms can act as potential agents to control blood glucose in proper manner. It is a natural product, safe to use and will also provide proper nutrients that might help to decrease complications associated with this metabolic disease. *P. florida* is much effective in reducing blood glucose level and hence can be use as an alternative remedy for the treatment of diabetes mellitus and its complications. Mushrooms thereby provide a rich reservoir for the development of new therapeutic agents. However, a further study is needed to identify the active fractions responsible for antidiabetic activity and to clarify the mechanism of the effect.

Plate-A: Fruiting body of *Pleurotus florida*

ACKNOWLEDGEMENTS

The authors wish to thank Head, Department of Botany and Head, Department of Zoology, DDU Gorakhpur University, Gorakhpur for providing necessary facilities.

REFERENCES

1. Adam, Z., A. Ismail, S. Khamis, M. Hanaffi, M. Mokhtar, M., M. Hamid, (2011). Antihyperglycemic Activity of *F. deltoidea* ethanolic extract in normal rats. *Sains Malaysiana*, 40(5):489–495.

2. Adebayo, E.A., J. K. Oloke, A.A. Ayandele, C. O. Adegunlola, (2012). Phytochemical, antioxidant and antimicrobial assay of mushroom metabolite from *Pleurotus pulmonarius* –LAU 09 (JF736658). *J Microbiol Biotech Res.*, **2(2):** 366-374.

3. Badole, S., L. and S. L. Bodhankar, (2007). Interaction of aqueous extract of *Pleurotus pulmonarius* (Fr.) Quel.-Champ with acarbose in alloxan induced diabetic mice. *Journal of Applied Biomedicine*, 5:157–166.

4. Cui, J. and Y. Chisti, (2003). Polysaccharopeptides of *Coriolus versicolor*: physiological activity, uses, and production. *Biotechnology Advances*, 21:109 – 122.

5. Edward, C.A., I. T. Johnson, N. W. Read, (1988). Do viscous polysaccharides slow absorption by inhibiting diffusion or convection? *Eur J Clin.Nutr.*, 42:307-312.

6. Gallagher, A.M., P. R. Flatt, G. Duffy, Y. H. A. Abdei-Wahab, (2003). The effect of traditional antidiabetic plants on *in vitro* glucose diffusion. *Nutrition Research*, 23:413-424.

7. Ganeshpurkar, A., S. Kohli, G. Rai, (2014). Antidiabetic potential of polysaccharides from the white oyster culinary-medicinal mushroom *Pleurotus florida* (higher Basidiomycetes). *International Journal of Medicinal Mushroom.* 16(3): 207-217.

8. Goyal, R., R. B. Grewal, R. K. Goyal, (2006). Nutritional attributes of *Agaricus bisporus* and *Pleurotus sajor caju* mushrooms. *Nutrition and Health,* 18:179-184.

9. Kanagasabapathy, G., K. H. Chua, S. N. A. Malek, S. Vikineswary, U. R. Kuppusamy, (2014). AMP-activated protein kinase mediates insulin like and lipo-mobilising effects of β- glucan rich polysaccharides isolated from *pleurotus sajor-caju* (Fr.), Singer mushroom, in 3T3-L1 cells. *Food Chem.*, 145:198-204.

10. Kim, D.H., B. K. Yang, S.C. Jeong, J. B. Park, S. P. Cho, S. Das, J. W. Yun, C. H. Song, (2001). Production of a hypoglycemic, extracellular polysaccharide from the submerged culture of the mushroom, *Phellinus lintenus. Biotechnology letter*, 23:513-517.

11. Kong Won-Sik (2004). Description of commercially important *Pleurotus* species. Mushroom Grower's handbook 1(II) Oyster Mushroom, Chapter 4.

12. Pullaiah, T. and K. C. Naidu, (2003). Antidiabetic plants in India and herbal based antidiabetic research. Daya Books, Regency Publications, New Delhi ISBN 81-87498-67-6.

13. Soosairaj, S., P. Raja, A. Kala, P. K. Raj, (2012). The survey of macroscopic fungi from a few districts of Tamil Nadu. *The Bioscan*, 7:669-671.

14. Thimmaiah S. K., (1999). Standard methods of biochemical analysis, Kalyani publication, New Delhi.

15. Vijaykumar, T. P., P. Nazni, Jaganmohan (2007). Nutritive indices and PER study of Mushroom varieties using wistar strain male Albino. *Mushroom Research*, 16(1): 23-26.

16. Vishwakarma, P., P. Singh, V. B. Kushwaha, N. N. Tripathi, (2018). The effect of aqueous extract of some wild edible macrofungi on *in vitro* diffusion of glucose. *KAVAKA,* 50: 34-37

17. World Health Organization (2008). Global strategy on diet, physical activity and health: diabetes. Accessed on May 4, 2008. URL: http://www.who.int/dietphysicalactivity/publications/facts/diabetes/en/index.html

18. Yang Byung-Keun, Kim Guk-Nam, Jeong Yong-Tae, Jeong Hun, Mehta Pradeep and Song Chi-Hyun (2008). Hypoglycemic effects of exo-biopolymers produced by five different medicinal mushrooms in STZ-induced Diabetic rats. *Mycobiology,* 36(1):45-49.

Current Status of Researches in Biosciences (2020) : 485-492
Editor : Prof. P.C. Joshi, N. Joshi, Reshman Yasmin, and D. K. Mansotra
Today & Tomorrow's Printers and Publishers, New Delhi - 110 002

STUDY ON THE QUALITY FACTOR OF EQUINE FIBRE

Amit Kumar Ghoshal and Shyama Pada Ghoshal

Charaktala, Post-Duilya, Howrah-711302

ABSTRACT

Horses are pre-existed by million. The fibre from horse mule donkey are generally used in making brushes strings of musical instruments. Materials instead of nylon or cutgut are now used for switechering in surgery. 22 samples (12 from horse & 10 from donkey) were studied. Following parameter were studied for the fibres.

In the case of horsetail, the colour was white brown tan and mixed. In the case of donkey, only mixed colour was found. In horsetail fibres, soft and coarse fibre were found but in case of donkey, only coarse fibres were found. Foreign materials such as sand and vegetable matter ranging between 0.1% to 0.5% in horse length varies between 19 cm to 53 cm. In case of donkey the length varies between 5.6 cm to 9.6 cm. Moisture percentage varies between 7.8 to 11.3 in case of horsetail and 12 to 15.8 in case of donkey's tail.

KEYWORDS: Equine, Fibre, Livestock, Economic welfare

INTRODUCTION

Horses constitute an important group of livestock animal not only for the enormous influence they had in moulding world history but also for the many roles they play in the economic welfare of mankind. Since prehistoric time horses have been in use both during war and peace times.

Horses have pre-existed mankind on earth by two million years. Modern breeds of horse are probably descendants of a prehistoric ancestor, originally found to occur both in the eastern and western hemisphere. The first equine ancestor, *Eohippus* lived approx 60 million years ago (early territory Eocene era) in North America. With the replacement of swamp by forests and grasslands, horses acquired many features, such as long legs, and shorter toes, and raised head. This growth was a handicap to horses as they could not conceal themselves from enemies, and therefore, developed the power of running of speedily to escape capture. There is an evolution of the present-day horse is due to the several adaptations brought about by the environmental changes.

The horse of today belongs to the order Perissodactyla. Family Equidae and the genus *Equus linn.* This genus represents four groups, viz. The horse group (*Equus*) with unstripped bodies shorter ears narrower hoofs, the ass group (*Asinus*) with unstripped bodies, longer ears and narrower hoofs, the Grevy's Zebra group (*Dolichohippous*) with stripped bodies and broad ears and the Zebra group (*Hippofigris*) with stripped body, head with narrow ears and elongated neck. While the former two groups of animals are represented in India, the latter two are found only in other parts of Asia and Africa.

Among the horses, there are two distinct types. The northern dunspecles which is still represented by the wild pony of Western Mongolia, known as the Przewalski's horse. *Equus Przewalskil* Pollakov occurs in the Gobi desert. It is the only true wild horse still living in the world.

It is about 12 hands (one hand =10.1 cm) in height with ungainly head, short erect mane and no forelock. It is yellowish dun in colour with a light nuzzle and narrow ass-like feet. Indian wild ass, *Equus leminous* khus lesson, is reported to occur in the Rann of Kutch (Gujarat).

The other type of E.L. Onager Boddaert of the Caspian and Mediterranean areas. it is a thin-skinned animal, less heavy quicker in movements, more intelligent and darker in colour than *E. Przeualskil*. The horse is reported to be the prehistoric forerunner of the Arab barb and Turk breeds. The modern Indian horse is from the Mediterranean areas.

Indian breeds: Kathiawari or Kutchi is one of India's best breeds found in Rajasthan. it is handy, vigorous, well known for its pace and speed and possesses good endurance power under tropical heat and intense cold. Animals between 12 and 15 hands (1.2 - 1.5m) in height and 1.37 - 1.52 m in girth. It has a close resemblance to Marwari breed, both of which possibly had a common ancestor, the Arab Stallion, which lived in the forests of Kathlawar and Bombay subsequent to the wreckage of the ship carrying the stud breeds on the west coast of India.

Marwari of Mallani is Rajasthani horse found in Marwar, Jaipur, Jodhpur and Udaipur. It is very tough and sturdy, possessing considerable staying powers and having an early pace and galt and somewhat uncertain temperature. Most prevalent colours are chestnut and bay.

Manipuri pony has been bred in Manipur state, Assam for many centuries. This breed is noted for its elegance, endurance and speed. Though small in size, the animal possesses a proportionate body and is sturdy and surefooted, it measures 11-13 hands (1.1- 1.4m) in height and its body weight is about 295 kg. The animals are utilised as poto ponies, racehorses and as pack animals. They are also valued as military pack transport ponies.

Bhutani or Bhutia is breed in the lower regions of the Himalayan Mountains from Punjab to Bhutan where most of the breeding work is undertaken by hill tribes. The prominent features of the animal are: compact body, broad forehead, short thick back, broad chest, straight shoulders, strong back, good bone, well ribbed up barrel, round muscular quarters, course hairy legs and long tail and mane. Height of a good pony is about 13.0- 13.2 hands (1.31 -1.33 m). Weight varies from 270 to 360 kg. Majority of the animals are gledings which are popular for riding and also as pack animals.

Spiti pony in inhabits the Spiti valley lies within the Kulu sub-division of the Kangra district. It is particularly hardy sure-footed and averages about 12 hands in height. The body is well developed.

Another type is Chummarti, hailing from the Chummarti valley of Tibet, is frequently met with in Kinnaur district and the neighbouring areas in Himachal Pradesh. There is very little difference between Spiti and Chummarti in body conformation (hight at wither) 1.27 -1.29m, length 1.34 -1.36 m, girth 1.34 -1.42m). The mares of this breed are cross-breed with stallions of imported Connemara, a native pony Ireland.

Donkeys: Donkey and mules are closely related to horses but for one or two minor differences, viz. the hind legs never have a trace of the chestnut, which is typically found below the hock on the hind legs in all horses. The period of gestation in the case of the donkey is about 12 months, one month more than that of a horse. The mules are sterile.

The weight of loads carried by mules excluding the saddle is 145 kg. Big, heavy type of mules are necessary for carrying artillery loads. The shape of the back must be straight from withers to croup or inclined to be roach-backed. It should be well covered with muscle, broad and level on the top and short, but sufficiently long to carry the pack saddle.

Use of equine fibres: The fibre from the horse, mules or donkeys are generally used in making various types of brushes. Besides that strings of the musical instrument are also used. Long back horse tall fibres used as stitching material in instead of nylon or cut-gut now.

Sample collection was found to be a big obstacle, in spite of contact with the number of institutions only 22 samples (12 horse and 10 donkeys) could be supplied by National Equine Centre (I.C.A.R) Hissar. Therefore the study is based on a very limited number of samples. The following parameters were considered for the study of fibres.

(A) **Colour:** Four colour generally found for horsetail fibres. They are (i) white (ii) brown and (iii) tan red and (iv) mixed colour whereas in case of donkey tail only mixed colour are generally found..

(B) **Texture:** Two types of fibres are generally found in the horsetail. They are soft and coarse. But in case of donkey only coarse type of tail fibres are found.

(C) **Foreign matter:** During the study, the following foreign materials were obtained from both the horse and the donkey tail. These are sand, dust and vegetable matter. The percentage of foreign matter varies from 0.1 to 0.5 per cent in case of horsetail and 0.1 to 0.6 per cent in case of donkey tail.

(D) Varies from 19 cm to 53 cm. In the case of horsetail and 5.6 cm to 9.3cm In case of donkey tail.

(E) **Moisture percentage:** Varies from 7.8 to 11.3 in case of horsetail sand 12.8 to 16.8 in case of donkey tail.

(F) **Foreign matter** :0.2%

Tentative specification of horsetail and donkey tail fibre:

1. **Scope:**This standard prescribed the grade designation and definition of quality of horse and Donkey tail fibres which are used as a fibre.

2. **Grade designation of horse and donkey tail fibres:** This grade indicates the characteristics and quality of horse and donkey tail fibres of specified trade description shall be those set out in :

   i) White

   ii) Brown / grey

   iii) Tan red and

   iv) Mixed

Table-(A) Analytical Results Of Horse Tail Fibres

S. NO	Name of the commodity/ code no.	Colour	Mean length (Length of whole bundle)	Foreign	Moisture matter	Texture
1	Pony Br. No.7	Brown	53 cms	0.2	7.83	Coarse
2	Pony Br. No.8	White	42 cms	0.3	10.72	Soft
3	Pony Br. No.8 (A)	Brown	60 cms	0.2	11.30	Coarse
4	Pony Br. No.9 (B)	Brown	26 cms	0.4	9.82	Coarse
5	Pony Br. No.10	Pale-Yellow mixed with black hair	20 cms	0.5	9.67	Coarse
6	Pony Br. No.11	Brown	20 cms	0.1	9.81	Coarse
7	Mare select	Black	33 cms	0.3	11.08	Coarse
8	Mare Br.No.6	50% white 50% Black Seprate bundles	34 cms	0.2	11.14	Coarse
9	Mare Krishna	Black with small bunch of pale yellow	20 cms	0.4	8.2	Coarse
10	Mare Rams	Tan red	19 cms	0.2	9.66	Soft
11	Mare Br.No.6	Black	19 cms	0.2	9.33	Soft
12	Mare Rani	Black	26 cms	0.1	8.90	Soft

Table - (B) Analytical results of donkey tail fibre

S. No.	Code	Colour	Texture matter	Foreign	Mean length (cm)	Moisture	Remarks
1	2	3	4	5	6	7	8
1	Long short	Pale yellow with	Coarse	0.4	3.16	12.85	All the samples
2	Long short	Mixed colours	Coarse	0.3	0.7	14.17	Contain two
3	Long short	Brown and pale yellow	Coarse	0.1	9.2	16.76	Different hair
4	Long short	Black yellowish black	Coarse	0.2	7.4	13.41	Bundle one short
5	Long short	Pale yellow with black hair	Coarse	0.6	5.6	12.78	and one long with
6	Long short	Yellowish brown pale yellow	Coarse	0.3	8.7	14.32	different colours
7	Long short	Mixed type pale yellow with black	Coarse	0.2	7.5	13.72	
8	Long short	Black Black with pale yellow	Coarse	0.5	6.8	14.18	
9	Sample contain burrs	Black and Tan red colour	Coarse	0.3	8.3	13.54	
10	Long short	Pale yellow mixed colours	Coarse	0.2	9.3	14.41	

3. **Texture**

   I) Soft

   II) Coarse

4. **Foreign matter**: 5%

5. **Moisture percentage**: 12%

6. **Length**: Length must be uniform in size with the variation of 0.5 cm and one bale /package contain single colour, texture and similar length. Average length shall be Indicated on the bale/package.

Note:

1. "Lot" means the total quantity of horsetail fibre offered for grading. It shall be of one colour, one type only.

Grade designation and special characteristics of Indian horse tail fibre

(A) Coarse (Texture)

Grade designation	Colour of fibres	Maximum foreign matter	Moisture percentage	Remark
White	White	.5%	10%	
Black	Black to brown			
Mixed	Mixed			

Note: Colour shall be indicated on the bale/package

(B) Soft

Grade designation	Colour of fibres	Maximum foreign matter	Moisture percentage	Remark
1	2	3	4	5
White	White	5%	10%	
Black	Black to brown			
Mixed	Mixed			

Note: Colour shall be indicated on the bale/package

Grade designation and special characteristics donkey tail hair fibre

Coarse Variety (texture)

Grade designation	Colour of fibres	Maximum foreign matter	Moisture percentage	Remark
1	2	3	4	5
Black	Black	.5%	8%	
Mixed	Mixed colour			

Note: Colour shall be indicated on the bale/package

Current Status of Researches in Biosciences (2020) : 493-506
Editor : Prof. P.C. Joshi, N. Joshi, Reshman Yasmin, and D. K. Mansotra
Today & Tomorrow's Printers and Publishers, New Delhi - 110 002

ETHNOBOTANICAL AND PHARMACOLOGICAL ASPECTS OF *IPOMOEA CARNEA* JACQ. AND *CELOSIA CRISTATA* LINN.

Ramveer Singh[1], Navneet[2], Dalip K. Mansotra[3] and Akash Kumar[4]

[1,2,4]Department of Botany and Microbiology,

[3]Department of Zoology and Environmental Science, Gurukula Kangri Vishwavidyalaya, Haridwar

E.mail: ramveer.gangwar@outlook.com

ABSTRACT

Uttarakhand is rich in medicinal wealth which is widely used by local inhabitants. Medicinal plants have various pharmacological activities like antibacterial, antifungal, glycosidase inhibitory, anti-inflammatory, antioxidant, antidiabetic, wound healing, immunomodulatory, cardiovascular, embryotoxic effect, anticancer, and anxiolytic activities. Various ethnomedicinal plants were used in different traditional medicinal systems like Ayurveda, Siddha, and Unani since the time of the existence of life. Ethnobotany is a branch of science which deal with various aspects such as medicine, religious and cultural perspective of the plants. This chapter is focused on the ethnobotanical and pharmacological properties of *Ipomoea carnea* and *Celosia cristata.*

KEYWORDS: Ethnobotany, Medicinal plants, Ayurveda, Pharmacology, *Ipomoea carnea, Celosia cristata*

INTRODUCTION

In this developing world about 3.4 billion peoples depended upon the traditional knowledge of the medicinal plants. They are using medicinal plant based natural products for the treatment of different disorders and diseases (Doughari *et al.*, 2009). The treatment of microbiological diseases with the help of traditional medicine plants becomes the main focus of numerous studies (Bhavnani & Ballow, 2000; Chariandy *et al.*, 1999). Plants synthesize different phytochemicals as secondary metabolites such as tannins, alkaloids, glycosides, terpenoids and phenols which shows antimicrobial properties (Cowan, 1999; Das, 2010; Singh & Navneet, 2016). It is supposed that about 14-28% of plant species are used for medicinal purpose and about 74% of them have pharmacologically active compounds.

***Ipomoea carnea* Jacq.**

General Introduction

It is widely used medicinal plant belongs to the Convolvulaceae family and mainly distributed in Uttarakhand, Chhattisgarh and Madhya Pradesh in India. But this plant is found all over the world like America, Brazil, Pakistan, Nepal, and Sri Lanka, etc. (Dwivedi 1999).

I. carnea belongs to morning glory species. Leaves of this plant are 4.0 – 7.5 cm long with 2.5 – 3.0 mm diameter. Leaves are simple and have an alternate leaf arrangement. The stem is thick, erect, woody and hairy; with the cylindrical shape and greenish color (Sharma & Bachheti, 2013).

Ethnobotanical Uses: In the Indian sub-continent region, *I. carnea* is used as folk medicine by several communities. Leukoderma, skin related diseases and lesions were treated by the milky juice of *I. carnea* (Adsul *et al.*, 2009), in menstruation provocation boiled roots of *I. carnea* were used (Chowdhury *et al.*, 1997). *I. carnea* was also used in dysentery, hypertension, venereal diseases and immunodeficiency (Meira *et al.*, 2012). *I. carnea* are reported in Ayurveda, Siddha, and Unani. This plant is used for Leukoderma. It is also used as an aphrodisiac, cathartic and purgative (Phillips *et al.* 1994). A recent study shows that it is also used against hypertension and AIDS (Lamidi *et al.*, 2000; Woradulayapinij *et al.*, 2005).

Pharmacological Uses

Anti-inflammatory activity: Water extract of young leaves of *I. carnea* was tested for anti-inflammatory activity. The dose of the plant extract in the current study is 250 mg/kg and 500mg/kg body weight with etoricoxib 6mg/kg body weight. The result shows that leaves of *I. carnea* have a strong anti-inflammatory activity at 500mg/kg body weight as compared with etoricoxib 6mg/kg body weight (Khalid *et al.*, 2011).

Antioxidant activity: The antioxidant activity of methanol extract of *I. carnea* was determined by DPPH (2,2-diphenyl-1picrylhydrazyl) free radical scavenging activity, FRAP (Ferric reducing antioxidant power assay), total phenolics, and ferric thiocyanate assay. The highest antioxidant activity for DPPH, FRAP, and total phenolic content was shown by n-butanol, chloroform and ethyl acetate fraction respectively (Abbasi *et al.*, 2010).

Antidiabetic activity: Albino wister male rats were taken for carryout the experiments. All rats had 150-200 gm weight and 200mg/dl blood glucose level were selected for the experiment. Streptozotocin was injected intraperitoneally to induce type 2 diabetes in rats with dose of 50mg/kg body weight. 6 groups of rats were created with 6 rats in each group. Rats were treated with the standard (Glibenclamide 10mg/kg) and aqueous extract of *I. carnea* (250 mg/kg and 500mg/kg). Blood glucose level was determined by electronic glucometer at different intervals (0 hours i.e. before treatment, 0.5, 1, 2, 4, 8, 12, 16 and 24 hours). Maximum reduction of blood glucose level for Glibenclamide (10mg/kg body weight), *I. carnea* (250mg/ml body weight) and *I. carnea* (500mg/kg body weight) were found 48.64±0.09 at 12 hr, 32.40±0.01 at 12 hr, 37.54±0.14 at 12 hr respectively (Adsul *et al.*, 2012)

Anticancer activity: *In-vivo* anticancer activity was carried out by Ehrlich's Ascites Carcinoma (EAC) cells. Albino mice have weight between 25 to 30g were taken and injected with EAC cells suspension (1X10^6 cells/ml) intraperitoneally in peritoneal cavity. After 15 days these cells were washed by normal saline and centrifuged at 1500 rpm for 15 min. These cells were suspended again in normal saline and repeat this process three times. 5 groups of male Swiss albino mice were taken with 12 mice in each group. All mice in groups were treated with EAC cells (2X10^6 cells/mouse) intraperitoneally (except normal group). 1st group treated with Normal with sodium Suspension (0.1%), 2nd group treated with Induced EAC cell (2 X10^6) with sodium Suspension (0.1%), 3rd group treated with Induced EAC cell (2X10^6) with 5fluorouracil (20mg/kg i.p), 4th group treated with Induced EAC cell (2X10^6) with HAEIC (250mg/kg p.o) with sodium suspension (0.1%), and 5th group treated with Induced EAC cell (2X10^6) with HAEIC (500mg/kg p.o) with sodium suspension (0.1%). Anticancer activity of the plant extract was measured in viable count, tumor weight, tumor volume, percentage of viable cell, non-viable cell count, percentage of non-viable cells, increase in life span (ILS), change in food intake and average increase in body weight. The tumor weight, tumor volume and viable cell count were increased significantly ($p < 0.001$) and food intake, non-viable cell count, increase in body weight were decreased significantly ($p < 0.001$). Mean survival time was increased in 4th and 5th group of mice. All these observation clearly suggest that the HAEIC showed anticancer activity significantly

(Anand *et al.*, 2013).

Antibacterial activity: Extraction of leaves done by different solvents i.e. n-hexane, ethanol, and acetone. Growth of *Proteus vulgaris* and *Salmonella typhimurium* were inhibited by crude acetone extract, while *Pseudomonas aeruginosa* was inhibited by ethanol extract (Adsul *et al.*, 2012).

Antibacterial activity of *I. carnea* stem extract was tested against *Klebsiella pneumoniae*, *Proteus mirabilis*, and *Pseudomonas aeruginosa* via agar disc diffusion method. *P. aeruginosa* showed large zone inhibition and found most susceptible to *I. carnea* stem extract (Khatiwora *et al.*, 2012).

Silver nanoparticles were synthesized with the water extract of *I. carnea* and check the antimicrobial activity with disc diffusion method against *Staphylococcus aureus*, *Bacillus cereus*, *Bacillus subtilis*, *Klebsiella pneumoniae*, *Aeromonas hydrophila*, *Salmonella typhi*, *Proteus vulgaris,* and *Pseudomonas aeruginosa*. Zone of inhibition (ZOI) at 2μg concentration for the above pathogen were 10.1±0.32, 10.5±0.28, 11.0±0.28, 11.06±0.34, 11.0±0.17, 11.2±0.38, 10.0±0.34, and 11.0±0.23mm respectively (Daniel *et al.*, 2014).

Antifungal activity: Antifungal activity of *I. carnea* leaves extract was studied against *Alternaria alternata* and *Curvularia lunata* (Agarwal & Uppadhay, 1997). *I. carnea* was tested against *Colletotrichum gloeosporioides* and *Cladosporium cucumerinum* growth, and spore germination of *A. alternata* and *A. porri*. For *Colletotrichum gloeosporioides*, percent mycelial growth inhibition shown by n-hexane, ethyl acetate, methanol, active fraction, phenol extractives of *I. carnea* were 0.0, 24.2±0.8 20.4±1.4, 68.5±0.7, and 78.3±0.4 respectively.This study shows that a fraction namely octadecyl p-coumarate was responsible for antifungal activity in a dose dependent manner (Nidiry *et al.*, 2011).

Wound healing activity: Wound healing activity of ethanol extract of *I. carnea* flower was studied. Kaempferol and 3-O-β-D glucoside found in *I. carnea* flower extract which helps in fibroblast proliferation and collagen fiber formation and scar drying (Ambiga *et al.*, 2007).

Immunomodulatory activity: *I. carnea* contain toxic component like indolizidine alkaloid swainsonine and nortropane alkaloid catystegines B1, B2, B3 and C1 (Cook, 1987; Sahayaraj & Ravi, 2008). Swainsonine toxic effects were studied on female rats, after the treatment with *I. carnea,* fewer changes occur in rat's body like body weight reduction, change in spleen/body weight ratio and thymus/body weight ratio. When *I. carnea* tested on pregnant rats with the dose of 7gm/kg body weight, all the progeny died after birth (Rosenberg, 1999; Lunardi, *et al.*, 2000).

Cardiovascular activity: Cardiovascular activity of *I. carnea* was checked in isolated frog heart. The initial blockade was 5-10 second when water extract was introduced and the blockade timing was increased with the dose of plant extract. This study shows a positive inotropic effect by releasing intracellular calcium or sodium extrusion (Bachhav *et al.*, 1999).

Embryotoxic effect: Aqueous leaf extract of *I. carnea* was introduced in rats with a dose of 0.0, 0.7, 3.0 and 15.0 mg/kg body weight on a daily basis in gestational days. Malformation, skeletal and visceral abnormalities were observed in this study. Which confirm that *I. carnea* act on maternal homeostasis and shows embryotoxic effects (Hosomi *et al*, 2008).

Nervous system effects: *Ipomoea* spp. intoxication was studied in goats. Fresh leaves, flower, and stem were fed to goats for up to 60 days. Depression, ataxia, hirsute coat, irregular muscular movements, abnormal postural reactions, spastic paresis, intense tremors, nystagmus, head tilting, and hypersensitivity to sound were observed in all treated goats. It noticed that cerebellum is the main affected organs in *Ipomoea* intoxication (Barbosa *et al.*, 2006; Rios *et al.*, 2008).

Anxiolytic Activity: The anxiolytic activity was studied in mice by using elevated plus maze open field test and hole board test model. The aqueous and methanol extract of *I. carnea* were used at 32.50 mg/ml and 16.25 mg/ml concentration respectively and diazepam used as a positive control. As compared with diazepam, both extracts (methanol and aqueous) showed higher anxiolytic effect. While the intraperitoneal (i.p) LD_{50} were found 325 mg/kg body weight for both the extracts. In hole board test model, observations show that the anxiolytic activity of *I. carnea* methanol and the aqueous extract is based on dose dependent manner when compared with diazepam at 1mg/kg, 2mg/kg as a standard (Bidkar *et al.*, 2012).

Anti-convulsant activity: The anticonvulsant activity of *I. carnea* was studied in Swiss albino rats by using strychnine, picrotoxin and MES-induced seizure models. 10 groups of Swiss albino rats were created with 5 rats in each. Before 30 minutes of MES, Albinos were treated with aqueous plant extract (100 and 200 mg/kg body weight), phenytoin and vehicle control. No death was observed in animals after the MES (Rout & Kar, 2013).

Sedative activity: The sedative effect of *I. carnea* was studied by phenobarbitone-induced sleeping time and head dip test. 66 albino mice in eleven group (6 in each group) taken on the wooden box which has 16 evenly spaced holes. Animals were treated with the plant extract 100, 200 and 400 mg/kg body weight, vehicle 5ml/kg body weight and diazepam 4mg/kg body weight. Animal dipped the head into the hole of the box were noted. The number of head dip for alcoholic extract and aqueous extract were decreased with the dose 200mg/kg and 400mg/kg body weight respectively (Rout & Kar, 2013).

Teratological activity: The teratological effect of *I. carnea* was studied on rat pups and adult offspring by prenatal exposure of 0.0, 0.7, 3.0 and 15.0 mg/kg body weight. This study was conducted during gestational days and found that the body weight of maternal and offspring was unaffected, while postnatal mortality, delay in the opening of both ears, negative geotaxis and reversible hyperflexion were observed. This is also confirmed that *I. carnea* had organ specific toxicities and affect the liver, thyroid, kidney, and pancreas (Schwarz *et al.*, 2003).

Phytochemistry: Traditional usage of a medicinal plant indicate that the plant has bioactive phytochemicals; these phytochemicals are responsible for their pharmacological properties.

Latex of *I. carnea* contain D-fucose, L-rhamnose, D-chinovose(6-deoxy-D-glucose), jalapinolate (11-hydroxypalmiticacid), D-glucose, convolvulinolate(11-hydroxy-pentadecane acid), ipurolic acid (3,11-dihydroxy-tetradecane acid), 7-hydroxy-decane acid (Legler, 1965).

Acacetin-7-galactoside, flavone glycoside, and ipomotocin were reported in the latex of *I. carnea*. While alkaloids, glycosides, tannins and reducing sugar were found in the dried powder leaves of *I. carnea*. Polyphenols (30-70mg catechol equivalent/g dry material) and flavonoids (80-120mg quercetin equivalent/g dry material) were also found in leaves, stem, and flower of *I. carnea* (Tirkey *et al.*, 1988).

Amino acid, alkaloids, carbohydrates, phenolic compounds, proteins, terpenoids, sterols, and saponins were reported in the methanol extract of flower and leaves of *I. carnea*. 2-epi-lentiginosine, swainsonine, N-methyl-trans-4-hydroxy-L-prolineat, calystegines B1, calystegines B2, calystegines B3, and calystegines C1 were found in leaves, seeds, and flower in the chromatographic study. 13 compounds were also found in GS-MS (gas chromatography mass spectrometry) of *I. carnea* hexane extract. Stearic acid, hexadecanoic acid, hexatriacontane, 1,2-diethyl phthalate, tetraacontane, octacosane, n-octadecanol, and 3-diethylamino-1-propanol were reported in GS-MS study (Patel *et al.* 2007).

Celosia cristata Linn.

General Introduction

C. cristata is an ornamental plant which belongs to Amaranthaceae family. It shows different pharmacological activities like antimicrobial, antioxidant, hemostatic, cytotoxicity, hepatoprotective and anthelmintic activity due to their phytochemicals (Al-Snafi, 2015).

Botanical Description: *C. cristata* is an annual little branched herb from Amaranthaceae family. Plant's height is about 60cm feet and covers approx. 60cm feet area. Leaves are green sometimes red or purple in color with simple and pinnate venation. Flower of this plant varies in color like pink, yellow or orange. The seed of this plant is very small, black colored and kidney shaped (Al-Snafi, 2015).

Traditional Uses of Celosia Herb: *C. cristata* is a native herb of Asia and used as a medicine and food from ancient time. Their leaves and young shoot are used as a vegetable which gave spinach like taste. Its different dishes are served with pepper, garlic, fresh lime, and red palm oil. *C. cristata* is used in bloody stool, uterine bleeding, skin burn and wound healing. Flower tops of the plant have been used for amenorrhea, dysentery, spitting up blood, hemorrhoids, leucorrhoea, and urine blooding. It is also effective against *Trichomonas*. Moreover, seeds are hypotensive and ophthalmic (Kumar, 2018).

The seed of *C. cristata* was used as antipyretic, alterative, ophthalmic, astringent and vulnerary. All parts of the plant have been used for coughs, bloody coughs, dysentery, abnormalities in menstruation cycle, mouth sores, intestinal bleeding, lungs bleeding, amenorrhea, hemorrhoids, retinal hemorrhages, conjunctivitis, eye diseases, urinary tract pathogen and liver problems (Surse *et al.*, 2014).

Physicochemical Characteristics: Different parameters of physicochemical characteristics were studied. Total ash, water soluble ash, acid insoluble ash, LOD (loss on drying), sulfated ash, and moisture content were tested. Analysis result of the above parameters is 5.54, 2.78, 1.14, 5.58, 0.8 and 5.54 % w/w respectively (Surse *et al.*, 2014).

Fluorescence Analysis: Fluorescence analysis of *C. cristata* was studied under visible and ultra violet light. Dried powder of plant was taken on a clean glass slide with their solvents. 1N NaOH (Alcohol), 1N NaOH (Aq), 1N HCl, 1% H_2SO_4, and 1% HNO_3 were used as a solvent. The dark brown color was observed for 1N NaOH (Alcohol) and 1N NaOH (Aq) and greenish brown for 1N HCl, 1% H_2SO_4 and 1% HNO_3 under UV light. Reddish Brown color was observed for 1N NaOH (Alcohol) and 1N NaOH (Aq), dark brown for 1N HCl, and 1% H_2SO_4, brown for 1% HNO_3 under visible light (Surse *et al.*, 2014).

Phytochemical Analysis: Phytochemical study of *C. cristata* showed the presence of amino acids, proteins, carbohydrates, alkaloids, flavonoids, saponins, gum, tannins, phenolic compounds, steroids, and triterpenoids, while oils and fats were found absent (Surse *et al.*, 2014).

Alkaloids, mucilages, flavonoids, saponins, tannins, triterpenoids, and phenolic compounds were present while oil, fats, carbohydrates, glycosides, and sterols were absent in aqueous, methanol and chloroform leaf extract of *C. cristata* (Rubini *et al.*, 2012).

Six compounds were isolated from ethanolic extract of *C. cristata*, these were identified as quercetin, β-sitosterol, 4-hydroxyphenethyl alcohol, kaempferol, stigmasterol, and 2-hydroxy octadecanoic acid (Xiang *et al.*, 2010). CCP-25 and CCP-27 two antiviral glycoprotein were isolated from *C. cristata* leaves (Gholizadeh & Kapoor 2004).

Celosin I, celosin II, celosin A, celosin B, cristatain, kaempferol, p-hydroxyphenylethanol, quercetin, β-sitosterol, stearic acid, stigmasterol, sphingosine, n-hexacosoic acid, palmitinic acid, and daucosterol compounds were isolated in *C. cristata* (Wang, 2012).

Pharmacological Activity:

Cytotoxicity: Cytotoxicity of *C. cristata* was studied in Fibroblast cell (Cos7) and in Cancer cell lines (SK-Hep1, HepG2, HeLa and 174T). IC_{50} was determined for aqueous extract against SK-Hep1, HepG2, HeLa, Cos7 and LS 174T. IC_{50} value for aqueous extract were 180.0, 200.0, 2773.5, 263.9 and >200.0 μg/ml respectively. IC_{50} for MeOH extract were 77.2 and 499.8 μg/ml against Cos7 and HeLa cell lines respectively. And IC_{50} for CH_2Cl_2 extract were 136.0 and 472.0 μg/ml against Cos7 and HeLa cell lines respectively (Herrmann *et al.*, 2011).

Hemostatic effect: Hemostatic effect of *C. cristata* was studied on rabbits and mice. Both animals were treated with plant extract with the dose of 1.7g/kg body weight and bleeding time (BT), plasma recovery time (PRT), coagulation time (CT), and prothrombin time (PT) were determined. PRT, CT, and PT were reduced in rabbits and BT reduced in mice as compared with a control group of rabbits and mice (Kumar, 2018).

Antimicrobial effects: The antimicrobial properties of different solvent extracts of *C. cristata* were determined against *Bacillus subtilis, Salmonella typhimurium, Staphylococcus aureus, Pseudomonas aeruginosa, Escherichia coli,* and *Candida albicans*. Minimal inhibitory concentration (MIC) was also determined by using a broth micro-dilution method. Methanolic and ethanolic extracts showed good antimicrobial activity against *C. albicans*, *B. subtillus*, and *S. aureus*; and the MIC value for above pathogens were 1mg/ml, 0.5mg/ml and 0.125mg/ml reported respectively (Yun *et al*., 2008). Petal aqueous extract of *C. cristata* was tested against *Proteus vulgaris, E. coli, S. aureus, Bacillus magaterium* and *B. subtilis*. Maximum zone of inhibition showed against *B. subtilis*. *B. subtilis* showed 8mm zone of inhibition while other bacteria showed no zone of inhibition (Sridevi *et al*., 2011). Flower extract of *C. cristata* was studied against *B. subtilis, S. aureus, E. coli, Klebsiella pneumoniae, P. aeruginosa, Proteus vulgaris, Aspergillus* sp., *Penicillium* sp., *Fusarium* sp. and *Rhizopus* sp. The inhibition zone of the above microorganisms were 9±0.5, 8±0.6, 5±0.6, 9±0.4, 7±0.6, 6±0.8, 12±0.6, 19±0.5, 14±0.7 and 23±0.9mm respectively (Rao *et al*., 2015).

Hepatoprotective effects: The hepatoprotective effect was determined in mice by analyzing the carbon tetrachloride (CCl_4)-induced hepatotoxicity. Semenoside A was isolated from *C. cristata* and Hepatoprotective effect of semenoside A was determined with an oral dose of 1.0, 2.0 and 4.0 mg/kg respectively. The results showed that it had remarkable hepatoprotective effects ($p<0.01$) (Sun *et al*., 2011).

Antioxidant effects: Antioxidant activity of *C. cristata* was determined by 2,2-diphenyl-1-picrylhydrazyl (DPPH) and 2,2'-azino-bis(3-ethylbenzothiazoline-6-sulphonic acid) (ABTS) radical scavenging method. Total polyphenol, flavonoid, and tannin were analyzed in methanolic extract of *C. cristata* by spectrophotometric methods. The total flavonoids, tannin, and polyphenol were found 2.34, 6.23 and 6.80 mg/g extract residue respectively in the methanolic extract. The DPPH and ABTS radical scavenging activities were 52.43 and 107.01 mg Trolox equivalent antioxidant capacity per gram extract residue respectively (Wen *et al*., 2006). Antioxidant activity of *C. cristata* was determined by using ferric reducing antioxidant power (FRAP) assay. *C. cristata* contains antiviral proteins (CCP-25 and CCP-27), FRAP assay of those proteins indicates that *C. cristata* has strong antioxidant activity (Gholizadeh & Kapoor, 2004).

Anthelmintic effects: The anthelmintic activity of aqueous, chloroform and methanol leaves extracts of *I. cristata* were analyzed. Adult earthworms

(*Pheretima posthuma*) were collected from the moist soil. All earthworms are 3-5 cm long. Fifty four earthworms were taken into nine groups, six earthworms in each group. Two concentrations (100 and 200mg/ml) of plant extract and two concentration of albendazole (100 and 200mg/ml) were prepared. Each group of earthworms treated with plant extracts, albendazole, and distilled water. Albendazole was marked as positive control and distilled water used as a negative control. Paralysis time for aqueous, chloroform, methanol extract and albendazole is 6.70±0.59, 59.67±0.88, 10.67±0.33, and 5.67±0.33 minute for 100mg/ml concentration respectively and 6.33±0.33, 59.67±1.45, 9.00±0.58, and 4.67±0.33 minute for 200mg/ml concentration respectively. Death time for aqueous, methanol extract and albendazole is 10.67±1.33, 12.67±0.88 and 7.67±0.33 minute for 100mg/ml concentration respectively and 6.00±0.00, 10.67±0.88 and 5.67±0.33 minute for 200mg/ml concentration respectively. The maximum anthelmintic effect was shown by aqueous extract in dose dependent manner (Rubini *et al.*, 2012).

CONCLUSION

This chapter focused on identifying the ethnobotanical and pharmaceutical aspect of *I. carnea* and *C. cristata*. Both plants contain many chemical constituents and bioactive compounds which are responsible for their medicinal properties. Many different pharmacological activities like antimicrobial, anti-inflammatory, wound healing, antidiabetic, antifungal, embryotoxic, anticancer, hepatoprotective and anxiolytic activity, etc. figured out with their respective plant.

REFERENCES

1. Abbasi, M.A., Zafar, A., Riaz, T., Rehman, A., Arshad, S., Shahwar, D. ... Ajaib M. (2010). Evaluation of comparative antioxidant potential of aqueous and organic fractions of *Ipomoea carnea*. *Journal of Medicinal Plants Research*, 4(18), 1883-1887.

2. Adsul, V.B., Khatiwora, E. & Deshpande, N.R. (2012). Evaluation of antioxidant activity of *Ipomoea carnea* leaves. *Journal of Natural Products and Plant Resources*, 2(5), 584-588.

3. Adsul, V.B., Khatiwora, E., Kulkarni, M., Tambe, A., Pawar, P., & Deshpande, N. (2009). GCMS study of fatty acids, esters, alcohols from the leaves of *Ipomoea carnea*. *International Journal of Pharm Tech Research*, 1, 1224–1226.

4. Adsul, V.B., Khatiwora, E., Torane, R., & Deshpande, N.R. (2012). Antimicrobial activities of *Ipomoea carnea* leaves. *Journal of Natural Products and Plant Resources*, 2, 597-600.

5. Agarwal, R.K., & Uppadhay, R.K. (1997). Antimicrobial activity of metal complexes

prepared from the leaf proteins of *I. carnea* Jacq. *Indian Drugs*, 14(2), 23-25.

6. Al-Snafi, A.E. (2015). The chemical constituents and pharmacological importance of *Celosia cristata* – A review. *Journal of Pharmaceutical Biology*, 5(4), 254-261.

7. Ambiga, S., Narayanan, R., Gowri, D., Sukumar, D. & Madhavan, S. (2007). Evaluation of wound healing activity of *Ipomoea carnea* Jacq. *Ancient Science of Life*, 26(3&4), 45-51.

8. Anand, G., Sumithira, G., Chinna, R.R., Muthukumar, A., & Vidhya, G. (2013). *In vitro* and *in vivo* anticancer activity of hydroalcoholic extract of *Ipomoea carnea* leaf against ehrlich ascites carcinoma cell lines. *International Journal of Advance Pharmarmaceutical Genuine Research*, 1(1), 39-54.

9. Bachhav, K.V., Burande, M.D., Rangari, V.D. & Mehta J.K. (1999). Effect of aqueous extract of *Ipomoea carnea* leaf on isolated frog and mouse heart. *Indian Journal of Experimental Biology*, 37(11), 1080-1084.

10. Barbosa, R.C., Riet-Correa, F., Medeiros, R.M., Lima, E.F., Barros, S.S., Gimen, E.J., … Gardner, D.R. (2006). Intoxication by *Ipomoea sericophylla* and *Ipomoea riedelii* in goats in the state of Paraiba, North-eastern Brazil. *Toxicology*, 47, 371-379.

11. Bhavnani, S.M., & Ballow, C.H. (2000). New agents for Gram-positive bacteria. *Current Opinion Microbiology*, 3, 528–534.

12. Bidkar, J.S., Bhujbal, M.D., Ghanwat, D.D., & Dama, G.Y., (2012). Anxiolytic activity of aqueous and methanolic extracts of *Ipomoea carnea* leaves. *International Journal of Universal Pharmaceutical Bio Science*, 1(2), 1-11.

13. Chariandy, C. M., Seaforth, C.E., & Phelps, R. H. (1999). Screening of medicinal plants from Trinidad and Tobago for antimicrobial and insecticidal properties. *Journal of Ethnopharmacolgy*, 64, 265–270.

14. Chowdhury, A.K., Ali, M.S., & Khan, M.O. (1997). Anti-microbial activity of *Ipomoea fistulosa* extractives. *Fitoterapia*, 68, 379–380.

15. Cook, C.D.K. (1987). *Ipomoea fistulosa*: A new problem for India. *Aquaphyte Journal*, 7(1), 12.

16. Cowan, M.M. (1999). Plant products as antimicrobial agents. *Clinical Microbiology Reviews*, 12(4), 564-582.

17. Daniel, S.C.G.K., Banu, B.N., Harshiny, M., Nehru, K., Ganesh, P.S., Kumaran, S. & Sivakumar, M. (2014). *Ipomoea carnea* based silver nanoparticle synthesis for antibacterial activity against selected human pathogens, *Journal of Experimental Nanoscience*, 9(2), 197-209.

18. Das, K., Tiwari, R.K.S., & Shrivastava, D.K. (2010). Techniques for evaluation of medicinal plant products as antimicrobial agent: Current methods and future trends. *Journal of Medicinal Plants Research*, 4 (2), 104-111.

19. Doughari, J. H., Human, I. S., Bennade, S., & Ndakidemi, P. A. (2009). Phytochemicals as chemotherapeutic agents and antioxidants: Possible solution to the control of antibiotic resistant verocytotoxin producing bacteria. *Journal of*

Medicinal Plants Research, 3(11), 839-848.

20. Dwivedi, S. N. (1999). Traditional health care among tribals of Rewa district of Madhya Pradesh with special reference to conservation of endangered and vulnerable species. *Journal of Economic and Taxonomic Botany*, 23(2), 315-320.

21. Ekka, N. R., & Dixit, V. K. (2007). Ethno-pharmacognostical studies of medicinal plants of Jashpur district (Chhattisgarh). *International Journal of Green Pharmacy*, 1(1), 2-4.

22. Gaur, K., Kori, M. L., Tyagi, L. K., Nema, R. K., Sharma, C. S., & Tripathi, P. (2009). In-vitro antioxidant activity of leaves of *Ipomoea fistulosa* Linn. *Academic journal of plant sciences*, 2(2), 60-64.

23. Gholizadeh, A., & Kapoor, H. C. (2004). Modifications in the purification protocol of *Celosia cristata* antiviral proteins lead to protein that can be N-terminally sequenced. *Protein and peptide letters*, 11(6), 555-561.

24. Herrmann, F., Romero, M. R., Blazquez, A. G., Kaufmann, D., Ashour, M. L., Kahl, S., ... Wink, M. (2011). Diversity of pharmacological properties in Chinese and European medicinal plants: cytotoxicity, antiviral and antitrypanosomal screening of 82 herbal drugs. *Diversity*, 3(4), 547-580.

25. Hosomi, R. Z., de Souza, H., Górniak, S. L., & Habr, S. F. (2008). Embryotoxic effects of prenatal treatment with *Ipomoea carnea* aqueous fraction in rats. *Brazilian Journal of Veterinary Research and Animal Science*, 45(1), 67-75.

26. Kadiyawala, A.L.A., Prasad, A.K., Kumar, S., Iyer, S.V., Patel, H.A., & Patel, J.A. (2012). Comparative antidiabetic studies of leaves of *Ipomoea carnea* and *Grewia asiatica* on streptozotocin induced diabetic rats. *International Journal of Pharmaceutical and Biological Archives*, 3(4), 853-857.

27. Khalid, M.S., Singh, R.K., Kumar, S.J., Suresh, D.K., Srinivas, R.K., & Raddy, N.I. (2011). Anti-diabetic activity of aqueous extract of *Ipomoea carnea* leaves in streptozotocin induced diabetic rats. *International Journal of Pharmacology and Bio Science*, 5, 45–54.

28. Khatiwora, E., Adsul, V.B., Kulkarni, M., Deshpande, N.R., & Kashalkar, R.V. (2012). Antibacterial activity of dibutyl phthalate: A secondary metabolite isolated from *Ipomoea carnea* stem. *Journal of Pharmacy Research*, 5(1), 150-152.

29. Kumar P.S. (2018). Phytochemical and pharmacological overview of *Celosia cristata* and future perspective as potential phytotherapeutic agent. *European Journal of Biomedical and Pharmaceutical sciences*, 1(5), 829-834.

30. Lamidi, M., Rondi, M.L., Ollivier, E., Faure, R., Ekekang, L.N., & Balansard, G. (2000). Constituents of *Ipomoea fistulosa* leaves. *Fitoterapia*, 71, 203-204.

31. Legler, G. (1965). Die bestandteile des giftigen glykosidharzes aus *Ipomoea fistulosa* Mart. ex Choisy. *Phytochemistry*, 4, 29–41.

32. Lunardi, C., Bason, C., Navone, R., Millo, E., Damonte, G., Corrocher, R., & Puccetti, A. (2000). Systemic sclerosis immunoglobulin G autoantibodies bind the human cytomegalovirus late protein UL94 and induce apoptosis in human endothelial cells. *Nature medicine*, 6(10), 1183-1186.

33. Meira, M., Silva, E. P. D., David, J. M., & David, J. P. (2012). Review of the genus *Ipomoea*: traditional uses, chemistry and biological activities. *Revista Brasileira de Farmacognosia*, 22(3), 682-713.

34. Nidiry, E. S. J., Ganeshan, G., & Lokesha, A. N. (2011). Antifungal activity and isomerization of octadecyl p-coumarates from *Ipomoea carnea* subsp. *fistulosa*. *Natural Product Communications*, 6(12), 1889-1892.

35. Patel, A. K., Singh, V. K., & Jagannadham, M. V. (2007). Carnein, a serine protease from noxious plant weed *Ipomoea carnea* (morning glory). *Journal of Agricultural and Food Chemistry*, 55(14), 5809-5818.

36. Phillips, O., Gentry, A. H., Reynel, C., Wilkin, P., & Gálvez Durand B, C. (1994). Quantitative ethnobotany and Amazonian conservation. *Conservation biology*, 8(1), 225-248.

37. Rao, S., Ekanath, M., Chandrakala, G., Rao, D.H., Venkatesh, R., & Kiran, P. (2015). Evaluation of pharmacological activity of selected flowers used in Bathukamma - State Festival of Telangana. *American Journal of Phytomedicine and Clinical Therapeutics*, 3(11), 666-678.

38. Rios, E., Cholich, L. A., Silva, J., & Acosta de Pérez, O. (2008). Histopathological lesions on central nervous system intoxication of goats by *Ipomoea carnea* spp *fistulosa* (Convolvuláceae). *Revista Veterinaria* 19(2), 130-134.

39. Rosenberg, A. (1999). Bones, joints, and soft tissue tumours. (Eds. Cotran, R.S., Kumar, V. and Colins, T.) Robbins Pathologic Basis of Disease. Pennsylvania: WB Saunders Company, pp.1215-1268.

40. Rout, S. K., & Kar, D. M. (2013). Sedative, anxiolytic and anticonvulsant effects of different extracts from the leaves of *Ipomoea carnea* in experimental animals. *International Journal of Drug Delivery Research*, 5(2), 232-243.

41. Rubini, D., Sudhahar D., & Anandarajagopal, K. (2012). Phytochemical investigation and anthelmintic activity of *Celosia cristata* leaf extracts. *International Research Journal of Pharmacy*, 3(5), 335-337.

42. Sabogal, A., & Borkowski, D. (2007). Estado actual de la investigación sobre *Ipomoea carnea*: toxicidad en ganado caprino. *Revista de Qumica*, 21(1-2), 29-35.

43. Sahayaraj, K., & Ravi, C. (2008). Preliminary phytochemistry of *Ipomea carnea* Jacq. and *Vitex negundo* Linn. leaves. *International Journal of Chemical Sciences*, 6(1), 1-6.

44. Schwarz, A., Górniak, S. L., Bernardi, M. M., Dagli, M. L. Z., & Spinosa, H. S. (2003). Effects of *Ipomoea carnea* aqueous fraction intake by dams during pregnancy on the physical and neurobehavioral development of rat offspring. *Neurotoxicology and teratology*, 25(5), 615-626.

45. Sharma, A., & Bachheti, R.K. (2013). A review on *Ipomoea carnea*. *International Journal of Pharma and Bio Sciences*, 4(4), 363-377.

46. Singh, A., & Navneet (2016). A review on medicinal plants and herbs of Uttrakhand (India): its traditional, ethanomedicinal and antimicrobial potential. *Nature and Science*, 14(12), 90-107.

47. Sridevi, A., Vijayalakshmi, M., & Narasimha G. (2011). Antibacterial activity of some higher plant floral petals. *Natural Products an Indian Journal*, 7(4), 168-170.

48. Sun, Z. L., Gao, G. L., Xia, Y. F., Feng, J., & Qiao, Z. Y. (2011). A new hepoprotective saponin from Semen *Celosia cristatae*. *Fitoterapia*, 82(4), 591-594.

49. Surse, S.N., Shrivastava. B., Sharma. P., Sharma J., & Gide P.S. (2014). Pharmacognostic standardisation of whole plant of *Celosia argentea* var. cristata (L). *International Journal for Pharmaceutical Research Scholar*, 3(I-3), 2277-7873.

50. Tirkey, K., Yadava, R. P., Mandal, T. K., & Banerjee, N. C. (1988). Pharmacological study of *Ipomoea carnea*. *Indian Veterinary Journal*, 65(3), 206-210.

51. Wang Y. (2012). Ph.D. thesis, Second Military Medical University. Chemical constituents and bioactivities of *Celosia cristata* L. and *Penthorum Chinense* Pursh.

52. Wen, Y., Islam, M. T., & Tahara, S. (2006). Phenolic constituents of *Celosia cristata* L. susceptible to spinach root rot pathogen *Aphanomyces cochlioides*. *Bioscience, biotechnology, and biochemistry*, 70(10), 2567-2570.

53. Woradulayapinij, W., Soonthornchareonnon, N., & Wiwat C. (2005). *In vitro* HIV type 1 reverse transcriptase inhibitory activities of Thai medicinal plants and *Canna indica* L. rhizomes. *Journal of Ethnopharmacology*, 101, 84-89

54. Xiang, C., Li, G.M., Hui. S., & Yadan C. (2010). Study on chemical consituents of *Celosia cristata* seed. *Journal of Jilin Agricultural University*, 32(6), 657-660.

55. Yun, S. M., Choi, B. H., Ku, H. O., Lee, M. H., Nam, H. M., Lee, K. J., ... & Son, S. W. (2008). Antimicrobial activities of the flower extract of *Celosia cristata* L. *Planta Medica*, *74*(09), PA31.

Current Status of Researches in Biosciences (2020) : 507-515
Editor : Prof. P.C. Joshi, N. Joshi, Reshman Yasmin, and D. K. Mansotra
Today & Tomorrow's Printers and Publishers, New Delhi - 110 002

ROLE OF SPECIES DISTRIBUTION MODELS IN BIODIVERSITY CONSERVATION

Gaurav Pant*, Dalip K. Mansotra, Shivalika Sharma, Meera Goswami and P. C. Joshi

Department of Zoology and Environmental Science,Gurukul Kangri University, Haridwar – 249404

**Email: rs.gauravpant@gkv.ac.in*

ABSTRACT

Climate change is becoming a threat to the Biodiversity at the global level, becoming the cause for extinction of different species of flora and fauna. Species distribution models (SDMs) plays an important role in biodiversity conservation from its initial phase of development to the present time through the linkage between sciences to decision process and policy making. These models gives the future scenarios of the different landscapes on the basis of known and projected ecological parameters. While there are many cautionary things exist, the persuasive capacity of the models to the non-scientific community about the effects of environmental change are overwhelming. Scientists should explain the ambiguity of the futures illustrated in their models, but also to involve the different stakeholders who will shape those future conditions. Stakeholders can assess the natural resources, raise their voices in the environmental policy making and articulate the array of solution, which they are keen to consent. Creating alternate scenarios is an academic exercise unless it is related to real viable decisions concerning important resources. The full potential of SDMs is reached by the combined involvement of public stakeholders, scientists and policy makers and can be used as an adaptive management tool for understanding the long and short-term changes which are undergoing in any complex landscape.

KEYWORDS: SDM, Biodiversity conservation, Environment policy, Ecological parameters.

INTRODUCTION

Emission of Greenhouse gases caused increase in temperatures, increased sea level, change in rainfall patterns and frequent and intense extreme events and causing species extinction. (IPCC, 2014; Cahill *et al.,* 2013). Many examples proves that climate change is creating an alarming situation for the Biodiversity causing extinction of different floral and faunal species. Several studies have investigated the climate change impacts on the future of biodiversity. Many studies for future prediction indicates worst case future scenarios, leading to the extinction of many species (Bellard *et al*., 2012). Climate change affects biodiversity from gene to biome level, resulting in the loss of habitat, range shifts and extinction of species. The response of species affected by climate change responses are in three ways, whether that species will change themselves, become extinct or move to the other area with the same conditions as their habitat (Rinawati *et al*., 2013; Pecl *et al.,* 2017). IUCN Red list shows that more than 28000 species have been comes under the threatened with extinction category, which are 27% of the total assessed species (https://www.iucnredlist.org/).

Species Distribution Models: Species distribution models (SDMs) plays an important role in biodiversity conservation from its initial phase of development to the present time through the linkage between sciences to decision process and policy making. Species distribution models gives the future scenarios of the different landscapes on the basis of known and projected ecological parameters.

In SDM, the following steps are usually taken (Austin, 2002):

(1) Location of species occurrence or other phenomena are complied

(2) Values of environmental predictor variables (such as climate) as these location are extracted from spatial databases

(3) The environmental models are used to fit a model to estimate to the sites of occurrence, or another measures such as abundance of the species

(4) The model is used to predict the variables of interest across the region of interest (and perhaps of the future or past climate).

The precise environmental layers together with affordable and efficient computers has led to a rapid growth in the development of predictive distribution modelling of species environmental and geographic distributions. The occurrence data of species presence or absence is important factor, which allows the use of various statistical modeling techniques. But the availability of occurrence data is not available for all the species. Depending on the availability of occurrence data, the selection of specific models is necessary to get the better results (Phillips *et al.,* 2006). Several ecological models are available depending on the availability of occurrence data. Some models viz. Maximum entropy (MaxEnt) model (Phillips *et al.*, 2006), Genetic Algorithm for Rule-set Production (GARP) (Stockwell, 1999) etc. which uses the presence only data and Generalised linear model (GLM) (McCullough and Nelder, 1989) and Generalised Additive models (GAM) (Hastie and Tibshirani, 1990) etc. uses presence/absence data to give the predictive distribution of species.

CONSERVATION APPLICATIONS

SDMs are used as a major tool now a days in a wide area for the conservation of different floral and faunal species. Several studies have been done for habitat suitability analysis and identification of a niche of different species which are prone to extinction.Applications for Entomological conservation also have been done using the SDMs for some endangered insects, which are important for the ecosystem functions. The distribution of a threatened saproxylic Stag Beetle (*Lucanus cervus*) in Belgium was studied using 8 different models i.e. artificial neural network, classification tree analysis, generalized linear models, mixture discriminant analysis, multiple adaptive regression splines and random forests and the model performance was evaluated by using AUC (Area under Curve), Sensitivity, (% correctly predicted presences in the evaluation set) and specificity (% correctly predicted absences in the evaluation set) and Kappa statistics. The predicted distribution map consist only 11% area was under conservation priorities in Natura 2000 scheme. After that the rest of the area have been conserved in three conservation areas (Thomaes *et al.,*2008).

Wildlife conservation is an important aspect with respect to the Biodiversity. Manel *et al.,* (1999) has done species distribution modeling for predicting the potential distribution of of common river bird, Plumbeous Redstart (*Rhyacornis fuliginosus*), in Indian and Nepal Himalaya by using multiple discriminant analysis (MDA), logistic regression (LR) and artificial neural networks (ANN). A total of 32 variables defining stream, altitude, slope, habitat structure and invertebrate abundance were used. Model performance was determined from the two methods: 1) Leave-one-out approach, in which LR is more accurate than other and 2) Holdout procedure, in which all methods performed similarly. True absence was predicted much better than true presence in all methods. Plumbeous Redstarts are one of the most common Himalayan river birds occurring only at 36% of the sites. It is important to test SDMs with independent data, and to use the array of criteria for assessing model performance. For analyzing bird distributions conventional multivariate methods does not have major advantages. LR and MDA were more efficient in the use of computer time than ANN and more accurate in providing testable hypotheses about environmental effects on presence.

Some plant studies have also been done using the SDMs for predicting their potential niches. The separation of the bioclimatic spaces of Himalayan Rhododendron tree species was studied using the suitability analysis. The distributions of two members of this species were predicted using environmental data for current condition (1950-2000). An Ecological Niche Modelling ensemble technique was used to map the suitable climatic condition based on 217 occurrence points with least correlated variables. Ensemble models showed high predictive power for Rhododendron species with bioclimatic variables, separating the climatic envelopes of the two species and the distribution range found for Rhododendron was wide. The most suitable area is western Yunnan, China with the least suitability in west and east. Its wide range of climatic spaces with highly dissected habitats tells about the adaptive capability of the species, and it might be useful in near future for the conservation management in the area, where these species are listed in threatened category. Models based on the group of variables gives good fits to data for prediction at the region level (Ranjitkar *et al.,*2014).

Rhododendron arboretum Ssp. *Nilagiricum* (Ericaceae), a species endemic to southern Western Ghats was modelled using the GARP modelling technique for predicting its potential spatial distribution. The results of the model suggests that its distribution was related to local climatic conditions i.e. annual temperature, rainfall, ad diurnal temperature ranges and not much dependent on elevation. The species distribution area was confined to Nilgiri, Anamalai, Palni and Meghamalai hills of Western Ghats. The predictive results gives the idea to identify key areas for the *in situ* conservation of this species. (Giriraj *et al.,* 2008)

MaxEnt modelling is the most used method with the use of presence only data, providing the suitable prediction. *Justicia adhatoda,* a medicinal plant species found in lesser Himalayan foothills was mapped using the MaxEnt Model. Its distribution was predicted using the bioclimatic layers and some other variables which are important for its distribution i.e. slope, aspect, elevation and land use/ land cover data. Maxent model gives highly accurate results with the significant AUC value of 92.3. Thus this modelling technique was found suitable for predicting the potential distribution of medicinal plant species and so it can be used for restoration and conservation planning of the species (Yang *et al.,* 2013). The distributional study on the Himalayan medicinal plant *Berberis aristata*was done using 3 different algorithms i.e. MaxEnt, GARP and Bioclim with the bioclimatic and topographical variables. Results shows the MaxEnt model has the higher accuracy when compared to present distribution of species (Ray *et al.,* 2011). The geographical distribution of *Pterocarpus santalinus* (Red sandars), a endemic and endangered species found only in the southern portion of Eastern Ghats in Andhra Pradesh, India; was studied using Ecological Niche Modeling (ENM) based on detailed distributional information, climatic and topographical data. Hierarchical Partitioning (HP) was done before using the climatic variable for the selection of best variable for better prediction. Among the Three models i.e. GARP, MaxEnt and Bioclim; GARP model gives the better prediction accuracy than others. The model success shows that ENM based techniques provide a better tool for understating the wide array of questions related conservation and biogeography (Babar *et al.,*2012).

The geographic distribution of *Aglaia bourdillonii* (Meliaceae), an endemic and threatened plant found in southern portion of Western Ghats was studied using ENM (GARP model) (Irfan-Ullah *et al.*, 2006). This species was found in a special habitat conditions and associated with some specific plants to sustain and maintaining populations (Pascal, 1988). It also requires safety from emergent species (e.g., *Cullenia exarillata* and *Palaquium ellipticum*) for survival (Ganesh *et al.,* 1996; Pascal, 1988) and also vulnerable to anthropogenic activities (such as logging) and its population is decreasing (Giriraj, 2005; Pascal, 1988).It is found only in flatter areas, and rare in deep valleys (Giriraj, 2005).A total of 53 occurrence points, 19 bioclimatic variables and slope, aspect, elevation and compound topographic index were used for SDMs using GARP model. The prediction of this species was good given current distribution of the species.Similar study, ENM for a critically endangered species *Gymnocladus assamicus* endemic to NE India using GARP and MaxEnt was done. This species was used by the local peoples for different uses, so being considered extinct because of the population decline, However a survey has been done before 2004-2007 for its population estimation and found 14 discrete populations. Due to the rugged terrain and remoteness of the area, there are some problems on field survey. So by prediction of the potential distribution of the species, further field inventories areas was targeted. ENM identified 26 sites found highly suitable for the species occurrence and the reconnaissance survey at the 14 most accessible area among the predicted sites was completed. Out of the 14, new population was found in 5 sites. So it can be concluded that Niche modelling can be used as a tool for locating new populations of rare and threatened species and enhance the current efforts for *in situ* conservation of the endangered plants (Menon *et al.,*2010).

Invasive species are also a major issue for the biodiversity loss replacing the local native species of a particular area. *Mikania micrantha* suitability for invasion was studied, which is a fast growing species using GARP, MaxEnt and Bioclim. Among the three models, Bioclim and GARP performed better than MaxEnt. Jackknife test showed Bio 13 (Precipitation of wettest period) and Bio 3 (Isothermality) having high percentage contribution in the expansion of *M. micrantha.* There is a significant difference showed in distribution frequency with increasing altitude and different climatic spaces. So it is obvious the decrease in the frequency of point locations with increasing altitude. (Rameshprabu *et al*., 2015).

Reintroduction of a species in a different area with suitable habitat conditions can be done using SDMs. For the reintroduction of *Ilex khasiana* (Critically endangered species) confined to Khasi hills of Meghalaya, MaxEnt modelling was used for habitat suitability modelling using 16 point locations and 16 environmental layers including Enhanced Vegetation Index (EVI) and Elevation data. The model predicted the species was confined to an area of H" 500 km^2 and influenced by EVI and elevation layers from April-May (flowering phenophase). The study reveals that the species potential habitat is confined to higher altitude Khasi hills within the current home range for its reintroduction (Adhikari *et al.,*2012).

LINKAGE BETWEEN SCIENCES TO DECISION PROCESS AND POLICY MAKING

The conservationists are not only the end user of Species distribution models output. For fulfilling the conservation goals the involvement of stakeholders, policymakers and managers is important (Euliss *et al*., 2011). From the literature survey it is being conclude that SDMs outputs have not been conveyed to the decision makers sufficiently. Even there are limited research have been done which gives the idea to work from theory to practice. More work should be done on the uses of SDMs in the making of public policy, stakeholder scenarios analysis and applied conservation practices (Driscoll *et al*., 2012). The better understanding of benefits of the different modelling techniques is necessary for the conservation community with respect to their ability to influence the policy making and their management.

CONCLUSION

The purpose of this review is to show the importance of different Species Distribution Models in the conservation of global biodiversity. The paper also encompasses some organisms which required a specific environmental conditions and also some species which required unrestricted movement across a landscape with different environmental envelopes and simulation studies at landscape level to know the changes in its ecosystem services. I conclude with the note that the conservation community should use the species distribution models throughout the planning process with the involvement of different stakeholders in the discussions of future scenarios and the decisions needed for getting the desired results of conservation.

REFERENCES

1. Adhikari, D., Barik, S. K., & Upadhaya, K. (2012). Habitat distribution modelling for reintroduction of *Ilex khasiana* Purk., a critically endangered tree species of northeastern India. *Ecological Engineering*, 40, 37-43.

2. Austin, M. P. (2002). Spatial prediction of species distribution: an interface between ecological theory and statistical modelling. *Ecological modelling, 157*(2), 101-118.

3. Babar, S., Amarnath, G., Reddy, C. S., Jentsch, A., & Sudhakar, S. (2012). Species distribution models: ecological explanation and prediction of an endemic and endangered plant species (*Pterocarpus santalinus* Lf). *Current Science*, 1157-1165.

4. Bellard, C., Bertelsmeier, C., Leadley, P., Thuiller, W., & Courchamp, F. (2012). Impacts of climate change on the future of biodiversity. *Ecology letters, 15*(4), 365-377.

5. Cahill, A. E., Aiello-Lammens, M. E., Fisher-Reid, M. C., Hua, X., Karanewsky, C. J., Yeong Ryu, H., ... & Wiens, J. J. (2013). How does climate change cause extinction?. *Proceedings of the Royal Society B: Biological Sciences, 280*(1750), 20121890.

6. Driscoll, C. T., Lambert, K. F., Chapin III, F. S., Nowak, D. J., Spies, T. A., Swanson, F. J., ...& Hart, C. M. (2012). Science and society: The role of long-term studies in environmental stewardship. *BioScience, 62*(4), 354-366.

7. Euliss Jr, N.H., Smith, L.M., Liu, S., Duffy, W.G., Faulkner, S.P.,Gleason, R.A. & Eckles, S.D. (2011) Integrating estimates of ecosystem services from conservation programs and practices into models for decision makers. *Ecological Applications* 21: S128–S134.

8. Ganesh T, Ganesan R, Devy MS, Davidar P, Bawa KS (1996) Assessment of plant biodiversity at a mid-elevation evergreen forest of Kalakad-Mudanthurai Tiger Reserve, Western Ghats, India. *Curr. Sci.* 71:379–392

9. Giriraj A (2005). Spatial characterization and conservation priortization in tropical evergreen forests of Southern Western Ghats using Geoinformatics. Ph.D. thesis, Barathidasan University, Thirunelveli

10. Giriraj, A., Irfan-Ullah, M., Ramesh, B. R., Karunakaran, P. V., Jentsch, A., & Murthy, M. S. R. (2008). Mapping the potential distribution of *Rhododendron arboreum* Sm. ssp. *Nilagiricum* (Zenker) *Tagg* (Ericaceae), an endemic plant using ecological niche modelling. *Current Science*, 1605-1612.

11. Hastie, T., & Tibshirani, R. (1990). Generalized Additive Models. Chapman Hall & CRC. *Monographs on Statistics & Applied Probability*. Chapman and Hall/CRC, 1.

12. https://www.iucnredlist.org/

13. Irfan-Ullah, M., Amarnath, G., Murthy, M. S. R., & Peterson, A. T. (2006). Mapping the geographic distribution of *Aglaia bourdillonii* Gamble (Meliaceae), an endemic and threatened plant, using ecological niche modeling. In *Plant Conservation and Biodiversity* (pp. 343-351). Springer, Dordrecht.

14. IPCC, 2014. Summary for Policymakers. In: field, c.b., Barros, V.R., Dokken, D.J., Mach, K.J., Mastrandrea, M.D., Bilir, T.E., Chatterjee, M., Ebi, K.L., Estrada, Y.O., Genova, R.C., Girma, B., Kissel, E.S., Levy, A.N., MacCracken, S., Mastrandrea, P.R., White, L.L. (Eds.), Climate change 2014: Impacts, adaptation, and vulnerability. Part A: Global and sectoral aspects. Contribution of working Group II to the Fifth Assessment Report of the Intergovernmental Panel on climate change. Cambridge University Press, Cambridge, United Kingdom, and New York, NY, USA, pp. 1–32.

15. Manel, S., Dias, J. M., & Ormerod, S. J. (1999). Comparing discriminant analysis, neural networks and logistic regression for predicting species distributions: a case study with a Himalayan river bird. *Ecological modelling,* 120(2), 337-347.

16. McCullough, P., Nelder, J.A., (1989). Generalised Linear Models, 2nd edition. Chapman and Hall, London.

17. Menon, S., Choudhury, B. I., Khan, M. L., & Peterson, A. T. (2010). Ecological niche modeling and local knowledge predict new populations of *Gymnocladus assamicus* a critically endangered tree species. *Endangered Species Research*, 11(2), 175-181.

18. Pascal JP (1988). Wet evergreen forests of the Western Ghats of India, ecology, structure, floristic composition and succession, 1st edn. Institutfrancais de Pondichery, p 345

19. Pecl, G. T., Araújo, M. B., Bell, J. D., Blanchard, J., Bonebrake, T. C., Chen, I. C., ...& Falconi, L. (2017). Biodiversity redistribution under climate change: Impacts on ecosystems and human well-being. *Science*, 355(6332), eaai9214.

20. Phillips, S. J., Anderson, R. P., & Schapire, R. E. (2006). Maximum entropy modeling of species geographic distributions. *Ecological modelling*, *190*(3-4), 231-259.

21. Rameshprabu, N., & Swamy, P. S. (2015). Prediction of environmental suitability for invasion of *Mikania micrantha* in India by species distribution modelling. *Journal of Environmental Biology*, 36(3), 565.

22. Ranjitkar, S., Kindt, R., Sujakhu, N. M., Hart, R., Guo, W., Yang, X., ... & Luedeling, E. (2014). Separation of the bioclimatic spaces of Himalayan tree rhododendron

species predicted by ensemble suitability models. *Global Ecology and Conservation*, 1, 2-12.

23. Ray, R., Gururaja, K. V., & Ramchandra, T. V. (2011). Predictive distribution modeling for rare Himalayan medicinal plant *Berberis aristata* DC. *Journal of Environmental Biology*, 32(6), 725.

24. Rinawati, F., Stein, K., & Lindner, A. (2013). Climate change impacts on biodiversity—the setting of a lingering global crisis. *Diversity*, 5(1), 114-123.

25. Stockwell, D. R. (1999). Genetic algorithms II. In Machine learning methods for ecological applications (pp. 123-144). *Springer,* Boston, MA.

26. Thomaes, A., Kervyn, T., & Maes, D. (2008). Applying species distribution modelling for the conservation of the threatened saproxylic Stag Beetle (*Lucanus cervus*). *Biological conservation*, 141(5), 1400-1410.

27. Yang, X. Q., Kushwaha, S. P. S., Saran, S., Xu, J., & Roy, P. S. (2013). Maxent modeling for predicting the potential distribution of medicinal plant, *Justicia adhatoda* L. in Lesser Himalayan foothills. *Ecological Engineering*, *51*, 83-87.